AF667862

3D Printing in Oral Health Science

Prabhat Kumar Chaudhari
Dinesh Bhatia • Jitendra Sharan
Editors

3D Printing in Oral Health Science

Applications and Future Directions

Springer

Editors
Prabhat Kumar Chaudhari
All India Institute of Medical Sciences
New Delhi, India

Dinesh Bhatia
North Eastern Hill University
Shillong, India

Jitendra Sharan
All India Institute of Medical Sciences
Bhubaneswar, India

ISBN 978-3-031-07368-7 ISBN 978-3-031-07369-4 (eBook)
https://doi.org/10.1007/978-3-031-07369-4

© The Editor(s) (if applicable) and The Author(s), under exclusive license to Springer Nature Switzerland AG 2022, corrected publication 2022
This work is subject to copyright. All rights are solely and exclusively licensed by the Publisher, whether the whole or part of the material is concerned, specifically the rights of translation, reprinting, reuse of illustrations, recitation, broadcasting, reproduction on microfilms or in any other physical way, and transmission or information storage and retrieval, electronic adaptation, computer software, or by similar or dissimilar methodology now known or hereafter developed.
The use of general descriptive names, registered names, trademarks, service marks, etc. in this publication does not imply, even in the absence of a specific statement, that such names are exempt from the relevant protective laws and regulations and therefore free for general use.
The publisher, the authors, and the editors are safe to assume that the advice and information in this book are believed to be true and accurate at the date of publication. Neither the publisher nor the authors or the editors give a warranty, expressed or implied, with respect to the material contained herein or for any errors or omissions that may have been made. The publisher remains neutral with regard to jurisdictional claims in published maps and institutional affiliations.

This Springer imprint is published by the registered company Springer Nature Switzerland AG
The registered company address is: Gewerbestrasse 11, 6330 Cham, Switzerland

This book is dedicated to my dear mother ***Late Smt. Sulochana Devi*** *and my loving father* ***Late Shri LalJi Chaudhari****, whose hardships in life and desire to see their son educated brought me here as a Faculty in India's Premier Medical Institute.*

Although my parents have attained the heavenly abode, their loving spirits nurture me still and inspires me to always do my best, making them proud and take nothing for granted.

Prabhat Kumar Chaudhari

My sincere gratitude to my dear parents ***Mr. Ashok and Mrs. Kanchan Bhatia*** *for their unconditional love, support and blessings for the completion of this book. I also acknowledge the constant support and cooperation of my wife and children during the preparation of this book.*

Dinesh Bhatia

For all those who encouraged me to pursue my dreams, especially my parents and teachers. Life is a crazy ride, and it is a privilege to go through it with my wife and lovely daughter while working on this project.

Jitendra Sharan

Preface

The edited book *3D Printing in Oral Health Science: Applications and Future Directions* has been written to bring out the latest developments in the field of 3D printing in Oral Health Sciences and its future developments. 3D printing is an emerging and revolutionizing technology which has undergone phenomenal expansion in recent years, impacting various aspects of daily life activities. The book has 14 chapters contributed by astute clinicians and stalwarts researchers from both academia and industry from Albania, Canada, India, Israel, Kingdom of Saudi Arabia, Nepal, New Zealand, Peru, South Korea, Taiwan, the USA and Vietnam. The book would present new insights and technological advancement not only for budding and experienced clinicians but also for the researchers from technological backgrounds interested in interdisciplinary research. The book starts with introductory concepts on 3D printing, principles and applications in different fields. It gives a broad overview of the technological revolution in oral health science with rapid technological advancements. The second chapter provides an in-depth practical understanding of the principles and applications of various 3D scanning methods for 3D digital data acquisition of the orofacial region. The next two chapters have been written to provide state of the art in-depth understanding about the 3D printing technology and about options of 3D printable materials choices for various oral healthcare applications. Fifth chapter on computer-aided design (CAD) for dental applications, which has become a significant part of modern dentistry as a diagnostic and treatment planning tool, has been included in the book to give readers a wider perspective. The chapters from 6 to 12 have focused on the 3D printing applications in Maxillofacial Surgery, Periodontics, Prosthodontics, Bioprinting for Craniofacial Regeneration, Customized 3D Metal Printed Implants for orofacial region and Endodontics. The second last chapter discusses the Limitations, Safety and Regulatory Considerations. The final chapter of the book talks about the future of 3D printing in Oral Health Science. Hence, it is a one-stop comprehensive book that would provide the latest technological advancements in the field of 3D printing in dentistry and healthcare, which would benefit the scientific, clinical and technological communities at large. We, the editors, thank Springer Nature-Switzerland (Publisher) for the opportunity to prepare this book and all the contributing authors for their contribution, support and cooperation despite the difficult time due to the pandemic situation and their other commitments to help us complete this task. We acknowledge their expertise, skills and knowledge shared in the course of preparing their respective chapters for the book. We thank our family members for their kind support and for allowing us sufficient time to complete the book as per standard guidelines. Finally, we thank the Almighty for the knowledge and wisdom to overcome all challenges in completing the book.

New Delhi, India — Prabhat Kumar Chaudhari
Shillong, Meghalaya, India — Dinesh Bhatia
Bhubaneswar, Odisha, India — Jitendra Sharan

Contents

Editors and Contributors

About the Editors

Prabhat Kumar Chaudhari is currently working as Additional Professor at the Division of Orthodontics and Dentofacial Deformities, Centre for Dental Education and Research, All India Institute of Medical Sciences (AIIMS), New Delhi, India. He has a broad background in Dental Surgery, with expertise in Orthodontics and Dentofacial Orthopaedics. He is a Member of the Faculty of Dental Surgery of the Royal College of Physicians and Surgeons (MFDS RCPS) of Glasgow, the Royal College of Surgeons (MFDS RCS) of England and the Faculty of Dental Trainers of the Royal College of Surgeons of Edinburgh. He is also a member of the National Academy of Sciences, India (NASI), and the National Academy of Medical Sciences (India). He is a part of the Cochrane India Network and Member of topic group 'Dental Diagnostics and Digital Dentistry' (TG-Dental) as part of the *International Telecommunication Union* and the World Health Organization (ITU/WHO) Focus Group on Artificial Intelligence for Health (FG-AI4H). Dr. Chaudhari is an Associate Editor of BMC Oral Health (digital dentistry section), Frontiers in Dental Medicine journal and Section Editor for the Journal of Oral Biology and Craniofacial Research. He has 11 years of teaching and research experience. Currently, he is involved in teaching and training of Cleft and Craniofacial Orthodontic Fellowship programme, Master of Dental Surgery (MDS) Students, Bachelor of Science in Dental Operating Room Assistance (BSc. DORA) programme and Bachelor of Medicine and Bachelor of Surgery (MBBS) students. His current research areas are the application of 3D printing, non-ionizing, non-invasive surface imaging in oral health science and artificial intelligence-based deep neural networks for dental application. He is committed to extending his research and training future generation of clinicians and researchers in these areas.

Dinesh Bhatia pursued his PhD in Biomechanics and Rehabilitation Engineering from MNNIT, Allahabad, India, in 2010 with Bachelor's (2002) and Master's degree (2004) in Biomedical Engineering from Mumbai University. He completed his MBA (Dual Specialization) from IMT

Ghaziabad in 2007. He is currently working as Associate Professor in the Department of Biomedical Engineering, North-Eastern Hill University (NEHU), Shillong, Meghalaya, India. He is the recipient of the 'Young Scientist Award (BOYSCAST 2011–12)' by the Government of India to pursue research in osteoarthritis (OA) for 1 year at Adaptive Neural Systems Laboratory, Biomedical Engineering Department, Florida International University, Miami, Florida, USA, where he was leading a multidisciplinary team of researchers. He is also the recipient of 'INAE Fellowship Award' in 2011 by the Indian National Academy of Engineering. He was selected as one of the 12 young biomedical scientists by the Indian Council of Medical Research (ICMR), Govt. of India, to pursue research fellowship (2014–2015) in the field of sensory prosthetics at the University of Glasgow, Scotland, UK. He has attended Biomechanics and Human Gait training at Munich, Germany, in March 2017 and training of use of Neuro-Diagnostics Equipment(s) at Ivanovo, Russia, in September 2017. He delivered an Invited Talk on Gait and Osteoarthritis in Kuala Lumpur, Malaysia, in August 2018. He has several research papers in reputed journals, conference, seminars and symposia with teaching and research experience of more than 18 years. He is an invited panel member of several professional bodies, editorial boards, committees, societies, and forums. He has worked on various funded projects on physically challenged, disabled and paralyzed persons and few projects are still ongoing. He has published 9 books and 21 books chapters till date and supervised several UG, PG and doctoral students. He has several patents, copyrights and designs to his name. His research focuses on understanding muscle mechanics, joint kinematics and dynamics involved in performing locomotion and routine tasks and undermining its effects during an injury or disease. His areas of interest are medical instrumentation, biomechanics and rehabilitation engineering, medical informatics, signal and image processing, marketing and international business.

Jitendra Sharan did his graduation in Dental Sciences and post-graduation in Orthodontics and Dentofacial Orthopaedics from Manipal University. After post-graduation, he joined Manipal University as faculty for 2½ years. In the pursuit of a better understanding of research insight, he joined the Centre for Dental Education and Research, All India Institute of Medical Sciences (AIIMS), New Delhi, India, as a Doctoral student. He worked on a novel technique for surface modification of Grade V titanium alloy in collaboration with the Indian Institute of Technology-Delhi, Jawaharlal Nehru University (JNU), and the Council of Scientific and Industrial Research—National Physical Laboratory (CSIR-NPL). He received his PhD degree from AIIMS-New Delhi, in 2018 and was the first to get a PhD in Dental Surgery (Orthodontics) from the institution. Since January 2018, he has been attached to the Department of Dentistry, AIIMS, Bhubaneswar, as faculty. He has published his research in reputed journals related to biomaterial science, orthodontics and interdisciplinary areas.He has written three books, out of those *MCQs in Orthodontics* (second Edition) was written for dental undergraduate students preparing for post-graduate entrance exams. It also helps post-graduate students of orthodontics who are preparing for the Indian Board of Orthodontics Level 2 exam. The book *Orthodontic Enquest* has been written for orthodontic post-graduates.At present, he is working on a novel orthodontic miniscrew implant design, bioconjugation of titanium alloys with biomolecules and nanoparticle doping of biomolecules for dental applications along with patient care.

Contributors

Bhaskar Agarwal Columbia-Asia Hospital, Ghaziabad, India

Dror Aizenbud Department of Orthodontics and Craniofacial Anomalies, School of Graduate Dentistry, Rambam Health Care Campus, Technion—Israel Institute of Technology, Haifa, Israel

Oral Biology Research Laboratory, Faculty of Medicine, Technion—Israel Institute of Technology, Haifa, Israel

Mohammad Khursheed Alam Orthodontic Department, College of Dentistry, Jouf University, Al-Jawf, Kingdom of Saudi Arabia

Ahmad Al-Hassiny Institute of Digital Dentistry, Wellington, New Zealand

Kelly Kathering Achachao Almerco Department of Pediatric Stomatology, Dentistry School, National University of San Marcos, Lima, Peru

Swaroop Anil Pratap Dental Clinic and Implant Centre, Allahabad, Uttar Pradesh, India

Manish Bajracharya Orthodontic Unit, Dental Department, National Academy of Medical Sciences (NAMS), Bir Hospital, Kathmandu, Nepal

Nikhil Belsure Imaginarium India Pvt. Ltd., Mumbai, India

Ravi Bhallamudi Mechanical Engineering Department, Indian Institute of Technology Bombay, Mumbai, India

Dinesh Bhatia Biomedical Engineering Department, North Eastern Hill University, Shillong, Meghalaya, India

Tal Capucha Department of Oral and Maxillofacial Surgery, Rambam Health Care Campus, Haifa, Israel

Prabhat Kumar Chaudhari Division of Orthodontics and Dentofacial Deformities, Centre for Dental Education and Research (CDER), All India Institute of Medical Sciences (AIIMS), New Delhi, India

Amrita Chawla Division of Conservative Dentistry and Endodontics, Centre for Dental Education and Research, All India Institute of Medical Sciences, New Delhi, India

Laishram Bijaya Devi Department of Orthodontics and Dentofacial Orthopedics, Dental College, Regional Institute of Medical Sciences (RIMS), Imphal, India

Harnoor Dhillon Division of Orthodontics and Dentofacial Deformities, Center for Dental Education and Research, All India Institute of Medical Sciences, New Delhi, India

Kunaal Dhingra Periodontics Division, Centre for Dental Education and Research, All India Institute of Medical Sciences, New Delhi, India

Isha Duggal Division of Orthodontics and Dentofacial Deformities, Center for Dental Education and Research (CDER), All India Institute of Medical Sciences, New Delhi, India

Ritu Duggal Division of Orthodontics and Dentofacial Deformities, Center for Dental Education and Research (CDER), All India Institute of Medical Sciences, New Delhi, India

Omri Emodi Department of Oral and Maxillofacial Surgery, Rambam Health Care Campus, Haifa, Israel

Adil Gandevivala Mahatma Gandhi Mission Dental College and Hospital, Navi Mumbai, India

Bhanupratap Gaur Mechanical Engineering Department, Indian Institute of Technology Bombay, Mumbai, India

Rupesh Ghyar Mechanical Engineering Department, Indian Institute of Technology Bombay, Mumbai, India

Bhumika Gumber Periodontics Division, Centre for Dental Education and Research, All India Institute of Medical Sciences, New Delhi, India

Shalini Gupta Division of Oral Medicine and Radiology, Centre for Dental Education and Research (CDER), All India Institute of Medical Sciences (AIIMS), New Delhi, India

Shankar Iyer Malo Smile USA, Elizabeth, NJ, USA

Department of Periodontics and Restorative Dentistry, Rutgers School of Dental Medicine, Newark, NJ, USA

Veena Jain Division of Prosthodontics, Centre for Dental Education and Research, All India Institute of Medical Sciences (AIIMS), New Delhi, India

Nayansi Jha Department of Orthodontics, Asan Medical Center, University of Ulsan College of Medicine, Seoul, South Korea

Mathew T. Kattadiyil Advanced Education Program in Prosthodontics, School of Dentistry, Loma Linda University, Loma Linda, CA, USA

Harsimran Kaur Manav Rachna Dental College and Hospital, Faridabad, Haryana, India

Yoon-Ji Kim Department of Orthodontics, Asan Medical Center, University of Ulsan College of Medicine, Seoul, South Korea

Dheeraj Koli Division of Prosthodontics, Centre for Dental Education and Research, All India Institute of Medical Sciences (AIIMS), New Delhi, India

Vijay Kumar Division of Conservative Dentistry and Endodontics, Centre for Dental Education and Research, All India Institute of Medical Sciences, New Delhi, India

Rong-Fu Kuo Department of Biomedical Engineering, National Cheng Kung University, Tainan, Taiwan

Medical Device Innovation Center, National Cheng Kung University, Tainan, Taiwan

Yu-Sheng Lin Department of Mechanical Engineering, Southern Taiwan University of Science and Technology, Tainan, Taiwan

Ajay Logani Division of Conservative Dentistry and Endodontics, Centre for Dental Education and Research, All India Institute of Medical Sciences, New Delhi, India

Kanika Makker Periodontics Division, Centre for Dental Education and Research, All India Institute of Medical Sciences, New Delhi, India

Hagai Hazan Molina Department of Orthodontics and Craniofacial Anomalies, School of Graduate Dentistry, Rambam Health Care Campus,, Technion—Israel Institute of Technology, Haifa, Israel

Oral Biology Research Laboratory, Faculty of Medicine, Technion—Israel Institute of Technology, Haifa, Israel

Nitika Monga Non Communicable Disease (NCD) Division, Indian Council of Medical Research (ICMR), New Delhi, India

Suraj Naik Mechanical Engineering Department, Indian Institute of Technology Bombay, Mumbai, India

Aditi Nanda Division of Prosthodontics, Centre for Dental Education and Research, All India Institute of Medical Sciences (AIIMS), New Delhi, India

Srivalli Natrajan Mahatma Gandhi Mission Dental College and Hospital, Navi Mumbai, India

K. S. Naveen Division of Orthodontics and Dentofacial Deformities, Centre for Dental Education and Research (CDER), All India Institute of Medical Sciences (AIIMS), New Delhi, India

Anh-Thu Nguyen Department of Dental Technology-Faculty of Odonto-Stomatology, University of Medicine and Pharmacy, Ho Chi Minh, Vietnam

Nakul Parasharami Lokmanya Hospital for Special Surgery, Pune, India

Sagar Parekh Imaginarium India Pvt. Ltd., Mumbai, India

Hieu M. Pham New York University College of Dentistry, New York University, New York, NY, USA

Ujjwal Pyakurel Kantipur Dental College Teaching Hospital & Research Centre, Kathmandu, Nepal

Private Practice, Shangrila Dental Clinic, Kathmandu, Nepal

Andis Qendro Oral and Maxillofacial Unit, Dental Implantology, Faculty of Dental Medicine, University Hospital Centre, Tirana, Albania

Adi Rachmiel Department of Oral and Maxillofacial Surgery, Rambam Health Care Campus, Faculty of Medicine, Technion—Israel Institute of Technology, Haifa, Israel

Ajoy Roychoudhury Department of Oral and Maxillofacial Surgery, AIIMS, New Delhi, India

Samrat Sagar Mechanical Engineering Department, Indian Institute of Technology Bombay, Mumbai, India

Jitendra Sharan Unit of Orthodontics and Dentofacial Orthopedics, Department of Dentistry, All India Institute of Medical Sciences (AIIMS), Bhubaneswar, Odisha, India

Sidhartha Sharma Division of Conservative Dentistry and Endodontics, Centre for Dental Education and Research, All India Institute of Medical Sciences, New Delhi, India

Dipti Shastri Department of Orthodontics and Dentofacial Orthopaedics, Faculty of Dental Sciences, King George's Medical University, Lucknow, Uttar Pradesh, India

Dekel Shilo Department of Oral and Maxillofacial Surgery, Rambam Health Care Campus, Haifa, Israel

Suraj Sinha Department of Craniofacial Orthodontics, Chang Gung Memorial Hospital, Taoyuan City, Taiwan

Ramandeep Kaur Sokhi Division of Orthodontics and Dentofacial Deformities, Centre for Dental Education and Research (CDER), All India Institute of Medical Sciences (AIIMS), New Delhi, India

Nimesh Soni Imaginarium India Pvt. Ltd., Mumbai, India

Owen Tao Dr. Gerald Niznick College of Dentistry, Rady Faculty of Health Sciences, University of Manitoba, Winnipeg, MB, Canada

Simon D. Tran Craniofacial Stem Cells and Tissue Engineering Laboratory, Faculty of Dental Medicine and Oral Health Sciences, McGill University, Montreal, QC, Canada

Flora Verma Viral Research and Diagnostic Laboratory, Department of Microbiology, Atal Bihari Vajpayee Institute of Medical Sciences and Dr. Ram Manohar Lohia Hospital, New Delhi, India

David T. Wu Department of Oral Medicine, Infection, and Immunity, Harvard School of Dental Medicine, Harvard University, Boston, MA, USA

Harvard John A. Paulson School of Engineering and Applied Sciences, Harvard University, Cambridge, MA, USA

Wyss Institute for Biologically Inspired Engineering, Harvard University, Boston, MA, USA

Craniofacial Stem Cells and Tissue Engineering Laboratory, Faculty of Dental Medicine and Oral Health Sciences, McGill University, Montreal, QC, Canada

Kevin Y. Wu Faculty of Medicine, University Laval, Quebec, QC, Canada

Tai-Hua Yang Department of Biomedical Engineering, National Cheng Kung University, Tainan, Taiwan

Edlira Zere Department of Orthodontics and Craniofacial Anomalies, School of Graduate Dentistry, Rambam Health Care Campus and Faculty of Medicine, Technion—Israel Institute of Technology, Haifa, Israel

Alon Zvirin Geometric Image Processing Lab, Faculty of Computer Science, Technion—Israel Institute of Technology, Haifa, Israel

An Introduction to Three-Dimensional (3D) Printing in Oral Health Science

1

Ritu Duggal and Isha Duggal

Contents

1.1 What Is 3D Printing (3DP)?

Oxford dictionary describes 3D printing as "The process of making a physical object from a 3D 'digital model file' by laying down successive thin layers of a material" [1].

Dentists are familiar with the traditional substractive manufacturing in which a block of material is removed to form an object using tools like lathes and milling machines [2]. It can be understood with the example of carving of wax block to make a tooth in the preclinical undergraduate laboratory course work. This method however, had shortcomings like inability to reproduce complex objects entirely and wastage of material. Thus, there was evolution of another way called additive manufacturing, i.e., adding material layer by layer to build up the 3D object. This technology allows creation of three-dimensional parts from computer-aided design (CAD) models by successively adding materials layer by layer until formation of physical object.

1.2 How Does 3D Printing Process Work?

A 3D printing process begins with feeding a CAD model to a specific software for preparation for 3D printing. Depending on the technology used by the 3D printer, the printing might be done layer by layer by using solidifying resin or sintering powder. The parts then undergo post-processing for the specific application. The

R. Duggal (✉) · I. Duggal
Division of Orthodontics and Dentofacial Deformities, Center for Dental Education and Research (CDER), All India Institute of Medical Sciences, New Delhi, India
e-mail: ritudug@gmail.com

© The Author(s), under exclusive license to Springer Nature Switzerland AG 2022

P. K. Chaudhari et al. (eds.), *3D Printing in Oral Health Science*,
https://doi.org/10.1007/978-3-031-07369-4_1

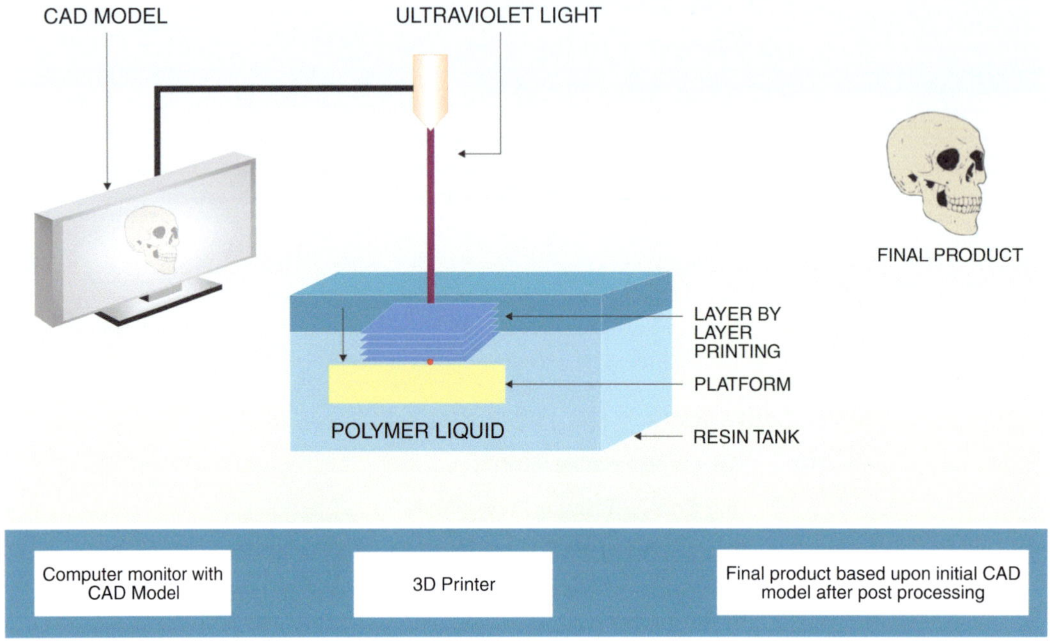

Fig. 1.1 Representation of the working process of a 3D printer

process of 3D printing essentially requires following steps (Fig. 1.1).

1. 3D digital model acquisition of desired object to be printed in standard tessellation language (STL) file format
2. Image processing by exporting the STL file to the 3D printer software to slice the digital models into layers
3. 3D printing
4. Postprocessing

Apart from these steps, the 3D printable materials including materials for support structures are also required.

1.2.1 3D Digital Model Acquisition

3D digital model can be acquired by non-nvasive surface imaging like desktop or intraoral scanning, computed tomography scanning, or any other suitable sources. Using specifically designed computer-aided design (CAD) softwares, the 3D data is saved as 3D digital models.

1.2.2 3D Digital Image Processing

In this step, CAD software further processes and prepares the STL file by breaking down the 3D model into thin layers of 16–300 μm each, known as "build layers" for 3D printing [3].

1.2.3 The 3D Printing

This step involves using a computer workstation to set up the print job; a build tray for fabricating the model; and 3D printable materials [3]. These 3D printers may use a laser light to cure liquid resin to make an object or fuse small particles of metal powder at high temperatures to build parts.

1.2.4 Post-processing

The support resin used with every printer prevents any deflection or movement. It also enables the printing of complex objects. Isopropyl alcohol (IPA) is required to rinse and to remove any uncured resin from the printed surface. Based on the technology and materials, the printed parts may also require post-curing to stabilize mechanical properties, manual work to remove support structures, or cleaning with compressed air or a media blaster to remove excess powder [4].

1.3 Brief History of 3D Printing Development

The 3D printing technology is not new when it comes to architectural designs, aerospace, defence, art, and engineering. Its use in medical science gained popularity especially in areas of precision medicine around 1990s. In the past 20 years, there has also been a sharp rise in the use of 3D printing in various domains of oral health science.

Chuck Hall of United States was the first person to develop the concept of 3D printing when he was working on "*Apparatus for production of three-dimensional objects by stereolithography*" on August 8, 1984 [5]. He used to design tabletop coatings through this process. In 1986, he established the 3D Systems Company to market the first machine for rapid prototyping of the tabletops, which was named as stereolithography (SLA).

In 1988, Scott Crump developed a similar variant of object formation by fusing 2D images and named it as "fused deposition modeling" (FDM), which was commercialized by Stratasys in 1990 [6].

Created by Larry Hornbeck of Texas Instruments in 1987, digital light projection (DLP) is identical to SLA except that a projector is used to cure an entire layer at a time. The DLP technology uses digital micromirror device, containing thousands of tiny mirrors that are able to move in two directions, on and off, thousands of times per second. It is advantageous in many ways, for example, faster printing speed and ability to produce more complex 3D objects with excellent resolution. It also uses lesser material and reduced waste, thus making it economical [3].

A third variant of 3D printing came from the developer of PolyJet photopolymer (PPP) printing in 1998. A variety of 3D printers—SLA, DLP, FDM, and PPP—are now commercially available for oral health applications.

1.4 Contemporary Applications and Limitations of 3D Printing in Oral Health Science

The versatility of 3D printing has enormous potential to revolutionize oral health science in terms of research, clinical care, and education. Further, there is a need to establish new, printable materials for dentistry, which would enhance the clinical applications of 3D printing. 3D printing has widespread applications in different disciplines of dentistry. Uses of 3D printing include the production of physical 3D printed models for prosthodontics, orthodontics and oral surgery, the production of surgical guides for dental implants, surgical splints for maxillofacial surgery applications, manufacture of dental, craniomaxillofacial and orthopedic implants, and the fabrication of copings and frameworks for implant and dental restorations [7].

To study the trend of the number of studies based upon use of 3D printing in journals indexed in PubMed over the last two 15 years (2006–2020), a search strategy was adopted and entered in PubMed. On the production of results, an excel sheet was made to analyze the data obtained in a chronological order. An overall upsurge in number of publications was observed during this time frame and the same is produced in Fig. 1.2. Out of the current total of 14,874 studies searched using the string "3D Printing"[MeSH Terms] OR ("3D Printing"[All Fields]), 4813 accounted for the field of medical sciences ("3D Printing"[All Fields] AND "Medicine"[All Fields]) and 1503 were dentistry-related studies ("3D Printing"[All Fields] AND "Dentistry"[All Fields]).

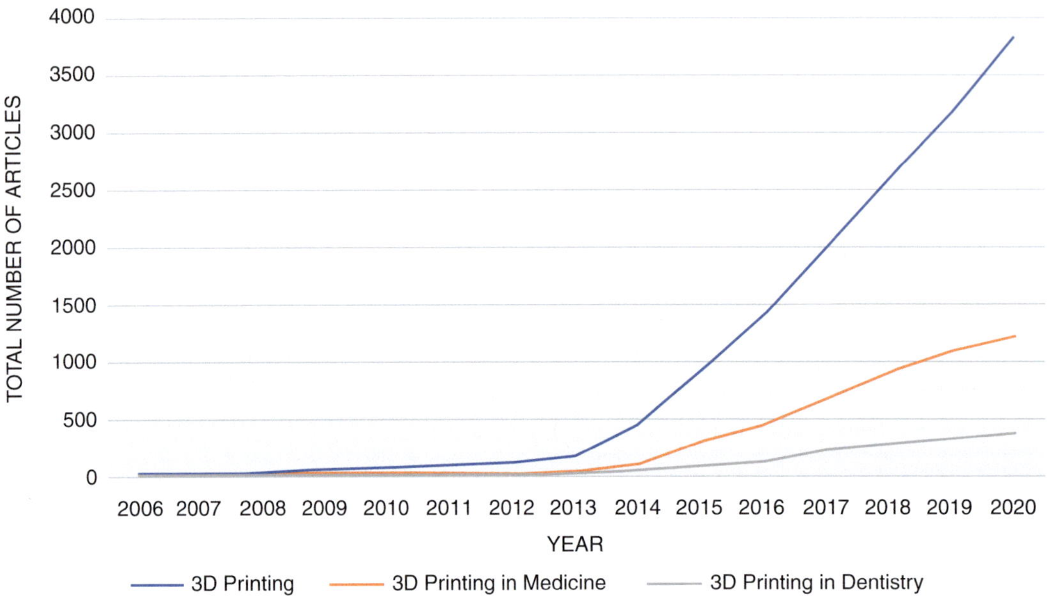

Fig. 1.2 Trends in publications related to 3D printing in medicine and dentistry in last 15 years

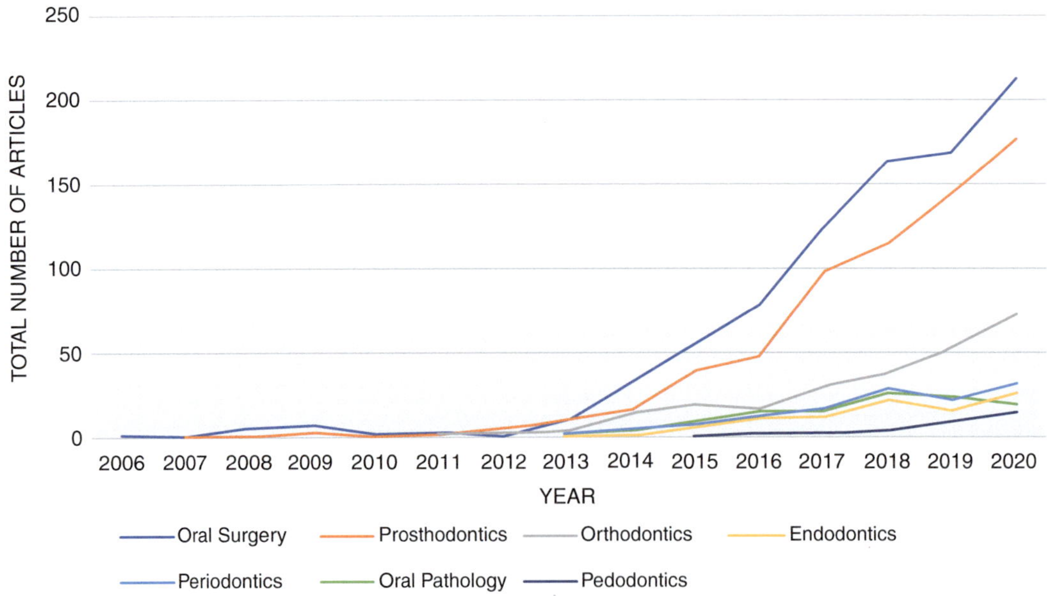

Fig. 1.3 Trends in publications related to 3D printing in various branches of dentistry in last 15 years

Similarly, on considering the fields of dentistry, the applications of 3D printing in dentistry have been shown in Fig. 1.3.

Amongst these 1503 studies related to 3D printing applications in dentistry ("3D Printing"[All Fields] AND "Dentistry"[All Fields]), majority were from the field of oral surgery, followed by prosthetics and consecutively by orthodontics.

Nonetheless, like every coin has two sides, 3D Printing also has certain limitations. It requires one to be skilled in computer softwares as perfect planning is usually required. Some of the materials used for printing surgical guides may not be biocompatible, thus restraining their use. Hence, factors like printing material availability, its properties, printing speed, and the desired resolution of the printed object must all be considered. In addition, the printing accuracy is mainly dependent upon the quality of the original scan which remains compromised when considering intraoral scanners for full arch or irregular surfaces.

1.5 Summary

The culmination of 3D printing technology and advanced materials has captivated the world of dentistry. Since with great power comes great challenges and responsibility, as of now a dentist's main trial would be to integrate these new technologies and equipment into the routine practice while ensuring the patient's highest standard of care, health, and safety.

References

1. Liu Q, Leu MC, Schmitt SM. Rapid prototyping in dentistry: technology and application. Int J Adv Manuf Technol. 2006;29:317–35.
2. Savini A, Savini GG. A short history of 3D printing, a technological revolution just started. In: 2015 ICOHTEC/IEEE International History of High-Technologies and their Socio-Cultural Contexts Conference (HISTELCON); 2015. p. 1–8.
3. Groth CH, Kravitz ND, Jones PE, Graham JW, Redmond WR. Three-dimensional printing technology. J Clin Orthod. 2014;48(8):475–85.
4. Bogdan E, Michorczyk P. 3D printing in heterogeneous catalysis-the state of the art. Materials (Basel). 2020;13(20):4534.
5. Hull CW. Apparatus for production of three-dimensional objects by stereolithography. U.S. Patent No. 4,575,330; 1986. https://patents.google.com/patent/US4575330A/en. Accessed 10 Apr 2020.
6. Gupta V, Nesterenko P, Paull B. An introduction to 3D printing. In: 3D printing in chemical sciences: applications across chemistry. London, UK: The Royal Society of Chemistry; 2019. p. 1–21.
7. Dawood A, Marti Marti B, Sauret-Jackson V, Darwood A. 3D printing in dentistry. Br Dent J. 2015;219(11):521–9.

Principles and Applications of Various 3D Scanning Methods for Image Acquisition for 3D Printing Applications in Oral Health Science

2

Yoon-Ji Kim, Nayansi Jha, Shalini Gupta, Alon Zvirin, Andis Qendro, Edlira Zere, Ramandeep Kaur Sokhi, K. S. Naveen, Jitendra Sharan, Suraj Sinha, Swaroop Anil, and Prabhat Kumar Chaudhari

Contents

Y.-J. Kim · N. Jha
Department of Orthodontics, Asan Medical Center, University of Ulsan College of Medicine, Seoul, South Korea
e-mail: yn0331@ulsan.ac.kr

S. Gupta
Division of Oral Medicine and Radiology, Centre for Dental Education and Research (CDER), All India Institute of Medical Sciences (AIIMS), New Delhi, India

A. Zvirin
Geometric Image Processing Lab, Faculty of Computer Science, Technion—Israel Institute of Technology, Haifa, Israel

A. Qendro
Oral and Maxillofacial Unit, Dental Implantology, Faculty of Dental Medicine, University Hospital Centre, Tirana, Albania

E. Zere
Department of Orthodontics and Craniofacial Anomalies, Faculty of Medicine, Rambam Health Care Campus, Technion—Israel Institute of Technology, Haifa, Israel

R. K. Sokhi · K. S. Naveen · P. K. Chaudhari (✉)
Division of Orthodontics and Dentofacial Deformities, Centre for Dental Education and Research (CDER), All India Institute of Medical Sciences (AIIMS), New Delhi, India
e-mail: dr.prabhatkc@aiims.edu, dr.prabhatkc@gmail.com

J. Sharan
Unit of Orthodontics and Dentofacial Orthopedics, Department of Dentistry, All India Institute of Medical Sciences (AIIMS), Bhubaneswar, Odisha, India

S. Sinha
Department of Craniofacial Orthodontics, Chang Gung Memorial Hospital, Taoyuan City, Taiwan

S. Anil
Pratap Dental Clinic and Implant Centre, Allahabad, Uttar Pradesh, India

© The Author(s), under exclusive license to Springer Nature Switzerland AG 2022
P. K. Chaudhari et al. (eds.), *3D Printing in Oral Health Science*,
https://doi.org/10.1007/978-3-031-07369-4_2

2.1 Noninvasive, Nonionizing Surface Imaging

2.1.1 Intraoral Scanning

2.1.1.1 An Introduction to Intraoral Scanners Including Its Advantages and Disadvantages

The intraoral scanner was first introduced in 1985, which allowed optical scanning of a tooth cavity and restoration made by milling using computer-aided design (CAD) and computer-aided manufacturing (CAM) technology [1, 2]. Since then, advancements in digital scanning and CAD/CAM have led to the use of intraoral scanners for digital impressions for restorative, surgical, and orthodontic indications. The accuracy of intraoral scanning technology has increased, and it is gradually replacing the conventional impression technique. Currently, the accuracy of full-arch intraoral scanned images exceeds that of irreversible hydrocolloid alginate and is comparable to that of rubber impression materials, such as polyvinyl siloxane and polyether [3]. In addition, the reliability of linear measurements taken from digital models has been confirmed [4–6]. As a result, various prostheses, such as inlays, crowns, fixed partial dentures, full dentures, and orthodontic appliances, such as customized prescription brackets [7], three-dimensional (3D) printed brackets [8], and clear aligners [9], are fabricated. Additionally, digital simulations of dental implant surgery and orthognathic surgeries are performed, and 3D-printed surgical guides provides better precision.

Advantages and Disadvantages

Intraoral scanners have been used effectively for recording of digital impression with several advantages. First, digital impression data can be easily customized to provide efficient and precise patient treatment. Most intraoral scanners support the conversion of scanned data to a standard tessellation language format (STL), which is one of the most native formats for 3D printers and CAD software. It facilitates data transfer for collaborative work in appliance design and 3D printing. In addition, appliance design and surgical planning can be combined with other types of image data such as 3D cone beam computed tomography, for precise planning and designing of an appliance. Second, it is patient-friendly, as it eliminates the process of using conventional materials to take an impression. It is especially convenient for children and those with a gag reflex and anxiety. Further, it reduces chair time [10]. However, a major disadvantage of the intraoral scanner is its high cost. The price ranges approximately USD 30,000–60,000, and usually, there are annual fees for software updates and maintenance. In addition, as there are advancements in hardware and software, clinicians may need to invest in a new intraoral scanner almost every 3–5 years, which can be an economic burden. Additionally, some scanners have a large head, which makes it difficult to scan the most distal tooth for patients with limited mouth opening.

2.1.1.2 Various Intraoral (IOS) Scanning Technologies and Principles

Tabletop vs. Intraoral Scanners (Fig. 2.1)

Digitization of tooth impression began with scanning the gypsum models using a tabletop model scanner [11]. Tabletop model scanners calculate the positions of the object three dimensionally, based on a known distance between the camera and light source [12]. Model scanners have a large field of view, and the scanner head moves along a precise linear or rotation axis [13]. In contrast, intraoral scanners have a small field of view and are handheld and thus move freely in many directions during the scan procedure. Numerous fragments of 3D scan data are acquired and superimposed to create the final 3D surface data. Therefore, a greater error may occur during the process of integrating the raw images. When larger objects are scanned, the amount of error may increase because a larger number of raw images need to be processed, possibly resulting in dimensional inaccuracies of the scanned data.

It is known that the accuracy of digital impressions obtained from a single tooth is equivalent or even superior to that of conventional impressions [14–16]. However, the accuracy of full-arch

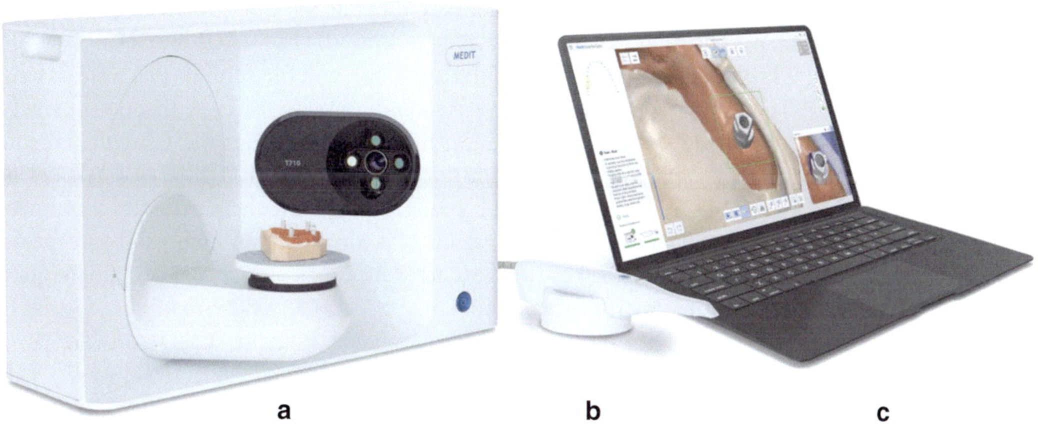

Fig. 2.1 (**a**) A tabletop scanner scans dental models of patients, and (**b**) an intraoral scanner directly scans patient's teeth. (**c**) Both tabletop and intraoral scanner need a computer software for data analysis and storage

intraoral scan data may vary depending on various factors [14, 17, 18]. A full-arch scan starts from the most distal molar and ends at the most distal molar of the contralateral side; therefore, errors in superimposition accumulate, leading to deformation of the arch shape and errors in the arch dimensions, particularly in the molar region [10–21]. Consequently, tabletop model scanners have been shown to have greater accuracy than intraoral scanners [22]. However, a tabletop scanner does not eliminate the process of taking the conventional impression and fabrication of a stone model, which explains why intraoral scanners are gaining attention. Despite the technical difficulties in obtaining an accurate intraoral scan, there have been significant developments in software algorithms for processing raw scanned images and constructing accurate 3D data that are clinically acceptable.

An intraoral scanner system consists of three components: (1) a handheld camera for scanning, (2) a computer connected to the intraoral scanner for processing raw data and storing scanned files, and (3) a software for constructing the captured images into 3D scanned data (Fig. 2.2). Moreover, there are three main technologies for acquiring the 3D scanned data of an object.

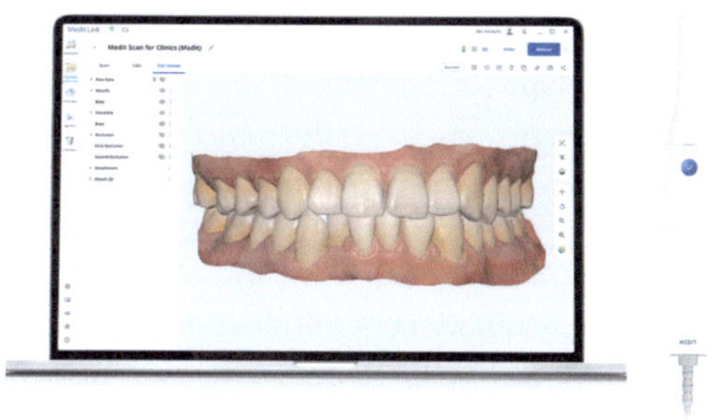

Fig. 2.2 An intraoral scanner system consists of a handheld scanner, a computer, and a software

1. **The confocal microscopy** technique is based on the theory of confocal microscopy, which was originally developed by Marvin Minsky. The key principle behind this technology is to differentiate between the "focus" (good data) and "out-of-focus" light (bad data). Laser light is projected through a filtering pinhole, and it falls on the target tissue. A sensor is placed in the confocal plane, and there is a small aperture that filters out any light from above or below the plane of focus. This eliminates bad data, thereby only allowing focused light to reach the sensor for processing; hence, the entire process improves the accuracy of the scan. This process is also called the point-and-stitch reconstruction because it takes thousands of slices and stitches to reconstruct the final image [23]. The most popular intraoral scan systems using this technique are iTero and Trios.
2. **The optical triangulation** method is based on the mathematical formula for a triangular

surface, where the position of a point can be calculated by viewing it from two different positions, with the angle and position of those two points in mind. The known variables in the equation are the "distance between the source of laser and sensor" and the "angle between the sensor and laser." Light is reflected from the surface, and the system measures the angle of reflection. With these values, the distance to the object from the laser source can be easily calculated. An opaque surface is usually ideal for uniform light dispersion, and to achieve this, a thin opaque powder coating is also applied to the target area [23]. Both active and passive lighting environments are supported in this technique. In the passive technique, the camera relies on ambient light for illumination; hence, there is some dependency on the actual shape and texture of the object. However, active techniques are less dependent on the shape of the object because the camera projects different colored lights to capture images or videos. Equipment utilizing passive lighting can be small, lightweight, and easy to produce because they do not require other hardware. Some of the systems utilizing this technique are Omnicam, FastScan™, DirectScan, etc. [24].

3. **Active wavefront sampling** (AWS) is a surface imaging technique that processes data from a digital camera to generate 3D geometries. The key principle of this technique is that it uses a special AWS module, which is an off-axis aperture rotating around the optical axis and capturing the rotation of target regions in a circular format on the image plane. Each point produces a circular pattern, and the depth can be calculated based on the radius of the pattern. Intelligibly, it can be seen as having multiple cameras, as it captures from different viewpoints, thereby increasing measurement sensitivity. The Lava™ intraoral scanner was developed using this technique [24].

After the scanned data are acquired and reconstructed, they are stored in several formats to facilitate them as an input for rapid prototyping or 3D printing. STL, wavefront 3D object file (OBJ), and polygon file format (PLY) are among the standard formats. STL is the most widely used format which defines the triangulated surface encoding by the unit normal and vertices of the triangle based on the Cartesian coordinate system. It is supported by most software packages, as it is quick and simple to implement, hence has wider adoption. Since STL format is readily available from the CAD software, it is often used to import data to CAM systems. STL format only describes the 3D surface geometry without showing the color, texture, or other CAD attributes [25].

A higher model fidelity requires a greater number of triangles; thus, the file size increases exponentially.

Other formats, such as OBJ, are richer and slightly more accurate as they contain more information. The OBJ format represents the 3D geometry, including geometric vertices, texture coordinates, vertex normals, and polygonal faces. As the OBJ has color and texture information, it is appropriate for cases of multicolor and multimaterial 3D printing, as well as complex cases that require further modeling and editing [26]. However, STL is preferred when sharing designs intended for 3D printing mainly because of its simplicity and small file size, making it ideal for storage and quick sharing.

Another commonly used format specifically designed for 3D scanners is PLY or the Stanford Triangle Format, which was originally designed by the Stanford Graphics Lab to store 3D data from scanners. In some cases, it can also be used as an alternative for STL.

2.1.1.3 Accuracy of Intraoral Scanning Technology

Methods for testing accuracy: Trueness and Precision (Fig. 2.3).

The accuracy of digital impressions can be expressed in terms of trueness and precision. According to ISO 5725-1, trueness is defined as "the discrepancy between a test object and a reference model" [26]. It refers to the dimensional accuracy of the acquired data compared to the

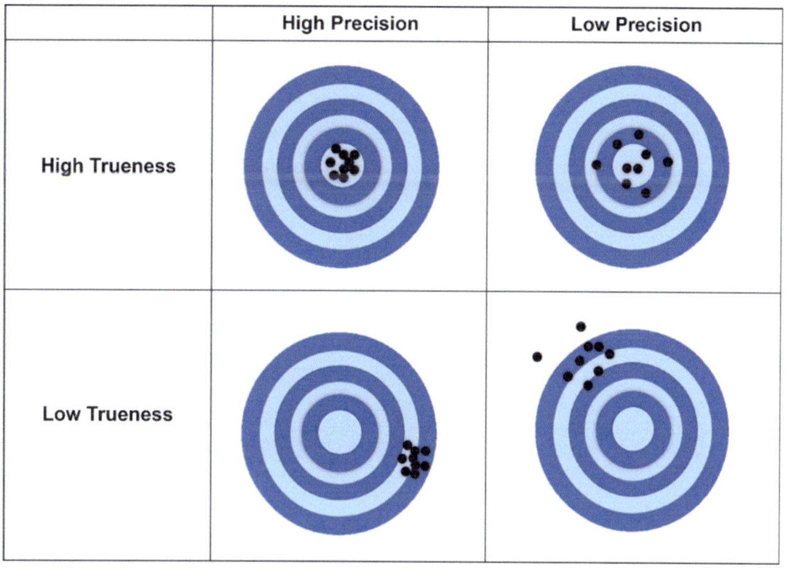

Fig. 2.3 Trueness and precision are methods for measuring accuracy of a system

reference or the "truth." Precision is the difference among the multiple acquisitions of data under the same prescribed conditions. It refers to the repeatability of data acquired from the same object [27, 28]. Both trueness and precision are vital for clinical efficiency and treatment outcomes, as they are directly related to the dimensional accuracy of prostheses and/or appliances delivered to the patient (Fig. 2.4).

In Vitro Trueness (Fig. 2.5a)

In vitro trueness refers to testing accuracy using reference data ("true" data) acquired from a dental cast or a standard dental model. One method is to compare the intraoral scan data acquired from a patient with the scan data of the gypsum model using a model scanner. Another method is to use a standard master model and perform scans of the model using an intraoral scanner and a model scanner and then compare the two scans. The limitation of the former method is that the reference scan data for analyzing trueness are the dental cast scan data, and errors may arise from the impression material. Dimensional changes occur in the impression material, which leads to inaccuracies in the dental cast. In contrast, the latter method is free of errors arising from the dimensional changes of the impression material; however, scans using intraoral scanners are taken on a model that is different from the actual patients' teeth. The intraoral environment is dark and humid with the tongue and cheek, limiting the movement of the scanner head. In addition, the teeth have significant differences in transparency and optical reflections compared to the dental models [29]. Therefore, errors in intraoral scan data may be greater when a patient, instead of a dental cast, is scanned. Therefore, a dental model scan is more accurate than that of a real patient, which leads to an underestimation of the error in trueness. Nevertheless, it is the most common method of testing accuracy in trueness, because it is easy to obtain reference scan data, and many studies have conducted in vitro evaluations of trueness using dental casts.

In Vivo Trueness (Fig. 2.5b)

In vivo trueness requires measuring the accuracy of the intraoral scan data with reference data taken directly from the patients' teeth. Obtaining accurate reference data for the analysis of trueness is difficult because dental arches cannot be scanned with tactile or other high-precision optical laboratory scanners. Previous studies on trueness used dental casts obtained from conventional impressions as a reference, and errors may have occurred from the dimensional stability of the impression material.

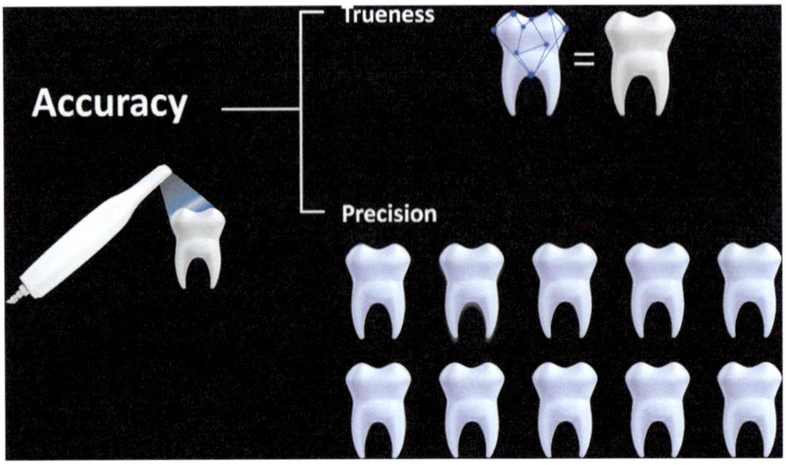

Fig. 2.4 Trueness refers to the dimensional error compared to the "truth" or the reference data. Precision refers to the repeatability of the acquired data

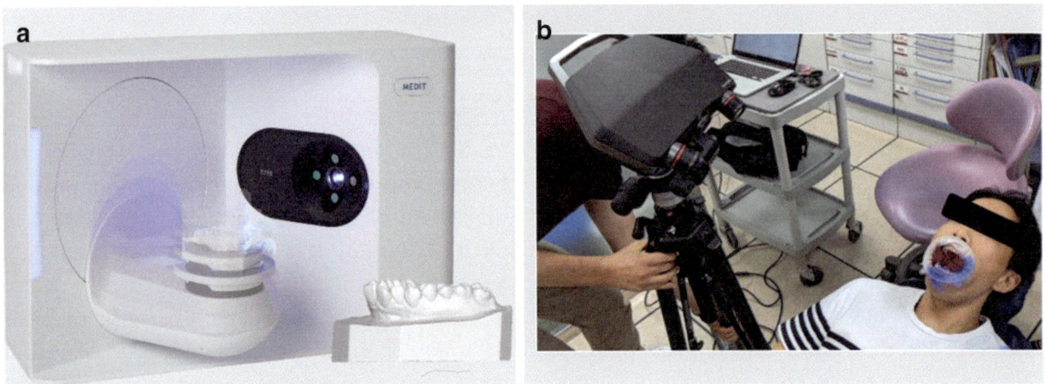

Fig. 2.5 Reference data for trueness can be acquired from a model scan (**a**) or from a direct scan of the patient's teeth (**b**)

An industrial optical scanner may be used to obtain reference scan data directly from the patient's mouth. Industrial-grade scanners have a large field of view; thus, fewer images are superimposed on each other. It has been reported to have an accuracy of less than 10 μm, which is more accurate than conventional impressions (Fig. 2.6, Table 2.1) [30]. Therefore, it may be used as a reference for measuring the trueness of full-arch digital impressions in vivo. In one study, the labial and buccal surfaces of the incisors and premolars were captured using an industrial scanner, and the trueness of the incisors and premolar area was assessed [31]. Another study evaluated full-arch trueness and precision using an industrial-grade scanner for reference scans [30]. As it is not possible to scan the whole arch using industrial scanners, reference spheres were attached to the canines and molars, and the industrial scanner captured the occlusal and surface of the dental arch, including the reference spheres, for analyzing dimensional accuracy (Fig. 2.7).

Table 2.1 Machine specifications of Solutionix C500 industrial scanner

Solutionix C500 (MEDIT, Seoul, Korea)	
Camera resolution	2 × 5.0 MP
Point spacing	0.028 ~ 0.157 mm
3D scanning area (FOV)	90/175/350/500 mm
3D scanning principle	Phase-shifting optical triangulation
Accuracy	7–10 μm
Light source	Blue LED
Mount	Detachable scanner head

Precision

Precision is analyzed from the repeated acquisition of data. It can be obtained from dental models (in vitro) or from a patient (in vivo). Similar to trueness, in vitro analyses underestimate errors due to the ease of acquiring scans outside the oral cavity [32–35] (Fig. 2.8).

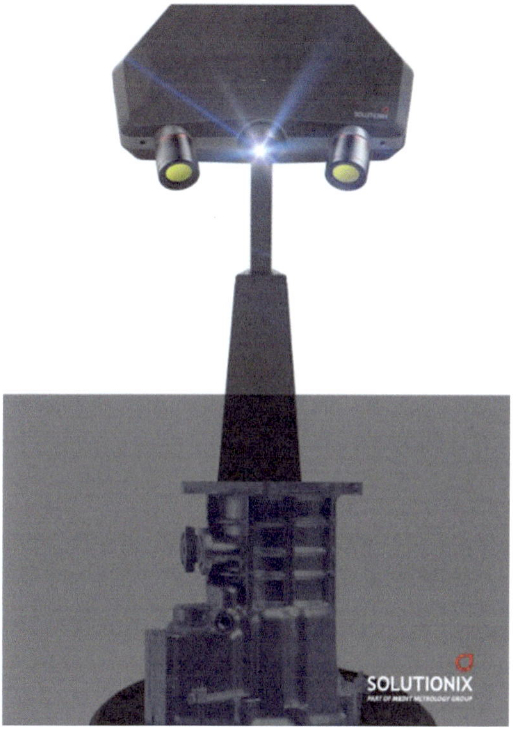

Fig. 2.6 An industrial scanner, Solutionix C500 (Medit, South Korea), with an accuracy of 7–10 μm

Effect of Software Versions on Intraoral Scanner Accuracy

The software version of intraoral scanners has a considerable effect on the accuracy of intraoral scan data. Software algorithms for processing scanned images are crucial for constructing accurate 3D scan data. Manufacturers continuously develop scanner software to process the raw data captured from the scanner with greater accuracy; therefore, clinicians must undergo software updates when available. Additionally, calibration should be performed regularly. Several studies have been conducted on the accuracy of scan data based on software updates. It has been reported that the marginal adaptation of crowns fabricated using digital impressions improved when the scanner used updated software [36]. Other studies showed increased trueness and precision after software updates. There may be differences in the improvements in accuracy according to the scanner type.

2.1.1.4 The Various Commercially Available Intraoral Scanners

Intraoral scanners have been developed by several companies in the digital device industry, such as iTero, Omnicam, and Trios. Europe is one of the largest consumers of intraoral scanners,

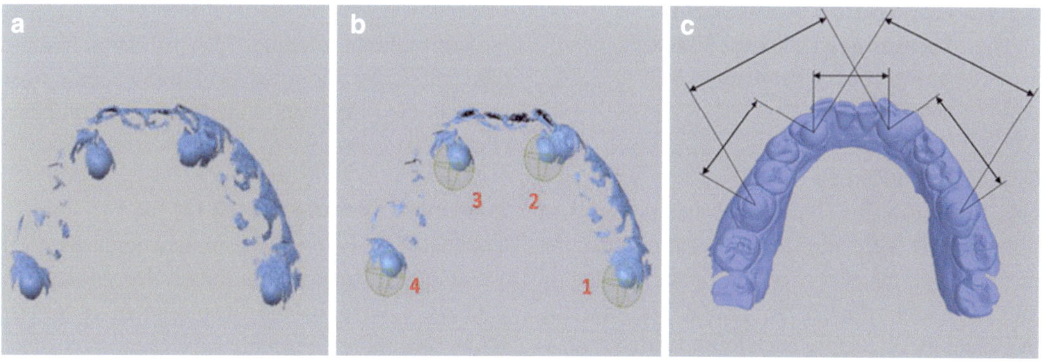

Fig. 2.7 (**a**) Intraoral reference scan acquired using an industrial scanner. (**b**) The four spheres are automatically detected, and the distance between the centers of the spheres is measured using the Geomagic Control X (Evatronix SA, Bielsko-Biala, Poland). (**c**) Same procedures are done in the intraoral scan data to measure the distance between the spheres

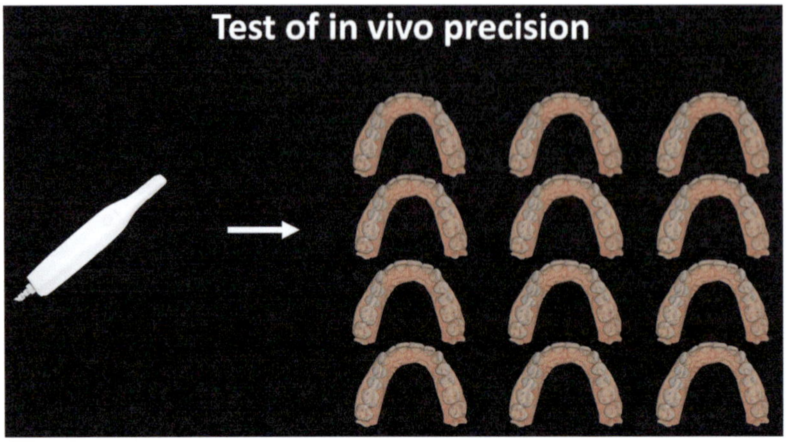

Fig. 2.8 Multiple intraoral scan data are compared to determine precision

followed by North America. There are several advantages and limitations of different types of scanners, and they mostly depend on their clinical use, such as in orthodontics, dental implants, restoratives, and clinician's preference.

Intraoral scanners with a lightweight and ergonomic design, including a small scanner head, facilitate better manipulation. However, a small head size may have a small field of view, which is associated with accuracy. In contrast, scanners with a large head might be difficult for patients with limited mouth opening or those with temporomandibular joint diseases. Currently, there are many different types of intraoral scanners available in the market, and below is the compilation of popular intraoral scanners (Table 2.2) [23, 37].

2.1.1.5 Application of Intraoral Scanning in Various Fields of Dentistry Including 3D Printing

Digital Orthodontics

One application of CAD/CAM technologies in orthodontics is the digital tooth setup. After the diagnosis of malocclusion, the patient's dental plaster model is used to simulate orthodontic treatment by dissecting each tooth from the plaster model and rearranging them into the desired position. This process provides orthodontists insights into the possibilities and limitations of tooth movement. However, this process is time-consuming and labor-intensive. Instead of using the gypsum model, intraoral scan data can be manipulated to obtain a tooth setup. Initially, the teeth from the intraoral scan data are individually segmented. An intraoral scan data contain all teeth as a single unit. Although each tooth has its boundary and teeth are not connected, the interdental areas are not captured by the intraoral scanner, which is called an "occlusion" area where no data are acquired. The software automatically fills the area of missing data [38]. Therefore, the tooth boundary and long axis of the clinical crown should be designated, which is used for tooth segmentation (Fig. 2.9).

After segmentation, the separated teeth are moved to the desired position using a software. The digital setup is more efficient than the conventional methods, as it eliminates the process of making a plaster model, and it is as accurate as the conventional methods [39, 40]. Digital setup data can be used to make orthodontic appliances such as clear aligners and customized brackets (Fig. 2.10).

3D Printed Dentures (Fig. 2.11)

Conventional removable complete dentures (CD) are standard treatment protocols for edentulous patients. Owing to the development of CAD/CAM technology, digitally fabricated removable CDs have gained attention in recent years. Compared to conventional CDs, it requires fewer visits, as low as two visits for final delivery. At the first visit, impressions are obtained using

Table 2.2 Commonly used commercially available intraoral scanners

Model	iTero	Trios®	CS 3500	Planmeca Emerald	Lythos Digital Impression System	3M True Definition	CEREC Omnicam	Medit i700
Marketed By	Align Technology Inc., USA	3 Shape, Denmark	Carestream Dental, USA	Planmeca, Finland	Ormco, USA	3M ESPE Dental, USA	Dentsply Sirona, USA	Medit, South Korea
Technology	Confocal laser scanner microscopy	Confocal microscopy and ultrafast optical scanning	Confocal laser scanner microscopy	Triangulation	Accordion Fringe Interferometry	3D-in-motion video technology	Confocal microscopy and triangulation	Triangulation
Work station configuration	Cart; portable	Cart; portable; mounted on dental chair	Portable	Portable; mounted on dental chair	Portable	Cart	Cart	Portable
File type	Open STL	Open STL	Open STL	Open STL	Open STL	Open STL	Open STL	Open STL
Tips	Disposable	Autoclavable	Autoclavable	Chemical disinfection or autoclaving	Disposable	The entire wand is cold disinfected	Metal with sapphire scratch-resistant lens: HLD (ProCide/Cidex OPA), dry heat sterilizable	Autoclavable
Tip sizes available	One size	3.4″H × 0.8″W × 0.8″D	Small and large	Small and large	One size	16.2 mm × 14.4 mm	4.5″ × 0.63″ × 0.63″	One size
Color	Yes	Yes	Yes	No	No	No	Yes	Yes
Powdering	No	No	No	Rarely	No	Yes	No	No
Unit price	………………	\$22,000 to \$42,000	……………	–	\$19,999	\$15,995	\$39,995	\$20,000
Annual software fee	First year included. \$330/month thereafter	\$1900 to \$4100	……………	–	\$0	Two data plans available. Standard: \$199/month. Advanced: \$329/month	N/A. No data transmission or per scan fees	\$0
Cloud storage	No	No, transfer via 3shape communicate	No	Yes, Planmeca Romexis Cloud	Yes	Yes	Yes	Yes

(continued)

Table 2.2 (continued)

Model	iTero	Trios®	CS 3500	Planmeca Emerald	Lythos Digital Impression System	3M True Definition	CEREC Omnicam	Medit i700
Unique features	Real-time 3D visualization; color scanning; advanced wand features with built-in controls, 20× faster scan speed	Real colors HD image capture; instant auto-occlusion capture; stop and resume scans at any point	Part of an open system; workflows for restorative dentistry, orthodontics, and implantology	Ergonomic, improved user interface and software; slimline tip for accessing posterior teeth; transillumination tip for caries detection	100% open platform with no fees	Seamless integrations to a range of workflows; new wand is the smallest in the industry	Battery backup ensures uninterrupted use; camera provides robust optical solution with no delicate moving parts	Smaller and lighter than Medit i500; easy to use, and ergonomic; full-arch accuracy 11 μm; powerful software with fast updates

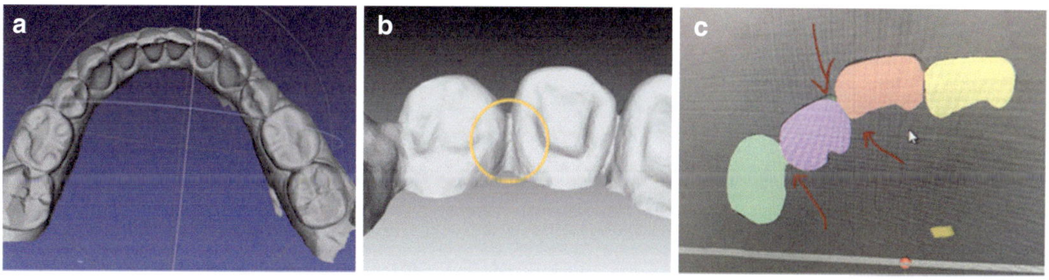

Fig. 2.9 (**a**) An intraoral scan data of a complete arch. (**b**) Scan data are not captured in the interdental area due to occlusion. (**c**) Tooth segmentation is needed for digital tooth setup

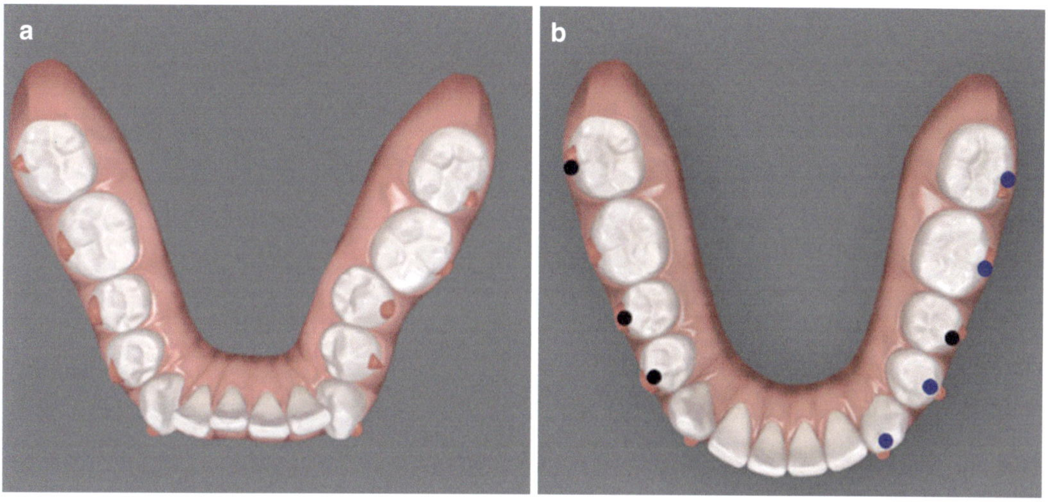

Fig. 2.10 (**a**) Intraoral scan data of a patient. (**b**) After digital setup

intraoral scanners. Then, the denture is designed virtually using a computer software. At the second visit, the dentures are provided to the patient.

There are two types of dentures with a digital protocol: milling and 3D printing. Both types of CAD/CAM dentures have been shown to have advantages over conventional CDs, such as better retention and superior mechanical and surface properties [41]. In addition, the digital denture protocol has decreased chair time, laboratory cost, and overall cost [42]. Regarding accuracy, CAD/CAM CDs have improved fit and trueness similar to those of conventional CDs [43]. Therefore, the digital protocol for CAD/CAM CDs may be especially beneficial for elderly and/or compromised edentulous patients, because it minimizes the treatment burden by reducing clinical procedures, time spent, and costs.

Dental Implants (Fig. 2.12)

The digital protocol for dental implants also saves chair time. For single and double abutments, the total mean time is less than 6 min, while conventional impressions require at least 20 min [44]. Additionally, digital impressions for dental implants eliminate the chances of infection, tray selection procedure, impression material distortion [11], and use of impression copings. In particular, screw-retained impression copings, which are preferred because of their higher accuracy, require a wide mouth opening for loosening of screw retentions on top of the impression trays. However, using an intraoral scanner, optical impressions are sufficient because only scans are needed to record dental implant positions, preventing the patients from opening their mouths wide for a long time [45].

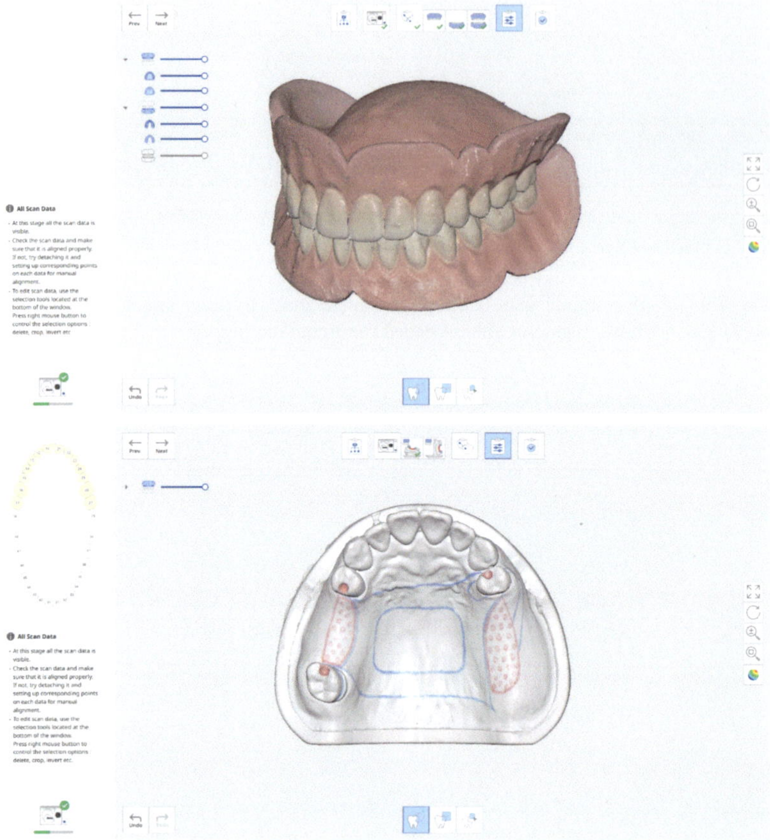

Fig. 2.11 A digital complete denture and a removable partial denture using the Medit Link software (ver 2.4, Medit, South Korea)

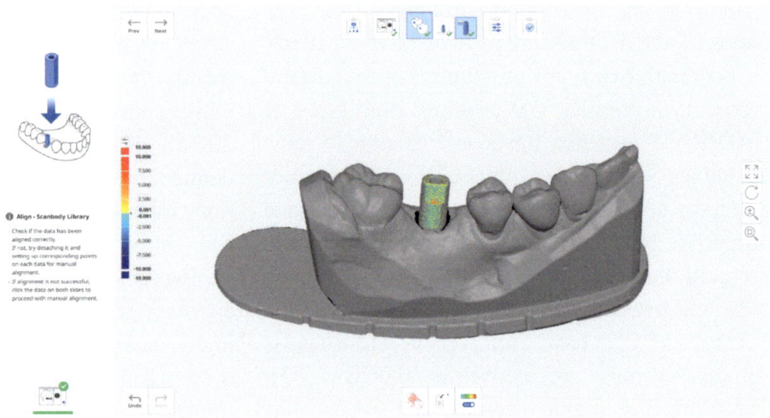

Fig. 2.12 A 3D model of implant scan body can be registered to the digital impression for accurate implant prosthesis

The workflow for an implant prosthesis using an intraoral scanner involves the following steps:

- Place the intraoral scanner in the region of the scan body (abutment/analog) in the oral cavity.
- Choose the arch (maxilla/mandible) to obtain an impression.
- Place the scanner approximately 5 mm away from each tooth surface and perform a scan of the occlusal tooth surface.
- Rotate the wand approximately 45°–90° to scan the lingual side, followed by the buccal side.
- Scan the surrounding proximal surface to which the scan body is connected.

Using the CAD/CAM system, abutment and prostheses are designed and manufactured (sintering and crown milling). Polished and finished products are then delivered in the patient's mouth [11].

When considering implant-supported restorations, passive fit of superstructures is important, which is directly related to the implant impression accuracy [45]. Misfit in fixed partial dentures can result in undue forces on the underlying teeth, which can still adapt well because of the presence of periodontal ligaments in the tooth structure. In contrast, implants are devoid of any periodontal ligaments, and such forces can result in screw loosening or implant fracture and other prosthetic complications [11]. A misfit of 30–150 μm between prosthetic framework and implant abutment can result in mechanical and biological complications [46]. To prevent long-term clinical complications and promote even distribution of masticatory forces [46], one should aim for a maximum possible passive fit [11].

The accuracy of digital impressions for implants depends on several factors, including implant-abutment connection, prosthetic components, clinical experience, placement depth, scanning technique [10], scanner type, and scanner speed. Thorough understanding of the digital workflow and good collaboration between the clinician and dental technician is essential for long-lasting treatment results.

2.1.2 Desktop Scanning

2.1.2.1 An Introduction of Desktop Scanning Including Advantages and Disadvantages

About 30 years ago, 3D scanning of the dental arch was introduced for use with computer-aided design and computer-aided manufacturing (CAD/CAM) technology to make dental restorations [47]. Its use in dentistry has now expanded significantly throughout this time. Desktop/benchtop scanners/extraoral scanners are devices used in dentistry to digitalize traditional impressions or plaster casts/models [48]. The scanning process of these scanners uses a nondestructive laser beam and numerous digital cameras, to reproduce high-resolution three-dimensional digital data of the target's surfaces [49]. Impressions and plaster dental models are placed inside a scanning chamber platform that rotates and tilts automatically during scanning, ensuring that the model's geometry is covered from all angles [49]. The target object is projected with laser light, and the cameras capture the object's mirror image. Following the completion of the scanning process, a rendered digital stereolithographic (STL) model is formed obviating the requirement for physical storage of dental models [49].

In simplest term, desktop scanners employ a nondestructive laser beam to illuminate the target object and several digital cameras to reproduce high-resolution images of the target's surfaces [50]. Many authors have suggest plaster model scanning as a potential solution for overcoming existing limitations in handling traditional plaster casts along with added advantages in diagnosis, treatment planning, and communication with other specialist including laboratory [50–57]. These digital dental models can be integrated with other diagnostic aids like cone beam computed tomography scan data for treatment planning, treatment outcome prediction, and creation of digital 3D printable files of planed devices or appliances like crown and bridge and surgical guides [58]. One of the important disadvantages of these desktop scanners is that most of the benchtop scanners can only scan 65–70% of impressions. These scanners have issues getting all of the digital impressions data because of the negative dynamic topography (i.e., undercuts) [58]. They have difficulty in capturing the areas of undercuts, especially in cases of mandibular anterior crowding.

The first step of the digital workflow in dentistry is 3D data acquisition of the oro-dental tissue either using desktop scanning of plaster dental models or using direct intraoral scanning [59]. The data acquisition by desktop scanners includes defining of surface topography and recording the surface data points with a high degree of accuracy and precision [60]. The recorded digital scan is then transformed into a universal file format, i.e., the STL (Standard Tessellation Language) file. Based on various scanning technologies and software processing techniques, the desktop scanners provide datasets of varied quality [61]. In a recent systematic review, authors found that the desktop scanners from different manufacturers of the same generation have nearly comparable accuracy [62].

2.1.2.2 Technical Characteristics of Some Commercially Available Desktop Scanners

A variety of desktop scanners are available, allowing dentists to scan and digitalize traditional impressions and models. Technical details of various commercially available desktop scanners have been presented in Table 2.1. The detailed description of some scanners presented below.

1. **Maestro 3D Dental Scanner**

   The Maestro dental scanner system (AGE Solutions, Pisa, Italy) has a LED projector with two digital cameras, which capture scans with 0.07 mm resolution and 10 μm accuracy (Fig. 2.13) [49]. The Maestro 3D extraoral scanner comes with several modules: Easy Dental Scan software for inspection and editing; Ortho Studio software for tooth, arch, overjet, and overbite measurements, cross sectioning, and occlusion inspection; Virtual Setup module for tooth movement, distance and collision evaluation, attachment management, modeling, and export for 3D printing. Scanning of the plaster dental model using Maestro easy dental scan software is completely automated. The acquisition of digital dental model and formation of virtual base from a plaster model using Maestro easy dental scan software and Maestro 3D dental scanner has been shown in Figs. 2.14, 2.15, 2.16, 2.17, 2.18, 2.19, 2.20, 2.21, 2.22, 2.23 and 2.24.

2. **OrthoInsight 3D Laser Scanner**

   In 2012, Ortho Insight 3D™ (Motion View Software, LLC, Chattanooga TN) was launched. The Ortho Insight 3D® desktop laser scanner allows to convert physical models and impressions into 3D digital models quickly and effortlessly [63]. The scanner uses three lasers which provide a high-resolution, robotic scan with a

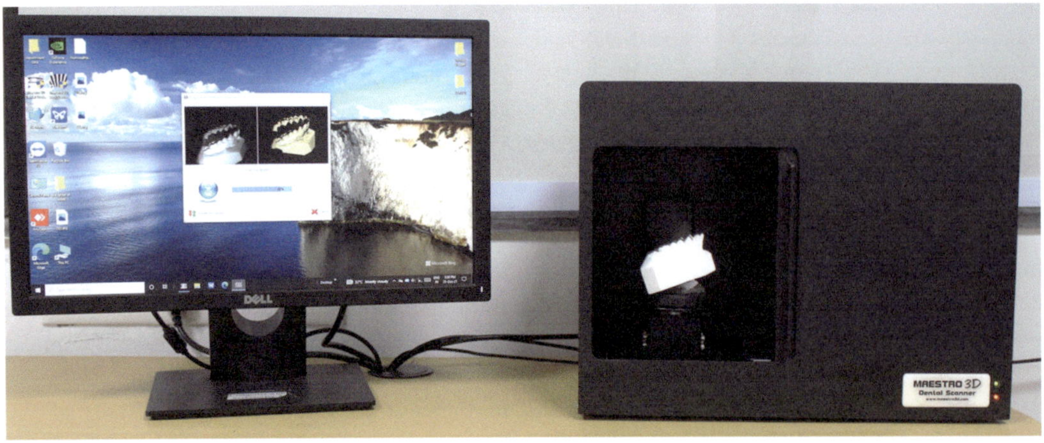

Fig. 2.13 Maestro 3D desktop scanner

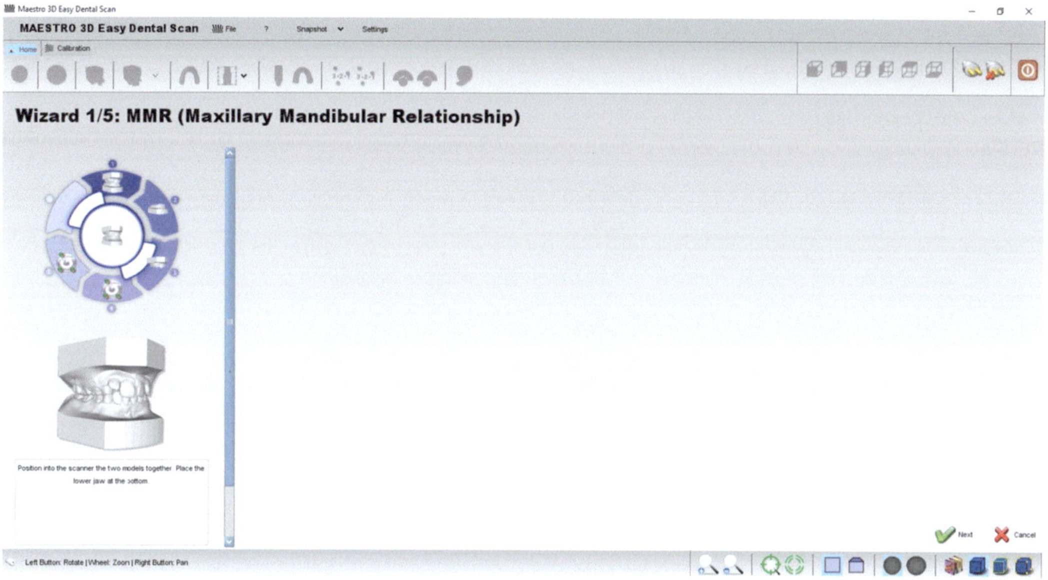

Fig. 2.14 Scanner instructing the placement of both maxillary and mandibular models in occlusion

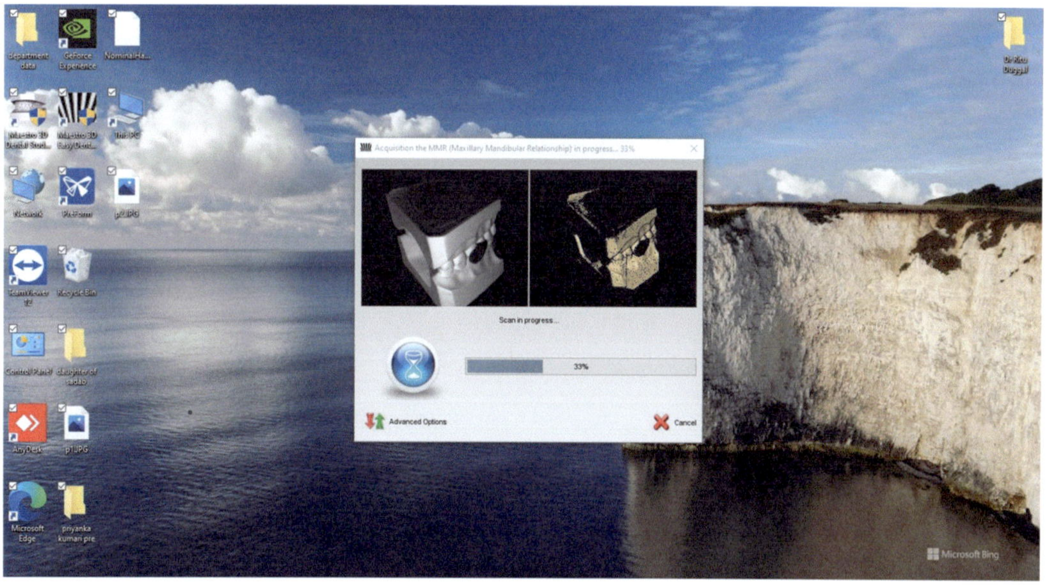

Fig. 2.15 Scanning of maxillary and mandibular models in occlusion

40–200 μm accuracy [49]. The scanning output can be saved in different file formats like free open file formats, i.e., STL (Standard Triangle Language/Standard Tessellation Language) file, OBJ (Wavefront OBJect) file, and PLY file (Polygon File Format or the Stanford Triangle Format) [64]. The software is able to reconstruct casts along with measuring the distance between teeth and measuring size of the teeth as well. By using the software, it enables us to digitize teeth, detect all the

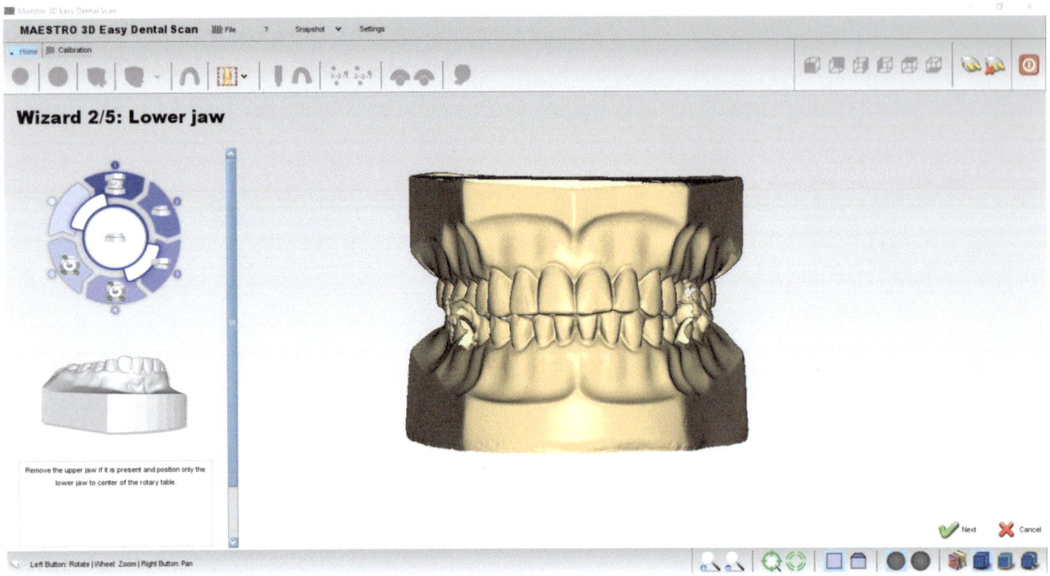

Fig. 2.16 Scanned digital models of maxillary and mandibular models in occlusion

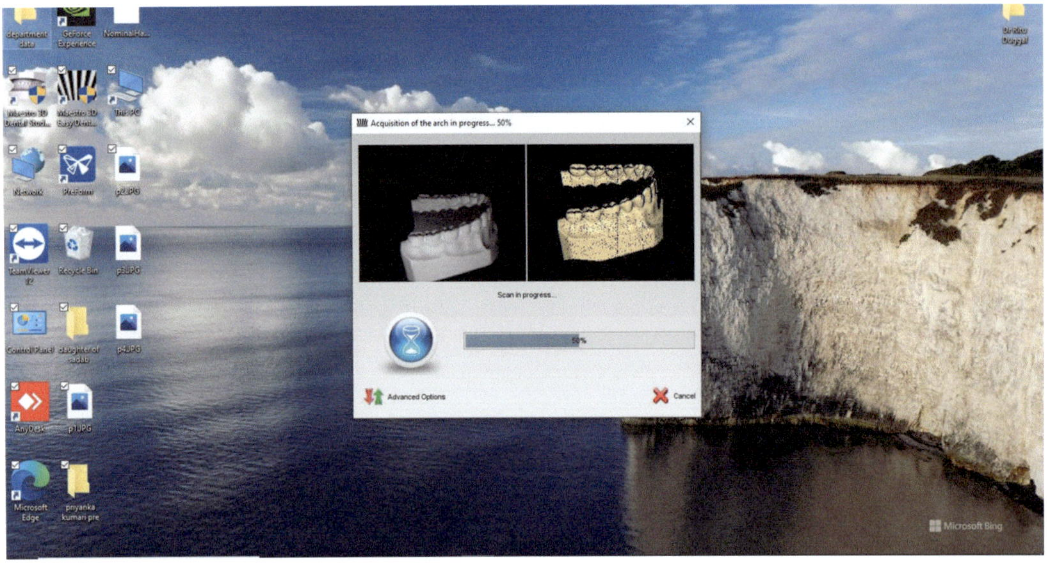

Fig. 2.17 Next step after removal of upper model, scanning of mandibular model starts

various landmarks, establish a facial axis, bring the teeth in occlusion, separation of the teeth from occlusion, measuring the mesio-distal distance of each tooth.

3. **3Shape Company Desktop Scanner**

   The 3Shape company desktop scanner (Copenhagen, Denmark) can be used for scanning of impressions and dental casts having applications in different specialties of dentistry. They have a wide range of available scanners (Table 2.3). The R500 and R700 series use red light laser technology with two 1.3-megapixel digital cameras which ensure 20 μm accuracy [49]. The advertised R500 series scanning time is 2 min and 20 s for a plaster model and 6 min and 40 s for an impression. The advertised

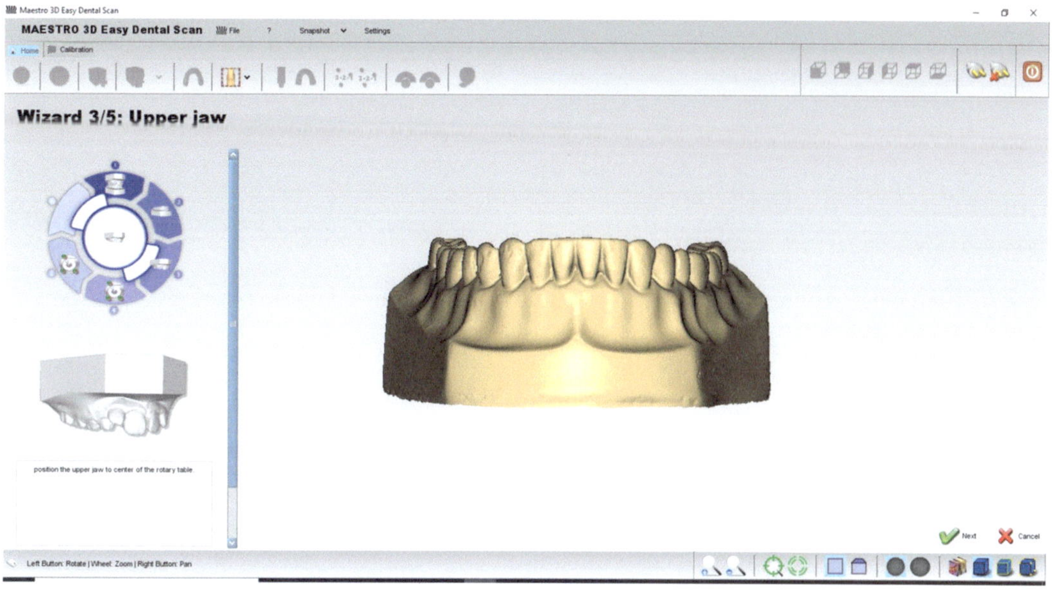

Fig. 2.18 Scanned digital model of mandibular model

Fig. 2.19 Lower model is removed, and the upper model is placed in the scanner for scanning

R700 series scanning time is 1 min and 30 s for a plaster model and 7 min for an impression which makes the scanner suitable for medium-sized orthodontic offices and laboratories. Ortho Analyzer™ is the 3Shape imaging and digital model software package which features sculpt and rebase applications with collision control, tooth movement simulation, superimposition of study models with photographs or DICOM data originating from CBCT scanners, and digital manufacture of appliances or dental restorations [49].

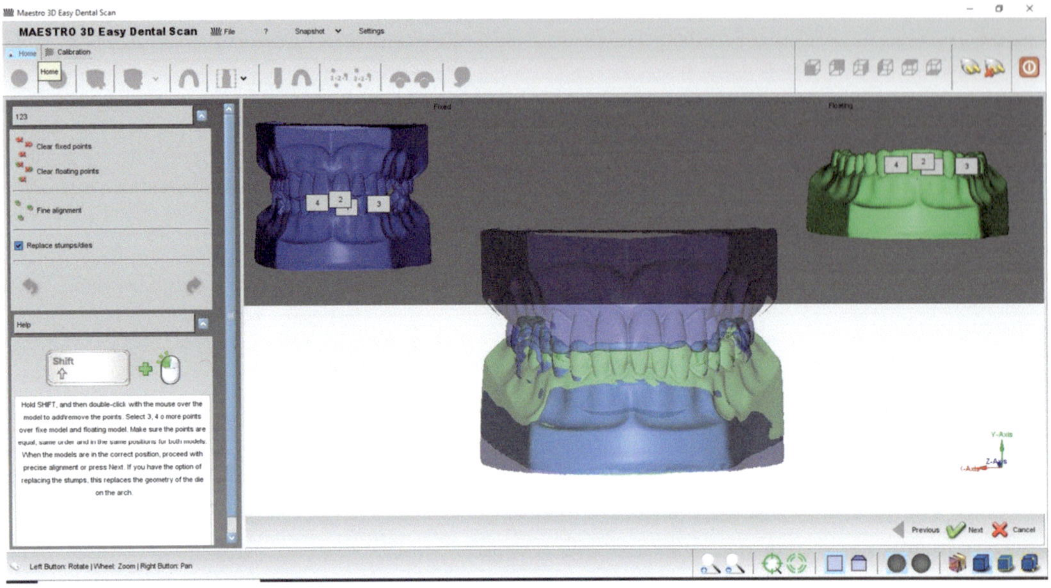

Fig. 2.20 Orientation of scanned digital mandibular model

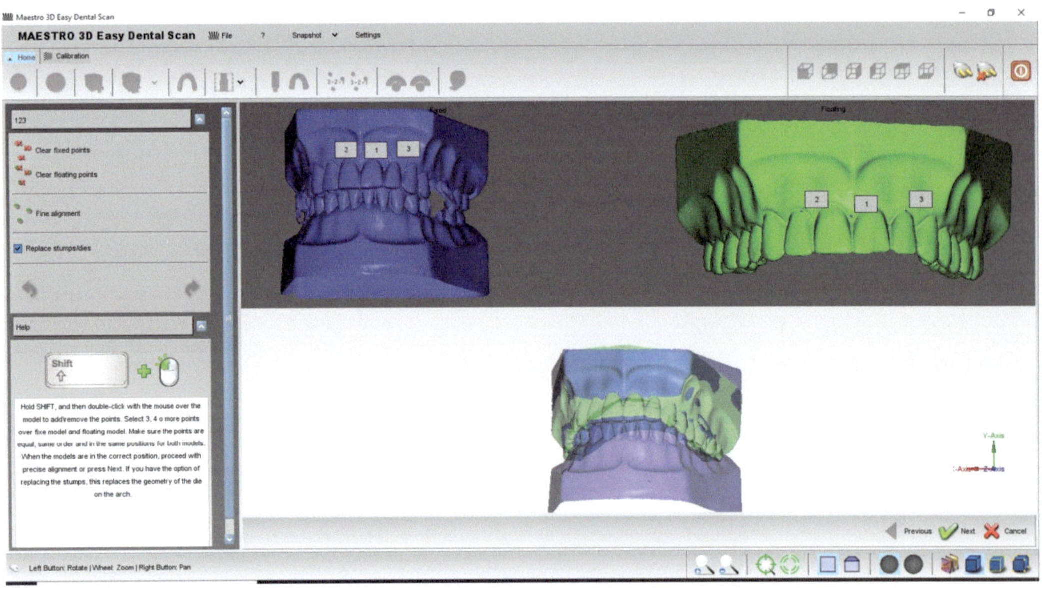

Fig. 2.21 Orientation of scanned digital maxillary model

2.1.2.3 Application of Desktop Scanning in Various Field of Dentistry Including 3D Printing

3D scanning is an innovative method for customizing dental appliances utilizing various types of software. On the digital scans of dental arch (Fig. 2.25) of the maxilla or mandible, a various dental prostheses like dental crowns, bridges, and orthodontic appliances can be designed digitally. These digital file of prosthesis or appliances can be saved in STL file format, and then, these can be 3D printed for clinical and laboratory applications using 3D printers and suitable 3D printable

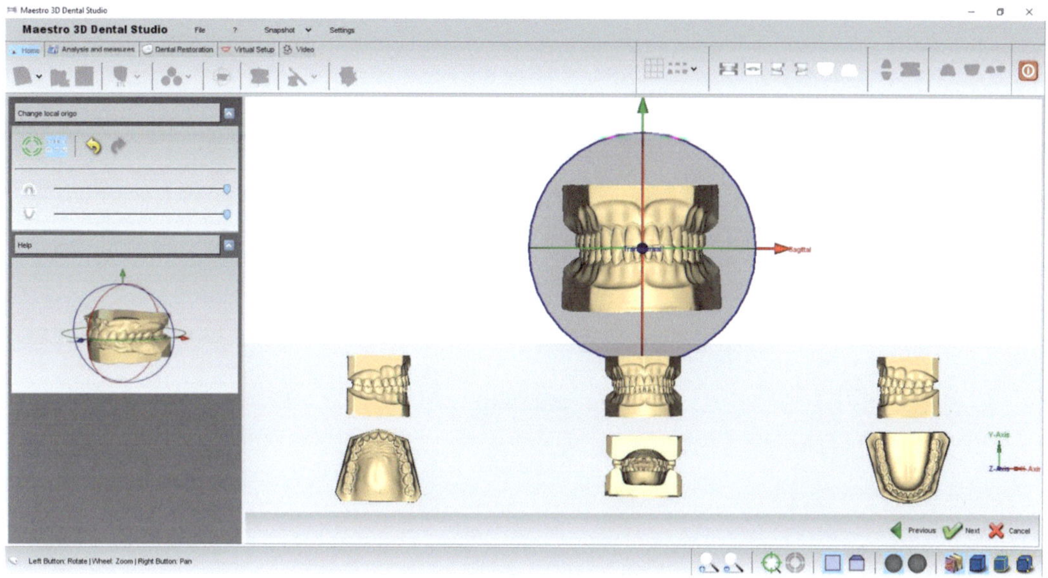

Fig. 2.22 Orientation of maxillary and mandibular models in occlusion

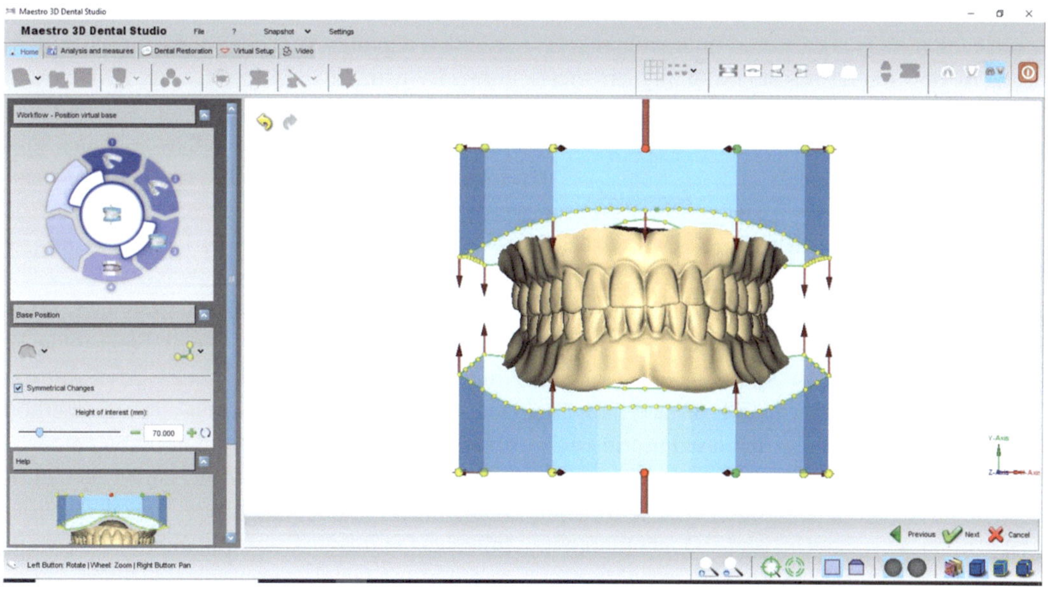

Fig. 2.23 Orientation of virtual bases on maxillary and mandibular models in occlusion

materials. 3D scanning technology makes it simple to replicate a patient's teeth, jaws, and other dental devices as closely as possible to the original. In the digitized file, a surgeon can make changes to tools and implants at any moment. The desktop scanners are already being utilized to build precise product/medical models in numerous disciplines of engineering, medicine, and dentistry. It finds broad application in dentistry, allowing for the creation of customized dental models and drill guides for dental implants [65]. They are also used in the field of maxillofacial prosthesis for planning surgeries. In orthodontics, digital models can be used to

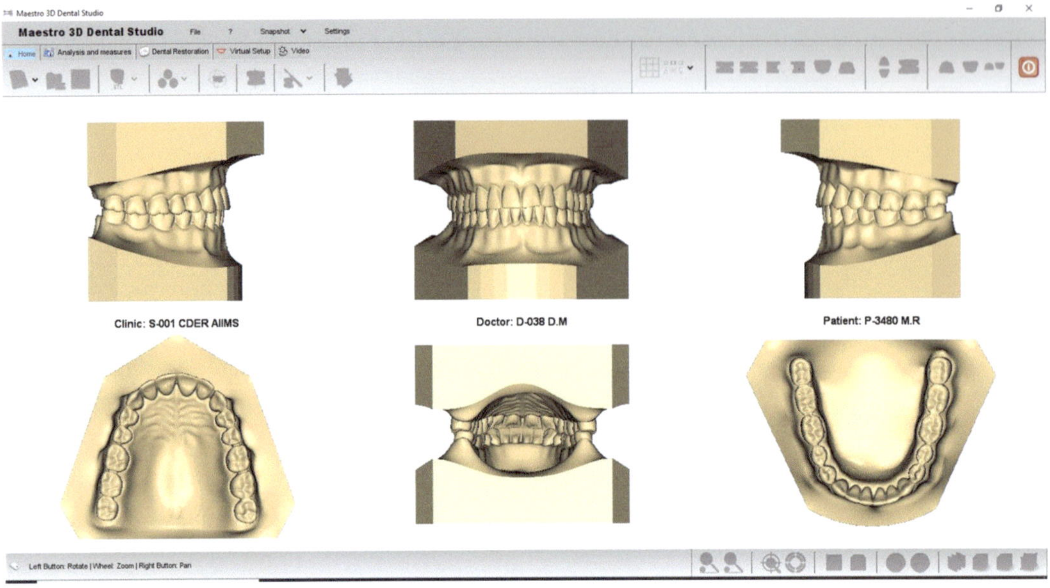

Fig. 2.24 Final maxillary and mandibular digital models

design and 3D print a wide range of orthodontic appliances such as customized indirect brackets, dental models for various stages of aligners thermoforming (Fig. 2.26), and direct printable retainers or aligners [49].

2.1.3 Facial Scanning

2.1.3.1 An Introduction to Facial Scanning Including Advantages and Disadvantages

The rapid developments in 3D surface imaging technology have led to its increased use in clinical dentistry helping to overcome the shortcomings of direct facial anthropometry, traditional usage of plaster models, and 2D photography [66–68]. 3D facial surface imaging technology or facial scanning allows to take a digital photograph that records the human facial surface in three dimensions by using 3D facial scanners. Facial scanners are noncontact devices that create a digital map of the facial surface by collecting its 3D shape and size data displayed on a 2D computer monitor giving the depth perception of the image to appear in 3D. Facial surface information is captured in the form of an array of data points, or a "point cloud," stored as *x*, *y*, *z* coordinates in a computer file. The "point cloud" describes a surface that captures the finer details and local features. By using a specific software, a 3D virtual image of the face consisting of discrete data points series is created, as well as polygons connecting these points, overlaid by the color mapping. The data of this virtual facial surface represent the original facial surface, recreating its best possible alignment [69].

Facial scanning is a fast, non-contact, high-resolution, non-invasive, and non-ionizing imaging technology that captures facial soft tissues in 3D. Being a non-contact method improves patient safety by minimizing potential contamination and discomfort. Performed without contacting the facial tissues, facial scanning prevents the image distortion; thus, it is an accurate tool for facial measurements (linear, angular, surface, and volumetric) that can be performed using the stored virtual facial data by the clinician and researchers [69, 70]. Using the appropriate software, facial scans can be rotated, zoomed, and translated showing a realistic perception and enable mathematical computations of soft tissues analysis to quantify the treatment outcomes. Further, facial scans can be merged with the images of hard tissues obtained from computed tomography and intraoral scans to

Table 2.3 Technical characteristics of some commercially available desktop scanners [37, 49]

Scanner name	Company	Country of origin	Model/ impression scanning both?	Data acquisition technology	Accuracy (μm)	Scanning time	Orthodontic software
R500™	3 Shape	Denmark	Both	Red laser	20	Plaster 2 min 20 s Impression 6 min 40 s	Ortho System Standard™
R700™	3 Shape	Denmark	Both	Red laser	20	Plaster 1 min 30 s Impression 7 min	Ortho System Standard™
R750™	3 Shape	Denmark	Both	Blue LED	10	Plaster 65 s Impression 180 s	Ortho System Premium™
R850™	3 Shape	Denmark	Both	Blue LED	7	Plaster 60 s Impression 160 s	Ortho System Premium™
R900™	3 Shape	Denmark	Both	Blue LED	15	Plaster 1 min 20 s Impression 2 min 10 s	Ortho System Standard™
R900L™	3 Shape	Denmark	Both	Blue LED	7	Plaster 40 s Impression 120 s	Ortho System Premium™
R1000™	3 Shape	Denmark	Both	Blue LED Multi line technology	5	Plaster 30 s Impression 75 s	Ortho System Premium™
R2000™	3 Shape	Denmark	Both	Blue LED Multi line technology	5	Plaster 30 s Impression 75 s	Ortho System Premium™
Maestro 3D Dental Scanner—MDS400	AGE solutions	Italy	Both	LED	10	2–3 min	Ortho Studio
7Series	Dental Wings	Canada	Both	Laser	15	15-min scan and design	DWOS™ Orthodontic Archiving
3Series	Dental Wings	Canada	Plaster	Laser	15	10-min scan and design	DWOS™ Orthodontic Archiving
iSeries	Dental Wings	Canada	Both	Laser	15	*	DWOS™ Orthodontic Archiving
Ortho Insight 3D Digital Scanner	Motion view	USA	Both	Laser	40–200	5-min scan and occlusion	Ortho Insight 3 D

reconstruct the 3D virtual patient. Facial scanning technologies and their computer-based software analyses provide an efficient tool for patient data collection for research purposes, which are less prone to human mishandling, and offer the benefits of reproducibility, ease of archive, retrieval, and inter-office transferability for clinicians [70, 71].

However, commercial facial scanning systems require expensive hardware and software, and designated room lead to their affordability only by specialized research centers or hospitals,

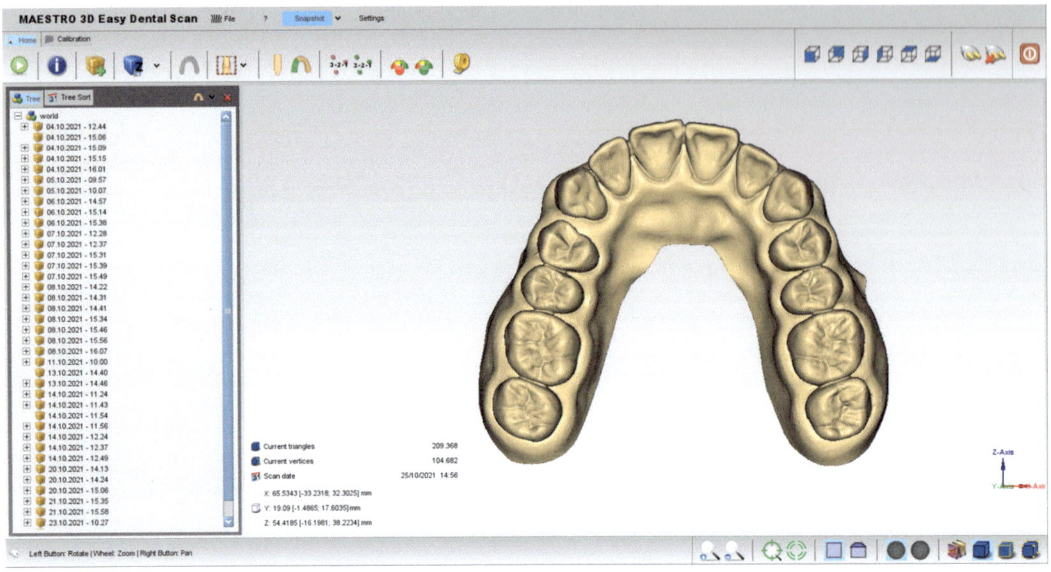

Fig. 2.25 Final digital file of the maxillary dental arch (scanned by Maestro 3D dental scanner)

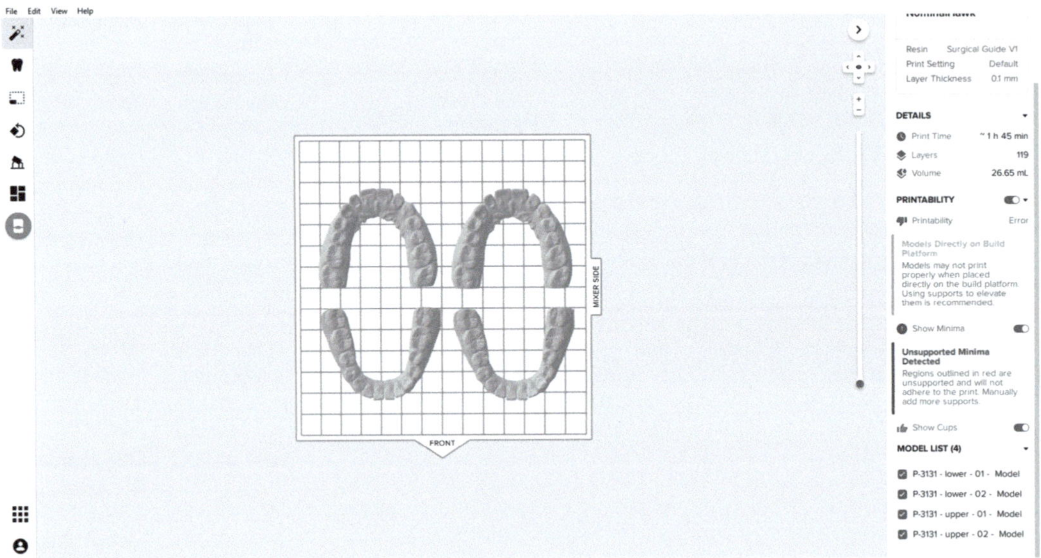

Fig. 2.26 Final STL file (digital file) showing different treatment stages for 3D printing of dental models for aligner thermoforming

remaining cost-prohibitive in health care including cleft centers or the dental clinicians in clinical routine [72]. Some 3D systems are affected by light conditions, for example, fluorescent light (having a frequency not synchronized with camera's framerate) or strong light such as direct sunlight, resulting in over-saturation of images. Camera calibration is required once in a while, especially if the system is moved to a different location. It is important to note that most scanning systems are not harmful to humans, but some laser-based systems can be used only if projected on static objects but harmful to human eyes, especially if in the non-visible spectrum.

Further, proper trained personnel and continued education regarding 3D facial imaging techniques and their software are mandatory. For these reasons, inexpensive facial scanning systems with increased portability such as a handheld facial scanner or that allows patients to take 3D images by themselves are entering the market as a tendency of the future facial scanning revolution in dentofacial settings [73, 74]. Also, it can be expected that high-quality 3D scans will be available on mobile devices in the near future and freely accessible to all.

2.1.3.2 Various Facial Scanning Technologies

Understanding the science behind different scanning methods is crucial for clinicians to select the appropriate facial scanning system for their dental practice or medical institution. The main 3D facial scanning methods are laser, structured light, and stereophotogrammetry [75–77].

The acronym LASER stands for light amplification by stimulated emission of radiation. Basically, laser is an energy form consisting of a narrow amplified light beam, characterized by spatial and temporal coherence. Laser scanning typically uses one or more laser beams projected onto the facial surface, complemented by custom-built sensors acquiring the reflected light and then calculating the x, y, z coordinates of the surface by triangulation [78, 79].

The laser facial scanning or contact digitization is noninvasive, done within a few seconds, accurate to approximately 0.5 mm, with some advantages (portability, speed, cost of laser scanners) in comparison with other types of 3D imaging technology for creating 3D images [80–83]. Minolta Vivid 910 3D Scanner (Konica Minolta Sensing, Inc., Osaka, Japan) with their recording intervals of 2.5 s per shot, high resolution, and the opportunity to link several scanners together proved to be practical, handy, precise, and accurate enough to assess facial morphology, its symmetry, and changes due to growth or clinical intervention, in plastic surgery and further clinical application [80, 83].

Structured light technology is based on the structured light patterns of white light being projected onto the subject. To acquire a single image of the face, one or two digital cameras at a known distance calibrated with the specifics of the projected light pattern from a video projector are used to catch 3D data using a standard triangulation method. The system's software captures the distortion of the light pattern, generates the shape and registers the accurate information of color and texture, and reconstructs the final 3D image [84]. 3D structured light scanners can obtain the facial geometric, color, and texture information without radiation and in high resolution, together with advantages of portability, decreased cost, ease of operation, and short scan time [85–89]. Basically, two options exist for pattern projection, a single pattern (lower accuracy, shorter capture time) versus multipatterns (higher accuracy, longer capture time). The single pattern is better suited for dynamic objects, while multipattern is better for static objects. To obtain a highly accurate and dense model, it is needed to illuminate the face with random patterns of light several times which needs patient collaboration and increases the capture time. Also, because it is difficult for one camera to provide an ear-to-ear capture, it may be necessary to use several cameras. A structured light 3D scanner using projected light patterns and a single camera, providing facial scans with extreme accuracy and a high resolution [87, 88] (Fig. 2.27), offers a cost-effective approach as stated by Amirav et al. [86, 87].

FaceScan3D (3D-Shape, Inc., Erlangen, Germany) and Artec EVA (Artec EvaTM; Artec Group, Luxembourg) are commercially used scanners to accurately capture and evaluate soft tissue changes in clinical dentistry [85, 90, 91].

Stereophotogrammetry is a 3D facial scanning technique used to capture topographic surface data based on the triangulation and fringe projection method. An image can be obtained by means of one or more simultaneous stereo pairs of photographs [92]. A 3D camera can capture the correct geometry and texture information of facial soft tissue surface which its corresponding software uses further to reconstruct the virtual 3D facial images. Fast acquisition time of stereophotogrammetry enables to eliminate inaccuracies

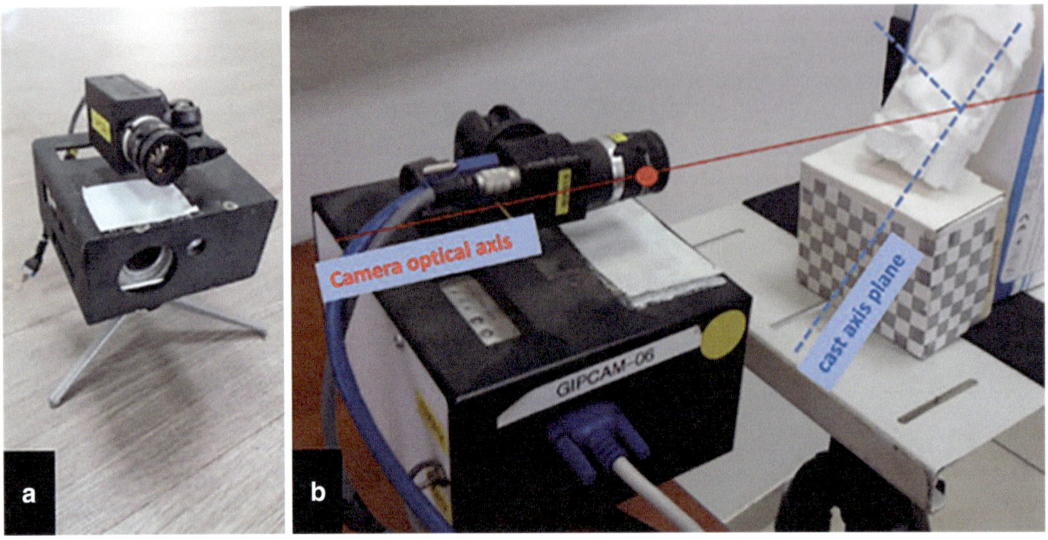

Fig. 2.27 (**a**) A noncontact structured light 3D surface scanning technology developed by the Geometric Image Processing (GIP) laboratory, Faculty of Computer Science, Technion—Israel Institute of Technology, Haifa, Israel. It consists of a DLP projector (Toshiba TDP-FF1A Projector, 800 × 600 pixels); digital video camera PTGrey Flea 3/CCD Camera (Point Grey Research® Inc.), gray scale (1280 × 960 pixels), synchronization switch and cable between camera and projector, and aluminum projector cage. (**b**) Structured light 3D surface imaging technique setup during moulage scanning process. (*Courtesy* of GIP Lab, Faculty of Computer Science, Technion—Israel Institute of Technology, Haifa, Israel; with permission)

related to motion artifacts, making it clinically advantageous in comparison to structured light, especially for children, together with its short calibration process, and proved accurate landmark detection process [76]. Vectra H1 (Canfield Scientific, Parsippany, NJ, USA) that merges 3D stereophotogrammetry and stereophotometry, Scanify (Fuel 3D Technologies, Chinnor, UK) and 3dMD technology (3dMD, Atlanta, USA) that integrates stereo and structured light (usually projecting a single constant light pattern in the near-infrared range being used to assist the stereo correspondence algorithm) are widely used 3D surface imaging system [93].

2.1.3.3 Accuracy of Facial Scanning Technology

The facial scanners' accuracy in clinical practice is a major focus of current research due to the evolution of 3D surface imaging technology and the need to clinically use it for diagnostic evaluation with reliability. Facial scanners must accurately and precisely quantify facial surface topography as well as or better than traditional anthropometry with its soft tissue and craniofacial measurement standards [76, 84, 94–103]. The nominal accuracy (NA) of a scanner represents the best measurement accuracy of the scanner as obtained by measuring standard geometric entities being provided by manufacturers. However, the clinical performance of a facial scanner can be affected by many factors and reduce its final accuracy. Also, human patients are not static as geometric objects used for calibration and accuracy estimation, and also human skin absorbs more light than the calibration objects. The difference between a real person's face and its scanned image with a more complex shape makes the practical accuracy (PA) of facial scanners to be different from NA and more relevant for clinical use [99, 100]. The nominal accuracy of 3dMD Face System (3dMD, Atlanta, USA) based on stereophotography is of 0.2 mm. Ye et al. (2016) found that the structured light scanning system and stereophotogrammetry scanning systems have high accuracy, reliability, and reproducibility in scanning and quantifying the human faces [103].

The literature increasingly reports that data derived from laser scanning are highly accurate when compared to direct anthropometry and laser facial scanners can be used in dentistry with a high degree of accuracy, precision, and reliability for soft tissue evaluation. Using the laser scanning, a constructed patchwork of triangles represents the facial contour with a precision of 0.5 mm. The Minolta Vivid700 3D surface laser scanner (Minolta USA, Ramsey, NJ) has shown to be accurate, offering great research potential and applied to evaluate orthodontic treatment changes, growth, and surgical simulations [104]. Also, the reliability and accuracy of certain 3D stereophotogrammetry systems for soft tissue facial measurements have been reported [76, 86, 90–93].

2.1.3.4 Technical Characteristics of Some Commercially Available Facial Scanning Devices

A 3D facial scanner represents a noncontact optical measuring device able to capture during a short scanning process (typically less than 1 s) a scanned 3D facial model with real skin texture color in open data format. Most of the facial scanners are mainly laser-based, 3D-stereophotogrammetric, and structured light-based. Currently, a variety of commercially available 3D surface imaging systems and facial scanners are in use in the dental market [73, 105–109] (Table 2.4).

One of the most efficient stereo photogrammetric systems used is the 3dMD technology (3dMD LLC, Atlanta, GA, USA). With 3dMD, 3D imaging protocols are no longer dependent on capturing that one vital 3D image in a single point-and-click setup. Temporal 3dMDface and 3dMDtrio Systems for ear-to-ear face capture, in contrast to other commercially available static 3D surface scanners that take one scan at a time over a duration of time, capture a progressive sequence of dense-surface 3dMD images at 1–120 3D frames per second. With a progressive sequence of 3dMD images, researchers now measure anatomical shape change in relation to particular facial functions and expressions. Due to its sophisticated algorithmic usage of a multi-camera system able to achieve optimal 3D surface coverage of the designated craniofacial complex, rapid image acquisition, and ear-to-ear capture with high facial geometric accuracy of <0.2 mm RMS (root mean square), the 3dMD technology reduces the artifacts and is capable of linear, angular, and volumetric measurements of 3D surfaces (Fig. 2.28) [96, 98].

2.1.3.5 Application of Facial Scanning in Various Fields of Dentistry

3D imaging of the face plays a crucial role in different specialties including maxillofacial prosthodontics, orthodontics, craniofacial surgery, the pediatric craniofacial area as an essential tool for diagnosis, treatment planning and outcome evaluations of craniofacial morphology, growth of young children with and without facial deformities, and human facial expression and aging due to ability to record changes in facial morphology [69, 75–78, 110–118]. 3D facial imaging is useful in monitoring the effects of treatments such as functional appliances and assessing the effects of extractions as part of orthodontic fixed appliance therapy [110–112] and postoperative facial swelling following orthognathic surgery due to the ability to record the volumetric changes [113]. 3D digital surface imaging enables a relatively accurate registration of growth and 3D facial surface morphology in children and adults, including those born with cleft lip and palate (CLP) enabling an accurate registration of 3D soft tissue morphology for preliminary diagnostic record, surgical lip closure, rhinoplasty, and orthognathic cleft surgeries at their late stages of final rehabilitation [114]. 3D data acquired by facial scanning can also be used for rapid prototyping to create facial prosthesis to reconstruct the lost facial tissue caused by trauma or malignancy [115]. 3D imaging technique can be used in prosthodontics, for example, for a better and more comprehensive aesthetic treatment planning in patients with severe tooth wear due to ability to visibly detect the changes in facial appearance caused by an artificial increase of vertical dimension of occlusion [116].

In craniofacial genetic research, facial scanning helps to describe anatomical variations and

Table 2.4 Commercially available 3D surface imaging systems and facial scanners in use in dental market

S.No	Name	Company	Technology	Accuracy	Output formats	Compatibility	Source
1	3dMDface system	3dMD, Atlanta, GA, USA	Stereophotogrammetry	0.2 mm	Supports OBJ as the default, as well as all other industry-standard open formats		https://3dmd.com/products/#https://3dmd.com/products/#!/face
2	EinScan Pro 2X	Shining 3D Tech. Co., Ltd. Hangzhou, China	Structured light	Up to 0.1 mm	OBJ; STL; ASC; PLY; P3; 3MF	Win7, Win8, Win10 (64bit)	https://www.einscan.com/handheld-3d-scanner/einscan-pro-2x/einscan-pro-2x-specs/
3	Artec EVA	Artec Group, Luxembourg	Structured light	Up to 0.1 mm	3D Mesh: OBJ, PLY, WRL, STL, AOP, ASC, PTX, E57, XYZRGB CAD: STEP, IGES, X_T	Windows 7, 8 or 10 x64	https://www.artec3d.com/portable-3d-scanners/artec-eva
4	PlanMeca ProFace (integrated into Planmeca's 3D imaging units)	Planmeca, Helsinki, Finland	Laser scanning and digital cameras	Up to 1 mm	OBJ, STL, PLY, and PRO (Planmeca's own format)	Win 10 Pro/Win 8.1 Pro/Win 7 Pro and iOS	https://www.planmeca.com/imaging/3d-imaging/planmeca-proface/
5	Structure Sensor Pro (Mark II)	Occipital, Moscow, Russia	Structured light	4 mm	Depends on the application used (STL, PLY, OBJ, and VRML)	iOS	https://support.structure.io/
6	Vectra imaging systems (H1, H2, XT, M3)	Canfield Scientific, Inc.	Stereophotogrammetry	0.2–0.3 mm^2	Files are analyzed using the MIRROR software by Canfield	iOS and Windows 10 Professional, 64 bit	https://www.canfieldsci.com/imaging-systems/categories/aesthetic-systems/
7	Obiscanner	FifthIngenium S.R.L.S., Milano, Italy	Infrared (IR) Laser Projector	–	STL, OBJ, PLY	Windows 10 integrated support	https://obiscanner.com/
8	Minolta Vivid 910 3D Scanner	Konica Minolta Sensing, Inc., Osaka, Japan	Laser	Good enough	STL	Windows 2000 Professional SP4, Windows XP2	http://www.minolta3d.com http://konicaminolta.us/3d
9	GIP 3D video camera	Geometric Image Processing Lab, Technion, Israel Institute of Technology	Structured light	0.1–0.2 mm	Custom video format + exported stl, obj	Windows 7,8,10	https://gip.cs.technion.ac.il/

Fig. 2.28 Imaging acquisition with the 3dMDface System. (*Courtesy* of 3dMD LLC, Atlanta, GA, USA; with permission. ©2021 3dMD. All rights reserved)

craniofacial growth patterns in order to perform quantitative genetics based on morphometric traits in understanding the aetiology of craniofacial anomalies [117]. Recently, facial scanning is gaining a great interest in studying facial morphology and facial expression while aging, also to better understand the sexual dimorphism of human faces and accurately categorize human faces by sex using 3D facial images [118–121].

The recent advancement of deep learning methods adapted to 3D models has an impact in this field also. Current trends and applications include classification and segmentation of teeth, palatal and dental shape, and dental occlusion and are summarized in a few surveys [122–125]. As deep neural networks exhibit ever more proficiency, specialized 3D datasets continue to accumulate, and more effort is bestowed from industrial applications as well as academic research, this trend is only expected to continue.

2.2 Ionizing Surface Imaging

2.2.1 Cone Beam Computed Tomography Imaging

Cone beam computed tomography (CBCT) was first introduced for dental use in 1998 by Mozzo et al. [NewTom 9000 (Quantitative Radiology, Verona, Italy)] [126]. It overcomes the limitations of two-dimensional (2D) conventional imaging in dentistry like superimpositions, distortions, and magnifications by offering high-resolution, three-dimensional image accuracy with multiplanar reconstructions at reduced radiation exposure than conventional CT [126]. The rapid development of improved, fast, cost-effective computer technology and software applications in dentistry has led to its incorporation in mainstream dentistry today. Its applications in dentistry have diversified from implant planning and orthodontic/orthognathic treatment planning in the early years to almost every discipline in dentistry [127].

2.2.1.1 Basic Principle of CBCT Imaging

CBCT is a form of computed tomography where the region of interest (ROI) is scanned by a cone-shaped X-ray beam in a single rotation of 180°–360° around a vertical axis of the patient's head and the imaging data are captured by detectors and processed by speciality software to construct 2D and 3D images in multiple anatomic planes [127]. The X-ray generation is at 1–5 mA and 90–120 kVp with an exposure time ranging from 5 to 40 s. The exposure can be pulsed or continuous with pulsed X-ray systems exhibiting improved spatial resolution due to reduced motion effect. The gantry of CBCT usually allows for the patient to be seated and/or standing with wheelchair access, while some old models had supine positioning. The positioning of the patient's head in the unit is usually assisted by lasers, while immobilization of the patient's head during exposure is achieved by chin cups, bite forks, or head restraints (Fig. 2.29) [126]. The size of the scanned object volume in ROI is called the field of view (FOV) and varies between different CBCT units (small, medium, large). The range of FOVs available is from 3.0 cm (H) × 3.0 cm (D) to 24 cm (H) × 16.5 cm (D). The FOV depends on the type and size of the detector and the degree of beam collimation of the X-ray tube head [127]. Most CBCT machines have collimators with predefined openings according to FOV sizes [128]. The detectors are image intensifier tube (IIT)/charge-coupled device (CCD) combination or flat panel detectors (FPDs). The FOV of FPDs is cylindrical, while IIT detectors have spherical FOV [127]. FPDs have a higher

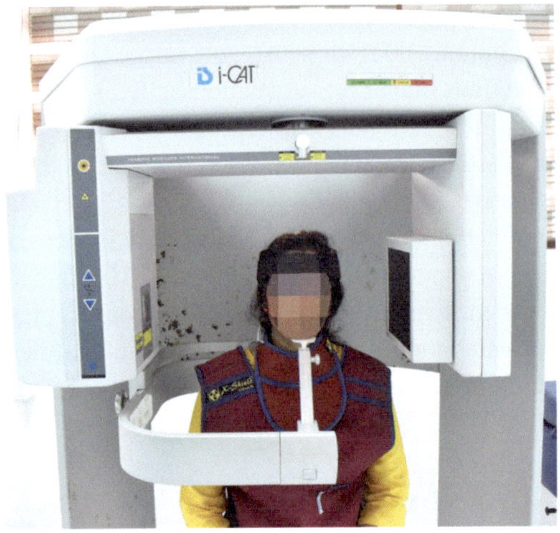

Fig. 2.29 Showing acquisition of CBCT scanner using i-CAT CBCT scanner

spatial resolution than IIT/CCD, and recently, complementary metal–oxide–semiconductor (CMOS) detectors with large FOV, high resolution, and low electronic noise have become available [128]. The individual volume elements in the acquired CBCT volumetric data are called voxels, and they are isotropic, i.e., equal in all three dimensions and can vary from 0.4 to 0.076 mm^3. The voxel size determines the number of basis images acquired during exposure [127]. The captured 2D images undergo preprocessing for correction of inherent pixel imperfections, variations in sensitivity, and uneven exposure before cone beam reconstruction with a filtered back projection called the Feldkamp, Davis, and Kress (FDK) algorithm. The reconstructed volume can be visualized in 1:1 ratio as various 2D cross-sectional images (axial, coronal, sagittal) and multiplanar reformation (MPR) (oblique, curved planar reformation and serial transplanar reformation) [126, 128]. Ray sum function adds up adjacent voxels to display thickened MPR slices like panoramic or cephalometric projections. Volume rendering function selectively displays voxels within a dataset to visualize volume using direct or indirect volume rendering tools. Maximum intensity projection (MIP) is the most commonly used direct volume rendering tool in which voxels with highest density values within a particular thickness are displayed, while voxels below the arbitrary threshold are excluded [126, 128]. The CBCT volumetric data are always in standardized file format, i.e., Digital Imaging and Communications in Medicine (DICOM 09v11dif) compliant with ISO 12052. Most of the CBCT scanners have their own built-in proprietary software which convert the file formats into DICOM which makes it easy for telecommunication and to use with other third-party software applications like CS3D, Dolphin 3D, Easy Guide, InVivo Dental, On Demand 3D, OsiriX, and Procera [126, 127]. Salient characteristics of various commercially used CBCT machines have been presented in Table 2.5.

2.2.1.2 3D Printing with CBCT Imaging

CBCT scans offer adequate resolution of images at lower radiation exposure for 3D rendering and reconstruction of dentomaxillofacial region for various applications in dentistry. 3D rendering involves segmentation or virtual separation of the anatomical region of interest from the surrounding structures for better visualization and analysis. Segmentation can be performed manually or semiautomatically using computer-aided approach (hybrid). The manual approach is time-consuming and is user-dependent involving slice by slice pooling of all slices to reconstruct the desired 3D volume but is also considered as the gold standard. The semi-automatic method is performed with the help of software, is faster, and is not user-dependent, thus useful for clinical and research purposes. Segmentation allows for conversion of the stacked 2D DICOM images to Standard Triangulation Language (STL) data format, a stereolithography file format commonly used for 3D printing. There are many commercial (fee-based) and open-source (free) software packages for segmenting DICOM images to STL format like 3D slicer, 3D views, Image J, InVesalius3, Mimics, OsiriX Lite, Dolphin 3D, and ITK-Snap. The construction of digital 3D model teeth and jawbone from CBCT of patient with multiple impacted anterior teeth was done with Materialize Mimics version 22.0 (Materialize NV, Belgium). The region of interest, selected in segmentation process to generate

Table 2.5 Salient characteristics of various commercially used CBCT machines

S. No	CBCT machine	Manufacturer	Field of View available (height × diameter in cm)	Focal spot (in mm)	Voxel size (mm)	Scan time	Availability of panoramic X-ray and cephalometric X-ray	Included software
1.	Veraviewepocs 3D R100	J. Morita, Japan	4 × 4, 4 × 8, 8 × 5, 8 × 8, 10 × 5, 10 × 8	0.5	0.125	Panoramic: 7.4–15 s 3D scan: 9.4 s	Yes	i-Dixel 2.0
2.	GALILEOS® COMFORT^{PLUS}	Dentsply Sirona Imaging, United States	15.4 spherical	0.5	0.25 0.125	2–14 s	Yes	SIDEXIS
3.	i-CAT FLX Cone Beam 3D Imaging Solution	Imaging Sciences Int'l, USA	**Standard:** 4 × 16, 6 × 16, 8 × 8, 8 × 16, 10 × 16, 11 × 16, 13 × 16 **Extended field of view:** 17 × 23	–	0.4, 0.3, 0.25, 0.2, 0.125	4.8 s, 8.9 s, 14.7 s, 17.8 s, 26.9 s	Yes	Tx STUDIO
4.	OP 3D™ Pro	KaVo, Germany	5 × 5, 6 × 9, 9 × 11 and (optional) 9 × 14 cm	–	0.85–0.42	1.2–8.7 s	Yes	DTX Studio™
5.	NewTom 5G	QR, Inc. Verona, Italy	6 × 6, 8 × 8, 12 × 8, 15 × 5, 15 × 12, 18 × 16	0.3	0.075–0.3	18–36 s	Yes	NNT software
6.	PreXion3D Eclipse 3D	PreXion, Inc.	8.1 × 7.5, 11.3 × 7.2	0.2	0.15	8–18 s	Yes	PreXion 3D Viewer, 3D Movie Maker, Customizable Implant Library
7.	Planmeca ProMax 3D s	Planmeca OY, Helsinki, Finland	5 × 5.5, 10 × 5.5, 10 × 9, 13 × 5.5, 13 × 9, 10 × 13, 13 × 13, 13 × 16, 23 × 16, 23 × 26	0.6	0.075–0.6	9–40 s	Yes	Planmeca Romexis
8.	CS 9300	Carestream Health, Rochester, USA	5 × 5, 8 × 8, 10 × 5, 10 × 10, 17 × 6, 17 × 11, 17 × 13.5	–	0.09–0.5	12–20 s	Yes	CS Imaging Software
9.	Gendex GXDP-70	Gendex Dental Systems, USA	6.1 × 4.1, 6.1 × 7.8	0.5 × 0.5	0.2 (standard)	2.3–12.6 s	Yes	Gendex™ VixWin™ Imaging Software
10.	Papaya 3D Plus	Genoray America Inc.	4 × 5, 7 × 7, 8 × 8, 14 × 8, 14 × 14	0.5	0.075–0.4	7.7 s in fast scan mode	Yes	Papaya 3D Plus Operation software

(continued)

Table 2.5 (continued)

S. No	CBCT machine	Manufacturer	Field of View available (height × diameter in cm)	Focal spot (in mm)	Voxel size (mm)	Scan time	Availability of panoramic X-ray and cephalometric X-ray	Included software
11.	Hyperion X9 CBCT Scanner	MyRay	5 × 5, 8 × 5, 8 × 8, 11 × 5, 11 × 8, 11 × 13	0.5	0.075	Pan: 7.5–13 s Ceph: 3.4 s 3D: 18 s	Yes	MyRay 3D Imaging Suite and Implant Library
12.	PaX-Reve3D Plus	Vatech, Korea	5 × 5, 8 × 6, 12 × 8, 15 × 15, 15 × 19	0.5	0.08–0.25	CBCT: 15/24 s Pano: 3–9.7 s Ceph: <0.3 s	Yes	EZ3D 2009 3D Image Viewer, EasyDent Panoramic, Cephalometric Image Viewer, Any Gateway DICOM File Interface Program

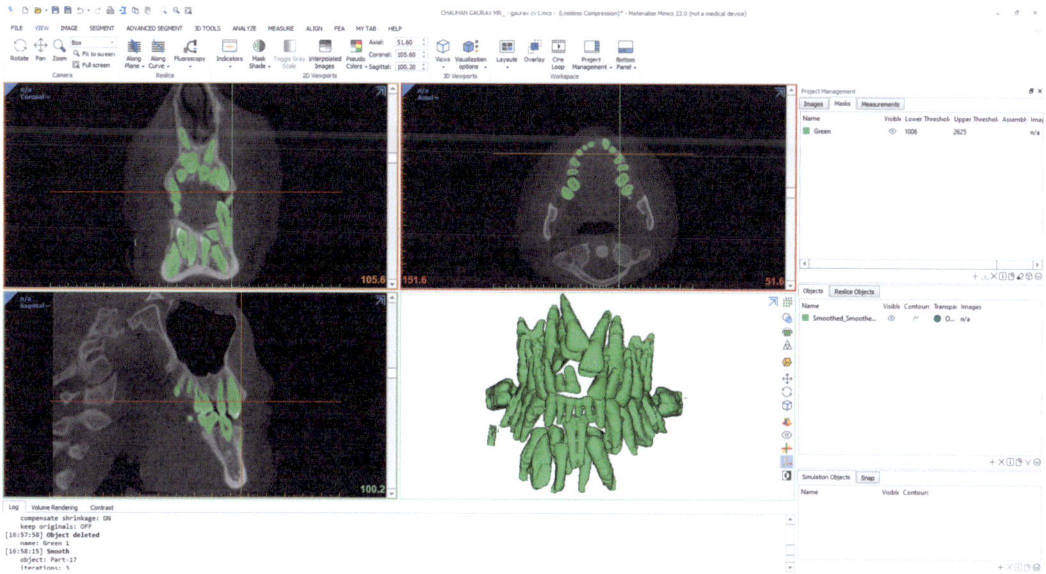

Fig. 2.30 3D model of maxillary and mandibular teeth after segmentation process then exported and saved in STL format (Mimics Innovation Suit software)

a digital 3D model of impacted teeth. The 3D model after segmentation process was exported and saved in STL format (Fig. 2.30). Cleaning the mesh of STL file was performed with Meshmixer (Autodesk) (Fig. 2.31). The final STL file of segmented teeth had been used for printing 3D models using FormLabs 3B 3D printer (Somerville, Massachusetts, United States) using surgical guide-V1 resin by FormLabs (Figs. 2.32 and 2.33). The model was printed in 4 h 30 min, and it took total 37.27 mL of resin. Use of artificial intelligence through deep learning and convolutional neural network (CNN) has shown promising results in automated segmentation of anatomical structures in CBCT [129, 130]. Diagnocat (https://eu.diagnocat.com/) is an AI analysis of dental X-rays, which can be used to create the STL—creation of 3D models from CBCT data. The same CBCT data were used to create the 3D model of maxilla and mandible including teeth in STL file format (Fig. 2.34) and then segmented of the maxillary and mandibular teeth (Fig. 2.35) using AI powered Diagnocat software. The easy operation and speed are the main advantages of AI powered Diagnocat software over the Materialize Mimics. The accuracy of the CBCT-derived 3D models depends on voxel size, FOV, patient positioning, artifacts, beam inhomogeneity of CBCT scanner, and the software packages used. Voxel sizes of 200 and 300 μm were found to underestimate 3D CBCT volumetric data which were significant for 300 μm as compared to voxel sizes of 76 μm and micro-CT (41 μm) [131, 132]. Smaller the voxel size, higher the resolution of the DICOM images and greater the accuracy of the 3D models.

2.2.1.3 Advantages of CBCT Over CT

CBCT has several advantages over conventional CT as it is cheaper and requires less space and lower effective radiation dosage to patient. The effective dose (E) depends on the category and model of CBCT scanner and ranges from 29 to 477 μSv which can be further reduced by varying patient positioning (tilting the chin) and supplementary personal protection like lead aprons and thyroid collars. The dose reduction for oral and maxillofacial imaging is 98.5–76.2% as compared to conventional CT. Some CBCT units are hybrid type as they also offer panoramic and cephalometric imaging. The CBCT scanners can collimate the X-ray beam to selected FOV and thus reduce radiation exposure to the patient and minimizes scattered radiation that degrades

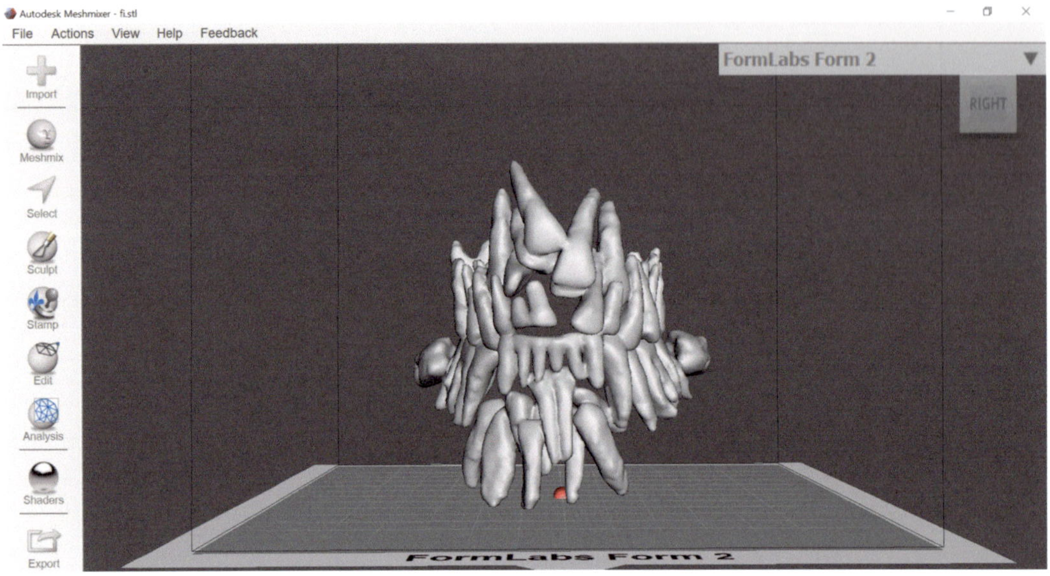

Fig. 2.31 STL file of 3D model of maxillary and mandibular teeth transferred to the Meshmixer (Autodesk) software for cleaning

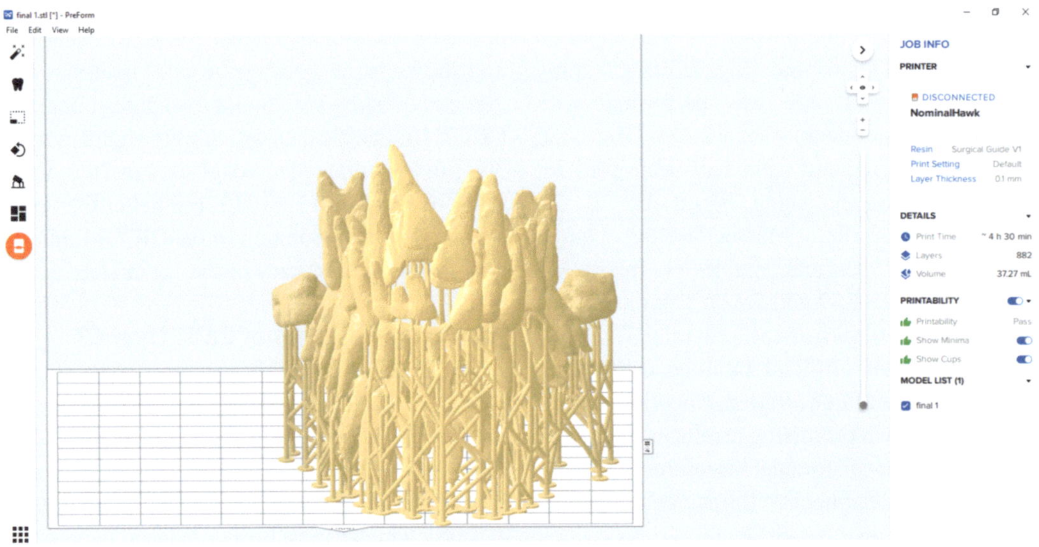

Fig. 2.32 Final STL file of 3D model of maxillary and mandibular teeth ready to 3D print using FormLabs printer

image quality. The voxels in CBCT are isotropic in all three proportions and produce a 1:1 image with submillimeter resolution (0.4–0.076 mm). Measurements calculated on screen are accurate with no distortion or magnification. The single rotation used in CBCT scanners reduces the scan time ranging from 5 to 40 s, thus reducing patient motion artifacts. The display modes for maxillofacial imaging in CBCT are very unique as it provides images in orthogonal planes and non-orthogonal (MPR) planes like oblique, curved planar (simulated panoramic images) and serial cross-sectional reformation. 3D visualization can be seen in ray sum, MIP, and 3D

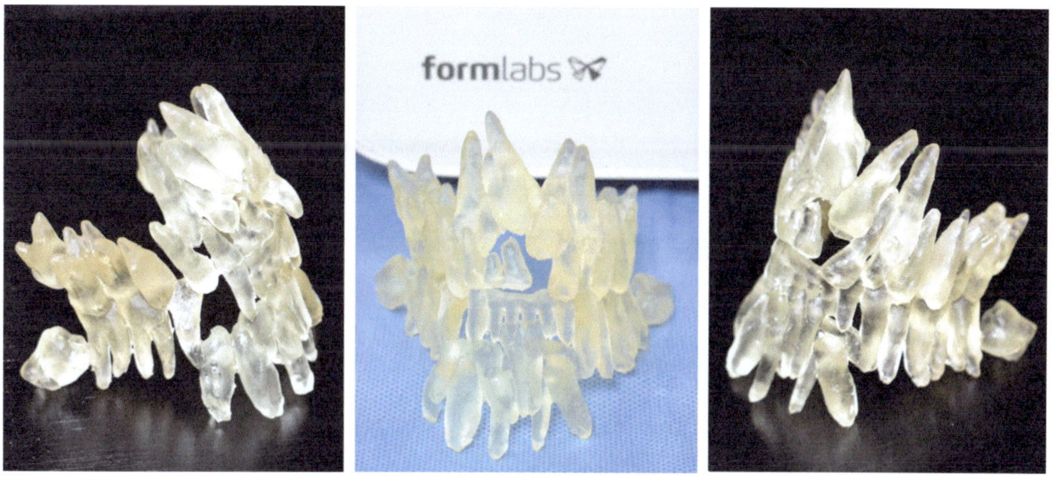

Fig. 2.33 Front and side (right and left) view of 3D printed model of maxillary and mandibular teeth using FormLabs printer

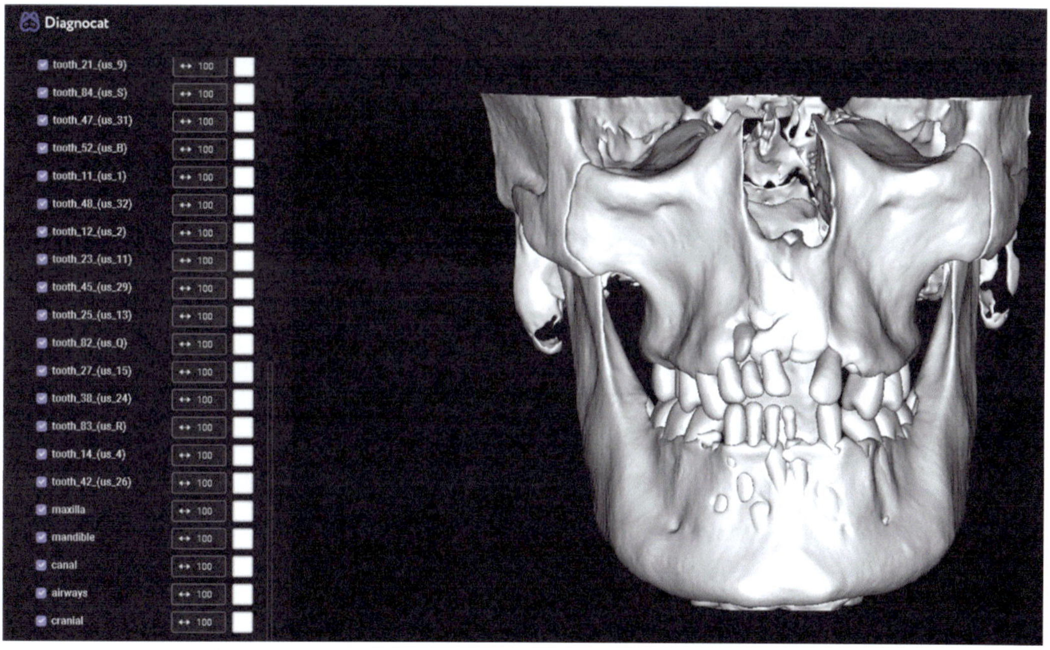

Fig. 2.34 CBCT data were used to create the 3D model of maxilla and mandible including teeth in STL file format using AI powered Diagnocat software

computer-generated models. Various artifact suppression algorithms are now available to reduce the metal artifacts in secondary reconstructions [126, 133].

2.2.1.4 Limitations of CBCT

The disadvantages of CBCT as compared to conventional 2D imaging in dentistry are the high radiation dosage to the patient. Various guidelines and recommendations for application of CBCT in dentistry have been issued to limit unnecessary radiation exposure to patients [134, 135]. The principles of ALARA (as low as reasonably achievable) and ALADA (as low as diagnostically acceptable) have been recommended for rationalizing use and referral for CBCT [136]. The clarity of CBCT images is affected by the cone beam projection geometry, detector sensi-

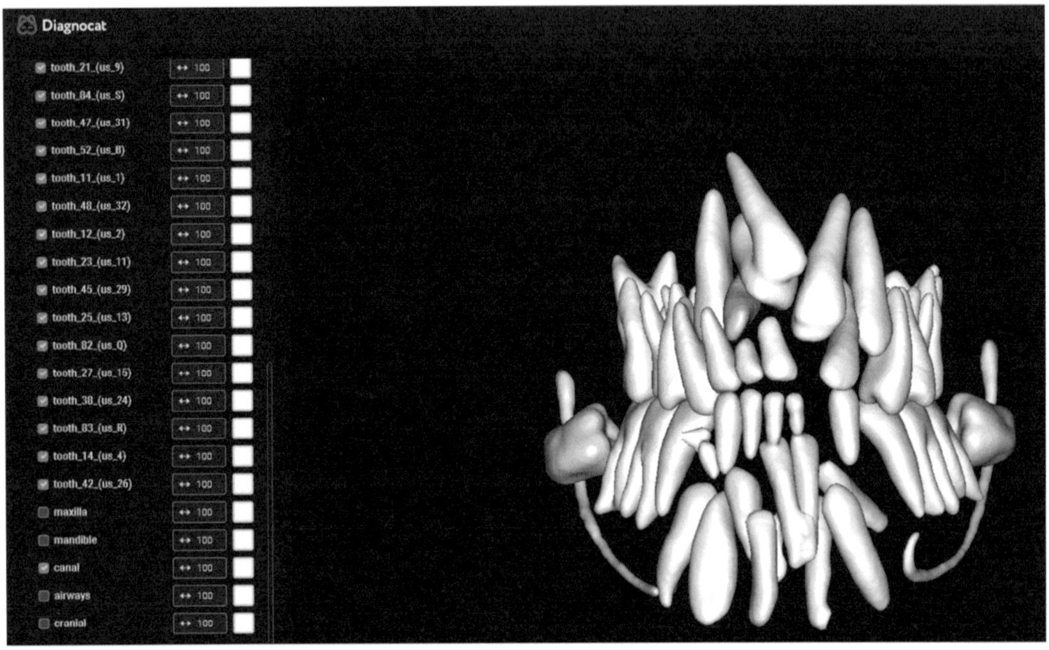

Fig. 2.35 CBCT data were used to create segmented 3D model of the maxillary and mandibular teeth using AI powered Diagnocat software

tivity, contrast resolution leading to noise, artifacts, and poor soft tissue contrast. Artifacts in CBCT limit adequate visualization of structures in ROI and are caused by beam hardening (cupping artifacts, streaks, and dark bands), patient movement (unsharpness, blurring), scanner-related artifacts (circular-/ring-shaped), and cone beam-related artifacts (partial volume averaging, undersampling and cone beam effect). Large volume of object during scanning causes scattered radiation and image noise leading to degradation of image quality. CBCT images have poor soft tissue contrast due to increased image noise, divergence of the X-ray beam, and numerous FPD-based artifacts. There is also limited correlation with Hounsfield units for standardized quantification of bone density [126, 133].

2.2.1.5 Application of CBCT Data for 3D Printing in Various Field of Dentistry

The first 3D printing technology was introduced by Charles Hull in 1986, and since then, it has been widely used in the industry, design, engineering, and manufacturing fields. It has also been used for surgical planning, custom surgical devices, and patient–physician communication in the fields of traumatology, cardiology, neurosurgery, plastic surgery, and cranio-maxillofacial surgery. Its use in dentistry ranges from prosthodontics, oral and maxillofacial surgery, oral implantology, orthodontics, endodontics, periodontology, and restorative dentistry. The various 3D printing technologies can accept computer-aided design (CAD) data based on 3D imaging, and hence, various applications have come into practice that allow for a low-cost and personalized service with simplification of the workflow related to digital dentistry. It is especially helpful where mechanical processing is inconvenient for products with complex and fine structures using multiple laboratory procedures [137, 138].

Acknowledgement The authors are grateful to the Indian Council of Medical Research (ICMR), New Delhi, for providing extramural research grant for research project (code I-1069), which helped to generate data for this book chapter.

Authors are thankful to Mr. Netrapal (Dental technician at Division of Orthodontics and Dentofacial Deformities, CDER, AIIMS, New Delhi) for assistance in generation of images for this chapter, Dr. Vandita Vats and

Dr. Okram DevDas Singh for providing inputs from general dental practice viewpoint needs of 3D imaging, and Mr. Eugene Shumilov, (Diagnocat, Inc., San Francisco, USA) for providing complementary access to generate data for this chapter.

In addition, authors are thankful to 3dMD Company (3dMD LLC, Atlanta, GA, USA) and GIP Lab (GIP Lab, Faculty of Computer Science, Technion—Israel Institute of Technology, Haifa, Israel) for assistance in providing the images for the facial scanning on this chapter.

References

1. Duret F, Blouin JL, Duret B. CAD-CAM in dentistry. J Am Dent Assoc. 1988;117:715–20.
2. Mormann WH. The evolution of the CEREC system. J Am Dent Assoc. 2006;137(Suppl):7S–13S.
3. Zimmermann M, Koller C, Rumetsch M, Ender A, Mehl A. Precision of guided scanning procedures for full-arch digital impressions in vivo. J Orofac Orthop. 2017;78:466–71.
4. Aragón ML, Pontes LF, Bichara LM, Flores-Mir C, Normando D. Validity and reliability of intraoral scanners compared to conventional gypsum models measurements: a systematic review. Eur J Orthod. 2016;38:429–34.
5. Vecsei B, Joós-Kovács G, Borbély J, Hermann P. Comparison of the accuracy of direct and indirect three-dimensional digitizing processes for CAD/CAM systems–an in vitro study. J Prosthodont Res. 2017;61:177–84.
6. Nawi N, Mohamed AM, Nor MM, Ashar NA. Correlation and agreement of a digital and conventional method to measure arch parameters. J Orofac Orthop. 2018;79:19–27.
7. Brignardello-Petersen R. No differences important to patients between orthodontic treatment with customized fixed appliances and conventional appliances. J Am Dent Assoc. 2017;148:e194.
8. Bae G-S, Kim Y-I, Kim S-S, Park S-B, Son W-S. 3D-printed double-wire bracket for anterior alignment. J Clin Orthod. 2017;51:377–81.
9. Ojima K, Dan C, Kumagai Y, Schupp W. Invisalign treatment accelerated by photobiomodulation. J Clin Orthod. 2016;50:309–17.
10. García-Gil I, Cortés-Bretón-Brinkmann J, Jiménez-García J, Peláez-Rico J, Suárez-García MJ. Precision and practical usefulness of intraoral scanners in implant dentistry: a systematic literature review. J Clin Exp Dent. 2020;12(8):e784–93.
11. Seo KS, Kim S, Kwon JH, Chang JS. Implant digital impressions with intraoral scanners: a literature review. Korean Acad Oral Maxillofac Implantol. 2017;21(1):2–13.
12. Marfell-Jones M, Stewart A, Olds T. Kinanthropometry IX: Proceedings of the 9th International Conference of the International Society for the Advancement of Kinanthropometry. Milton Park, UK: Taylor & Francis; 2006.
13. Flugge TV, Schlager S, Nelson K, Nahles S, Metzger MC. Precision of intraoral digital dental impressions with iTero and extraoral digitization with the iTero and a model scanner. Am J Orthod Dentofac Orthop. 2013;144:471–8.
14. Güth JF, Keul C, Stimmelmayr M, Beuer F, Edelhoff D. Accuracy of digital models obtained by direct and indirect data capturing. Clin Oral Investig. 2013;17:1201–8.
15. Boeddinghaus M, Breloer ES, Rehmann P, Wöstmann B. Accuracy of single-tooth restorations based on intraoral digital and conventional impressions in patients. Clin Oral Investig. 2015;19:2027–34.
16. Berrendero S, Salido MP, Valverde A, Ferreiroa A, Pradíes G. Influence of conventional and digital intraoral impressions on the fit of CAD/CAM-fabricated all-ceramic crowns. Clin Oral Investig. 2016;20:2403–10.
17. Atieh MA, Ritter AV, Ko CC, Duqum I. Accuracy evaluation of intraoral optical impressions: a clinical study using a reference appliance. J Prosthet Dent. 2017;118:400–5.
18. Ender A, Mehl A. Influence of scanning strategies on the accuracy of digital intraoral scanning systems. Int J Comput Dent. 2013;16:11–21.
19. Kuhr F, Schmidt A, Rehmann P, Wöstmann B. A new method for assessing the accuracy of full arch impressions in patients. J Dent. 2016;55:68–74.
20. Rhee YK, Huh YH, Cho LR, Park CJ. Comparison of intraoral scanning and conventional impression techniques using 3-dimensional superimposition. J Adv Prosthodont. 2015;7:460–7.
21. Sousa MV, Vasconcelos EC, Janson G, Garib D, Pinzan A. Accuracy and reproducibility of 3-dimensional digital model measurements. Am J Orthod Dentofac Orthop. 2012;142:269–73.
22. Reich S, Vollborn T, Mehl A, Zimmermann M. Intraoral optical impression systems – an overview. Int J Comput Dent. 2013;16:143–62.
23. Kravitz ND, Groth C, Jones PE, Graham JW, Redmond WR. Intraoral digital scanners. J Clin Orthod. 2014;48(6):337–47.
24. Logozzo S, Zanetti E, Franceschini G, Kilpela A, Makynen A. Recent advances in dental optics-Part 1: 3D intraoral scanners for restorative dentistry. Opt Lasers Eng. 2014;54:203–21.
25. Ciobata ND. Standard tessellation language in rapid prototyping technology. Mater Mech. 2012.
26. ISO5725-1. Accuracy (trueness and precision) of measurement methods and results – Part 1: General principles and definitions. 1994; ISO 5725-1.
27. Goracci C, Franchi L, Vichi A, Ferrari M. Accuracy, reliability, and efficiency of intraoral scanners for full-arch impressions: a systematic review of the clinical evidence. Eur J Orthod. 2016;38:422–8.
28. Güth J-F, Edelhoff D, Schweiger J, Keul C. A new method for the evaluation of the accuracy of full-arch digital impressions in vitro. Clin Oral Investig. 2016;20:1487–94.

29. Li H, Lyu P, Wang Y, Sun Y. Influence of object translucency on the scanning accuracy of a powder-free intraoral scanner: a laboratory study. J Prosthet Dent. 2017;117:93–101.
30. Kwon M, Cho Y, Kim DW, Kim M, Kim YJ, Chang M. Full-arch accuracy of five intraoral scanners: in vivo analysis of trueness and precision. Korean J Orthod. 2021;51(2):95–104.
31. Nedelcu R, Olsson P, Nyström I, Rydén J, Thor A. Accuracy and precision of 3 intraoral scanners and accuracy of conventional impressions: a novel in vivo analysis method. J Dent. 2018;69:110–8.
32. Ender A, Attin T, Mehl A. In vivo precision of conventional and digital methods of obtaining complete-arch dental impressions. J Prosthet Dent. 2016;115(3):313–20.
33. Flügge TV, Schlager S, Nelson K, Nahles S, Metzger MC. Precision of intraoral digital dental impressions with iTero and extraoral digitization with the iTero and a model scanner. Am J Orthod Dentofac Orthop. 2013;144(3):471–8.
34. Grünheid T, McCarthy SD, Larson BE. Clinical use of a direct chairside oral scanner: an assessment of accuracy, time, and patient acceptance. Am J Orthod Dentofac Orthop. 2014;146(5):673–82.
35. Sun L, Lee JS, Choo HH, Hwang HS, Lee KM. Reproducibility of an intraoral scanner: a comparison between in-vivo and ex-vivo scans. Am J Orthod Dentofac Orthop. 2018;154(2):305–10.
36. Shim JS, Lee JS, Lee JY, Choi YJ, Shin SW, Ryu JJ. Effect of software version and parameter settings on the marginal and internal adaptation of crowns fabricated with the CAD/CAM system. J Appl Oral Sci. 2015;23:515.
37. Martin CB, Chalmers EV, McIntyre GT, Cochrane H, Mossey PA. Orthodontic scanners: what's available? J Orthod. 2015;42:136–43.
38. Chen BT, Lou WS, Chen CC, Lin HC. A 3D scanning system based on low-occlusion approach. In: Proceedings of the Second International Conference on 3-D Digital Imaging and Modeling (Cat. No. PR00062), Ottawa, ON, Canada, 8 October 1999. Piscataway, NJ: IEEE; 1999. p. 506–15.
39. Barreto MS, Faber J, Vogel CJ, Araujo TM. Reliability of digital orthodontic setups. Angle Orthod. 2016;86:255–9.
40. Im J, Cha JY, Lee KJ, Yu HS, Hwang CJ. Comparison of virtual and manual tooth setups with digital and plaster models in extraction cases. Am J Orthod Dentofac Orthop. 2014;145:434–42.
41. Srinivasan M, Kamnoedboon P, McKenna G, Angst L, Schimmel M, Özcan M, Müller F. CAD-CAM removable complete dentures: a systematic review and meta-analysis of trueness of fit, biocompatibility, mechanical properties, surface characteristics, color stability, time-cost analysis, clinical and patient-reported outcomes. J Dent. 2021;113:103777.
42. Srinivasan M, Schimmel M, Naharro M, O'Neill C, McKenna G, Müller F. CAD/CAM milled removable complete dentures: time and cost estimation study. J Dent. 2019;80:75–9.
43. Herpel C, Tasaka A, Higuchi S, Finke D, Kühle R, Odaka K, Rues S, Lux CJ, Yamashita S, Rammelsberg P, Schwindling FS. Accuracy of 3D printing compared with milling-a multi-center analysis of try-in dentures. J Dent. 2021;110:103681.
44. Patzelt SB, Lamprinos C, Stampf S, Att W. The time efficiency of intraoral scanners: an in vitro comparative study. J Am Dent Assoc. 2014;145:542–51.
45. Sawase T, Kuroshima S. The current clinical relevancy of intraoral scanners in implant dentistry. Dent Mater J. 2020;39(1):57–61.
46. van der Meer WJ, Andriessen FS, Wismeijer D, Ren Y. Application of intra-oral dental scanners in the digital workflow of implantology. PLoS One. 2012;7(8):e43312.
47. Mörmann W, Brandestini M, Ferru A, Lutz F, Krejci I. Marginal adaptation of adhesive porcelain inlays in vitro. Schweiz Monatsschr Zahnmed. 1985;95:1118.
48. Joós-Kovács G, Vecsei B, Körmendi S, Gyarmathy VA, Borbély J, Hermann P. Trueness of CAD/CAM digitization with a desktop scanner - an in vitro study. BMC Oral Health. 2019;19(1):280.
49. Taneva E, Kusnoto B, Evans CA. 3D scanning, imaging, and printing in orthodontics. In: Issues in contemporary orthodontics, vol. 148(5). Rijeka, Croatia: IntechOpen; 2015. p. 862–7. Available from: https://www.intechopen.com/chapters/48165.
50. Correia GD, Habib FA, Vogel CJ. Tooth-size discrepancy: a comparison between manual and digital methods. Dental Press J Orthod. 2014;19(4):107–13. https://doi.org/10.1590/2176-9451.19.4.107-113.oar.
51. Zilberman O, Huggare JAV, Parikakis KA. Evaluation of the validity of tooth size and arch width measurements using conventional and three-dimensional virtual orthodontic models. Angle Orthod. 2003;73(3):301–6.
52. Garino F, Garino GB. Comparison of dental arch measurements between stone and digital casts. World J Orthod. 2002;3(3):1–5.
53. Palmer NG, Yacyshyn JR, Northcott HC, Nebbe B, Major PW. Perceptions and attitudes of Canadian orthodontists regarding digital and electronic technology. Am J Orthod Dentofac Orthop. 2005;128(2):163–7.
54. Stevens DR, Flores-Mir C, Nebbe B, Raboud DW, Heo G, Major PW. Validity, reliability, and reproducibility of plaster vs digital study models: comparison of peer assessment rating and Bolton analysis and their constituent measurements. Am J Orthod Dentofac Orthop. 2006;129(6):794–803.
55. Marcel T. Three-dimensional on-screen virtual models. Am J Orthod Dentofac Orthop. 2001;119(6):666–8.
56. Mayers M, Fitestone AR, Rashid R, Vig KWL. Comparison of peer assessment rating (PAR) index scores of plaster and computer-based digital models. Am J Orthod Dentofac Orthop. 2005;128(4):431–4.

57. Horton HMI, Miller JR, Gaillard PR, Larson BE. Technique comparison for efficient orthodontic tooth measurements using digital models. Angle Orthod. 2010;80(2):254–61.
58. Jansen CE. Understanding the potential of digital intraoral and benchtop scanning workflows. Compend Contin Educ Dent. 2015;36(10):726–31; quiz732.
59. Koch GK, Gallucci GO, Lee SJ. Accuracy in the digital workflow: from data acquisition to the digitally milled cast. J Prosthet Dent. 2016;115:749–54.
60. Vlaar ST, van der Zel JM. Accuracy of dental digitizers. Int Dent J. 2006;56:301–9.
61. Chen Z, Linghu C, Yu K, Zhu J, Luo H, Qian C, Chen Y, Du Y, Zhang S, Song J. Fast digital patterning of surface topography toward three-dimensional shape-changing structures. ACS Appl Mater Interfaces. 2019;11:48412–8.
62. Jedliński M, Mazur M, Grocholewicz K, Janiszewska-Olszowska J. 3D scanners in orthodontics—current knowledge and future perspectives—a systematic review. Int J Environ Res Public Health. 2021;18:1121. https://doi.org/10.3390/rs13030374.
63. https://www.motionviewllc.com/desk-top-scanner. Accessed 21 Oct 2021.
64. https://www.inn-soft.com/scanner. Accessed 21 Oct 2021.
65. Javaid M, Haleem A. Current status and applications of additive manufacturing in dentistry: a literature-based review. J Oral Biol Craniofac Res. 2019;9(3):179–85.
66. Farkas LG, Katic MJ, Forrest CR, et al. International anthropometric study of facial morphology in various ethnic groups/races. J Craniofac Surg. 2005;16:615–46.
67. Holberg C, Schwenzer K, Mahaini L, Rudzki-Janson I. Accuracy of facial plaster casts. Angle Orthod. 2006;76(4):605–11.
68. Zogheib T, Jacobs R, Bornstein MM, et al. Comparison of 3D scanning *versus* 2D photography for the identification of facial soft-tissue landmarks. Open Dent J. 2018;12:61–71.
69. Ireland AJ, McNamara C, Clover MJ, et al. 3D surface imaging in dentistry – what we are looking at. Br Dent J. 2008;205(7):387–92.
70. Da Silveira AC, Martinez O, Da Silveira D, Daw JL Jr, Cohen M. Three-dimensional technology for documentation and record keeping for patients with facial clefts. Clin Plast Surg. 2004;31(2):141–8.
71. Karatas OH, Toy E. Three-dimensional imaging techniques: a literature review. Eur J Dent. 2014;8(1):132–40.
72. Lekakis G, Claes P, Hamilton G, Hellings P. Three-dimensional surface imaging and the continuous evolution of preoperative and postoperative assessment in rhinoplasty. Facial Plast Surg. 2016;32(1):088–94.
73. Mai HN, Kim J, Choi YH, Lee DH. Accuracy of portable face-scanning devices for obtaining three-dimensional face models: a systematic review and meta-analysis. Int J Environ Res Public Health. 2020;18(1):94.
74. Chong Y, Liu X, Shi M, Huang J, Yu N, Long X. Three-dimensional facial scanner in the hands of patients: validation of a novel application on iPad/iPhone for three-dimensional imaging. Ann Transl Med. 2021;9(14):1115.
75. Hajeer MY, Millett DT, Ayoub AF, Siebert JP. Applications of 3D imaging in orthodontics: Part I. J Orthod. 2004;31(1):62–70.
76. Naini FB, Akram S, Kepinska J, Garagiola U, McDonald F, Wertheim D. Validation of a new three-dimensional imaging system using comparative craniofacial anthropometry. Maxillofac Plast Reconstr Surg. 2017;39(1):23.
77. Alshammery FA. Three dimensional (3D) imaging techniques in orthodontics-an update. J Family Med Prim Care. 2020;9(6):2626–30.
78. Arridge S, Moss JP, Linney AD, James DR. Three dimensional digitization of the face and skull. J Maxillofac Surg. 1985;13(3):136–43.
79. Baik HS, Lee HJ, Lee KJ. A proposal for soft tissue landmarks for craniofacial analysis using 3-dimensional laser scan imaging. World J Orthod. 2006;7(1):7–14.
80. Kau CH, Richmond S, Zhurov AI, et al. Reliability of measuring facial morphology with a 3-dimensional laser scanning system. Am J Orthod Dentofac Orthop. 2005;128(4):424–30.
81. Kovacs L, Zimmermann A, Brockmann G, et al. Accuracy and precision of the three dimensional assessment of the facial surface using a 3-D laser scanner. IEEE Trans Med Imaging. 2006;25(6):742–54.
82. Germec-Cakan D, Canter HI, Nur B, Arun T. Comparison of facial soft tissue measurements on three-dimensional images and models obtained with different methods. J Craniofac Surg. 2010;21(5):1393–9.
83. Djordjevic J, Toma AM, Zhurov AI, Richmond S. Three-dimensional quantification of facial symmetry in adolescents using laser surface scanning. Eur J Orthod. 2014;36(2):125–32.
84. Ma L, Xu T, Lin J. Validation of a three-dimensional facial scanning system based on structured light techniques. Comput Methods Prog Biomed. 2009;94(3):290–8.
85. Alkhayer A, Becsei R, Hegedűs L, et al. Evaluation of the soft tissue changes after rapid maxillary expansion using a handheld three-dimensional scanner: a prospective study. Int J Environ Res Public Health. 2021;18(7):3379.
86. Amirav I, Luder AS, Halamish A, et al. Design of aerosol face masks for children using computerized 3D face analysis. J Aerosol Med Pulm Drug Deliv. 2014;27(4):272–8.
87. Amirav I, Masumbuko CK, Hawkes MT, et al. 3D analysis of child facial dimensions for design of medical devices in low-middle income countries (LMIC). PLoS One. 2019;14(5):e0216548.
88. Kimmel R. Numerical geometry of images: theory, algorithms, and applications. New York: Springer; 2003.

89. Rubinstein O, Honen Y, Bronstein AM, Bronstein MM, Kimmel R. 3D-color video camera. In: 12th International Conference on Computer Vision Workshops, ICCV Workshops. New York: IEEE; 2009. p. 1505–9.
90. Meyer-Marcotty P, Stellzig-Eisenhauer A, Bareis U, Hartmann J, Kochel J. Three-dimensional perception of facial asymmetry. Eur J Orthod. 2011;33(6):647–53.
91. Modabber A, Peters F, Kniha K, et al. Evaluation of the accuracy of a mobile and a stationary system for three-dimensional facial scanning. J Craniomaxillofac Surg. 2016;44(10):1719–24.
92. Ras F, Habets LL, van Ginkel FC, Prahl-Andersen B. Quantification of facial morphology using stereophotogrammetry—demonstration of a new concept. J Dent. 1996;24(5):369–74.
93. Liu C, Artopoulos A. Validation of a low-cost portable 3-dimensional face scanner. Imaging Sci Dent. 2019;49(1):35–43.
94. Plooij JM, Swennen GR, Rangel FA, et al. Evaluation of reproducibility and reliability of 3D soft tissue analysis using 3D stereophotogrammetry. Int J Oral Maxillofac Surg. 2009;38(3):267–73.
95. Weinberg SM, Naidoo S, Govier DP, Martin RA, Kane AA, Marazita ML. Anthropometric precision and accuracy of digital three-dimensional photogrammetry: comparing the Genex and 3dMD imaging systems with one another and with direct anthropometry. J Craniofac Surg. 2006;17(3):477–83.
96. Aynechi N, Larson BE, Leon-Salazar V, Beiraghi S. Accuracy and precision of a 3D anthropometric facial analysis with and without landmark labeling before image acquisition. Angle Orthod. 2011;81(2):245–52.
97. Artopoulos A, Buytaert JA, Dirckx JJ, Coward TJ. Comparison of the accuracy of digital stereophotogrammetry and projection moire profilometry for three-dimensional imaging of the face. Int J Oral Maxillofac Surg. 2014;43(5):654–62.
98. Dindaroğlu F, Kutlu P, Duran GS, Görgülü S, Aslan E. Accuracy and reliability of 3D stereophotogrammetry: a comparison to direct anthropometry and 2D photogrammetry. Angle Orthod. 2016;86(3):487–94.
99. Zhao YJ, Xiong YX, Yang HF, Wang Y. Evaluation of measurement accuracy of three facial scanners based on different scanning principles. Beijing Da Xue Xue Bao. 2014;46(1):76–80.
100. Zhao YJ, Xiong YX, Wang Y. Three-dimensional accuracy of facial scan for facial deformities in clinics: a new evaluation method for facial scanner accuracy. PLoS One. 2017;12(1):e0169402.
101. Ghoddousi H, Edler R, Haers P, Wertheim D, Greenhill D. Comparison of three methods of facial measurement. Int J Oral Maxillofac Surg. 2007;36(3):250–8.
102. Ceinos R, Tardivo D, Bertrand MF, Lupi-Pegurier L. Inter- and intra-operator reliability of facial and dental measurements using 3D-stereophotogrammetry. J Esthet Restor Dent. 2016;28(3):178–89.
103. Ye H, Lv L, Liu Y, Liu Y, Zhou Y. Evaluation of the accuracy, reliability, and reproducibility of two different 3D face-scanning systems. Int J Prosthodont. 2016;29(3):213–8.
104. Kusnoto B, Evans CA. Reliability of a 3D surface laser scanner for orthodontic applications. Am J Orthod Dentofac Orthop. 2002;122(4):342–8.
105. Knoops PG, Beaumont CA, Borghi A, et al. Comparison of three-dimensional scanner systems for craniomaxillofacial imaging. J Plast Reconstr Aesthet Surg. 2017;70(4):441–9.
106. Rudy HL, Wake N, Yee J, Garfein ES, Tepper OM. Three-dimensional facial scanning at the fingertips of patients and surgeons: accuracy and precision testing of iPhone X three-dimensional scanner. Plast Reconstr Surg. 2020;146(6):1407–17.
107. 3-D digitizing-breakthrough in process innovation VIVID 910 PET: Polygon editing software, EAT: Easy alignment target-based registration Compatible with all major 3D software for Modeling and CAD, CAM and CAT. Konica Minolta Sensing. Inc. 2001, 2002. http://konicaminolta.us/3d.
108. Ritschl LM, Roth M, Fichter AM, et al. The possibilities of a portable low-budget three-dimensional stereophotogrammetry system in neonates: a prospective growth analysis and analysis of accuracy. Head Face Med. 2018;14(1):11.
109. Verhulst A, Hol M, Vreeken R, Becking A, Ulrich D, Maal T. Three-dimensional imaging of the face: a comparison between three different imaging modalities. Aesthet Surg J. 2018;38(6):579–85.
110. Sade Hoefert C, Bacher M, Herberts T, Krimmel M, Reinert S, Göz G. 3D soft tissue changes in facial morphology in patients with cleft lip and palate and class III mal occlusion under therapy with rapid maxillary expansion and delaire facemask. J Orofac Orthop. 2010;71(2):136–51.
111. Hoefert CS, Bacher M, Herberts T, et al. Implementing a superimposition and measurement model for 3D sagittal analysis of therapy-induced changes in facial soft tissue: a pilot study. J Orofac Orthop. 2010;71(3):221–34.
112. Erten O, Yılmaz BN. Three-dimensional imaging in orthodontics. Turk J Orthod. 2018;31(3):86–94.
113. Kau CH, Cronin A, Durning P, Zhurov AI, Sandham A, Richmond S. A new method for the 3D measurement of postoperative swelling following orthognathic surgery. Orthod Craniofac Res. 2006;9(1):31–7.
114. Staderini E, Patini R, Camodeca A, Guglielmi F, Gallenzi P. Three-dimensional assessment of morphological changes following nasoalveolar molding therapy in cleft lip and palate patients: a case report. Dent J (Basel). 2019;7(1):27.
115. Ciocca L, Scotti R. CAD-CAM generated ear cast by means of a laser scanner and rapid prototyping machine. J Prosthet Dent. 2004;92(6):591–5.
116. Lin WS, Harris BT, Phasuk K, Llop DR, Morton D. Integrating a facial scan, virtual smile design, and 3D virtual patient for treatment with CAD-CAM ceramic veneers: a clinical report. J Prosthet Dent. 2018;119(2):200–5.

117. Richmond S, Howe LJ, Lewis S, Stergiakouli E, Zhurov A. Facial genetics: a brief overview. Front Genet. 2018;9:462.
118. Lambros V, Amos G. Three-dimensional facial averaging: a tool for understanding facial aging. Plast Reconstr Surg. 2016;138(6):980e–2e.
119. Tanikawa C, Takata S, Takano R, Yamanami H, Edlira Z, Takada K. Functional decline in facial expression generation in older women: a cross-sectional study using three-dimensional morphometry. PLoS One. 2019;14(7):e0219451.
120. Tanikawa C, Zere E, Takada K. Sexual dimorphism in the facial morphology of adult humans: a three-dimensional analysis. Homo. 2016;67(1):23–49.
121. Tanikawa C, Akcam MO, Gokalp H, Zere E, Takada K. Population affinity and variation of sexual dimorphism in three-dimensional facial forms: comparisons between Turkish and Japanese populations. Sci Rep. 2021;11(1):16634.
122. Tian S, Dai N, Zhang B, Yuan F, Yu Q, Xiaosheng Cheng X. Automatic classification and segmentation of teeth on 3D dental model using hierarchical deep learning networks. New York: IEEE; 2019. p. 84817–28.
123. Lian C, Wang L, Wu TH, et al. Deep multi-scale mesh feature learning for automated labeling of raw dental surfaces from 3D intraoral scanners. IEEE Trans Med Imaging. 2020;39(7):2440–50.
124. Nauwelaers N, Matthews H, Fan Y, et al. Exploring palatal and dental shape variation with 3D shape analysis and geometric deep learning. Orthod Craniofac Res. 2021; https://doi.org/10.1111/ocr.12521.
125. Hung K, Yeung AWK, Tanaka R, Bornstein MM. Current applications, opportunities, and limitations of AI for 3D imaging in dental research and practice. Int J Environ Res Public Health. 2020;17(12):4424.
126. Venkatesh E, Elluru SV. Cone beam computed tomography: basics and applications in dentistry. J Istanb Univ Fac Dent. 2017;51(3 Suppl 1):S102–21.
127. Abramovitch K, Rice DD. Basic principles of cone beam computed tomography. Dent Clin N Am. 2014;58(3):463–84.
128. Pauwels R, Araki K, Siewerdsen JH, Thongvigitmanee SS. Technical aspects of dental CBCT: state of the art. Dentomaxillofac Radiol. 2015;44(1):20140224.
129. Kamio T, Suzuki M, Asaumi R, Kawai T. DICOM segmentation and STL creation for 3D printing: a process and software package comparison for osseous anatomy. 3D Print Med. 2020;6(1):17.
130. Lo Giudice A, Ronsivalle V, Grippaudo C, Lucchese A, Muraglie S, Lagravère MO, et al. One step before 3D printing—evaluation of imaging software accuracy for 3-dimensional analysis of the mandible: a comparative study using a surface-to-surface matching technique. Materials (Basel). 2020;13(12):2798.
131. Kim YH, Jung B-Y, Han S-S, Woo C-W. Accuracy evaluation of 3D printed interim prosthesis fabrication using a CBCT scanning based digital model. PLoS One. 2020;15(10):e0240508.
132. Effect of voxel size on the accuracy of 3D reconstructions with cone beam CT - PubMed [Internet]. [cited 2021 Oct 17]. Available from: https://pubmed.ncbi.nlm.nih.gov/23166362/
133. Kamburoğlu K. Use of dentomaxillofacial cone beam computed tomography in dentistry. World J Radiol. 2015;7(6):128–30.
134. Carter L, Farman AG, Geist J, Scarfe WC, Angelopoulos C, Nair MK, et al. American Academy of Oral and Maxillofacial Radiology executive opinion statement on performing and interpreting diagnostic cone beam computed tomography. Oral Surg Oral Med Oral Pathol Oral Radiol Endodontol. 2008;106(4):561–2.
135. Horner K, Islam M, Flygare L, Tsiklakis K, Whaites E. Basic principles for use of dental cone beam computed tomography: consensus guidelines of the European Academy of Dental and Maxillofacial Radiology. Dentomaxillofac Radiol. 2009;38(4):187–95.
136. Jaju PP, Jaju SP. Cone-beam computed tomography: time to move from ALARA to ALADA. Imaging Sci Dent. 2015;45(4):263–5.
137. Oberoi G, Nitsch S, Edelmayer M, Janjić K, Müller AS, Agis H. 3D printing—encompassing the facets of dentistry. Front Bioeng Biotechnol. 2018;6:172.
138. Tian Y, Chen C, Xu X, Wang J, Hou X, Li K, et al. A review of 3D printing in dentistry: technologies, affecting factors, and applications. Scanning. 2021;2021:e9950131.

3 Commonly Used 3D Printing Technologies in Oral Health Science

Ramandeep Kaur Sokhi, Nitika Monga, Kelly Kathering Achachao Almerco, Dipti Shastri, Manish Bajracharya, Ujjwal Pyakurel, Mohammad Khursheed Alam, and Prabhat Kumar Chaudhari

Contents

The original version of this chapter was revised. The correction to this chapter can be found at https://doi.org/10.1007/978-3-031-07369-4_15

R. K. Sokhi · P. K. Chaudhari (✉)
Division of Orthodontics and Dentofacial Deformities, Centre for Dental Education and Research (CDER), All India Institute of Medical Sciences (AIIMS), New Delhi, India
e-mail: dr.prabhatkc@aiims.edu

N. Monga
Non Communicable Disease (NCD) Division, Indian Council of Medical Research (ICMR), New Delhi, India
e-mail: nitika.monga@icme.gov.in

K. K. A. Almerco
Department of Pediatric Stomatology, Dentistry School, National University of San Marcos, Lima, Peru
e-mail: kelly.achachao.a@upch.pe

D. Shastri
Department of Orthodontics and Dentofacial Orthopaedics, Faculty of Dental Sciences, King George's Medical University, Lucknow, Uttar Pradesh, India

M. Bajracharya
Orthodontic Unit, Dental Department, National Academy of Medical Sciences (NAMS), Bir Hospital, Kathmandu, Nepal

U. Pyakurel
Kantipur Dental College Teaching Hospital & Research Centre, Kathmandu, Nepal

Private Practice, Shangrila Dental Clinic, Kathmandu, Nepal

M. K. Alam
Orthodontic Department, College of Dentistry, Jouf University, Al-Jawf, Kingdom of Saudi Arabia
e-mail: dr.mohammad.alam@jodent.org

© The Author(s), under exclusive license to Springer Nature Switzerland AG 2022, corrected publication 2022
P. K. Chaudhari et al. (eds.), *3D Printing in Oral Health Science*, https://doi.org/10.1007/978-3-031-07369-4_3

3.1 Introduction

Three-dimensional (3D) printing is becoming an integral part of dental and medical sciences, expanding to all the major aspects of oral health, beginning from diagnosis and treatment planning to treatment execution and follow-up. Improvement in technology has developed and brought up an integrated digital system that has the potential to be used extensively for clinical care of patients and research by healthcare professionals throughout the world.

The term "3D printing" is relatively new; however, 3D printing technologies and their concepts are not recent. 3D printing is generally used to describe a manufacturing approach that builds a three-dimensional object from a computer-aided design (CAD) model, one layer at a time, adding up to multiple layers to form an object. The process of 3D printing involves depositing the material layer by layer in horizontal cross sections to form a complete object. This process is more correctly described as additive manufacturing and is also referred to as rapid prototyping. For an object to be in 3D, it has to have three axes: the x-, y-, and z-axis (Fig. 3.1) where x-axis is horizontal axis, y-axis is vertical axis, and z-axis is the depth (height) axis. The z-axis or the depth axis is the one that gives a 3-dimensional meaning to the object [1, 2].

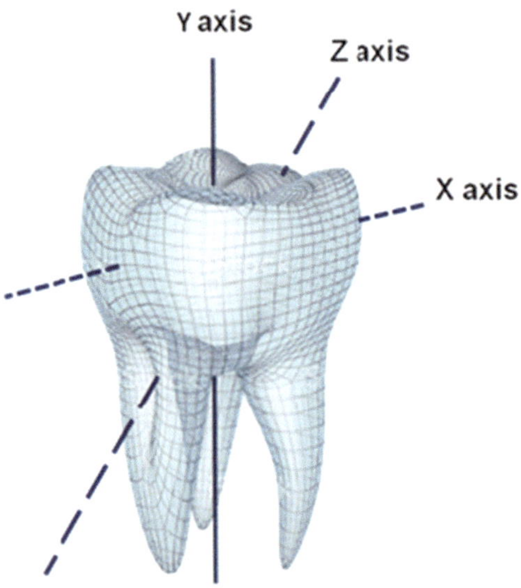

Fig. 3.1 An object depicting the 3 Dimensions in three planes i.e., x-, y-, z-axis

The materials used to print these solid 3D objects can be metal, resin, and other biocompatible variants. The various methods that apply the principle of additive manufacturing are stereolithography (SLA), fused deposition modeling (FDM), selective laser sintering (SLS), PolyJet printing, and bioprinting [3].

3D printing is a part of the digital workflow in the dental office. The digital workflow involves a series of sequential unidirectional patterns that involve 3D scanning, treatment planning, computed-aided designing (CAD), and 3D printing process. In a conventional orthodontic workflow, plaster models are made by taking physical impressions of the patient's oro-dental tissues. In contrast, in a digital workflow, 3D printing is used to make a model using the stereolithography (STL) file of the digital model generated through either intraoral or desktop scanning and CAD. STL has several acronyms such as "Standard Triangle Language" and "Standard Tessellation Language" [4].

This chapter will discuss the commonly used 3D printing technologies and their applications in dentistry.

3.2 Brief History

The concept of 3D printers was visualized almost four decades ago but has gained momentum in the last two decades in the dental field. The initiation of 3D printing technology dates back to 1980 when Hideo Kodama from Nagoya Japan, at Municipal Industrial Research Institute, conceptualized the single-beam method of curing. In his experiments, ultraviolet (UV) rays were used using Toshiba mercury lamp along with photosensitive resin. For each corresponding cross section, a black and a white film was used as a cover to shield and manage the region of exposure. These basics of solidification of polymer in thin consecutive layers are what later lead to development of stereolithography (SL) technology [4].

Three years later, in 1984, Chuck Hull made 3D printing a reality by inventing the first commercial rapid prototyping technology, i.e., stereolithography (SLA) and the STL file format. He developed the concept of 3D printing while using UV light to cure tabletop coating [5]. In 1984 and later in 1986, he received the US patent for the same [5]. Hull defined stereolithography as a method and apparatus for making solid objects by successively printing thin layers of the ultraviolet light-curable material one on top of the other. He also founded the company 3D systems to commercialize the first computer for quick prototyping, called SLA-1 stereolithography (SLA) 3D printer in 1986 (3D Systems, Rock Hill, SC) [5]. In 1988, Scott Crump invented another technique called fused deposition modeling (FDM), which was later commercialized by Stratasys (Stratasys Ltd., Minneapolis, MN; USA) in 1990. Crump also developed acrylonitrile butadiene styrene (ABS) material for FDM machines that are being used widely by the majority of FDM 3D printers today. In 1998, object geometries (a subsidiary of Stratasys) was founded that developed the PolyJet photopolymer (PPP) printing [6].

In 1984, Carl Deckard at the University of Texas conceptualized the idea of SLS (Selective Laser Sintering) method independent of SLA and FDM technology. While the SLA and FDM were useful in making plastic (synthetic resin) and nylon parts, the SLS was able to solidify metal powder on exposure to a laser beam to make the prototype in metal parts. In 1992, DTM (Desktop Manufacturing) Corporation was founded by Deckard to manufacture the first SLS printers. Later, DTM Corporation was acquired by 3D Systems.

EnvisionTEC (Envision TEC, Dearborn, MI; USA) was founded in 2002 by engineer entrepreneur Al Siblani. EnvisionTEC is the biggest digital light processing (DLP) printer's manufacturer. DLP is a registered trademark of Texas Instruments Incorporated, Dallas, TX, USA. It is similar to SLA except for the light source; in contrast to the SLA laser, a projector is used in DLP to cure the entire layer at a time.

Many industrial patents of the 3D System and Stratasys companies expired a few years ago. In 2005, Stratasys FDM Software patent expired, and the RepRap Project (The Replication Rapid-Prototype Project) of Dr. Adrian Bowyer launched an open-source project to develop a 3D printer that can design itself—or at least print most parts of it [7]. The most popular open-source 3D printer for dental applications are FormLabs printers. FormLabs was officially founded in 2011 and brought their first desktop-sized, easy-to-use, and affordable stereolithography 3D printers (Forms 1 and 2). In 2019, FormLabs developed the Form 3 and Form 3L. Form 3 uses low force stereolithography (LFS), a new SLA technology developed by FormLabs that promises smoother surface finish and more detailed prints.

At present, several companies are offering many variants of printers that use versions of SLA, FDM, PolyJet, and DLP technologies for dental applications. Over the year, 3D System and Stratasys have acquired other companies with different 3D printing technologies and have become the biggest and best 3D printing companies. Other popular manufactures of the 3D printer for dental applications are EnvisionTEC and FormLabs.

3.3 Basic Terminologies

3.3.1 2D Printing

Two-dimensional printing is similar to the process of printing like we do to print a picture using ink printers. These printings can be seen in two dimensions, only, i.e., *x*- and *y*-axis. Printing from left to right will be a horizontal axis, i.e., *x*-axis, while the top to bottom will be the vertical axis, i.e., *y*-axis [4].

3.3.2 3D Printing

It is a technology, where a 3D printer lays down the material in sequential layers, i.e., one layer at a time in order to print an object. Since the process involves adding the material one layer followed by subsequent layering, it is called as additive manufacturing. Along with the *x*- and *y*-axis, the *z*-axis is seen, which is called the depth axis. z-axis is the height of the object that gives it a 3D shape [1, 2]. Few examples of such printers are as follows: SLA, DLP, FDM, SLS, inkjet, EBM (electron-beam melting), and LOM (laminated object manufacturing).

3.3.3 Additive Manufacturing

In 3D printing, additive manufacturing is a process in which sequential material layers are put on top of each other and ultimately shaped into an object [4].

3.3.4 Subtractive Manufacturing (Computer-Aided Manufacturing or Milling)

This process is just the opposite of additive manufacturing. In this, a single block of material is carved out to produce the object, e.g., milling seen in CEREC (Chairside Economical Restoration of Esthetic Ceramics).

3.3.5 Intraoral Scanning

Intraoral scanner captures the optical images of the dental hard and soft tissues. The light source from the handheld scanner is directed over the area to be scanned, e.g., teeth, and gingiva etc. The image is captured by sensors, which is processed with the help of software by generating point clouds. The software then analyses these points, which help in creating a mesh framework giving the final virtual image of the scanned object [8].

3.3.6 Desktop Scanning

This type of scanning is used to scan the object physically, i.e., scanning of impression can be in alginate or polyvinyl siloxane or the gypsum cast [4].

3.3.7 STL File (Stereolithographic File or Standard Tessellation Language File)

STL is the global format for the 3D printing files. As the object is scanned with the scanner (intraoral scanner/desktop scanner), it is stored in the computer in STL file format [4].

3.3.8 Stereolithography (SLA)

It is a rapid prototyping technology that will be discussed in detail later in the chapter.

3.3.9 Stereophotogrammetry

It is a 3D camera technology where the cameras are arranged as a stereo-pair. Here, the 3D coordinate points are marked on an object (e.g., face) followed by photographs taken from different angulations and positions. The image is then calculated by collecting the points which are obtained along the X, Y, and Z system of coordinates [9].

3.3.10 Computer-Aided Design (CAD)

The CAD program is used to begin processing and preparation of the file for printing or milling. It is a software on the computer that prepares the STL file for 3D printing of the intended object [4].

3.3.11 Resolution

It is one of the way to measure the print quality of a 3D printer. It is of two types horizontal and vertical resolution [10].

3.3.11.1 Horizontal Resolution

It is the smallest movement that a printer's extruder makes with a layer along the x- and *y*-axis, i.e., 2 dimensionally. The lesser the value, the higher the details of the printed object, hence better the resolution.

3.3.11.2 Vertical Resolution

It is the movement that a printer's extruder makes with a layer along the *z*-axis. It can also be defined as the minimum thickness of the layer that can be printed in one pass by the printer head while the object is being printed. The smaller the layer height, the better will be the surface finish yielding more details and thereby longer time to print.

While choosing a print resolution for dental or orthodontic models, a maximum accuracy of 100 μm has been seen to be adequate to manufacture high-quality models for diagnostic purposes, retainers, and other appliances. But for a print with better resolution, it might take a longer time to print. According to the American Board of Orthodontics, for a model to have high quality, it can be printed with a layer height of 16 microns (16 μm or 0.016 mm). Conversely, for printing models at a higher speed with decreased surface quality, they can be printed with a layer height at approximately 30 microns (30 μm or 0.030 mm) [11].

3.3.12 Accuracy

In simple terms, accuracy is how closely the 3D printed model resembles the digital model. It determines the repeatability of a scanner. Two different terms that represent accuracy are "*Trueness* and *Precision*." *Trueness* is defined as the comparison between reference data and a test data set, while *precision* is the comparison between the numerous data sets that are obtained from the same object by using the same scanner (repeatability). Researchers measure potential deviations between the reference data set and the test data set, thereby determining the real accuracy of a scanner [12].

3.4 Various 3D Printing Technologies for Oral Health Applications: Principles and Applications

The process of 3D printing typically starts with a 3D model, which is digitally designed or obtained by physical object scanning.

The process of 3D printing can be broadly divided into three steps;

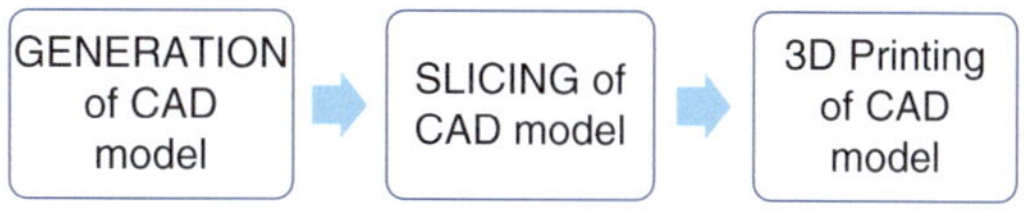

3D printing techniques can be broadly categorized based on different principles (Table 3.1),; e.g., SLA and DLP are based on the principle of vat polymerization; FDM is based on material extrusion whereas PolyJet photopolymerization is based on material jetting, and SLS is based on a powder-bed fusion of polymers. In contrast, SLM/DMLS/EBM are based on powder-bed fusion technology using metal powder, and LOM is based on sheet lamination [13, 14].

For printing any object using 3D printing technology, a digital workstation comprising of the computer system and a 3D printer is necessary. The computer system has software installed, which generates the CAD (computer-aided design) model. This CAD model is the STL file of the object that will be 3D printed. The software then breaks up the CAD model (which is in STL format) in thin cross-sectional slices ranging between 16 and 300 μm [4] (for better understanding, the thickness of a wool fiber is in the range of 25–100 μm) [15]. These layers are referred to as "build layers." The time required to 3D print plays an important role in the surface texture and accuracy. It is dependent on the vertical height of the object being printed. For example, to print one object having 20 μm layer, height will take much longer time to print as

Table 3.1 Additive manufacturing technologies with their principles and material used for 3D printing

Technique based on	Technology	Description	Material used
Vat polymerization	SLA DLP	They employ the use of the light source for curing the liquid resin in a vat of photopolymer	Photopolymer resin, ceramics
Material extrusion	FDM	Material is extruded out through a heated nozzle to build layers	Polymer
Powder-bed fusion	SLS SLM DMLS	Creates objects by using thermal energy to fuse regions of a powder bed	Polymer, ceramic, metal
Material jetting	Inkjet/PolyJet photopolymerization (PPP)	The object created by depositing small droplets of build material, which are then cured on exposure to light	Photopolymer, wax
Direct energy deposition	EBM	Objects are created using thermal energy to fuse regions of a powder bed	Metal: powder or wire
Sheet lamination	LOM	Parts are built by layering the sheets on each other (additive) followed by trimming sheets of material (subtractive)	Hybrids, metals, ceramic

compared to printing 20 objects with 1 μm layer height. The materials used in 3D printing may include a variety of them, depending on the particular type of printer. Materials can include ranging from plastics to metal to even clays and even human cells [4].

3.4.1 Stereolithography (SLA)

Also known as laser lithography [16], it is based on the technique of vat polymerization. Both SLA and DLP are vat polymerization techniques. *Vat polymerization* is a process that makes use of photopolymer resin (liquid) which gets solidified (cured) under the light source. The concept of stereolithography was conceptualized by Hull in the 1980s. The stereolithographic apparatus includes an ultraviolet (UV) laser light, a build platform on which the object is being fabricated, and a resin tank consisting of resin material from which the object will be printed [4, 14]. The UV light source placement, i.e., the laser beam, is right on top of the resin tank. The laser beam is then projected upon the top of photosensitive liquid polymer, drawing out cross section of the object. As the first layer is laid down and cured by the UV light, the build platform descends down into the resin tank to a distance which equals the layer thickness that has just been cured. This causes the uncured resin to flow over the cured layer. This uncured layer is once again exposed to the laser beam projected on it for the next layer to cure. This step is repeated n number of times until the object is formed. The drawback of steriolithography printers is that they require post-cure processing, thus affecting the accuracy of the printed object. Also, they take a longer time to print as the UV light first draws the whole layer and then cures it [17]. As the curing is not done fully, it requires a separate UV curing chamber for curing the printed model to 100%. This might lead to shrinkage of the object, having a greater impact along *z*-axis [18]. Post-cure processing is done because as the SLA printer is finished with printing while the object is still attached to the build platform, it is said to be in "green state." This means that the polymerization reaction is still not complete, which would further affect the mechanical properties, thus affecting the stability and strength of the 3D printed object [19].

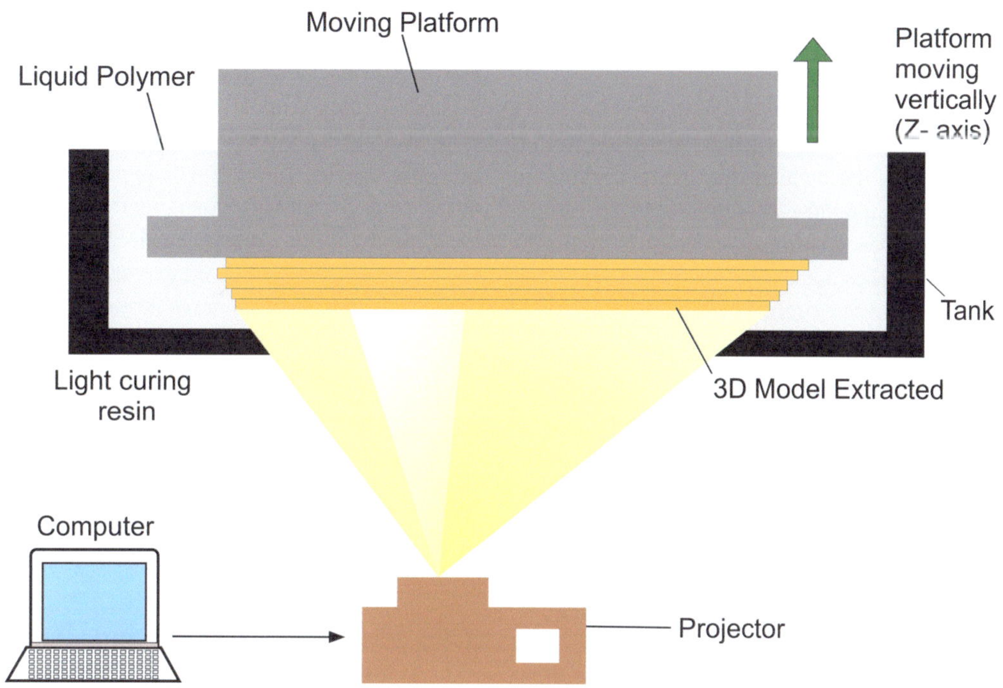

Fig. 3.2 Printer setup and mechanism of action of the digital light processor (DLP) printer. (Adopted from Groth CH, Kravitz ND, Jones PE, Graham JW, Redmond WR. Three-dimensional printing technology. J Clin Orthod. 2014;48:475–85. (To reproduce in publication, permission applied))

3.4.2 Digital Light Processing (DLP)

DLP is a vat polymerization technique based on similar technology as the SLA, with the difference being in the light source. The apparatus of a DLP printer (Fig. 3.2) consists of an LED screen and a digital microdevice (DMD). These two combine to make up the DLP chip. A DMD consists of thousands of small tiny micromirrors that help to direct the light and create the layer. In simple terms, a DMD is a projector that flashes an image (of the cross sections created by CAD software) on the top of liquid resin over the build platform, followed by simultaneous curing. Since the projector is a digital display, each layer's image is in the form of square pixels because the image is being flashed in 2 dimensions *X* and *Y*. When it continues to form layers on top of one another, it results in a three-dimensional layer made up of small rectangular cubes called voxels. To keep it simple, the voxel is a 3D pixel that can be stacked on top of each other, making up a layer. The voxels here in DLP are cube-shaped. In contrast to the SLA, DLP technology builds the model in voxels, resulting in better finish quality of the printed object. In this, only one type of resin material can be used at a time, and it also requires post-cure processing followed by washing and removal of supports by cutting them off [2]. Printers that use DLP technology are much quicker. They can build up objects with better resolutions with layer thickness reaching up to 30 μm. Few studies have described a range for the error that can be considered clinically acceptable. Between 0.2 and 0.5 mm, it has been considered to be an acceptable range of clinical accuracy [20–25]. EnvisionTEC, Dearborn, MI, USA, is one of the many producers of the DLP printer manufacturers.

3.4.3 Fused Deposition Modeling (FDM)

Stratasys Ltd. in the 1990s popularized the FDM technology. The material used here is thermoplastic in nature, which can be extruded out through a temperature-controlled heated nozzle. The nozzle head motion is monitored by a processor that traces and deposits the material on the build platform in extremely thin layers. This can be correlated to using a glue gun where hot glue is extruded through a heated nozzle. As the melted material is extruded out of the hot nozzle, it hardens immediately, helping in achieving a better accuracy [26]. Materials used are acrylonitrile butadiene styrene polymer (ABS), polyphenyl sulfones, waxes, polycaprolactone, polycarbonates, and polyamides. These are thermoplastic in nature and are available in spools [4, 26]. FDM is the most frequently used 3D print technology today and is also known as fused filament fabrication (FFF), such as MakerBot (Stratasys, Ltd., Eden Prairie, MN, USA).

3.4.4 Inkjet 3D Printing/PolyJet Photopolymerization (PPP)

This technology uses the same principle as the standard office inkjet printer. As the process begins by slicing of the CAD data, then it is transferred to the PolyJet printer that begins to construct the object one layer at a time. Similar to how the standard inkjet printer lays down ink pigment, the material is extruded out through hundreds of nozzle heads that deposit small amounts of UV curable material on the build platform, one layer after layer. The material is then cured with the help of UV light which is projected from a UV light device attached to the printer's head consisting of nozzles. Once the laying down of cross-sectioned layer is complete, the platform descends to make room for another layer to form. By the use of pressure and vibrations, small heat droplets of ink are forced through the nozzle helping in the adherence of one layer to the other. As these layers can be as thin as 16 μm, the surface quality is excellent for these printers, although there may be stratification lines present on the surface. With these types of printers, multiple materials can be used for the fabrication of a single model. However, it has been observed that there is more wastage in this kind of technology because the print heads need to be cleared and cleaned out to prevent clogging. To prevent clogging, little material is expelled out while changing resins resulting in wastage of the unused resin. This unused/wasted material is not recyclable and reusable. It is collected in a tank that has to be emptied at periodic intervals. Stratasys and 3D devices are currently the only PPP printer manufacturers [2, 4]. Object 30 Dental Prime (desktop 3D printer), J-700 (mainly used for the production of clear aligners), J-720 (full colored 3D printer), Objet 250, and Objet 250/500 are the few models by Stratasys Ltd. used for dental applications.

3.4.5 Selective Laser Melting (SLM) and Selective Laser Sintering (SLS)

Dr. Carl Deckard developed and patented this technology.

Sintering is a process of forming a solid mass of the material by using heat/pressure without actually melting it to the point of liquefaction.

Selective laser sintering (SLS) is the process which uses a high-power laser that basically fuses the particles together. SLS is a powder-bed fusion process for producing polymer parts. Here, the melting point of the material plays an important role. This process involves the printer to effectively lay down the material over and across the surface of a powder bed. Then with the help of an energy source heat/laser/pressure, the material is made to solidify by increasing the temperature of the material below its melting point. In this way, it is selectively solidifying the material on the powder bed. After the first layer is done, another layer of powder is laid down, followed by material layering on top of it for subsequent exposure by the laser. Hence, the material fuses with the previous layers. As this process is performed repeatedly, the build platform lowers down with each successive laser fusion. The materials used here are in powder form of nylon, ceramics, and even glass. Polyamides, polyethylene, polycaprolactone, hydroxyapatite, stainless steel, cobalt–chromium, and titanium alloys are also the choice of materials [2].

SLM (selective laser melting) and *DMLS (direct metal laser sintering)* are the metal-based additive manufacturing technologies based on powder-bed fusion technology as SLS. In here, melting metallic powder is used that fuses together [27]. The laser beam is so strong enough that can melt and weld the particles together to form a solid object.

Selective laser melting (SLM): The difference between SLS and SLM is that in SLS, the laser beam heats the powder material to just below the melting point, whereas in SLM, the material is heated above the melting point. SLM uses only those types of metal powders that have a single melting temperature, which would allow the full melting of the particles, while in DMLS, the powder is composed of materials having variable melting points. DMLS works mainly at a molecular level at elevated temperatures [28, 29].

Direct metal laser sintering (DMLS) is a laser-based additive manufacturing technology building the object layer by layer having thickness of 0.1 mm for each layer. The apparatus includes a powder of metal, radiant heaters, and a laser that is computer controlled. The process begins with adding incremental layers of the powder material on a movable platform of the machine. The high-energy beam laser is directed over the platform bed consisting of the metal powder particles which then fuses the metal powder present along the focal zone of the laser beam. The object is built layer by layer through programming of CAD file via software. After the first layer has formed, the platform then moves down and again through programming a film of powder material is laid down followed by exposure to laser-beam source, melting of powder particles and fusing them together and forming the next layer. This process continues till the whole object is 3D fabricated [30–34].

3.4.6 Electron-Beam Melting (EBM)

This technology works at higher temperatures of up to 1000 °C, leading to phase differences created through solidification, unlike other technologies. This process is similar to SLS, but instead of a laser, an electron beam is used for laying down layers successively and creating highly porous mesh-like parts. These parts are strong and dense, with no voids. The materials used for EBM are stainless steel, copper, and titanium. The main areas of their application are in orthopedic surgeries and also for dental maxillofacial surgeries providing better fixation and ingrowth of the bone [35].

3.4.7 Laminated Object Manufacturing (LOM)

This technology employs the principle of both additive and subtractive manufacturing to build the object. The additive technology part includes successively layering the material in sheets on top of each other and sticking them up together either by using pressure, heat, or even an adhesive. As this process is completed, the subtractive technology plays its part. With the help of a laser or a drilling machine, the object is cut into the desired object. The materials that could be used are metals, ceramics, plastics, composites, and paper. Due to the wide variety of material that could be used, 3D printing by this method yields to be a comparatively inexpensive method. From looking at the accuracy perspective, they are not as accurate as of the SLA and SLS technology printers. Their use is seen in casting, processing ceramics, concept modeling, and also for architectural application [36].

3.5 Commonly Used 3D Printers in Different Branches of Oral Health Science

A number of 3D printers have been used for dental applications. The most commonly used 3D printers in dentistry along with their application and materials used have been presented in Tables 3.2 and 3.3. The applications of 3D printing in different specialties of dentistry have been discussed below:

Table 3.2 Few of the commonly used 3D printers for application in oral health science

Model name	Objet 30 OrthoDesk	ProJet MJP 3600 Dental	ULTRA® 3SP™ Ortho	D4K DENTAL PRO	Form 3B	FORMIGA P 110	ZENITH	Creo® C5
Manufacturer	**Stratasys, Ltd., Eden Prairie, Minnesota**	**3D Systems, Rock Hill, South Carolina**	**EnvisionTEC, Gladbeck, Germany**	**EnvisionTEC, Gladbeck, Germany**	**Formlabs, Somerville, Massachusetts, United States**	**EOS, Munich, Germany**	**ZENITH U, Dong-gu, Daegu, Korea**	**Planmeca, Helsinki, Finland**
Technology based upon	PLP (PolyJet printing)	MultiJet Printing (MJP) technology	DLP (Digital Light Processing)	DLP (Digital Light Processing)	LFS (Low Force Stereolithography)	SLS (Selective Laser Sintering)	Stereolithography Apparatus (SLA)	DLP (Digital Light Processing)
Build volume	300 × 200 × 100 mm (11.81 × 7.87 × 3.94 in.)	11.75 × 7.3 × 8 in. (298 × 185 × 203 mm)	10.5 × 6.9 × 3 in. (266 × 175 × 76 mm)	148 × 83 × 110 mm (5.8 × 3.3 × 4.3 in.)	145 mm (w) × 145 mm (d) × 185 mm (h) 5.7″ (w) × 5.7″ (l) × 7.3″ (h)	200 × 250 × 330 mm (7.9 × 9.8 × 13 in.)	110 × 110 × 150 mm	131 × 81 × 131 mm (5.1 × 3.5 × 5.1 in.)
Minimum layer thickness	28 μm	25–30 μm	50–100 μm	15–150 μm	25–300 μm	50 μm	16, 50, 100 μm	50 μm
Cost/price	$10,000–$50,000	$50,000–$99,999	$10,000–$50,000	$15,000 approximately	The Form 3B Basic Package starts at $4999	$250,000	–	–
Weight	106 kg (234 lb)	299 kg (659.18 lb)	89.8 kg (198 lb)	–	17.5 kg (38.58 lb)	Approx. 600 kg (1.323 lb)	17.5 kg (38.58 lb)	31 kg (68.3 lb)
Source	https://www.stratasys.co.in/3d-printers/objet30	https://www.aniwaa.com/product/3d-printers/3d-systems-projet-mjp-3600-dental/	https://envisiontec.com/wp-content/uploads/2016/09/2017-ULTRA-3SP-Ortho.pdf	https://envisiontec.com/wp-content/uploads/2020/03/2020-D4K-Dental-.pdf	https://www.creat3d.shop/3d-printers/formlabs-form-3b.html	https://www.eos.info/en/additive-manufacturing/3d-printing-plastic/eos-polymer-systems/formiga-p-110-velocis	http://zenith3d.co.kr/eng/	http://publications.planmeca.com/Brochures/CAD_CAM/Creo_fly_en_low.pdf

Table 3.3 3D Printers and their material names available

Model name	Material used
Objet 30 OrthoDesk By Stratasys, Ltd., Eden Prairie, Minnesota	Materials compatible with Objet30 OrthoDesk: • *NEW Biocompatible MED625FLX*: Orthodontic indirect bonding trays and implant gingival masks are made of this flexible translucent material. • *Biocompatible MED610*: Surgical guides and cast partial frames are made of rigid clear plastic. • *Biocompatible VeroGlaze MED620*: Denture try-ins, diagnostic wax-ups, and custom trays (for biocompatible applications) are all done using this rigid, opaque material. • *VeroDent MED670*: It is suitable for orthodontic labs and clear aligner manufacture since it has high-quality detail, strength, and durability. • *VeroDentPlus MED690*: Produces fine details and finishes are achieved with strength, accuracy, and durability. It is available in a dark beige color that makes it ideal for crown and bridge work. *Source*: https://www.stratasys.com/en/materials/search?properties=c20833234e854a5ba5adef51a3e5b564&industries=8f9a6a1b80df473687572945f1ef8163&technologies=731e07a1a51b42419acd1cb75142dfe6&printers=48cf6e7bbd35438880d34c5d4d50d8d2&sortIndex=0
ProJet MJP 3600 Dental By 3D Systems, Rock Hill, South Carolina	The 3D Systems ProJet MJP 3600 Dental can 3D print in three different engineered plastics, from the VisiJet M3 Dental line of materials (with VisiJet S300 as a support material): *Rigid plastic printing material*: • *VisiJet M3 Dentcast*: A dark green printing material for castable and pressable wax-ups for the production of highly accurate smooth surface prosthetic devices. *Rigid translucent plastic printing material*: • *VisiJet M3 StonePlast*: Used for fabrication of medical-grade parts such as drill guides, working models and thermoforming models. *Rigid plastic with stone-like finish material*: • *VisiJet M3 PearlStone*: Used to create accurate and cost-effective crowns, bridges, orthodontic devices, implants, and partial dentures that have the appearance of dental stone.

(continued)

Table 3.3 (continued)

Model name	Material used
ULTRA® 3SP™ Ortho and Perfactory® Micro Ortho By EnvisionTEC, Gladbeck, Germany	*E-Aqua Model*: Material for orthodontic models that may be cleaned without the use of alcohol. *Key-Ortho IBT for EnvisionTec*: It is a tasteless and biocompatible material used for bonding indirect trays. *E-MFP (Coming Soon)*: This material is used in the production of ready-to-use partial denture without the requirement of casting and assembling. The dental professionals will be able to create the entire partial denture, including the framework and built-in teeth in a biocompatible 3D printing material having incredible strength and flexibility. *E-Denture Pro (Coming Soon)*: E-Denture Pro material may be used to create denture bases for entire digital denture solutions. It has great strength as well as a pleasant aesthetic appearance. It comes in five shades that closely resemble natural gingiva. *E-Dent 1000 (Coming Soon)*: High-quality monolithic dentures and try-ins are made with this material. The material is simple to finish and polish, and it can be stained with a variety of composite staining kits to achieve a beautiful, natural appearance. *E-SepFree*: EnvisionTEC's revolutionary new E-SepFree dental model material is the only 3D printable dental model material that meets two unique needs in the dental industry that were previously impossible to meet with 3D printed models. It is simple to release when making orthodontic appliances without the use of a separator. It is also simple to carve when contouring contact areas, opening interproximals, and a variety of other tasks. *E-Partial*: It is a castable material designed to meet a specific need in the dental industry: production of delicate partial frameworks with thin features and some casting flexibility. The clasp flexes without breaking due to the material's flexural strength. *Press-E-Cast*: It is a wax-filled photopolymer material that can be used to make full anatomical crowns and bridges with exceptional dimensional accuracy in *X*, *Y*, and *Z*, as well as a smooth surface finish. It is also possible to print partial frameworks with it. *NextDent C&B MFH*: It is a biocompatible class IIa material that is been designed for crowns and bridges. It is simple to finish and polish, and it can be stained with a variety of composite staining kits. The printed crowns blend in perfectly with the existing teeth due to the perfect balance of opacity and translucency. It complies with the following ISO 10993-1 biocompatibility standards. *NextDent Denture 3D+*: It is a class IIa biocompatible material that can be used to print all types of removable denture bases. When compared to standard PMMA (polymethyl methacrylate) denture base materials, this material shrinks significantly less. *E-Model Light DLP*: Model Light DLP printing material for DLP and cDLM 3D printers is the perfect choice for a wide variety of high-accuracy dental modeling needs. *E-Model HS*: It is used to quickly print orthodontic arch models on EnvisionTEC cDLM 3D printers. *KeySplint Soft for EnvisionTEC*: Splints and night guards are ideal for printing using this material. *E-OrthoShape*: It is recommended by orthodontic professionals for the low-cost mass manufacture of models on which clear thermoformed aligners can be made. *NextDent Gingiva Mask*: Used to print parts of the model that need a certain flexibility, such as Gingiva Masks on implant models. *E-Tray*: Individually customized impression trays are made out of this material. *NextDent Ortho IBT*: It is a biocompatible (class I) material designed for use in orthodontics for making indirect bonding tray. *E-Guard*: A biocompatible, crystal-clear material that can be used to make accurate orthodontic splints and retainers. *E-Guide*: A certified class I material used to manufacture high-precision surgical drill guides for implant surgery. *E-Model Beige*: It is the ideal material for thermoforming aligners over orthodontic models. *E-Denstone*: It is a material that was developed with the goal of producing high-quality, scannable dental models quickly. It has a matte texture and resembles typical gypsum models in appearance and feel. *Source*: https://envisiontec.com/3d-printing-materials/dental/

Form 3B By Formlabs, Somerville, Massachusetts, United States	*Dental LT Clear Resin (V2)*: A long-term biocompatible (second-generation) material used for fabrication of splints and occlusal guards. *Model Resin*: Used mainly for printing crown and bridge models with removable dies (has high precision, accuracy). *Class I biocompatible resin (EN-ISO 10993-1:2009/AC:2010, USP Class VI)./Dental SG Resin*: Surgical and pilot drill guides are the primary applications. *Class II biocompatible resin/Denture base resin*: Developed for biocompatible denture bases and teeth with fewer steps and less variability than traditional workflows. *Class II a biocompatible/Denture Teeth Resin*: Used to create biocompatible denture bases and teeth with fewer steps and less variability than traditional workflows. *Custom Tray Resin*: It is used to directly manufacture impression trays for implants, dentures, crowns and bridges, and many other comprehensive treatments. *Permanent Crown Resin*: This long-term biocompatible material with tooth-colored, ceramic-filled resin was invented for 3D printing dental crowns, inlays, on-lays, and veneers. *Temporary CB Resin*: Temporary CB resin is indicated for up to seven-unit bridges and comes in four VITA colors. It has excellent marginal adaptability, strength, and aesthetics. To repair restorations made using temporary CB resin, use standard temporary cements. *Soft Tissue Starter Pack*: This material will expand your digital capabilities with 3D printed gingiva masks and implant models. *IBT Resin*: For efficient and accurate dental bracket placement, this flexible biocompatible material has been developed. *BioMed Amber Resin*: Biocompatible parts that require short-term skin and mucosal membrane contact should be made of stiff, functionally robust medical-grade material. *BioMed Clear Resin*: Medical-grade material for rigid, non-brittle, biocompatible parts with long-term skin and mucosal membrane contact, as well as end use medical equipment. *Source*: https://support.formlabs.com/s/article/Choosing-the-Right-Material?language=en_US#dental

(continued)

Table 3.3 (continued)

Model name	Material used
FORMIGA P 110 By EOS, Munich, Germany	*Alumide®*: This material has a high-rigidity metallic finish and offers a variety of post-processing choices. It also has a high-temperature resistance, resulting in greater dimensional stability as the temperature rises. *PA 2200®white*: The robust white pieces manufactured of PA 2200 have a highly steady profile. The components are noteworthy for their strength, stiffness, and chemical resistance. *PA 2201*: This material is FDA-approved for contact with food (FDA, 21 CFR) and has similar features as PA 2200 in terms of strength, stiffness, and chemical resistance. *PA 3200 GF*: Parts manufactured of this white polyamide 12 powder packed with glass beads have a high degree of stiffness and elongation at break. They are also abrasion-resistant and thermally resilient. This material is largely utilized in vehicle engines and deep-drawing tool molding materials. *PrimePart®ST*: The material is white and has a halogen fire retardant component. The 3D printed pieces have a high elasticity and tensile strength. Parts used in the interior of aircraft, such as ventilation ducts and exhaust valves, are the most common. *PA 2105*: PA 2105 is a thin polyamide 12 powder that has been tinted with light skin colors. The most common application is for the direct construction of dental models with high accuracy and surface quality using digital technologies. *PA 1101*: PA 1101 is a white polyamide 11 powder that is intended for laser sintering applications. It is created from repurposed materials (castor oil). Elasticity and good impact resistance characterize the material. *PA 1102 black*: Except for its black color, it is identical to PA1102 white polyamide powder. *PrimeCast® 101*: It is a powdered polystyrene (PS) grade with a gray color. It has good dimensional accuracy, a low melting point, excellent surface quality, and good strength. Laser sintering and rapid prototyping can be used to process it. It is ideal for lost pattern casting in plaster and ceramic shells, as well as master patterns for vacuum casting. *Written from Technical Datasheet, Supplied by EOS *Source*: https://www.eos.info/en/additive-manufacturing/3d-printing-plastic/eos-polymer-systems/formiga-p-110-velocis
ZENITH By ZENITH U, Dong-gu, Daegu, Korea	ZMD-1000B TEMPORARY ZMD-1000B CLEAR-SG ZMD-1000B MODEL ZMD-1000B CASTABLE ZENITH has exclusive material (acrylate photopolymer resin) for various dental applications such as dental model surgical guide, clear aligners, splint, temporary crown and wax up pattern. *Source*: http://www.zenith3d.co.kr/eng/content/01products/01_03.php
Creo® C5 Planmeca Helsinki, Finland	*Class I biocompatible resin/FotoDent® Guide by Dreve*: Surgical guides are printed with this material. *FotoDent® Model by Dreve*: Resin of excellent quality for printing dental models and study models. *FotoDent® Setup by Dreve*: Fast printing resin for thermoforming aligner bases. *Source*: https://www.henryschein.co.nz/3d-printing/3d-printing-equipment/planmeca-creo-c5-printer

3.5.1 Oral and Maxillofacial Surgery

3D printing techniques can be used as a new approach to surgically prepare the patient before surgery, along with the 3D simulation. 3D Imaging using computed tomography (CT) scans helps us to create a 3D anatomical model by rapid prototyping, helping to get an overview and getting acquainted with the complex anatomical structures [37]. The surgeon, with the help of these 3D models, can view and plan the treatment better that helps in patient education and motivation before surgery and also for the precise location of the area where the surgery has to be performed. This has seen to reduce the inclusion of human errors significantly [29]. Anderl et al. in the early 1990s developed an acrylic model using CT-guided stereolithography that allowed successful preoperative treatment planning as well as intraoperative management in an 8-year-old patient having wide midline craniofacial cleft [38].

Rapid prototyping has also been used for the fabrication of customized reconstruction plates, and the 3D CT data can be utilized for further morphological reconstruction of the area having the bony defect which is mainly seen in cases of reconstruction surgeries and fractures. For planning orthognathic surgeries, personalized orthognathic surgical guides (POSG) can be prepared using the CAD software, which helps to predetermine the exact position of the bony segments so as to where precisely to drill holes and which points to mark for placing surgical guides [3, 39].

3D printed models also help in the development of personalized implant surgical guides. Commercial devices that are currently available using a 3D CT patient scan can help to assist in implant placement in cases that require full mouth rehabilitation. Such systems require the development of surgical guides to ensure that implants are precisely located, sized, and angled to restore the anatomy and function [40].

Apart from their clinical applications, the 3D printed models can be used as an educational tool for the students for a better understanding of the subject in-depth and also for the researchers for better visualization to carry out further research in this field [37].

3.5.2 Orthodontics

With the advancement in digital technology, 3D printing has emerged as a solution that could help us to achieve a precision of orthodontic hardware. It was Normando in 1980 who had introduced the idea of face scanning and 3D printing to correct the dental arch shape precisely. He also used it for the fabrication of orthodontic brackets [41]. By using the 3D CAD data, it has become very convenient for the practitioners to virtually modify teeth using the software into their ideal positions. This has also led to easy in-office printing of the model. In the current scenario, the most common application of the 3D printers is for clear retainers and aligner fabrication. Aligners are being a preferred choice for correction of various malocclusions, as it requires fewer patient visits, thus being a compliant treatment modality. The physical models are printed, and then, they are fabricated using a thermoplastic sheet, similar to what we see with Invisalign or various aligner manufacturing companies. In-house 3D printable biocompatible aligner material (Tera Harz TC-85 DAC resin, Graphy, Seoul, Korea) has been created by Korean researchers to overcome the limitation of thermoforming sheet type aligner. The material can be used to directly 3D print orthodontic aligners. Direct 3D printing of aligner is a newer treatment approach for treating orthodontic malocclusions. The dental professional can easily create aligners in one day time and deliver to the patient without the need to outsource remote location laboratory support. This can help the orthodontist produce large volume of aligners in a short time [42]. Also, much-advanced software are available that helps to shape up the model base and also guide in trimming. These software can also be used for designing bracket pads, hooks, etc. Full bracket customization is possible for each tooth that can be placed on the tooth by 3D printed guides [43]. Customized orthodontic archwire bending is also possible for customized brackets, thus making it possible for an orthodontist to achieve a higher level of patient care.

Incognito system utilizes an innovative software, which is used by an orthodontist to build a personalized smile-orientated fully customized lingual brackets and wires for the teeth. The system is available as a one-stop solution for the customized lingual appliance, where the bracket base, slot, and the face could be customized according to the treatment plan of the patient [44]. Fully customized orthodontic appliances provide a better outcome as compared to the conventional orthodontic system. It has also been beneficial in splint therapy for TMJ disorders by the rapid prototyping for the preparation of customized splints and also saving the laboratory time as well [3, 45, 46]. Rapid prototyping also permits the production of specific orthodontic appliances, including Digital Herbst appliance, mouth guards, retainers, expanders, and appliances for sleep apnea. For 3D printing active appliances, they have to be imported first via STL, and then, the 3D structure is made around it. Care should be taken to accurately position them at the correct height. The use of the DLMS technology [47] has been seen to have clear and smooth surfaces without any sharp edges making it to have better adaptation with the palatal gingivae [2, 3, 48, 49].

Surgical wafers, cutting guides, repositioning guides, and 3D models can also be printed with much more accuracy for orthognathic surgery. These surgical wafers and models are very crucial for the success of an orthognathic surgery. Various stents for cleft lip and palate can also be designed and printed easily. 3D models are created from the digital images. With virtual planning software, the digital planning of the various patient-specific device fabrication can be done and exported as an STL file. Error corrections (if any) are done on the STL file generated by the planning software to avoid any printing failures during the 3D printing of the device or appliances. Minimizing the error is very important as it directly affects the final outcome of the surgery. The necessary offset values are added for easy fit. The laboratory time is greatly reduced with increased accuracy. Dolphine, Mimics, Deltaface, Simplant OMS etc., are generally used for planning. CAD-CAM as well as direct resin print can be used for fabrication of devices and appliances. 3D printing and virtual planning have helped the orthodontist to plan for orthognathic surgery to higher degree of accuracy with ease. SLA, PolyJet, DLP printers, and medical-grade surgical guide resins are generally used.

3.5.3 Implantology and Prosthodontics

3D printing techniques have revolutionized the field of prosthodontics as fabrication of denture has become much easier and patient-friendly [50]. Kim et al. have suggested that the fabrication of removable prosthesis and removable partial dentures has been successful in patients with decreased mouth opening and also in patients with gag reflex and treatment can be done with ease [51]. Studies have shown that the 3D printed fixed and removable dentures had comparable clinical accessibility and physical property compared to the conventional ones. Most of the modern 3D printing methods used today are SLS, SLM, and SLA to produce porous structures, whereas inkjet printing makes dense ceramic structures complex [52].

3D printing, along with CT data, has been used for the preparation of surgical guides for dental implant placement. The data are superimposed on the software in STL format, and with proper planning, the clinician is able to prepare surgical guides with the right amount of dimensions along with accuracy and precision, making it less time-consuming during the surgical process.

DLP has been seen to be an efficient method for printing customized zirconia dental implants with adequate dimensional accuracy, though the mechanical properties have shown flexural strength near to traditionally produced ceramics. Customized dental implants printed using the SLM method showed increased strength, density, and sufficient dimensional accuracy. SLM has been proven to be an efficient means for 3D printing and fully dense customized implants. The additional oral tissue or bone donation, and the use of allographs are also avoided [53, 54].

Studies have shown that selective laser melting (SLM) and electron-beam melting (EBM) were used for the 3D printing process of metal-based implant prosthesis [55]. This has immensely cut down the amount of time that was spent by the dental technicians in the laboratory. They also yield a more precise framework when seen in comparison to the conventional frame.

In additive manufacturing, metallic and polymer-based material is more commonly used for the fabrication of crowns and other dental prostheses. Through increasing the digital workflow, these silicone prostheses can be printed directly, offering appropriate esthetics, and also reducing the patient's number of appointments [56]. With facial deformities and gag reflexes, additive manufacturing is an essential factor to consider for treat ment planning as it will increase patient comfort.

3.5.4 Restorative Dentistry

UV light-induced photopolymerization of resin was the primary mean of 3D printing in dentistry. The resins have shown some degree of shrinkage and dimension instability, which has affected its physical properties. Such changes in physical properties require further research [57]. A study has concluded that 3D printing with resins could create more accurate temporary crowns than traditional methods. Even though it is documented in the literature that 3D printing resins used for restorative purposes are clinically acceptable, resins still have a disadvantage because of shrinkage. For the fabrication of crowns, bridges, and metal copings, the restorations can either be milled from the 3D data, or they can be fabricated conventionally by printing out the model using 3D data [58]. 3D printing of master models can be done for the use in conventional aspects of fabrication, e.g., various types of restorations, adding material for veneers that can be used to educate and guide and increase awareness in patients about the various types of restorative treatment. With the advent of digitization and considering the pace with which it is growing in dentistry since the past few years, the combination of a cone-beam computer tomography (CBCT) along with digital impressions can become a standard in the coming future [59].

3.5.5 Endodontics

Endodontic treatment of a tooth with an obliterated pulp chamber can be a challenging task in locating the canal orifice. This is where 3D printing plays a role. For dealing with such clinical situations, a 3D printed guide of the tooth can be used by the clinician for locating the canal orifice. With the help of the guide, it would be much easier to place the bur at the correct location and angulation [54]. In cases of endodontic surgeries, where the location of the lesion is attached to the cortical plate, 3D printed guides are beneficial as they can help determine the correct amount of bone that needs to be resected in order to remove the lesion [60]. For placing grafts and implants, 3D guides can aid in better placement with accuracy keeping other factors in consideration like bone support, teeth, mucosa using patient-specific 3D printed guides. With 3D printing, the chairside time has dramatically reduced. However, research is still going on the properties of 3D printed ceramic restorations for their fracture toughness and strength.

3.5.6 Periodontics

The periodontium is a complex structure comprising mainly bone, gingiva, and cementum. All these tissues have different characteristics, and regeneration of periodontium is regulated by several cell types, signaling pathways, and interactions of various factors in the oral cavity. Accordingly, 3D prints frequently used in periodontology mainly focuses on regenerative periodontology and 3D recommendations for esthetic correction of gingiva. Hoang et al. used the term additive bio-manufacturing for the manufacture of 3D printed scaffolds for supporting tissue regeneration in areas of defects [61, 62]. A computed tomography (CT) scan of the patient's defect serves as a reference template for the fabrication of scaffold. Then, a print wax mold is designed to create a scaffold on the basis of the CT scan image [63] which can be utilized to restore the periodontal defect.

Adequate bone support is very essential for long-term survivability and functionality of the dental implant. Under pathological conditions where periodontal tissue integrity is compromised, placement of implant is always crucial; this can be due to inadequate support for osseointegration in the remaining tissue. To improve the quality and quantity of bone in the localized area, 3D printing guided bone regeneration is one of the most promising tools. The primary purpose of 3D printers is to have a more controlled slow growing tissue/bone regeneration [64] for preventing the ingrowth of rapidly regenerating tissues in the defect area, which has led to better integrity and functionality of the oral cavity [65].

Clinically, 3D printing has become more common in the anterior region of the oral cavity in gingival esthetic surgeries. For each patient, specific guides can be 3D printed for carrying out gingivectomy and smile designing procedures. These templates have proven accuracy and precision, leading to better surgical skills. From an educational perspective, the idea to print 3D models that represent gums, periodontal tissues, and tissue-related defects is helpful to learn proper proprioception and abilities [66].

3.6 Summary and Key Points

3D scanning, along with CAD technology, has been at the forefront, which has a definite positive impact on the different aspects of oro-dental health. With the advancement and development of these technologies, it has become much more comfortable to virtually modify and print 3D objects. Such sophisticated techniques have taken care of various aspects of day-to-day requirements such as storage space over the conventional dental practice. This has enhanced the efficiency of the clinical care of patients allowing for a better fit of appliance resulting in more satisfied patients, staff personnel, and a high level of oral care. 3D printing can also create complex geometrical forms from the digital data for treatment planning of difficult clinical situations. Furthermore, the increased usage of intraoral scanning devices has contributed to providing a better comprehensive analysis in the printing of 3D models. Digital models, 3D facial scans, and CBCT integration tools enable the simulation of procedures and creation of efficient patient communication tools. Although 3D printing is now relatively cost-effective, still for some printers the cost of the materials used and the maintenance of machinery remain a point of concern. It is necessary to recognize the need for well-trained operators, post-processing, and compliance to strict health and safety measures. With the technologies evolving (Fig. 3.3),

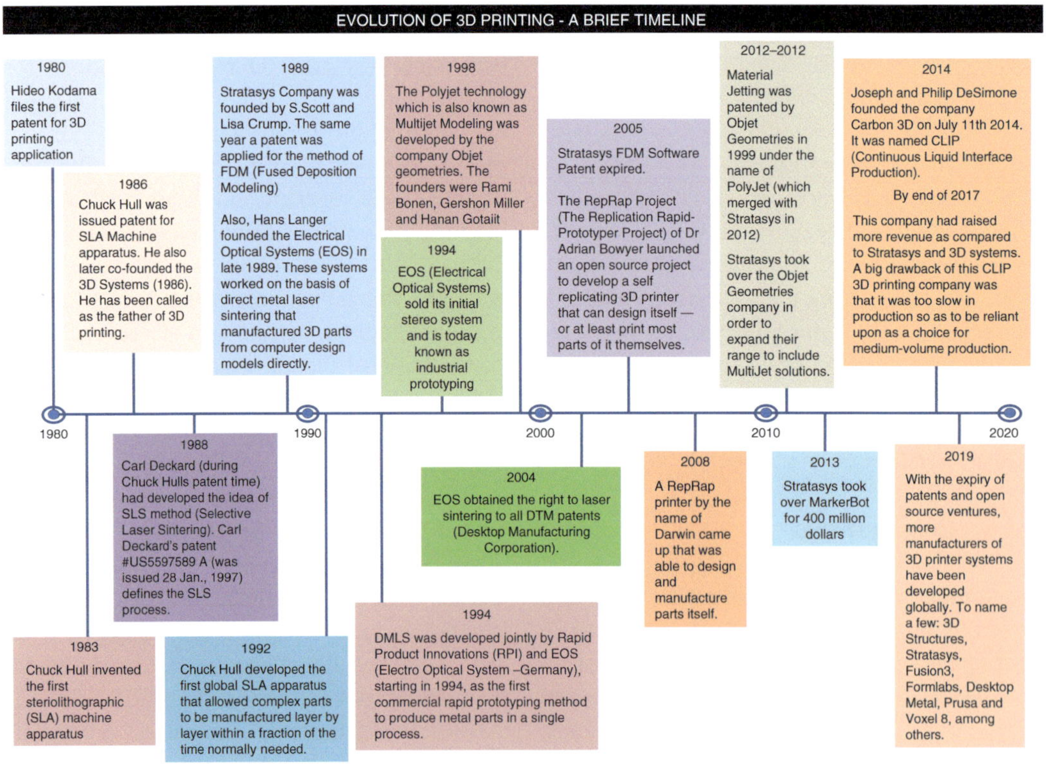

Fig. 3.3 Brief timeline of the evolution of 3D printing

both the dentist and the patient must take part in the changes which can be made in the future. To summarize the whole chapter, key points have been highlighted in Box 3.1.

Box 3.1: Key point summary

- 3D printing also known as additive manufacturing or rapid prototyping can create the design of an object using CAD software followed by fabrication of 3 dimensional solid objects from a digital file that is the STL file of the scanned dental hard and soft tissues.
- STL (an abbreviation of "stereolithography") also known as Standard Tessellation Language/Standard Triangle Language is the global standard format for 3D printing files.
- There are numbers of technologies based on additive manufacturing. Few of them are as follows: SLA, DLP, FDM, PolyJet photopolymerization, SLS, SLM, DMLS, EBM, LOM.
- SLA utilizes photopolymerization of resin by mean of UV light. As the layer is being formed, the built platform keeps descending to accommodate the object.
- DLP technology uses a different light source than SLA. It uses small micro-mirrors which act as projector flashing image of the cross sections; thus, the printing is much faster than SLA.
- FDM is based on material jetting where a hot thermoplastic material is extruded through a heated nozzle and is deposited layer by layer. As the material is extruded, it sets immediately, giving better accuracy.
- PolyJet photopolymerization is similar to standard in-office inkjet printer. These variants of printer use hundreds of nozzles which deposit the material, and the same is simultaneously cured by UV light.
- SLS/SLM/DMLS are based on powder-bed fusion technology. Here, melting point of material is of prime importance. SLS uses a laser beam to melt and solidify the material, while SLM and DMLS are metal-based additive manufacturing technologies which use metallic powder for fabrication of metal parts or objects.
- EBM is a very similar to SLS but instead of a laser light, it utilizes an electron beam for creation of the 3D printed object.
- LOM is a technology where both the additive and subtractive manufacturing processes take place. Its main use is seen in casting ceramics, architectural application, etc.
- Most commonly used printers for dental applications are SLA, DLP, and PolyJet printing.

Acknowledgments The authors are grateful to the Indian Council of Medical Research (ICMR), New Delhi, for providing extramural research grant for research project (code I-1069), which helped to generate data for this book chapter.

References

1. Hajeer MY, Millett DT, Ayoub AF, Siebert JP. Applications of 3D imaging in orthodontics: Part I. J Orthod. 2004;31:62–70.
2. Taneva E, Kusnoto B, Evans CA. 3D scanning, imaging, and printing in orthodontics. In: Bourzgui F, editor. Issues in contemporary orthodontics. London: InTech Open; 2015.
3. Oberoi G, Nitsch S, Edelmayer M, Janjić K, Müller AS, Agis H. 3D printing—encompassing the facets of dentistry. Front Bioeng Biotechnol. 2018;6. https://doi.org/10.3389/fbioe.2018.00172.
4. Groth CH, Kravitz ND, Jones PE, Graham JW, Redmond WR. Three-dimensional printing technology. J Clin Orthod. 2014;48:475–85.
5. Hull CW. Apparatus for production of three-dimensional objects by stereolithography, U.S. Patent No. 4,575,330; 1986. https://patents.google.com/patent/US4575330A/en. Accessed 4 Oct 2020.
6. Mouzakis DE. Advanced technologies in manufacturing 3D-layered structures for defense and aerospace. In: Lamination - theory and application. 1st ed. London: InTech Open; 2018. p. 89–113.
7. Campbell I, Diegel O, Kowen J, Wohlers T. 3D printing and additive manufacturing state of the industry: Annual Worldwide Progress Report. 2017.

8. Mangano F, Gandolfi A, Luongo G, Logozzo S. Intraoral scanners in dentistry: a review of the current literature. BMC Oral Health. 2017;17. https://doi.org/10.1186/s12903-017-0442-x.
9. Adams LP, Spirakis A. Stereo photogrammetry. In: Optical measurement methods in biomechanics. Boston, MA: Springer; 1997. p. 17–38.
10. Favero CS, English JD, Cozad BE, Wirthlin JO, Short MM, Kasper FK. Effect of print layer height and printer type on the accuracy of 3-dimensional printed orthodontic models. Am J Orthod Dentofac Orthop. 2017;152:557–65.
11. Digital Model and 3D Printing Requirements, Original Release 4.23.2013, Last Updated 30 Jun 2016. www.AmericanBoardOrtho.com. Accessed 4 Oct 2020.
12. Patzelt SB, Vonau S, Stampf S, Att W. Assessing the feasibility and accuracy of digitizing edentulous jaws. J Am Dent Assoc. 2013;144:914–20.
13. Quan H, Zhang T, Xu H, Luo S, Nie J, Zhu X. Photo-curing 3D printing technique and its challenges. Bioact Mater. 2020;5:110–5.
14. Manfredi D, Calignano F, Krishnan M, et al. Additive manufacturing of Al Alloys and aluminium matrix composites (AMCs). In: Monteiro WA, editor. Light metal alloys applications. London: InTech Open; 2014. p. 3–34.
15. Skals I, Gleba M, Taube M, Mannering U. Wool textiles and archaeometry: testing reliability of archaeological wool fibre diameter measurements. Danish J Archaeol. 2018;7(2):161–79.
16. Ruiz MD, Frías MÁ, Rider RM. Fundamentals of stereolithography, a useful tool for diagnosis in dentistry. ODOVTOS-Int J Dent Sci. 2015;17:15–21.
17. Kim SY, Shin YS, Jung HD, Hwang CJ, Baik HS, Cha JY. Precision and trueness of dental models manufactured with different 3-dimensional printing techniques. Am J Orthod Dentofac Orthop. 2018;153:144–53.
18. Brown GB, Currier GF, Kadioglu O, Kierl JP. Accuracy of 3-dimensional printed dental models reconstructed from digital intraoral impressions. Am J Orthod Dentofac Orthop. 2018;154:733–9.
19. https://formlabs.com/blog/ultimate-guide-to-stereolithography-sla-3d-printing/. Accessed 4 Dec 2020.
20. Hassan WN, Yusoff Y, Mardi NA. Comparison of reconstructed rapid prototyping models produced by 3-dimensional printing and conventional stone models with different degrees of crowding. Am J Orthod Dentofac Orthop. 2017;151:209–18.
21. Hazeveld A, Slater JJ, Ren Y. Accuracy and reproducibility of dental replica models reconstructed by different rapid prototyping techniques. Am J Orthod Dentofac Orthop. 2014;145:108–15.
22. Schirmer UR, Wiltshire WA. Manual and computer-aided space analysis: a comparative study. Am J Orthod Dentofac Orthop. 1997;112:676–80.
23. Hirogaki Y, Sohmura T, Satoh H, Takahashi J, Takada K. Complete 3-D reconstruction of dental cast shape using perceptual grouping. IEEE Trans Med Imaging. 2001;20:1093–101.
24. Halazonetis DJ. Acquisition of 3-dimensional shapes from images. Am J Orthod Dentofac Orthop. 2001;119:556–60.
25. Bell A, Ayoub AF, Siebert P. Assessment of the accuracy of a three dimensional imaging system for archiving dental study models. J Orthod. 2003;30:219–23.
26. Zein I, Hutmacher DW, Tan KC, Teoh SH. Fused deposition modeling of novel scaffold architectures for tissue engineering applications. Biomaterials. 2002;23:1169–85.
27. Rimell JT, Marquis PM. Selective laser sintering of ultra-high molecular weight polyethylene for clinical applications. J Biomed Mater Res. 2000;53:414–20.
28. https://www.3dhubs.com/knowledge-base/introduction-metal-3d-printing/. Accessed 4 Oct 2020.
29. Shilo D, Emodi O, Blanc O, Noy D, Rachmiel A. Printing the future—updates in 3D printing for surgical applications. Rambam Maimonides Med J. 2018;9:e0020.
30. Mullen L, Stamp RC, Brooks WK, Jones E, Sutcliffe CJ. Selective laser melting: a regular unit cell approach for the manufacture of porous, titanium, bone in-growth constructs, suitable for orthopedic applications. J Biomed Mater Res B Appl Biomater. 2009;89(2):325–34.
31. Stamp R, Fox P, O'Neill W, Jones E, Sutcliffe C. The development of a scanning strategy for the manufacture of porous biomaterials by selective laser melting. J Mater Sci Mater Med. 2009;20(9):1839–48.
32. Hollander DA, Von Walter M, Wirtz T, et al. Structural mechanical and in vitro characterization of individually structured Ti-6Al-4V produced by direct laser forming. Biomaterials. 2006;27(7):955–63.
33. Mangano C, Raspanti M, Traini T, Piattelli A, Sammons R. Stereo imaging and cytocompatibility of a model dental implant surface formed by direct laser fabrication. J Biomed Mater Res A. 2009;88(3):823–31.
34. Dabrowski B, Swieszkowski W, Godlinski D, Kurzydlowski KJ. Highly porous titanium scaffolds for orthopaedic applications. J Biomed Mater Res B Appl Biomater. 2010;95(1):53–61.
35. Van Noort R. The future of dental devices is digital. Dent Mater. 2012;28:3–12.
36. 3D printing process – what is laminated object manufacturing (LOM). https://thre3d.com/how-it-works/sheet-lamination/laminated-object-manufacturing-lom/. Accessed 4 Dec 2020.
37. Zaharia C, Gabor AG, Gavrilovici A, Stan AT, Idorasi L, Sinescu C, Negruțiu ML. Digital dentistry—3D printing applications. J Interdiscip Med. 2017;2:50–3.
38. Anderl H, Zur Nedden D, Mu W, Twerdy K, Zanon E, Wicke K, Knapp R. CT-guided stereolithography as a new tool in craniofacial surgery. Br J Plast Surg. 1994;47:60–4.

39. Li B, Shen S, Jiang W, Li J, Jiang T, Xia JJ, Shen SG, Wang X. A new approach of splint-less orthognathic surgery using a personalized orthognathic surgical guide system: a preliminary study. Int J Oral Maxillofac Surg. 2017;46:1298–305.
40. Papaspyridakos P, Lal K. Complete arch implant rehabilitation using subtractive rapid prototyping and porcelain fused to zirconia prosthesis: a clinical report. J Prosthet Dent. 2008;100:165–72.
41. Normando D. 3D orthodontics from verne to shaw. Dent Press J Orthod. 2014;19:12–3.
42. Chan E, Panayi N, Polychronis G, Papageorgiou SN, Zinelis S, Eliades G, Eliades T. In-house 3D-printed aligners: effect of in vivo ageing on mechanical properties. Eur J Orthod. 2021;5:cjab022. https://doi.org/10.1093/ejo/cjab022. Epub ahead of print
43. Kohli TMA. 3D printing in dentistry – an overview. Acta Sci Dent Sci. 2019;6:35–41.
44. Nguyen T, Jackson T. 3D technologies for precision in orthodontics. Semin Orthod. 2018;24:386–92.
45. Salmi M, Paloheimo KS, Tuomi J, Ingman T, Makitie A. A digital process for additive manufacturing of occlusal splints: a clinical pilot study. J R Soc Interface. 2013;10:20130203. https://doi.org/10.1098/rsif.2013.0203.
46. Salmi M, Tuomi R, Sirkkanen R, Ingman T, Mäkitie A. Rapid tooling method for soft customized removable oral appliances. Open Dent J. 2012;6:85–9.
47. Graf S. Direct printed metal devices-the next level of computer-aided design and computer-aided manufacturing applications in the orthodontic care. APOS Trends Orthod. 2017;7:253–9.
48. Farronato G, Santamaria G, Cressoni P, Falzone D, Colombo M. The digital-titanium Herbst. J Clin Orthod. 2011;45:263–7.
49. Al Mortadi N, Eggbeer D, Lewis J, Williams RJ. CAD/CAM/AM applications in the manufacture of dental appliances. Am J Orthod Dentofac Orthop. 2012;142:727–33.
50. Hu F, Pei Z, Wen Y. Using intraoral scanning technology for three-dimensional printing of Kennedy class I removable partial denture metal framework: a clinical report. J Prosthodont. 2019;28:e474–6.
51. Kim JE, Kim NH, Shim JS. Fabrication of a complete, removable dental prosthesis from a digital intraoral impression for a patient with an excessively tight reconstructed lip after oral cancer treatment: a clinical report. J Prosthet Dent. 2017;117:205–8.
52. Gan N, Ruan Y, Sun J, Xiong Y, Jiao T. Comparison of adaptation between the major connectors fabricated from intraoral digital impressions and extraoral digital impressions. Sci Rep. 2018;8:529. https://doi.org/10.1038/s41598-017-17839-4.
53. Osman RB, van der Veen AJ, Huiberts D, Wismeijer D, Alharbi N. 3D-printing zirconia implants; a dream or a reality? An in-vitro study evaluating the dimensional accuracy, surface topography and mechanical properties of printed zirconia implant and discs. J Mech Behav Biomed Mater. 2017;75:521–8.
54. Chen J, Zhang Z, Chen X, Zhang C, Zhang G, Xu Z. Design and manufacture of customized dental implants by using reverse engineering and selective laser melting technology. J Prosthet Dent. 2014;112:1088–95.
55. Revilla León M, Klemm IM, García-Arranz J, Özcan M. 3D metal printing - additive manufacturing technologies for frameworks of implant borne fixed dental prosthesis. Eur J Prosthodont Restor Dent. 2017;25:143–7.
56. Unkovskiy A, Spintzyk S, Brom J, Huettig F, Keutel C. Direct 3D printing of silicone facial prostheses: a preliminary experience in digital workflow. J Prosthet Dent. 2018;120:303–8.
57. Anadioti E, Kane B, Soulas E. Current and emerging applications of 3D printing in restorative dentistry. Curr Oral Health Rep. 2018;5:133–9.
58. Mai HN, Lee KB, Lee DH. Fit of interim crowns fabricated using photopolymer-jetting 3D printing. J Prosthet Dent. 2017;118:208–15.
59. Krastl G, Zehnder MS, Connert T, Weiger R, Kühl S. Guided endodontics: a novel treatment approach for teeth with pulp canal calcification and apical pathology. Dent Traumatol. 2016;32(3):240–6.
60. Garcia-Sanchez A, Mainkar A, Ordonez E, Sanchez S, Weinstein G. 3D-printed guide for endodontic surgery. Clin Dent Rev. 2019;3(1):13.
61. Hoang D, Perrault D, Stevanovic M, Ghiassi A. Surgical applications of three-dimensional printing: a review of the current literature and how to get started. Ann Transl Med. 2016;4:456. https://doi.org/10.21037/atm.2016.12.18.
62. Hung KC, Tseng CS, Dai LG, Hsu S. Water-based polyurethane 3D printed scaffolds with controlled release function for customized cartilage tissue engineering. Biomaterials. 2016;83:156–68.
63. Pilipchuk SP, Monje A, Jiao Y, Hao J, Kruger L, Flanagan CL, et al. Integration of 3D printed and micropatterned polycaprolactone scaffolds for guidance of oriented collagenous tissue formation in vivo. Adv Healthc Mater. 2016;5:676–87.
64. Carter SSD, Costa PF, Vaquette C, Ivanovski S, Hutmacher DW, Malda J. Additive biomanufacturing: an advanced approach for periodontal tissue regeneration. Ann Biomed Eng. 2017;45:12–22.
65. Bottino MC, Pankajakshan D, Nör JE. Advanced scaffolds for dental pulp and periodontal regeneration. Dent Clin N Am. 2017;61:689–711.
66. Li Z, Liu YS, Ye HQ, Liu YS, Hu WJ, Zhou YS. Diagnosis and treatment of complicated anterior teeth esthetic defects by combination of whole-process digital esthetic rehabilitation with periodontic surgery. Beijing Da Xue Xue Bao. 2017;4:71–5.

An Overview of 3D Printable Materials for Dental and Craniofacial Applications

4

Nikhil Belsure, Sagar Parekh, and Nimesh Soni

Contents

4.1 Introduction to 3D Printed Materials

Additive manufacturing (AM) also known as 3D printing is a technique for fabricating objects by printing successive layers of materials on top of each other. A wide range of structures and complex geometries can be fabricated via 3D printing using a 3D computer-aided design (CAD) model as the input. 3D printing has witnessed widespread applications in aerospace, engineering, dental, and biomedical applications.

N. Belsure (✉) · S. Parekh · N. Soni
Imaginarium India Pvt. Ltd., Mumbai, India
e-mail: sagar@imaginarium.io;
nimesh@imaginarium.io

Dental materials have a long history, with the Romans using gold for dental crowns and bridge restorations [1]. Since then, various materials including metals and alloys, polymers, and ceramics have been researched and applied in dentistry. Novel materials and technological developments have accelerated adoption of 3D printing in digital dentistry.

Among the greatest advantages of using 3D printing for dental applications is the possibility to offer patient-specific care and conformity to the patients' anatomy. 3D printing finds its applications throughout the dental care workflow in various specialties of dental medicine. Earlier, limited availability of materials which could be 3D printed posed a hindrance to widespread adoption for dental applications. However, recent

© The Author(s), under exclusive license to Springer Nature Switzerland AG 2022
P. K. Chaudhari et al. (eds.), *3D Printing in Oral Health Science*,
https://doi.org/10.1007/978-3-031-07369-4_4

advances in material sciences have enabled further acceleration and adoption of 3D printing for newer applications especially in dentistry. Extensive adoption of 3D printing has consequently led to the growth and acceptance of digital dentistry. Digital dentistry aids dental technologists and clinicians to produce various orthodontic appliances by combining oral 3D scanning, CAD designing, and 3D printing [2]. Crowns, bridges, implants, impression trays, gingiva masks are among the few patient-specific instruments made with 3D printing in digital dentistry. Various polymer, metal, and ceramic composite materials are used to produce these instruments.

Apart from the existing plethora of materials for digital dentistry, groundbreaking material research is paving the way for novel materials in dental applications. Biomaterials are under study to use cellular matrix as building blocks especially for regenerative medicine. Ceramics which are excruciatingly difficult to process are also being 3D printed. Novel composite polymers and metal alloys also have the potential to change the landscape of 3D printing for dental and craniofacial applications. This chapter delineates the various types of materials used in 3D printing with applications pertaining to digital dentistry and also the future scopes of material development.

4.2 Classification of 3D Printing Materials

The most fundamental way to classify 3D printing materials is on the basis of their state during the processing in the 3D printer [3]. Liquid resin-based materials are among the most commonly used applications in digital dentistry. These materials are usually processed in digital light projection (DLP), stereolithography (SLA), and PolyJet printing (PJP) techniques. Some materials are also used in the form of powders. Selective laser sintering (SLS), binder jetting, and Multi Jet Fusion (MJF) make use of polymers in powder form to produce parts. Direct metal laser sintering (DMLS) and selective laser melting (SLM) techniques make use of metal powders to produce parts. Fused deposition modeling (FDM) is a technique which uses solid-based materials in the form of polymer filaments to produce parts.

4.2.1 Liquid-Based Materials

Photocurable resins are the most common among the most widely used 3D printable materials for dental applications. SLA technology uses a scanning laser to cure the liquid resin for fabrication of objects in a layer-by-layer fashion. DLP technology makes use of LED (light-emitting diode) projectors to cure the resin. Liquid resin materials processed in SLA or DLP are used to produce surgical guides, dental models, custom impression trays, permanent and temporary crowns, and gingiva masks. Form 3B (FormLabs, Somerville, Massachusetts, USA) range of printers are a recent example of desktop-based LFS (low-force stereolithography) used for dental applications with a diverse range of materials available for all dental applications [4]. Among the prominent original equipment manufacturers (OEMs) for DLP-based 3D printers for dental applications are NextDent (NextDent, Soesterberg, the Netherlands), Asiga (Asiga, Alexandria, New South Wales, Australia), SprintRay (SprintRay, Los Angeles, California, USA), Rapid Shape (Rapid Shape, Heimsheim, Germany), EnvisionTEC (EnvisionTEC, Dearborn, Michigan, USA), and DWS (DWS, Thiene, Vicenza, Italy).

PolyJet printing (PJP) is another commonly used technique employing the use of liquid waxes and resins for 3D printing objects. PJP is used for anatomical models, implant drill guides, and some flexible materials can also be used to produce orthodontic splints. However, the equipment, materials as well as the running costs of the PJP technology are comparatively higher than other 3D printing techniques [3].

4.2.2 Powder-Based Materials

Powder-based materials can be sintered or melted using laser radiation. The materials spectrum ranges from polymers, metals to ceramics. Technologies such as selective laser sintering (SLS) and Multi Jet Fusion (MJF) use polyamide powder as a common material. Direct metal laser sintering (DMLS) and selective laser melting (SLM) use metal alloy powders like titanium, steel, cobalt chromium for 3D printing. In dentistry, metal 3D printing is mostly used for printing oral maxillofacial (OMF) implants and dental crowns [5]. Polymer powder-based 3D printing techniques are used to produce anatomical models and surgical guides. Both polymer and metal powders exhibit a high degree of biocompatibility [6, 7].

4.2.3 Solid-Based Materials

Solid-based materials in the form of filaments find application in fused deposition modeling (FDM) technology. The filament is heated and extruded through a hot nozzle and deposited in a layer-by-layer fashion to make an object. FDM technology finds limited applications in dentistry as the output is coarse, lacks precision, and requires a significant amount of finishing [8]. Polylactic acid (PLA) and acrylonitrile-butadiene-styrene (ABS) polymers are the most commonly used materials in FDM technology. This technology finds limited use for applications like anatomical models and dental impression trays. Functional parts or end use parts cannot be 3D printed with the FDM technique as the output has low mechanical strength and is not biocompatible [3].

4.2.4 AM Technologies and Materials Matrix

3D printing technologies use liquid resins, powdered polymers or metals and solid filaments for manufacturing of patient-specific dental applications. Various DLP and SLA machine manufacturers use liquid resins to produce parts. SLS and MJF technologies make use of polymer powders for 3D printing objects. DMLS and SLM processes make use of metal alloy powders primarily to produce dental and OMF implants. FDM technology uses solid polymer filaments extrusion to produce 3D printed parts. Resin-based 3D printing techniques are the most widely adopted for digital dentistry and dominate the landscape. Powder-based 3D printers are used for very specific use and have lesser material offerings as compared to resin-based techniques. FDM being a very primitive technique with fewer material options finds very limited use in dentistry [3].

Table 4.1 depicts the original equipment manufacturers (OEM), technology, and materials matrix for various 3D printing techniques.

4.3 Digital Dentistry with 3D Printed Materials: A Broad Perspective

Digital dentistry is augmented by the use of AM technology and advent of various materials for diverse dental specialties. Oral maxillofacial surgery, prosthodontics, orthodontics, endodontics, and periodontics are the dental specialties taking advantage of AM technologies. AM has already made its mark for almost three decades in the field of oral maxillofacial surgery [9]. OMF specialization utilizes AM for anatomical models, drilling, and cutting guides and patient-specific implants. Prosthodontics uses 3D printing for producing surgical guides for implant placement, dentures, impression trays, crowns, and bridges. Orthodontic devices like splints, aligners, and impression trays are also produced by using 3D printing technology, much faster than traditional CAM (computer-aided manufacturing) techniques [10]. Additive manufacturing in endodontics finds application in guided endodontic access cavity preparation by the use of surgical guides [11]. For periodontal applications, additive manufacturing is primarily used to produce gingiva masks and guides for implant placements.

Table 4.1 Technology and materials matrix of prominent 3D printing OEMs and their respective materials

Type	OEM	Technology	Printers	Material
Liquid materials	Formlabs (USA)	Low-Force Stereolithography (LFS)	Form 2 Form 3B Form 3BL	Photocurable resin
	DWS (Italy)	Stereolithography (SLA)	XFab Series, XPro Series, DFAB	Photocurable resin
	3D Systems (USA)	Digital Light Processing (DLP)	NextDent 5100	Photocurable resin
	EnvisionTec (USA)	Digital Light Processing (DLP)	Envision One	Photocurable resin
	Asiga (Australia)	Digital Light Processing (DLP)	MAX Series	Photocurable resin
	SprintRay (USA)	Digital Light Processing (DLP)	PRO Series, Moonray	Photocurable resin
	Rapid Shape (Germany)	Digital Light Processing (DLP)	D Series	Photocurable resin
	Stratasys (USA)	Material Jetting	Connex Series	Photocurable resins
Powder materials	EOS (Germany)	Selective Laser Sintering (SLS)	Formiga, P Series	Polymer powder
	EOS (Germany)	Direct Metal Laser Sintering (DMLS)	M series	Metal powder
	HP (USA)	Multi Jet Fusion (MJF)	HP MJF 5200 Series	Polymer powder
	3D Systems	Selective Laser Sintering (SLS)	sPRO Series, ProX	Polymer powder
	3D Systems (USA)	Direct Metal Laser Sintering (DMLS)	DMP Series	Metal powder
	SLM (Germany)	Selective Laser Melting	SLM series	Metal powder
Solid materials	Ultimaker (Netherlands)	FDM	Ultimaker S5, 3, 3+	Polymer filament

4.3.1 Prominent Use Cases for Dental Applications in Additive Manufacturing

4.3.1.1 Oral Maxillofacial Surgery

Oral maxillofacial and reconstructive surgeries rely on AM for precise surgeries and reducing the time required for surgical procedures [12]. 3D printed surgical guides, anatomical models, and patient-specific implants are the prominent used cases for AM in OMF specialization. SLA and DLP techniques use photocurable resins to print surgical guides and models. Formlabs (Formlabs, Somerville, Massachusetts, USA) 3D printer uses model resin material to print dental models which include anatomical and implant models. Formlabs portfolio also includes surgical guide resin for manufacturing anatomy-specific dental surgical guides. Likewise, major liquid-based 3D printing technologies make use of similar resins for anatomical models and surgical guides for OMF applications. Material jetting technologies are used for anatomical models. Table 4.2 depicts the material applicability matrix of the prominent material OEMs for various dental specialties.

Polymer powder-based 3D printing techniques SLS and MJF are suitable to print maxillofacial anatomical models. In OMF reconstructive surgeries, SLS technique is used for printing fibular grafting guides and anatomical models [13]. Metal powder-bed fusion techniques which include DMLS and SLM technologies are used to produce patient-specific implants (PSIs). These PSIs are usually manufactured in biocompatible titanium material. In clinical settings, FDM 3D printers have been used to make anatomical models [14]. FDM technique can be used to produce relatively low-cost models for dental applications.

4.3.1.2 Prosthodontics

AM is used extensively for fixed and removable prosthodontics. Surgical guides for dental implant placements are manufactured using photocurable resins via the SLA or DLP techniques [15]. Various liquid-based 3D printer OEMs offer proprietary surgical guide resins for guided implant placement. Table 4.2 enumerates the OEMs and their respective materials for this particular application. Complete dentures and try-in dentures are also manufactured using the SLA, DLP, or material jetting AM techniques. Newly introduced castable wax resins for SLA and DLP machines can be directly used to 3D print the sacrificial patterns and further casted into metal crowns by traditional lost-wax casting method [16]. However, 3D printing allows the direct printing of permanent and temporary crowns with the advent of the resins for direct crown making [17]. These resins are photocurable ceramic-filled hybrid resins for the 3D printing single crowns, bridges, inlays, on-lays, and veneers.

The traditional method of lost-wax casting can be eliminated by directly 3D printing the crowns in metal and other metal alloys. DMLS and SLM technologies directly 3D print the crowns in titanium and cobalt chromium alloy (CoCr) [18]. CoCr has been traditionally used as a compatible material for dental crowns. Studies also provide evidence for the use of titanium for 3D printing dental crowns and implants [19].

4.3.1.3 Orthodontics

The workflow for orthodontics has been simplified and augmented by the use of digital dentistry. Traditional orthodontics processes have observed a paradigm shift by the adoption of AM in the workflow. Digital dentistry has not only made the orthodontic process more comfortable for the patient but also efficient in practice [20]. SLA and DLP 3D printers offer several photocurable resin materials for producing models which can be used with the conventional vacuum forming process to make clear aligners. Some OEMs also offer dental grade clear resin materials to directly 3D print clear aligners and occlusal splints [21]. Dental impression trays are also manufactured by using specific resin materials. Resins for fabrication of bonding trays are the newest development in the photocurable resin. Commercial entities like Saratech (Saratech, Mission Viejo, California, USA) and Smile Direct Club (Nashville, Tennessee, USA) also use powder-based AM processes, specifically Hewlett Packard (HP) MJF process to 3D print mouth molds for clear aligner therapy [22, 23]. FDM technology is comparatively less precise than other printing techniques. However, it can be used to 3D print models for orthodontic applications [8]. 3D printing of occlusal splints and aligners is not possible with FDM due to material and accuracy constraints.

4.3.1.4 Endodontics

Use of 3D printing for endodontic applications is limited to training and education using anatomical models and guided access for cavity location and preparation through the use of surgical guides [24]. Proprietary surgical guide photocurable resins with SLA and DLP are used for printing the surgical guide. Dental models are printed with model resins in SLA and DLP, polymer powders in SLS, HP MJF, and FDM. The guided surgeries have an increased accuracy and also reduce significant risks. Anatomical models for endodontics also are suitable for education as well as simulation of the surgery.

4.3.1.5 Periodontics

3D Printing finds its use in periodontics for gingival aesthetic surgeries [25]. In periodontology, patient-specific gingiva masks are a popular application for 3D printing. Several photocurable resins offered with SLA and DLP 3D printers can be used to directly print patient-specific gingiva masks. Periodontics is among the unexplored frontiers of dentistry for 3D printing due to the limitation in the material options. However, with new material developments there could be solutions for other periodontic applications too.

Table 4.2 depicts the materials and applicability matrix of various OEMs for dental applications. It describes the materials portfolio of each OEM for dental appliances ranging from dental models to crown and implant materials.

Table 4.2 Additive manufacturing materials and applicability matrix

Applications → OEMs ↓	Surgical guides	Impression trays	Models	Implants	Temporary crown	Permanent crown	Castable material	Dentures	Gingiva masks	Splints/aligners	Bonding trays
Formlabs	Surgical guide resin	Custom tray resin	Model resin Draft resin	–	Temporary CB resin	Permanent crown resin	Castable wax resin	Denture teeth (FLDTA201) Denture Base (FLDBLP01)	Flexible 80A resin	Dental LT clear resin (V2)	IBT resin
DWS	DS3000	DS3500	Invicta (907,915) Precisa (RD096, RD097, RD-CO2)	–	Temporis	–	Fusia RF080 Fusia DC710	–	GL4000	Thermoforming models Therma (294, RD095)	–
3D Systems	NextDent SG	NextDent tray	NextDent Model (1.0, 2.0, Ortho)	LaserForm Ti Gr23 A	NextDent C&B MFH	LaserForm CoCr (C)	NextDent cast	NextDent denture 3D+ NextDent try-in	NextDent gingiva mask	NextDent OrthoClear NextDent Ortho rigid	NextDent Ortho IBT
EnvisionTec	E Guide	E Tray	E Aquamodel E Model HS E Sepfree E Orthoshape E Dentstone	–	NextDent C&B MFH	–	Press E-cast	NextDent denture 3D+	NextDent gingiva mask	Keysplint E guard	Keyortho IBT NextDent Ortho IBT
Asiga	DentaGUIDE	–	DentaMODEL	–	–	–	DentaCAST	DentaTRY DentaBASE DentaTOOTH	DentaGUM	–	DentalBT
SprintRay	SprintRay Surgical Guide 2	–	SpritnRay (Die and Model, Die and Model 2, Study Model)	–	–	–	SprintRay Castable 2	SprintRay try-in	–	SprintRay splint	SprintRay IDB
Rapid Shape	3Delta GuideS	3Delta tray	3Delta Model (320, SilverGray Ortho)	–	3Delta Etemp	–	3Delta cast M 3Delta cast P	3Delta try-in	3Delta gingiva mask	–	–
Stratasys	Clear Biocompatible (MED610)	–	VeroDent (MED670) VeroDentPlus (MED690)	–	–	–	–	Biocompatible VeroGlaze (MED620)	Flexible biocompatible (MED625FLX)	–	Flexible biocompatible (MED625FLX)

EOS	Polyamide PA 12	–	Polyamide PA 12	Titanium (Ti6AlV4) CobaltChrome SP2	–	CobaltChrome SP2	–	–	–	–	–
HP MJF	Polyamide PA 12	–	Polyamide PA 12	–	–	–	–	–	–	–	–
SLM	–	–	–	Titanium (Ti6AlV4)	–	SLM MediDent (CoCr)	–	–	–	–	–
FDM Printers	–	–	PLA, ABS	–	–	–	–	–	–	–	–

4.3.2 Factors for Adopting 3D Printed Materials in Digital Dentistry

Adoption of digital dentistry including the 3D printing process has been attributed to several advantages over the traditional CAD/CAM process. The quality of restorations coupled with the quick and easy fabrication are the advantages of adopting digital dentistry [2]. The possibility of providing mass customization and patient-specific dental appliances is also a major factor for the adoption of digital dentistry. Adopting 3D printing for orthodontic applications enables patient comfort during the process and is also efficient as compared to the conventional process. Dental implants, crowns, and bridges that are 3D printed in castable materials, casted through lost-wax casting process also improve the conventional workflow. The casting process can be completely eliminated by using metal 3D printing for high-quality custom implants and crowns using dental grade materials.

Advantages of 3D printing materials over the conventional materials used in the dental manufacturing workflow are dependent on several factors. Formulation of materials to deliver quicker turnaround times, enable faster printing, and better accuracy and resolution are important factors influencing the use of 3D printed materials in dentistry. Photopolymerization techniques using photocurable resin materials dominate the 3D printing processes used in dentistry. The possibility to print fast builds, good z-axis strength, and ability to print in clear materials make resin-based 3D printing processes an attractive proposition [26]. SLA and DLP processes can typically print in resin materials with the z-axis resolution of 50–200 μm. With MJP techniques, z-axis resolution of up to 20 μm can be achieved; however, due to material limitations, MJP printers are typically used to produce models. With powder-based techniques, z-axis resolution of 100 μm can be achieved. Conclusively, it can be claimed that resin-based 3D printing techniques are more accurate and quicker than powder-based techniques for producing parts.

3D printed materials are also used as patterns for lost-wax casting; hence, the castability of the materials and ash content of castable materials is also a crucial factor. Wax-like resins used in SLA and DLP techniques and liquid thermoplastic waxes used in the MJP technique are optimized to provide high resolution and low ash content for precise casting process [27].

Biocompatibility is a very important factor when considering 3D printed materials, especially for applications where the appliance comes in contact with the body tissues. Surgical guides, aligners, occlusal splints, dentures, crowns, and implants are some of the applications which necessitate biocompatibility. Dental surgical guide resins by Formlabs (FormLabs, Somerville, Massachusetts, USA) and other manufacturers comply with class I biocompatibility and ISO 10993-1:2018 standards [28]. Resins used for long-term dental devices like aligners, splints, dentures, and crowns comply with class IIA biocompatibility and ISO 10993-1 standards [29]. Powder-based polymers materials like polyamides (PA) are used for surgical guide applications, especially in oral maxillofacial surgery. These PA materials comply with USP (U.S. Pharmacopeia) class VI biocompatibility [30]. Cobalt chrome powder used to direct 3D print crowns is class IIA biocompatible. Titanium alloy powder (Ti6Al4V) used for manufacturing maxillofacial implants through DMLS and SLM techniques meets biocompatibility standards as per studies [31]. Titanium powder (Ti6AL4V) is non-toxic and meets USP class VI biocompatibility.

Mechanical properties of the 3D printed materials are an important factor of consideration for dental applications. For occlusal devices and aligners, mechanical strength is an important factor. Studies have shown that 3D printed devices for orthodontics do not meet the flexural strength of 65.0 MPa required in accordance with ISO 20795-1 [32]. Compared to milled or pressed occlusal devices, 3D printed devices have lower mechanical properties [32]. However, the accuracy of 3D printed devices is conclusively better than thermoformed devices, which may help in achieving accurate treatment outcomes [33]. The low mechanical performance of the 3D printed devices can be mitigated by optimizing build direction, post-curing, and appropriate material use [34]. Titanium metal powder used for 3D printing implants also needs to meet the mechanical strength of the bone and simultaneously also facilitate tissue and bone integration. Studies have shown, leverage of design and the mechanical properties of titanium powder to enhance the mechanical strength and mimic natural load-bearing structures by the use of intelligent design [35].

Resin-based 3D printing materials are also being offered with color compatibility. Permanent and temporary crown resins offered by several OEMs are available with the standard shades. Gingiva masks and dentures are also available with several colors which can be selected by matching it according to patient needs. Hence, color compatibility is an important factor of consideration.

Pricing is also a prime factor for adoption of 3D printing. The cost of biomedical materials is among a limiting factor for the adoption of 3D printing [36]. However, this is touted to change in the future due to wider use and falling prices. Entry-level desktop 3D printing solutions are also a driving factor for falling prices and adoption at the clinic level. Affordable materials coupled with compact entry-level 3D printing machines will be the key to the growth of digital dentistry.

4.4 Liquid-Based Materials

In liquid-based AM technologies, the most common materials used are photopolymers. Photopolymers are soluble, light-sensitive materials which undergo polymerization with light reaction; this light is mostly in the ultraviolet (UV) spectrum. The two major components of photopolymers are monomers and photoinitiators. Monomers are chemically bonded small molecules which join with other monomers, oligomers (formed when few monomers undergo polymerization), or polymers (formed when a large number of monomers undergo polymerization) in a repeated fashion to form new polymers. Polymers are basically large macromolecules that contain a large number of repeating units of monomers. Mostly, photopolymers consist of monomers based on acrylates or methacrylate [37].

Photoinitiators convert light energy into chemical energy by forming free radicals or cations upon UV exposure. They can break into two or more particles with UV reaction and at least one of the particles will react with monomers or oligomers and bind them together. Photoinitiators are sensitive to certain wavelengths of light and can be found to exist naturally or synthesized chemically.

4.4.1 Introduction to Liquid Material-Based 3D Printing Technologies

4.4.1.1 Stereolithography (SLA)

Stereolithography (SLA) is an additive manufacturing process that belongs to the vat photopolymerization family. In SLA, an object is created by selectively curing a polymer resin layer-by-layer using an UV laser beam. The materials used in SLA are photosensitive thermoset polymers that come in a liquid form [26] (Fig. 4.1).

4.4.1.2 Material Jetting (MJ)

Material jetting (MJ) is an additive manufacturing process which is similar to 2D (2 dimensional) printing. In material jetting, a printhead (similar to the printheads used for standard inkjet printing) dispenses droplets of a photosensitive material that solidifies under UV light, building the part layer by layer. Thermoset polymers (acrylates) in the liquid form are typically the materials which are used with MJ [26] (Fig. 4.2).

4.4.1.3 Digital Light Processing (DLP)

Digital light processing (DLP) is a 3D printing process where a projector is used as a light source to cure photopolymer resin. It is very similar to SLA where the only difference is that instead of a UV laser to cure the photopolymer resin, a safelight (light bulb) is used. Objects in DLP are created in the same way as SLA; however, a basic

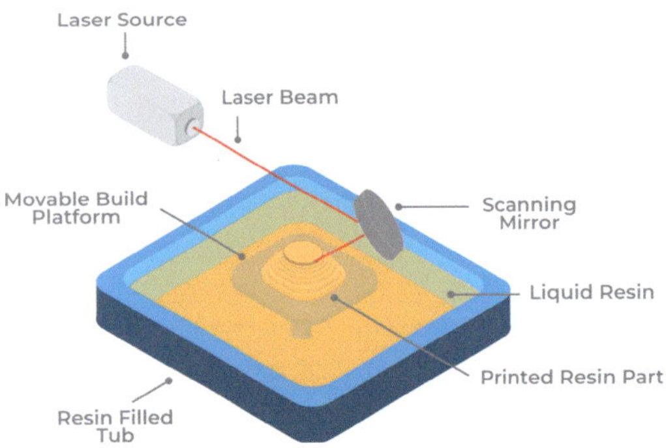

Fig. 4.1 Stereolithography (SLA) 3D printing process illustration. (Source: Imaginarium)

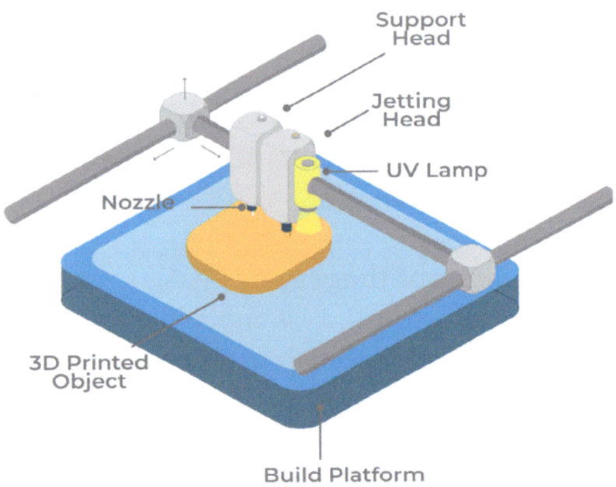

Fig. 4.2 Material jetting (MJ) 3D printing process illustration. (Source: Imaginarium)

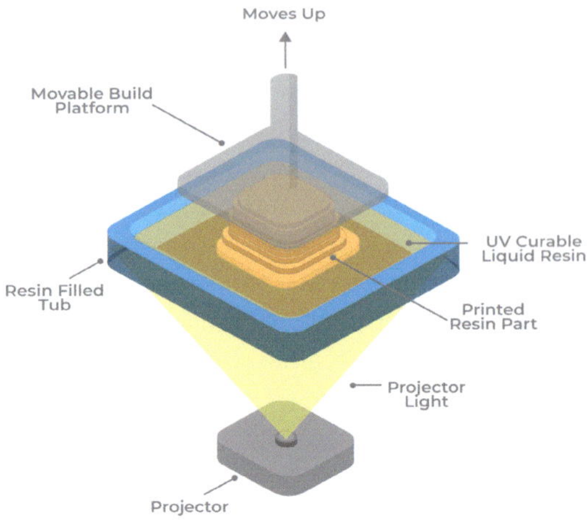

Fig. 4.3 Digital light processing (DLP) 3D printing process illustration. (Source: Imaginarium)

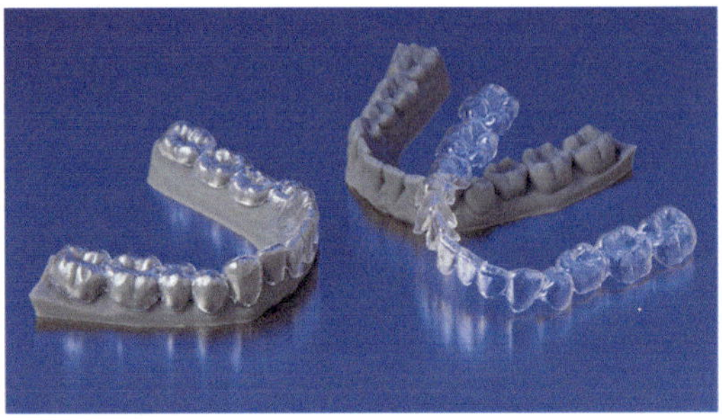

Fig. 4.4 Clear aligner. (Source: Formlabs)

difference is that DLP has an inverted build platform which results in the printing of part in an upside-down manner [26] (Fig. 4.3).

4.4.2 Dental Applications of Liquid-Based Materials

4.4.2.1 Clear Aligners

Clear aligners are transparent trays made of special material which are used to straighten teeth like braces. They use gentle and constant force to move the teeth in the required position without going through the hassles of metal wires and brackets. Clear aligners perform this alignment without the use of wires and brackets, with an added advantage of aesthetics. They are custom-made for each patient through a digital scan [9].

The manufacturing involves two steps, printing the digital teeth alignment model and thermoforming or vacuum forming of the polymer sheet on to the printed model. The model printing material does not require biocompatibility and is available with all the liquid material-based technologies mentioned below (Fig. 4.4):

- *Formlabs*—Model and Draft Resin
- *DWS*—Invicta, Precisa
- *3D System*—NextDent Model
- *EnvisionTec*—E Aquamodel, E Model HS
- *Asiga*—DentaMODEL
- *SprintRay*—Sprintray Die & Model
- *Rapid Shape*—3Delta Model 320
- *Stratasys*—Vero Dent, Vero DentPlus

4.4.2.2 Splints and Occlusal Guards

An occlusal appliance, also known as a night guard or splint, is worn over the occlusal, or biting, surface of either your upper or lower teeth. It is prescribed for many reasons, such as protec-

Fig. 4.5 Night guard. (Source: Formlabs)

tion from habitual tooth grinding or clenching, early treatment of TMJ, or temporomandibular joint pain [9].

The material requires biocompatibility and the list of OEMs with the available materials are mentioned below (Fig. 4.5):

- *Formlabs*—Dental LT Clear Resin
- *3D System*—NextDent Ortho Clear, NextDent Ortho Rigid
- *EnvisionTec*—Keysplint E Guard
- *SprintRay*—Sprintray Splint

4.4.2.3 Digital Dentures

In the fabrication of dentures in digital workflow, the computer-aided design and engineering tools are used to create perfectly-fitting dentures using digital impressions obtained through 3D scanning. These digital impressions are sent to a lab where the dentures are crafted using 3D modeling technology. The digital dentures are then manufactured using 3D printing technology [38]. The material requires biocompatibility. The list of OEMs with the available materials are mentioned below (Fig. 4.6):

- *Formlabs*—Denture Teeth
- *3D System*—NextDent Denture 3D+
- *EnvisionTec*—NextDent Denture 3D+
- *Asiga*—DentaTRY, DentaBASE, DentaTOOTH
- *SprintRay*—Sprintray Try-in
- *Rapid Shape*—3Delta Try-in
- *Stratasys*—Biocompatible VeroGlaze

4.4.2.4 Crown and Bridges

Crowns and bridges are prosthetic devices that are attached on to the teeth. Unlike removable devices such as dentures, which can be taken out and

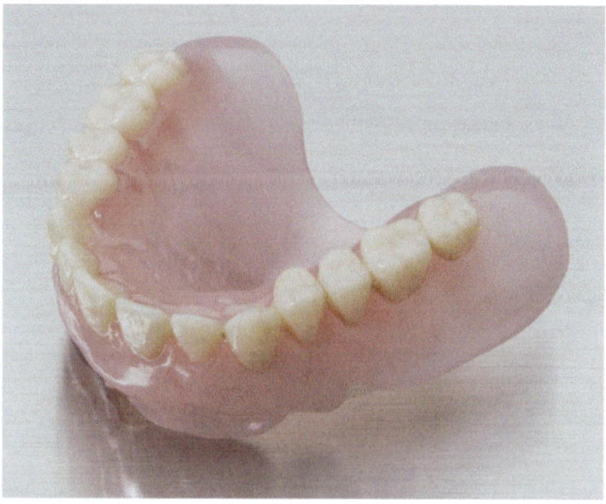

Fig. 4.6 Digital dentures. (Source: Formlabs)

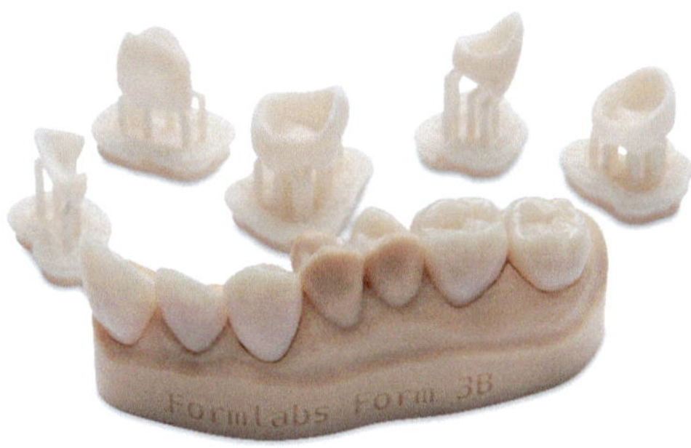

Fig. 4.7 Permanent crowns. (Source: Formlabs)

cleaned daily, crowns, and bridges are cemented and fixed permanently onto the existing teeth or implants. With digital dentistry, crowns and bridges can be easily manufactured by printing in direct biocompatible materials or can also be made in castable material which is later casted in metal through the lost-wax casting method [9].

The direct material requires biocompatibility and the list of OEMs with the available materials are mentioned below (Fig. 4.7):

- *Formlabs*—Temporary and permanent CB resin
- *DWS*—Temporis
- *3D System*—NextDent C&B MFH
- *EnvisionTec*—NextDent C&B MFH
- *Rapid Shape*—3Delta Etemp

4.4.2.5 Castable Wax

Castable wax material is typically used for manufacturing sacrificial patterns for lost-wax cast-

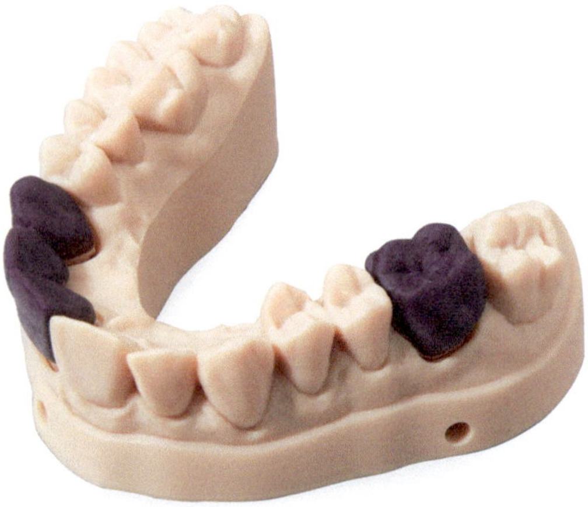

Fig. 4.8 Castable wax crowns. (Source: Formlabs)

ing [39]. These 3D printed patterns are then cast into dental grade metals. Castable wax materials are often referred to as thermoplastic waxes, as they were composed of part plastic and part waxes. The castable wax printing material does not require biocompatibility and is available with all the liquid material-based technologies (Fig. 4.8).

- *Formlabs*—Castable Wax Resin
- *DWS*—Fuchsia DC710, RF080
- *3D System*—NextDent Cast
- *EnvisionTec*—Press E-cast
- *Asiga*—DentaCAST
- *SprintRay*—Sprintray Castable 2
- *Rapid Shape*—3Delta Cast M, 3Delta Cast P

4.4.2.6 Impression Trays

Impression trays are used for taking impressions for implants, dentures, crowns and bridges, and other comprehensive cases. 3D printed impression trays offer consistent, accurate impressions for high-quality digital dentistry [40].

The material requires biocompatibility and the list of OEMs with the available materials are mentioned below (Fig. 4.9):

- *Formlabs*—Custom Tray Resin
- *DWS*—DS3500
- *3D System*—NextDent Tray
- *EnvisionTec*—E Tray
- *Rapid Shape*—3Delta Tray

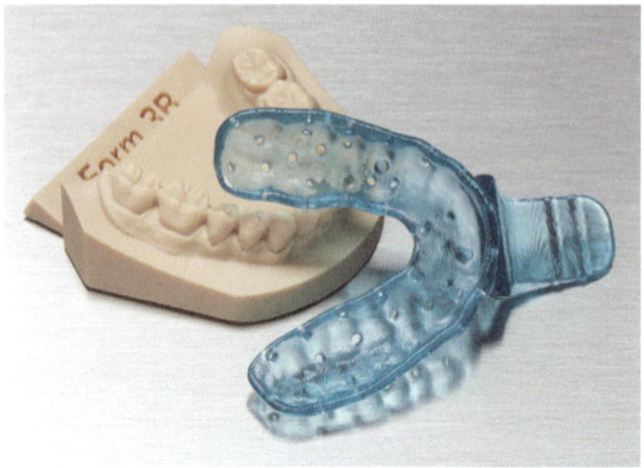

Fig. 4.9 Custom impression trays. (Source: Formlabs)

4.4.2.7 Surgical Guides

A surgical guide replicates the exact surfaces of the patient's intraoral setting and assists the surgeon to drill implants into the bone with optimal accuracy. Upon placement on the patient's jaw, the surgical guide uses sleeves to help guide the surgical instruments and implant to the proper location. A surgical guide is created by taking impressions on an intraoral scan of the desired surgical implant site as well as a computer-guided 3D implant planning system [41]. When properly used, surgical guides can increase the predictability of dental implant treatment outcomes. Any implant clinician faces the dilemma of whether to insert implants freehand or with the help of a surgical guide. With good technique, the use of surgical guides can be a confidence building and predictable method for implant placement. It can assist the practitioner in avoiding damage to anatomic structures, as well as limiting fenestration and dehiscence of the alveolar ridge at potential implant sites [42].

The material requires biocompatibility and the list of OEMs with the available materials is mentioned below (Fig. 4.10):

- *Formlabs*—Surgical Guide Resin
- *DWS*—DS3000
- *3D System*—NextDent SG
- *EnvisionTec*—E Guide
- *Asiga*—DentaGUIDE
- *SprintRay*—Sprintray Surgical Guide 2
- *Rapid Shape*—3Delta GuideS
- *Stratasys*—Clear Biocompatible MED610

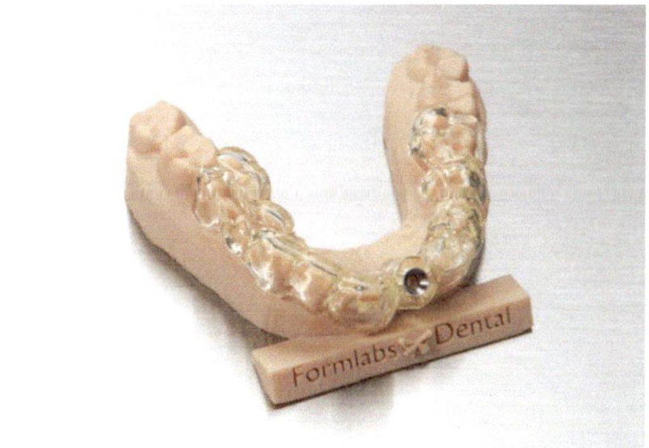

Fig. 4.10 Custom surgical guides. (Source: Formlabs)

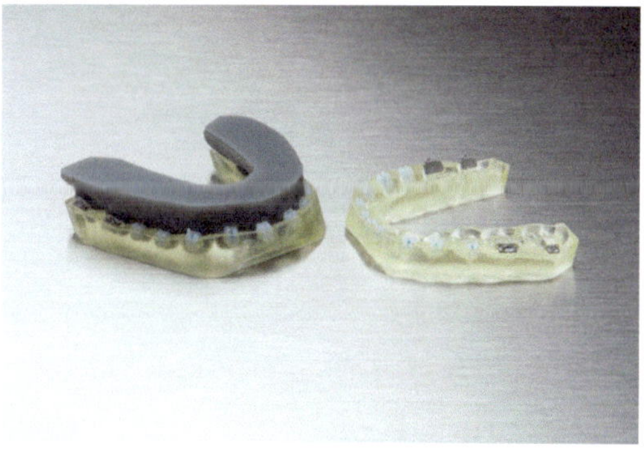

Fig. 4.11 Indirect bonding tray (IBT). (Source: Formlabs)

4.4.2.8 Indirect Bonding Trays

The indirect bonding technique optimizes fixed appliance installation at the orthodontic office, ensuring precise bracket positioning, among other advantages. In this laboratory clinical phase, material and methods employed in creating the transfer tray are decisive to accuracy [43]. The material requires biocompatibility and the list of OEMs with the available materials is mentioned below (Fig. 4.11):

- *Formlabs*—IBD Resin
- *3D System*—NextDent Ortho IBD
- *EnvisionTec*—NextDent Ortho IBD, Keyortho IBD
- *Asiga*—DentalIBD
- *SprintRay*—Sprintray IBD
- *Stratasys*—Flexible Biocompatible

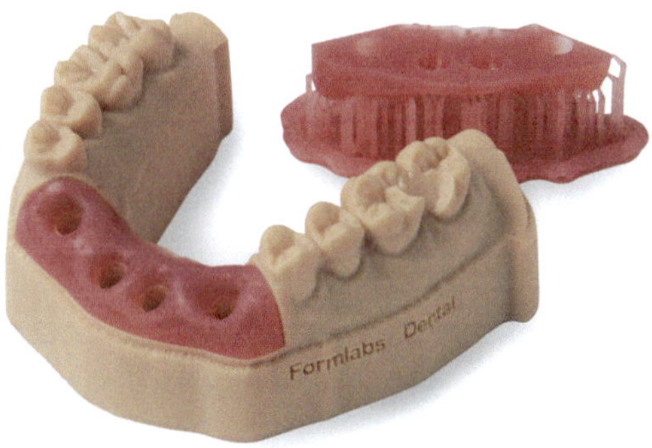

Fig. 4.12 Gingiva mask. (Source: Formlabs)

- *Rapid Shape*—3Delta Gingiva Mask
- *Stratasys*—Flexible Biocompatible MED625FLX

4.4.2.9 Gingiva Masks

A gingival mask is a removable artificial gum. The mask is convenient, removable, and functional. A gingival mask completely hides dental work and tooth roots. It rapidly restores a healthy-looking smile [44].

The material requires biocompatibility and the list of OEMs with the available materials is mentioned below (Fig. 4.12):

- *Formlabs*—Flexible 80A Resin
- *DWS*—DS4000
- *3D System*—NextDent Gingiva Mask
- *EnvisionTec*—NextDent Gingiva Mask
- *Asiga*—DentaGUM

4.5 Powder-Based Materials

Powder-based material for 3D printing is based on two categories, polymer powder and metal powders. Polymer powders are typically used in powder-bed fusion processes like selective laser sintering (SLS) and Multi Jet Fusion (MJF). Among the most commonly used polymer powders are polyamide (PA) and polyamide composites like PA—glass filled and PA—carbon fiber reinforced (CFR). Polyamide is also known as nylon. Polyamide material offers very high mechanical and thermal properties as well as high resistance to various chemicals. Among the most important characteristics of polyamide is

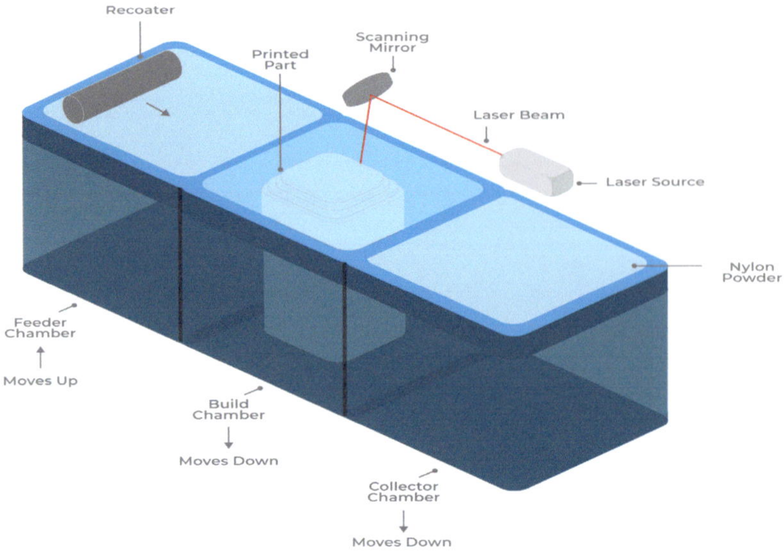

Fig. 4.13 Selective laser sintering (SLS) 3D printing process illustration. (Source: Imaginarium)

that it is biocompatible, which ensures that it can be safely used in contact with external and internal human tissues. Typically, the polyamide powder is semi-crystalline in nature with particles having sizes in the range of 60 μm. Polyamide powder is blended with glass beads or carbon fibers for high-strength applications [45]. Duraform PA 12 by 3D Systems (USA) and PA 2200 by EOS (Germany) are commonly used materials by the OEMs.

Metal powders are used with DMLS and SLM processes. Titanium (Ti6Al4V) powder and cobalt chromium powder are suitable for dental applications due to their strength, durability, and biocompatibility. Particle size distribution for metal powders usually ranges from 10 to 50 μm [46]. Cobalt chromium metal powder is used to produce crowns and copings through 3D printing. Titanium powder is used for 3D printing customized oral and maxillofacial implants.

4.5.1 Introduction to Powder Material-Based 3D Printing Technologies

4.5.1.1 Selective Laser Sintering (SLS)

Selective laser sintering (SLS) is an AM process that falls under powder-bed fusion category. In SLS, a laser selectively sinters the particles of a polymer powder, fusing them together and building a part layer by layer [5]. SLS uses thermoplastic polymers in a granular or powdered form (Fig. 4.13).

4.5.1.2 Multi Jet Fusion (MJF)

Multi Jet Fusion (MJF) by Hewlett Packard (HP) is an industrial 3D printing technology that belongs to the powder-bed fusion family. Parts are built by thermally fusing polymer powder particles layer by layer. The materials used in MJF are thermoplastic polymers (usually nylon) that come in a granular form. In MJF, an ink (fusing agent) is dispensed on the powder that promotes the absorption of infrared light. An infrared energy source then passes over the building platform and fuses the inked areas [47] (Fig. 4.14).

4.5.1.3 Direct Metal Laser Sintering (DMLS) and Selective Laser Melting (SLM)

Direct metal laser sintering (DMLS) and selective laser melting (SLM) are 3D printing technologies that directly create a metal part from its 3D computer model. This process makes use of powder metals to build parts [3]. The working material for this 3D printing process is finely powdered metal, with the size of the metal particles being 20–40 μm. The particle size and shape

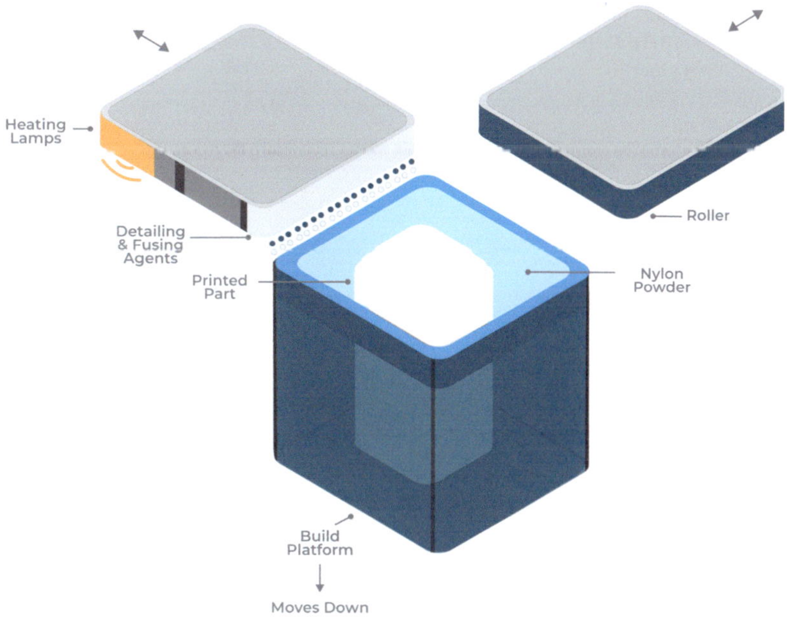

Fig. 4.14 Multi Jet Fusion (MJF) 3D printing process illustration. (Source: Imaginarium)

limit the detail resolution of the final part. Smaller metal particle size and less variation allow better resolution (Fig. 4.15).

4.5.2 Dental Applications of Powder-Based Materials

4.5.2.1 Crowns and Bridges

Both crowns and bridges are fixed prosthetic devices. Unlike removable devices such as dentures, which can be removed and cleaned daily, crowns and bridges are cemented and fixed permanently onto the existing teeth or implants. These crowns and bridges are manufactured in mainly cobalt chromium and also in gold [3]. The invisible teeth are directly made in metal, and the aesthetically visible teeth are made with a coating of ceramic or porcelain over metal. Using 3D printing, these crowns can be manufactured in cobalt chromium via DMLS, SLM, or electron-beam melting (EBM) process (Fig. 4.16).

4.5.2.2 Oral Maxillofacial Implants

Various prefabricated maxillofacial implants are used for the surgical treatment of patients. In addition to these prefabricated implants, customized CAD/CAM implants become increasingly important for a more precise replacement of damaged anatomical structures [48].

The basic requirements for successful outcome of material are, it should be biocompatible, corrosion resistant, must possess adequate mechanical property to withstand stress, produce least artifacts under imaging like computed tomography (CT) scan and MRI (magnetic resonance imaging), and interfere minimally in normal growth, remodeling, and development of bone. The materials used for implants at present are metals, ceramics, and polymers. Metallic components are exclusively used in dental implants because of their higher strength and contour-ability [49]. The materials used for manufacturing oral maxillofacial implants are titanium alloys and cobalt chromium (Fig. 4.17).

4.5.2.3 Dental and Aligner Models

Dental models are used for manufacturing aligners, for making removable crowns and bridges for checking their form and fit. They are also used for surgical planning in oral maxillofacial surgeries and for patient education. The material required may or may not be biocompatible. The most common material used for making the models is polyamide-12 which is class VI biocompatible material when printed via SLS technology

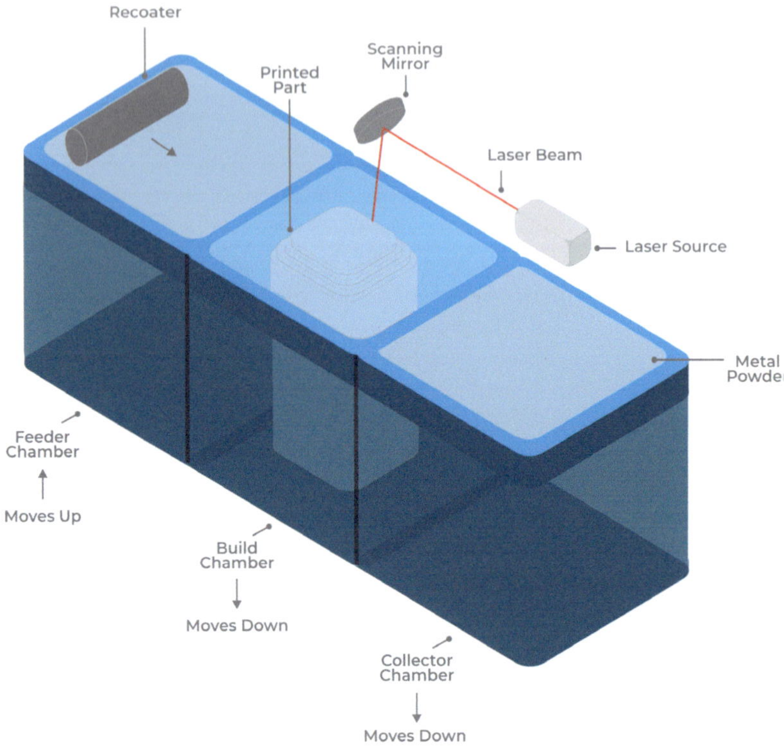

Fig. 4.15 Direct metal laser sintering (DMLS) 3D printing process illustration. (Source: Imaginarium)

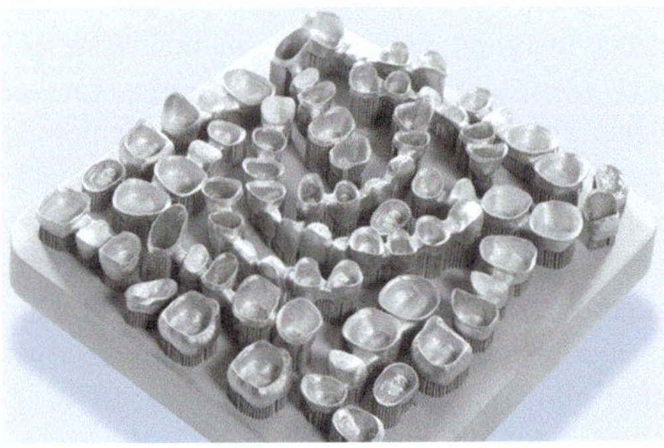

Fig. 4.16 Full mouth bridge and dental crowns directly printed in metal 3D Printing. (Source: GE Additive)

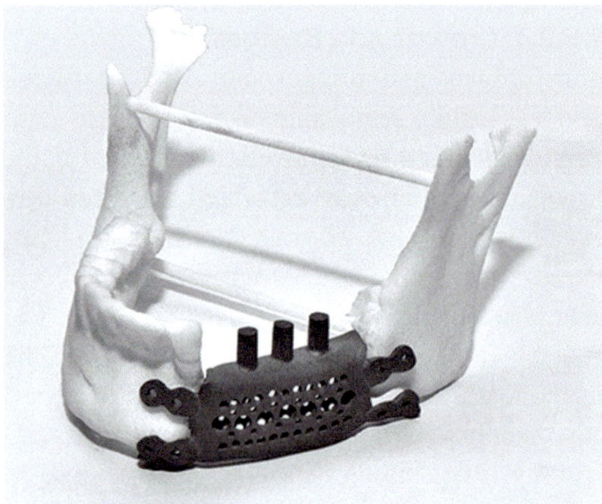

Fig. 4.17 Patient-specific mandibular implant. (Source: Imaginarium)

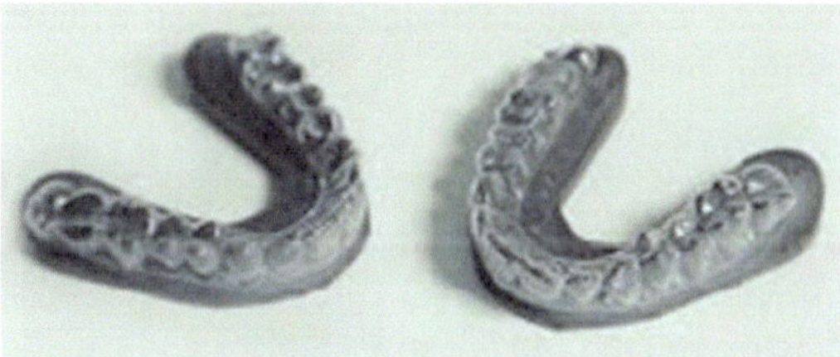

Fig. 4.18 Aligner dental models printed in MJF technology. (Source: HP Inc)

on 3D system or EOS printers and is an autoclavable biocompatible material when printed via HP MJF technology due to the addition of adhesives (Fig. 4.18).

4.6 Solid-Based Materials

In solid-based AM technologies, the material is in the form of filament. Different materials such as ABS, PLA, PETG (polyethylene terephthalate glycol), HIPS (high-impact polystyrene), PVA (polyvinyl alcohol), PP (polypropylene), and TPE (thermoplastic elastomer) can be printed. One of the most common materials printed is PLA.

PLA is a shiny, hard, and biodegradable substance which is a stronger and flexible material for a 3D printer. Chemically PLA contains lactic acid and lactides. This has a low melting temperature of 173–178 °C and tensile strength 2.7–16 GPa with several application areas including medical and dental.

4.6.1 Introduction to Solid Material-Based 3D Printing Technologies

4.6.1.1 Fused Deposition Modeling (FDM)

Fused deposition modeling (FDM) or fused filament fabrication (FFF) is an AM process that belongs to the material extrusion family. It is a method of additive manufacturing where layers of materials are fused together in a pattern to create an object. The material is usually melted just past its glass transition temperature and then extruded by a hot nozzle in a pattern next to or on top of previous extrusions, creating an object layer by layer [5]. In layman's terms, a typical FDM 3D printer takes a plastic filament and squeezes it through a hot end, melting it, and then depositing it in layers on the print bed. These layers are fused together, as they keep building up throughout the print, and eventually forming the finished part (Fig. 4.19).

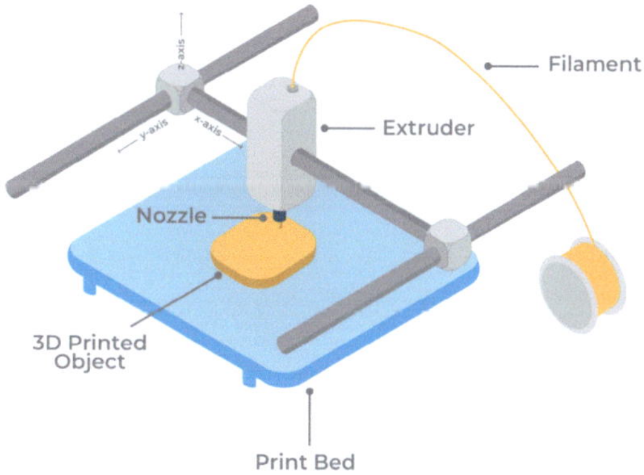

Fig. 4.19 FDM 3D printing process illustration. (Source: Imaginarium)

4.6.2 Dental Applications of Solid Materials

4.6.2.1 Dental and Anatomical Models

Dental models are used for manufacturing aligners, making removable crowns and bridges and for checking their form and fit. They are also used for surgical planning in oral maxillofacial surgeries and for patient education. The material required may or may not be biocompatible. FDM technology is suitable only for 3D printing of anatomical models [5]. The most common material used for printing the models using the solid-based AM technology is PLA, ABS, medical-grade PETG, and nylon (Fig. 4.20).

Fig. 4.20 Dental models printed in FDM technology

4.7 Advancements in Materials for Oral Healthcare

There are evidently few gaps in the material development for dental applications. As the use of additively manufactured materials in dentistry increases, concerns about their safety, compatibility, and durability are being addressed. Novel materials are being developed with these considerations in place. Biomaterials, ceramics, novel polymers, and metals are among the few materials being studied by researchers across the globe [26]. Maxillofacial surgery and reconstructive dentistry are poised to be the biggest beneficiaries of the new materials.

4.7.1 Biomaterials with AM for Dental Applications

Bioprinting is defined as a single approach combining a set of techniques incorporating cells, biologically active compounds such as growth factors, and extracellular matrix components within or onto a printed substrate. Different material delivery methods and technologies have been used, including contact bioprinting, contactless bioprinting and inkjet deposition, and other methods [50]. Inkjet printing methods have been extensively explored for biomaterials printing because the printing nozzles can print without getting in contact with the substrate and hence decreasing the possibility of cross-contamination. Inkjet printing also provides a platform on which multiple cell types can be constructed. Structural and conformal cell printing methods create a printed cell construct fabricated from the bottom upward (i.e., layer by layer or cell by cell), thereby producing a heterogeneous cell and biomolecular structure in 3 dimensions. Structural cell printing requires that the scaffold, cells, and biomolecules be simultaneously or sequentially printed. Conformal cell printing is a hybrid approach that allows biomolecules to be printed on top of thin layers of prefabricated scaffolding.

Polylactic-co-glycolic acid (PLGA) is a material used in the medical industry since its approval by FDA in 1986; however, it lacks the mechanical stability required for bone regeneration. PLGA can be mixed with other materials to increase its stability and offer additional reinforcement. PLGA and tricalcium phosphate (TCP) compounds have been successfully used for animal testing in research studies [51]. Other studies have also reported that the 3D β-tricalcium phosphate (3D-TCP) scaffold produced by 3D printing performed a fully controllable and customizable structure and had good biocompatibility [52]. PLGA/TCP and β-TCP scaffolds manufactured by 3D printing have a potentially broad dental application for personalized levels of treatments with the potential to improve oral and maxillofacial bone regeneration [53] (Fig. 4.21).

Current research hints at the use of biological materials like cells, proteins to 3D print biological implants. These can be referred to as 3D printed tissue or organs by using a bioprinter and materials consisting of scaffold and living cells. There is currently no biological implant available. However, there have been few examples showing how much organ printing has advanced. The printing of heart vessels and tissues, cartilage, skin, and liver tissue are some of the organs are in the works [54].

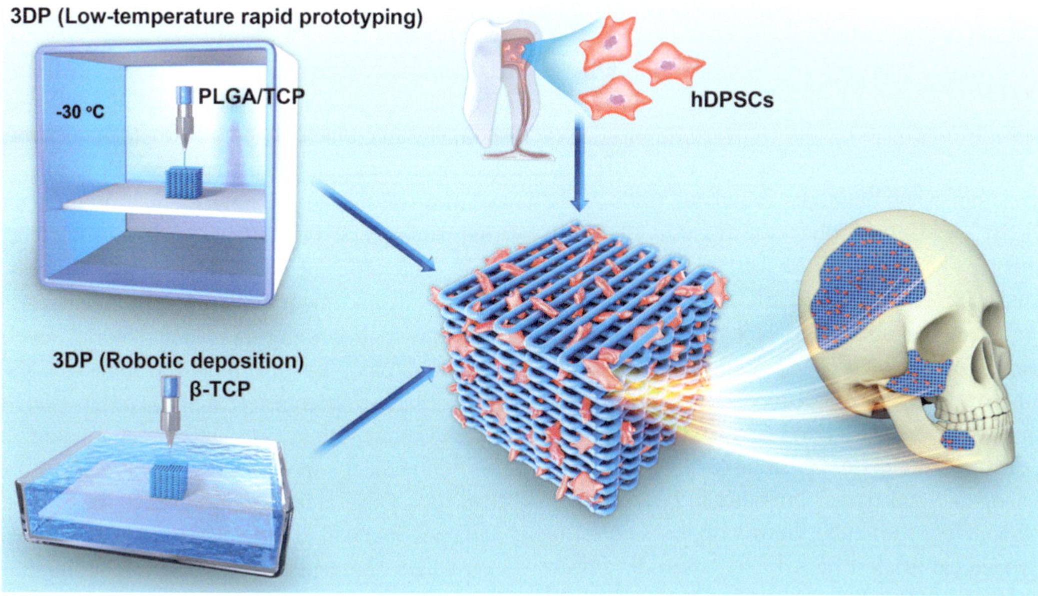

Fig. 4.21 Overview of the application of two types of 3DP technologies used to fabricate scaffold incorporated with human dental pulp stem cells (hDPSCs) for oral and maxillofacial bone reconstruction [53]

4.7.2 3D Printed Ceramics for Dental Applications

Ceramic materials have been challenging to produce conventionally due to their unique attributes like high melting temperature, high coefficient of thermal expansion, and high hardness. Ceramic materials have been extensively used in dentistry, for crowns, inlays, on-lays, bridgeworks, dentures, and veneers due to their close mimicry to natural teeth materials. Due to technical constraints of working with ceramics in 3D printing, feasibility studies have explored the use of binder jetting process with ceramic powders for dental applications [55].

Zirconia material for 3D printing is emerging as a promising material for dental applications. Several research studies have investigated the suitability of using zirconia for dental applications. These studies evaluate the density, crystalline structure, porosity, surface roughness, hardness, toughness, and biocompatibility of zirconia. Chewing simulation studies have also been performed to evaluate the tribological performance of 3D printed zirconia [56]. Zirconia can be printed with paste extrusion as well as DLP photopolymerization techniques. However, both the processes are capable of producing only green parts. Further post-processing steps including debinding and sintering are required to achieve the final properties of the hard zirconia. Studies exploring the use of DLP technique for zirconia 3D printing have concluded that zirconia dental implants can be printed with sufficient dimensional accuracy and have mechanical properties similar to conventionally produced zirconia implants [57].

4.7.3 Other Novel Materials: PEEK, Nitinol

Other novel materials are also emerging in the dental application sector. These include polymers and metal alloys. Among the polymers, polyether ether ketone (PEEK), which has traditionally been used as a medical-grade material with conventional manufacturing, has been re-introduced as a 3D printing material. Medical-grade PEEK powders and filaments have been recently commercialized by material science majors across the world. Another traditionally used material which is making the waves with 3D printing is Nitinol, a nickel titanium alloy.

In recent years, PEEK has been investigated widely in oral and maxillofacial surgery with possible applications in dental implants, skull implants, and bone replacement material for mandibular and maxillary reconstructions. CFR-PEEK or carbon fiber reinforced—PEEK is another variation of PEEK being investigated for use in implants. CFR-PEEK is considered as a promising candidate to replace metallic materials because of the inherent advantages of PEEK and improved mechanical properties [58]. Studies have also indicated FDM printed CFR-PEEK and PEEK materials to be suitable candidates for orthopedic and dental implants [59]. ApiumTec (ApiumTec, Karlsruhe, Germany) is also offering its medical-grade PEEK FDM-based 3D printer for printing PEEK implants commercially [60]. Evonik Industries, German material sciences giant is also offering an implant grade PEEK filament material for FDM 3D printing [61].

Shape memory alloys like Nitinol which is a nickel titanium (NiTi) alloy have also exhibited properties suited for making implants. Nitinol exhibits properties like shape memory effect, superelasticity, and high damping properties. In dentistry, it is already being used as a dental orthodontic wire. Other applications being explored are dental implants and dentures which exploit the shape memory effect. Blade-type implants made of NiTi have been used for patients who possess a comparatively narrow jaw bone structure in order to ensure tight initial fixation [62]. As NiTi is also relatively challenging to machine, additive manufacturing process has been touted as the suitable process to manufacture NiTi objects. SLM process has been demonstratively proved to use NiTi for printing parts successfully [63, 64].

Novel materials evidently pave the way for newer and unexplored territories of 3D printing in dental applications. Several application areas in dentistry which are so far untouched upon due to material constraints can be solved by further materials research.

4.8 Summary

3D printed materials for digital dentistry are transforming the conventional work flow of the dental appliance manufacturing. This digital workflow, right from capturing the patient anatomy to manufacturing of the dental devices, has brought an amalgamation of perfect patient care outcomes, speed, and efficiency to the treatment process. With the advancement in the additive manufacturing processes becoming stagnated, material development remains an unexplored frontier for the future. Material science research has progressed to develop safer, stronger, and biomimetic materials for biomedical applications. Treatments like Teeth-in-a-Day, clear aligners, etc., have been made possible with the development of new materials which augment the traditional dental manufacturing work flow. Biocompatible polymers and metal alloys have also made customized maxillofacial implants safer, achieving anatomical conformity through 3D printing. With a lot of ersatz for biological materials currently available in 3D printing, the next objective of materials research is to bioprint living cells and organs. Although bioprinting is touted to provide the proverbial "gold standard" treatment especially for oral and maxillofacial surgery and restorative dentistry in future, the winding path is riddled with many regulatory and scientific hurdles.

References

1. Xu X, He L, Zhu B, Li J, Li J. Advances in polymeric materials for dental applications. Polym Chem. 2017;8(5):807–23. https://doi.org/10.1039/C6PY01957A.
2. Zaharia C, Gabor AG, Gavrilovici A, Stan AT, Idorasi L, Sinescu C, Negruţiu ML. Digital dentistry—3D printing applications. J Interdiscip Med. 2017;2(1):50–3. https://doi.org/10.1515/jim-2017-0032.
3. Dawood A, Marti Marti B, Sauret-Jackson V, Darwood A. 3D printing in dentistry. Br Dent J. 2015;219(11):521–9. https://doi.org/10.1038/sj.bdj.2015.914.
4. High accuracy 3D printing materials for dental labs and practices [Internet]. Available from: https://dental.formlabs.com/materials/.

5. Kessler A, Hickel R, Reymus M. 3D printing in dentistry—state of the art. Oper Dent. 2019;45:30–40. https://doi.org/10.2341/18-229-L.
6. Tang JC, Luo JP, Huang YJ, Sun JF, Zhu ZY, Xu JY, Dargusch MS, Yan M. Immunological response triggered by metallic 3D printing powders. Addit Manuf. 2020; https://doi.org/10.1016/j.addma.2020.101392.
7. Rimington R, Capel A, Player D, Bibb R, Christie S, Lewis M. Feasibility and biocompatibility of 3D-printed photopolymerized and laser sintered polymers for neuronal, myogenic, and hepatic cell types. Macromol Biosci. 2018;2018:1800113. https://doi.org/10.1002/mabi.201800113.
8. Taczała J, Czepułkowska W, Konieczny B, Sokołowski J, Kozakiewicz M, Szymor P. Comparison of 3D printing MJP and FDM technology in dentistry. Arch Mater Sci Eng. 2020;101(1):32–40. https://doi.org/10.5604/01.3001.0013.9504.
9. Oberoi G, Nitsch S, Edelmayer M, Janjić K, Müller AS, Agis H. 3D printing-encompassing the facets of dentistry. Front Bioeng Biotechnol. 2018;6:172. https://doi.org/10.3389/fbioe.2018.00172.
10. Martorelli M, Gerbino S, Giudice M, Ausiello P. A comparison between customized clear and removable orthodontic appliances manufactured using RP and CNC techniques. Dent Mater. 2013;29(2):e1–10. https://doi.org/10.1016/j.dental.2012.10.011.
11. Connert T, Zehnder MS, Amato M, Weiger R, Kühl S, Krastl G. Microguided endodontics: a method to achieve minimally invasive access cavity preparation and root canal location in mandibular incisors using a novel computer-guided technique. Int Endod J. 2018;51(2):247–55. https://doi.org/10.1111/iej.12809.
12. Tarsitano A, Ciocca L, Scotti R, Marchetti C. Morphological results of customized microvascular mandibular reconstruction: a comparative study. J Craniomaxillofac Surg. 2016;44(6):697–702. https://doi.org/10.1016/j.jcms.2016.03.007.
13. Leiggener C, Messo E, Thor A, Zeilhofer HF, Hirsch JM. A selective laser sintering guide for transferring a virtual plan to real time surgery in composite mandibular reconstruction with free fibula osseous flaps. Int J Oral Maxillofac Surg. 2009;38(2):187–92. https://doi.org/10.1016/j.ijom.2008.11.026.
14. Kamio T, Hayashi K, Onda T, Takaki T, Shibahara T, Yakushiji T, Shibui T, Kato H. Utilizing a low-cost desktop 3D printer to develop a "one-stop 3D printing lab" for oral and maxillofacial surgery and dentistry fields. 3D Print Med. 2018;4(1):1–7. https://doi.org/10.1186/s41205-018-0028-5.
15. Anunmana C, Ueawitthayasuporn C, Kiattavorncharoen S, Thanasrisuebwong P. In vitro comparison of surgical implant placement accuracy using guides fabricated by three different additive technologies. Appl Sci. 2020;10(21):7791. https://doi.org/10.3390/app10217791.
16. Katreva I, Dikova T, Abadzhiev M, Tonchev T, Dzhendov D, Simov M, Angelova S, Pavlova D, Doychinova M. 3D-printing in contemporary prosthodontic treatment. Scripta Scientifica Medicinae Dentalis. 2016;2(1):7–11.
17. 3D printing permanent crowns with the Form 3B – White Paper [Internet]. 2020. Available from: https://3d.formlabs.com/permanent-crowns-application-guide/.
18. Venkatesh KV, Nandini VV. Direct metal laser sintering: a digitised metal casting technology. J Indian Prosthodont Soc. 2013;13(4):389–92. https://doi.org/10.1007/s13191-013-0256-8.
19. Traini T, Mangano C, Sammons RL, Mangano F, Macchi A, Piattelli A. Direct laser metal sintering as a new approach to fabrication of an isoelastic functionally graded material for manufacture of porous titanium dental implants. Dent Mater. 2008;24(11):1525–33. https://doi.org/10.1016/j.dental.2008.03.029.
20. Groth C, Kravitz ND, Shirck JM. Incorporating three-dimensional printing in orthodontics. J Clin Orthod. 2018;52(1):28–33.
21. 3D printing splints with Formlabs SLA 3D Printers, Formlabs [Internet]. 2020. Available from: https://dental.formlabs.com/indications/splints-and-occlusal-guards/guide-v2/.
22. Producing molds for clear dental aligners with HP Multi Jet Fusion 3D printing. HP Inc [Internet]. 2018. Available from: https://saratech.com/wp-content/uploads/2019/07/Producing-molds-for-clear-dental-aligners-with-HP-Multi-Jet-Fusion-3D-printing_SARATECH.pdf.
23. SmileDirectClub to 3D print 20M unique aligner molds a year with HP MJF [Internet]. 3D Printing Media Network. 2019. Available from: https://www.3dprintingmedia.network/smiledirectclub-3d-print-20-million-aligner-hp/.
24. Anderson J, Wealleans J, Ray J. Endodontic applications of 3D printing. Int Endod J. 2018;51(9):1005–18. https://doi.org/10.1111/iej.12917.
25. Li Z, Liu YS, Ye HQ, Liu YS, Hu WJ, Zhou YS. Diagnosis and treatment of complicated anterior teeth esthetic defects by combination of whole-process digital esthetic rehabilitation with periodontal surgery. J Peking Univ Health Sci. 2017;49(1):71–5.
26. Stansbury JW, Idacavage MJ. 3D printing with polymers: challenges among expanding options and opportunities. Dent Mater. 2016;32(1):54–64. https://doi.org/10.1016/j.dental.2015.09.018.
27. Marwah OMF, Sharif S, Zainol MA, Ibrahim M, Mohamad EJ. 3D printer patterns evaluation for direct investment casting. Appl Mech Mater. 2013;2013:1400–3.
28. Declaration of conformity for surgical guide resin. Formlabs 2019. [Internet]. Available from: https://dental-media.formlabs.com/filer_public/ca/1b/ca1bbd27-d09b-4bec-b851-2be3479c5ff2/sgv2-declaration_of_conformity.pdf.
29. Declaration of conformity for temporary CB resin. Bego 2020. Available from: https://media.form-

labs.com/m/6e1926d2c347bce0/original/-ENEU-Declaration-of-Conformity-for-Temporary-CB.pdf.
30. Biocompatibility: USP class VI and ISO 10993-1 information. 3D Systems 2020. Available from: http://infocenter.3dsystems.com/materials/classvi.
31. Wang H, Zhao B, Liu C, Wang C, Tan X, Hu M. A comparison of biocompatibility of a titanium alloy fabricated by electron beam melting and selective laser melting. PLoS One. 2016;11(7):e0158513. https://doi.org/10.1371/journal.pone.0158513.
32. Berli C, Thieringer FM, Sharma N, Müller JA, Dedem P, Fischer J, Rohr N. Comparing the mechanical properties of pressed, milled, and 3D-printed resins for occlusal devices. J Prosthet Dent. 2020;124(6):780–6. https://doi.org/10.1016/j.prosdent.2019.10.024.
33. Jindal P, Juneja M, Siena FL, Bajaj D, Breedon P. Mechanical and geometric properties of thermoformed and 3D printed clear dental aligners. Am J Orthod Dentofac Orthop. 2019;156(5):694–701. https://doi.org/10.1016/j.ajodo.2019.05.012.
34. Reymus M, Fabritius R, Keßler A, Hickel R, Edelhoff D, Stawarczyk B. Fracture load of 3D-printed fixed dental prostheses compared with milled and conventionally fabricated ones: the impact of resin material, build direction, post-curing, and artificial aging-an in vitro study. Clin Oral Investig. 2020;24(2):701–10. https://doi.org/10.1007/s00784-019-02952-7.
35. Tan XP, Tan YJ, Chow CSL, Tor SB, Yeong WY. Metallic powder-bed based 3D printing of cellular scaffolds for orthopaedic implants: a state-of-the-art review on manufacturing, topological design, mechanical properties and biocompatibility. Mater Sci Eng C Mater Biol Appl. 2017;76:1328–43. https://doi.org/10.1016/j.msec.2017.02.094.
36. Quan H, Zhang T, Xu H, Luo S, Nie J, Zhu X. Photocuring 3D printing technique and its challenges. Bioact Mater. 2020;5(1):110–5. https://doi.org/10.1016/j.bioactmat.2019.12.003.
37. Bagheri A, Jin J. Photopolymerization in 3D printing. ACS Appl Polym Mater. 2019;1:593–611. https://doi.org/10.1021/acsapm.8b00165.
38. Chung Y, Park J, Kim T, Ahn J, Cha H, Lee J. 3D printing of resin material for denture artificial teeth: chipping and indirect tensile fracture resistance. Materials. 2018;11:1798. https://doi.org/10.3390/ma11101798.
39. Tahayeri A, Morgan M, Fugolin A, Bompolaki D, Athirasala A, Pfeifer C, Ferracane J, Bertassoni L. 3D printed versus conventionally cured provisional crown and bridge dental materials. Dent Mater. 2018;34:192–200. https://doi.org/10.1016/j.dental.2017.10.003.
40. Sun Y, Chen H, Li H, Deng K, Zhao T, Wany Y, Zhou Y. Clinical evaluation of final impressions from three dimensional printed custom trays. Sci Rep. 2017;7:14958. https://doi.org/10.1038/s41598-017-14005-8.
41. Sahwil H. An introduction to surgical guides in dentistry. 19 Feb 2019. DDS Labs Blogs [Internet]. Available from: https://blog.ddslab.com/surgical-guides-in-dentistry.
42. Trobough KP, Garrett PW. Surgical guide techniques for dental implant placement [Internet]. 24 Jul 2018. Available from: https://decisionsindentistry.com/article/surgical-guide-techniques-for-dental-implant-placement/.
43. Nojima LI, Araújo AS, Alves M Jr. Indirect orthodontic bonding—a modified technique for improved efficiency and precision. Dental Press J Orthod. 2015;20(3):109–17. https://doi.org/10.1590/2176-9451.20.3.109-117.
44. Gingival masks for receding gums [Internet]. Available from: https://www.billdorfmandds.com/dental-procedures/gingival-mask/.
45. Ligon SC, Liska R, Stampfl J, Gurr M, Mülhaupt R. Polymers for 3D printing and customized additive manufacturing. Chem Rev. 2017;117(15):10212–90. https://doi.org/10.1021/acs.chemrev.7b00074.
46. Slotwinski JA, Garboczi EJ, Stutzman PE, Ferraris CF, Watson SS, Peltz MA. Characterization of metal powders used for additive manufacturing. J Res Natl Inst Stand Technol. 2014;16(119):460–93. https://doi.org/10.6028/jres.119.018.
47. Cai C, Tey WS, Chen J, Zhu W, Liu X, Liu T, Zhao L, Zhou K. Comparative study on 3D printing of polyamide 12 by selective laser sintering and multi jet fusion. J Mater Process Technol. 2021;288:116882. https://doi.org/10.1016/j.jmatprotec.2020.116882.
48. Memon AR, Wang E, Hu J, Egger J, Chen X. A review on computer-aided design and manufacturing of patient-specific maxillofacial implants. Expert Rev Med Devices. 2020;17(4):345–56. https://doi.org/10.1080/17434440.2020.1736040.
49. Sapru BL, Mohan SM, Roy SK. Implants in maxillofacial surgery. Med J Armed Forces India. 1999;55(4):285–6. https://doi.org/10.1016/S0377-1237(17)30348-9.
50. Farré-Guasch E, Wolff J, Helder MN, Schulten EA, Forouzanfar T, Klein-Nulend J. Application of additive manufacturing in oral and maxillofacial surgery. J Oral Maxillofac Surg. 2015;73(12):2408–18. https://doi.org/10.1016/j.joms.2015.04.019.
51. Chen SH, Lei M, Xie XH, Zheng LZ, Yao D, Wang XL, Li W, Zhao Z, Kong A, Xiao DM, Wang DP, Pan XH, Wang YX, Qin L. PLGA/TCP composite scaffold incorporating bioactive phytomolecule icaritin for enhancement of bone defect repair in rabbits. Acta Biomater. 2013;9(5):6711–22. https://doi.org/10.1016/j.actbio.2013.01.024.
52. Abarrategi A, Moreno-Vicente C, Martínez-Vázquez FJ, Civantos A, Ramos V, Sanz-Casado JV, Martínez-Corriá R, Perera FH, Mulero F, Miranda P, López-Lacomba JL. Biological properties of solid free form designed ceramic scaffolds with BMP-2: in vitro and in vivo evaluation. PLoS One. 2012;7(3):e34117. https://doi.org/10.1371/journal.pone.0034117.

53. Cao S, Han J, Sharma N, Msallem B, Jeong W, Son J, Kunz C, Kang HW, Thieringer FM. In vitro mechanical and biological properties of 3D printed polymer composite and β-tricalcium phosphate scaffold on human dental pulp stem cells. Materials (Basel). 2020;13(14):3057. https://doi.org/10.3390/ma13143057.
54. Ho CM, Ng SH, Yoon YJ. A review on 3D printed bioimplants. Int J Precis Eng Manuf. 2015;16(5):1035–46. https://doi.org/10.1007/s12541-015-0134-x.
55. Miyanaji H, Zhang S, Lassell A, et al. Process development of porcelain ceramic material with binder jetting process for dental applications. J Mater. 2016;68:831–41. https://doi.org/10.1007/s11837-015-1771-3.
56. Branco AC, Silva R, Santos T, Jorge H, Rodrigues AR, Fernandes R, Bandarra S, Barahona I, Matos APA, Lorenz K, Polido M, Colaço R, Serro AP, Figueiredo-Pina CG. Suitability of 3D printed pieces of nanocrystalline zirconia for dental applications. Dent Mater. 2020;36(3):442–55. https://doi.org/10.1016/j.dental.2020.01.006.
57. Osman RB, van der Veen AJ, Huiberts D, Wismeijer D, Alharbi N. 3D-printing zirconia implants; a dream or a reality? An in-vitro study evaluating the dimensional accuracy, surface topography and mechanical properties of printed zirconia implant and discs. J Mech Behav Biomed Mater. 2017;75:521–8. https://doi.org/10.1016/j.jmbbm.2017.08.018.
58. Devine DM, Hahn J, Richards RG, Gruner H, Wieling R, Pearce SG. Coating of carbon fiber-reinforced polyetheretherketone implants with titanium to improve bone apposition. J Biomed Mater Res B Appl Biomater. 2013;101(4):591–8. https://doi.org/10.1002/jbm.b.3286.
59. Han X, Yang D, Yang C, Spintzyk S, Scheideler L, Li P, Li D, Geis-Gerstorfer J, Rupp F. Carbon fiber reinforced PEEK composites based on 3D-printing technology for orthopedic and dental applications. J Clin Med. 2019;8(2):240. https://doi.org/10.3390/jcm8020240.
60. Apium medical PEEK 3D printing [Internet]. Available from: https://apiumtec.com/en/apium-m-series-medical-peek-3d-printing.
61. Evonik biomaterials for medical applications [Internet]. Available from: https://3d-printing.evonik.com/en/additive-manufacturing-materials/biomaterials-medical-applications/peek-filaments.
62. Nouri A. Titanium foam scaffolds for dental applications. Metallic Foam Bone. 2017;2017:131–60. https://doi.org/10.1016/B978-0-08-101289-5.00005-6.
63. Saedi S, Turabi AS, Andani MT, Haberland C, Elahinia M, Karaca H. Thermomechanical characterization of Ni-rich NiTi fabricated by selective laser melting. Smart Mater Struct. 2016;25(3):035005. https://doi.org/10.1088/0964-1726/25/3/035005.
64. Mitchell A, Lafont U, Hołynska M, Semprimoschnig C. Additive manufacturing - a review of 4D printing and future applications. Addit Manuf. 2018;24:606–26. https://doi.org/10.1016/j.addma.2018.10.038.

Fundamentals of Computer-Aided Design (CAD) in Dental Healthcare: From Basics to Beyond

5

Ahmad Al-Hassiny

Contents

5.1 Introduction

Computer-aided design and computer-aided manufacturing (CAD/CAM) encapsulate digital workflows carried out across many different industries. It was only a matter of time before digitalization started taking place in dentistry as well. CAD/CAM has now become a common occurrence in dental offices across the globe [1]. Digital technologies have transformed modern dentistry on all fronts, from communication and diagnosis to treatment [2]. The entire process is now much more efficient and predictable [3]. In this chapter, we will discuss how CAD/CAM has changed dentistry completely. We will cover its application in various disciplines, such as restorative dentistry, prosthodontics, orthodontics, implantology, and maxillofacial surgery, focusing on acquisition, design, and manufacturing.

A. Al-Hassiny (✉)
Institute of Digital Dentistry,
Wellington, New Zealand
e-mail: director@instituteofdigitaldentistry.com

5.1.1 A Brief History of Computer-Aided Designing and Computer-Aided Manufacturing (CAD/CAM)

Although CAD/CAM has only recently become a mainstream in the dental industry, its history started much earlier when Dr. François Duret proposed optical impressions and a rudimentary dental CAD/CAM system as early as 1973. In 1984, he demonstrated the first crown produced using a patented electro-optical scanner to take digital impressions and a simple chairside CAD/CAM system [4]. This technology was met with skepticism at the time. However, the need for

© The Author(s), under exclusive license to Springer Nature Switzerland AG 2022

P. K. Chaudhari et al. (eds.), *3D Printing in Oral Health Science*,
https://doi.org/10.1007/978-3-031-07369-4_5

tooth-colored restorative alternatives to amalgam or gold prevailed. Composite materials were not advanced enough at the time, and their polymerization shrinkage limited their application in chairside dentistry.

Recognizing the potential of this market, Dr. Werner Mörmann and Dr. Marco Brandestini developed a similar system in 1983 at the University of Zurich, independent of Dr. Duret's work, and they implemented it in the same year as the CEREC (CEramic REConstruction) 1 unit (Sirona Dental Systems GmbH, Bensheim, Germany). In 1985, the first chairside inlay using the CEREC 1 system was carried out [5]. It did not have sophisticated CAD software for dental anatomy as it simply copied the cavity and required the occlusion to be refined in the mouth. The milling process was performed by a single disk working on a one and a half axes. CEREC was developed with ceramic crowns in mind, as the name, an abbreviation for CEramic REConstruction, suggests. Ceramic materials available at this time (namely feldspathic ceramics) were not as strong or resilient as contemporary materials, so the demand for full-metal or porcelain-fused-to-metal restorations remained. This was the one reason for the relatively small market penetration of chairside CAD/CAM systems at the time. Laboratory CAD/CAM systems were adopted by dental technicians earlier as they were utilized for manufacturing metal frameworks or full-ceramic restorations in a greater volume [6].

It is without question that the marketing efforts of CEREC and the improvement in material sciences over time were two instrumental factors in the growth of dental CAD/CAM. The invention of a lithium disilicate material, IPS e.max CAD by Ivoclar Vivadent in 1998, contributed significantly to the growing popularity of chairside CAD/CAM and same-day dentistry [7]. This material is available in various aesthetic shades and levels of translucency and has a flexural strength similar to natural enamel. Since then, milling machines have been proven to be a valuable tool in dentistry, and now additive manufacturing in the form of 3D printers has started gaining mass popularity. Although dental professionals began considering 3D printers for their businesses in the mid-2010s, the technology had already been around for nearly 30 years [8]. The growing competition has also driven prices lower, making them more affordable for the masses. Digital dentistry is without a doubt the future of the profession, and the evolution of the industry is happening right now.

5.2 CAD/CAM Workflow in Dentistry

In its entirety, the CAD/CAM workflow consists of three parts [5]: acquisition, design, and manufacturing. This chapter will cover all three aspects, including what they involve and their capabilities and limitations.

5.2.1 Open vs. Closed Systems

CAD/CAM dentistry is not a one-size-fits-all modality, and there are now many options on the market to find a solution tailored to individual needs and budgets. An open system or architecture essentially allows communication and data exchange between software and hardware of different brands and companies [9]. Open systems now dominate the market, with most devices enabling the export output data in the ubiquitous standard tessellation language (STL) file type. All design programs on the market can read scans taken with an open system, giving you the flexibility to customize your CAD/CAM workflow [10].

It is not necessary for those starting with digital dentistry to switch from conventional methods to a fully digitalized workflow at once. Investing in an intraoral scanner can be the first step. An open system is vital if you plan to start with a scanner only and intend to have the restoration manufactured elsewhere (by a laboratory or a milling center). Thankfully, these days all scanners on the market are outwardly open. Meaning they enable exporting of files that can be used with any software. This was not the case six years ago.

Open systems result from the fact that many CAD/CAM manufacturers do not cater to every part of the workflow. Some companies (such as Medit or Carestream Dental) make scanners without fully-fledged designing capabilities or CAD software. In contrast, others (e.g., 3Shape, Dentsply Sirona) offer scanners with comprehensive CAD software that can fabricate various dental prosthetics. For those companies that do not supply CAD software or manufacturing units, it is necessary to rely on third parties to provide the missing links. Therefore, the data must be interchangeable within the work chain [9].

Closed systems are also found across the entire industry. This is especially true for companies that sell complete end-to-end CAD/CAM systems with manufacturing units included (e.g., Zirkonzahn, CEREC) or design their workflow so that you can only send scan data to be manufactured by their specific milling centers exclusively (e.g., Nobel Biocare). In closed systems, the formatted output data obtained from scanning is a proprietary file that can only be used in corresponding software or associated systems. They may also prevent the import or export of files, scans, or designs [9]. These systems are not necessarily worse than open systems as they provide you with complete workflow, but they also often offer little flexibility and limited usage of third-party libraries or materials. The market trend seems to be more inclined toward open systems [2]. Prominent dental companies have switched from closed to open systems in the past, notably Straumann abandoning their rigid system and adopting Dental Wings Operating System (DWOS) by Dental Wings in 2012. CEREC also opened its system in 2017 due to what appeared to be market pressure, enabling STL export from their scanners.

5.2.2 3D File Formats

No matter what system you choose, the data you create either through scanning or designing will be stored in some form of 3D file format. There are over 100 types of 3D file formats, but the most common across industries is standard tessellation/triangle language (STL) [11]. This format contains information about the 3D geometry of the object without any information about surface color or texture. It does not include information a milling machine might require, such as margin lines, but this is typically detected manually by CAD software. If you want to transfer your scan data with surface color information, you must export the scan information in the correct format. The file types used in dentistry that include color are the Polygon File Format (PLY) and Wavefront.OBJ. Including STL, these are the three most commonly used file types in the industry [12] (Fig. 5.1).

As mentioned above, some companies also develop their own proprietary file formats that are unique for their products. For example, TRIOS by 3Shape can export data in STL, PLY, and DCM, a unique 3Shape format. DCM contains information about geometry, color, measurements, HD photographs, annotations, patient information, or a margin line as the user sets it. This file type can only be read and used by other 3Shape products. Similar file types exist for other major CAD/CAM companies like CEREC which exclusive to their products.

5.2.3 Chairside vs. Laboratory CAD/ CAM

The simplest way to classify dental CAD/CAM systems is into CAD/CAM for chairside and laboratory use [13]. In the subsequent sections, we will discuss acquisition, designing, and manufacturing units and how they differ depending on whether they are made for dental offices or laboratories. Keep in mind that this does not mean they cannot be used in both scenarios. In simplified terms, dental office CAD/CAM systems tend to be more compact and offer a somewhat rigid, streamlined workflow. On the other hand, laboratory systems tend to have more flexibility with materials, tools, and manufacturing capabilities and focus on detail rather than efficiency [14].

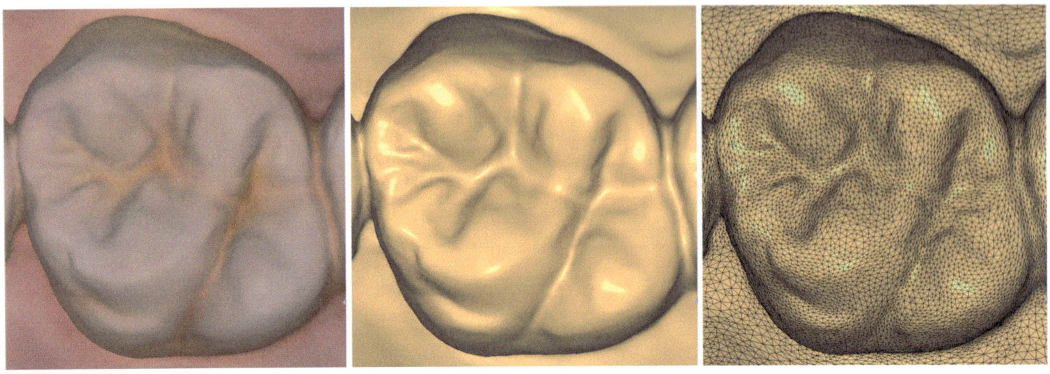

Fig. 5.1 An example of an intraoral scan exported in PLY (colour), STL (monochrome), and visualized tessellated mesh. (Source: Dr. Ahmad Al-Hassiny)

5.3 3D Surface Data Acquisition for Computer-Aided Designing

To design any dental prosthetic for a patient, you first need to obtain data. This is done using an acquisition unit, one of the three pillars of a complete CAD/CAM system [5]. Using a scanner, you can acquire a digital copy of the patient's dental and orofacial structures and the relationship of the jaws [2]. There is also an option of a pre-operative scan which is used for different indications.

Multiple studies have compared the accuracy of conventional impression techniques using impression material with digital impressions. The latest studies concluded that digital scans have comparable accuracy [15] or are even superior [16] to physical impressions. Research findings also show that CAD/CAM-fabricated dental restorations are of excellent quality and accuracy when compared to conventionally manufactured dental restorations [1, 17–19]. However, it is essential to note that there are differences in the accuracy between scanners. Some scanners can struggle with maintaining accuracy when scanning more complicated structures such as edentulous arches or when scanning for full-arch prosthetics. Therefore, not all scanners are suitable for digital impression in full-arch implant-supported fixed dental prostheses [20, 21].

There are many other benefits of digital impressions, such as rarely provoking a gag reflex and making the treatment more comfortable for patients [2, 22]. One of the most significant advantages of digital impressions is immediate feedback for the clinician. You can examine the scans on your computer during the procedure. These scans can be rotated and enlarged, and some systems highlight seemingly inadequate areas for your convenience, for example, insufficient tooth reduction during crown preparation. All this information is readily available immediately after scanning and has been shown to improve the quality of tooth preparations [23]. Lastly, another advantage is that scans and design data are stored on a computer and can be easily reaccessed in the future, and are easily transferred. The technique in which these digital impressions are obtained differs significantly, depending on whether your acquisition device is for a chairside or laboratory use.

5.3.1 Laboratory Desktop Scanners

Laboratory scanners are desktop devices that acquire digital impressions from conventionally taken impressions or, more commonly, from physical models of said impressions. Mechanical scanners (e.g., NobelProcera by Nobel Biocare) were used in the past, but now optical scanners have quickly become an industry standard [23].

Impressions or models are placed on a scanning holder or a plate and retained by a fixture or adhesive putty. The construction of these labora-

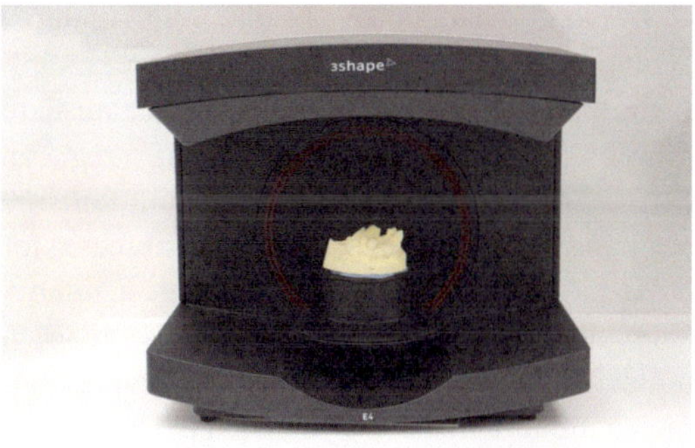

Fig. 5.2 A tabletop laboratory scanner, E4 by 3Shape. (Source: Institute of Digital Dentistry Laboratory, Wellington, New Zealand)

tory scanners varies in size, design, and mechanics, but it has become common for the holders to be attached to a mechanical arm that moves on at least one axis. The holder itself also rotates and tilts during operation. These simultaneous multi-axis movements allow the software to capture as much data in as little time as possible. Nowadays, scanning a single model using a latest generation laboratory scanner can be carried out within seconds (Fig. 5.2).

Laboratory scanners are classified as optical scanners as they use a static light that is projected onto the object's surface, which is then used to create a digital impression. A digital sensor registers the change in depth of this projected light. This data is calculated, analyzed and used by software to create the 3D image [24]. In earlier generations of desktop scanners, it was necessary to coat surfaces in a **spray powder**. This spray contained a fine powder that would render the surface matte, limiting the light scattering and making scanning process easier. Nowadays, this is usually unnecessary and only reserved for scanning exceptionally lustrous, transparent, or pitch-black objects. This advancement also paved the way for the development of desktop scanners capable of capturing color.

The **bite scan** is usually obtained by scanning the upper and lower models held together in centric occlusion. This is one of the reasons why it is beneficial to provide a laboratory with full-arch digital impressions—even if you provide a bite registration, full-arch models make evaluating the impressions' accuracy more reliable. The sign of correctly taken digital impressions or bite registrations is when the models do not rock in any direction while occluding, given the occlusion is cleaned of any interferences (e.g., bubbles, pulls). In the case of quad- or triple-tray impressions, models often rock buccal-lingually as they lack adequate occlusal stops from the opposing half of the arch. This often makes evaluation, mounting, and scanning of models more difficult.

Narrow, undercut spots, especially in areas of crown preparations, can be difficult to scan for laboratory-based systems. For this reason, dental laboratories can turn prepared teeth into **removable dies**. These dies are similar to traditional techniques and have the captured soft tissue removed to reveal the finishing line of the preparation fully [24]. Modern scanner advancements have led to the ability to capture these difficult areas more predictably. Removable dies can lead to an inaccurate vertical or horizontal position of dies/preps within an arch resulting in tight or open contacts. Although multiple model sectioning systems have been developed in the past, e.g., "The Carrot model" by Willi Geller or Giroform by Amann Girrbach, this issue prevails to some degree [25]. For this reason, control, unseparated models are often also poured to check the contacts during a final seating of the restoration.

There are many benefits of scanners compared to traditional techniques. Dental technicians often faced issues, such as distortion, pulls, and bubbles with conventional impressions. All these frustrating issues with traditional impressions do not exist with digital scans [26]. Furthermore, the COVID-19 pandemic has raised awareness of the cross-infection risks that come with handling physical impressions. Everyone in the dental team, from dental assistants to technicians, who are handling physical impressions, can be potentially exposed to viruses or bacteria if they are not adequately disinfected. The harsh chemicals used in this disinfection process can also deteriorate the surfaces of impressions or stone models if misused. These are all issues that are completely avoided with digital impressions fabricated with intraoral scanners [27]. IOS are designed to be safely wiped down, scanner tips, and carts included. Many scanners on the market now also offer an option of removable scanning heads, either single-use or reusable, the latter type being capable of withstanding multiple autoclave cycles. When using digital technologies, there is no transfer of any physical impressions or materials. This enables ideal cross-infection control [28]. Unfortunately, laboratory desktop scanners are still prone to the risks of physical impressions. They are not as effective for cross-infection control as they still require handling of impressions by multiple people or departments. Proper disinfection of any impression and models is paramount in these cases [29].

5.3.2 Chairside Intraoral Scanners

Unlike laboratory scanners, **chairside or intraoral scanners** (IOS) have always been optical and they are designed to obtain digital impressions directly in a patient's mouth. The shape of the oral cavity has limited their design. The general appearance of intraoral scanners across all the different companies does not vary as much as laboratory scanners. For instance, all intraoral scanner heads are small enough to fit in a mouth, long enough to capture the last molar's distal aspect, and light enough to allow operation with one hand [30] (Fig. 5.3).

Intraoral scanners have been through multiple evolutions and improvements over the years [5]. Almost all modern IOS have some mechanism to prevent condensation while scanning, e.g., an external or in-built heater or an in-built fan. The scanning tips are either wiped and disinfected for cross-infection control, covered with a removable single-use scanning head, or contain a removable scanning head capable of withstanding multiple autoclave cycles [31]. The market trend seems to be moving toward wireless, battery-powered scanners with removable scanning heads, rather than traditional wired ones, with fixed scanning heads that require cold disinfection.

The intraoral scanner itself is connected either to a **personal desktop or laptop computer** (e.g., i700 by Medit, CS3700 by Carestream, Emerald by Planmeca) or comes with a **custom build cart** purchased with the scanner (e.g., TRIOS MOVE by 3Shape, CEREC Primescan by Dentsply Sirona). The laptop/desktop IOS tends to be a cheaper, entry-level type of scanner. Companies that sell an IOS with a cart also often offer a cheaper laptop/mobile version, e.g., 3Shape's TRIOS 4 "Pod" and TRIOS 4 with a MOVE cart (Fig. 5.4).

When it comes to **scanning**, it is once again recommended to scan arches in their entirety. Although it is not as crucial as conventional

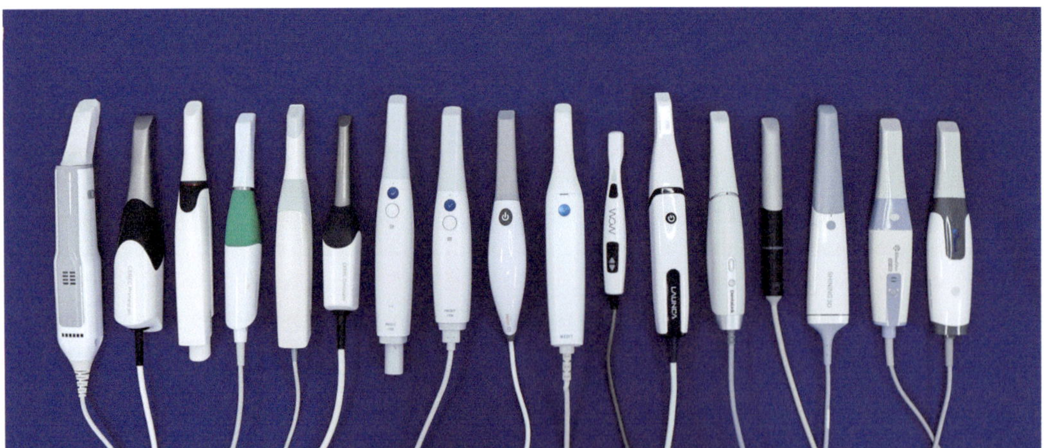

Fig. 5.3 Intraoral scanners (left to right): iTero Element 5D Plus (Align Technology), CEREC Primescan (Dentsply Sirona), TRIOS 4 (3Shape), Emerald S (Planmeca), TRIOS 3 (3Shape), CEREC Omnicam (Dentsply Sirona), i700 Wireless (Medit Corp), i700 (Medit Corp), Heron IOS (Heron), i500 (Medit Corp), WOW (Biotech Dental), DL-206 (Launca), S6000 (DentaLink), Virtuo Vivo (Straumann Digital), Aoralscan 3 (Shining 3D), Helios 600 (Eighteeth), AS100 (Alliedstar). (Source: Institute of Digital Dentistry, Wellington, New Zealand)

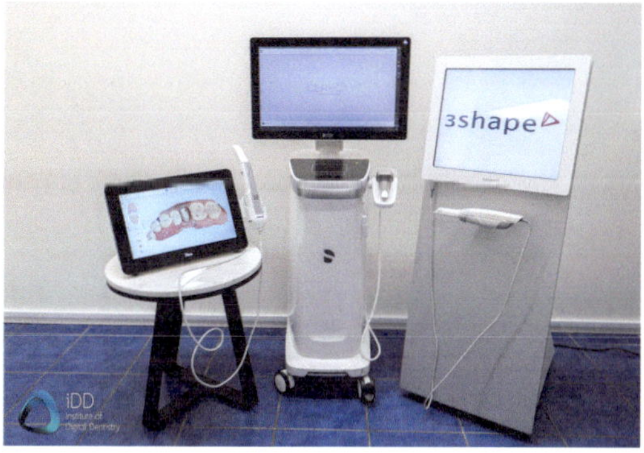

Fig. 5.4 An example of intraoral scanners integrated with mobile computers/carts. Left to right: iTero Element 5D Plus (Align Technology), CEREC Primescan (Dentsply Sirona), TRIOS 3 (3Shape). (Source: Institute of Digital Dentistry, Wellington, New Zealand)

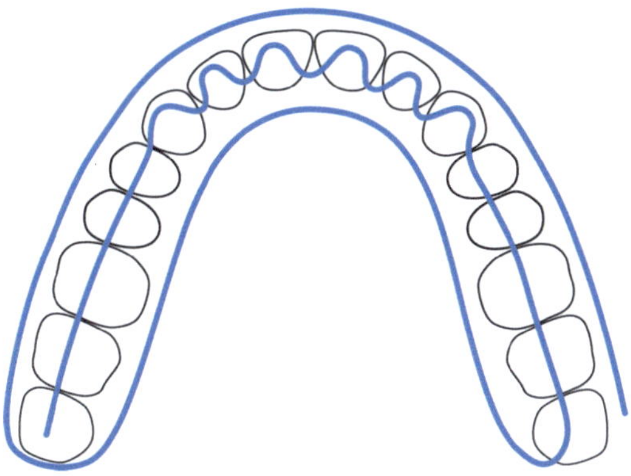

Fig. 5.5 Commonly recommended scanning strategy for dentate arches. (Source: Dr. Ahmad Al-Hassiny, Wellington, New Zealand)

impressions, it helps ensure an accurate evaluation of the relationship between the jaws and makes the virtual articulator function work more reliably [32]. For a simple same-day crown procedure or quadrant dentistry using an in-house digital workflow, simple quadrant scans can be utilized with excellent results [33]. When using an IOS, it is crucial to follow recommended **scanning protocols or scanning strategies** set out by the company [30]. The typical scanning protocol is occlusion first, then the buccal surfaces, and finally scanning the lingual aspect of the dental tissues (Fig. 5.5).

Following a scanning protocol is the most efficient way to scan and minimizes the risk of incorrect image stitching, resulting in a warped scan. Following prescribed scanning protocols also ensures optimum accuracy of the digital image [30]. **Bite registration** is a scan of the buccal aspect, often in centric occlusion. The jaw scans are aligned to this mesh, and they are assigned coordinates relative to each other, meaning that even exported scans appear in this position relative to each other (Fig. 5.6).

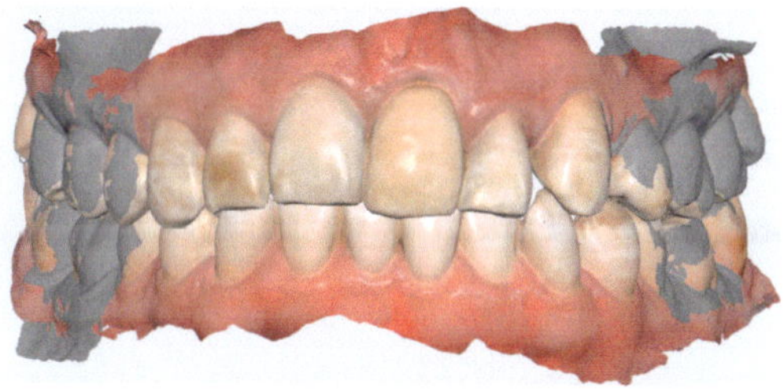

Fig. 5.6 An example of the bite or occlusion scans (in gray) with mandibular and maxillary scans aligned accordingly. Scanned with i700 (Medit), previewed in MeditLink (Medit). (Source: Dr. Ahmad Al-Hassiny, Wellington, New Zealand)

Like laboratory scanners, coating the oral cavity in scanning powder is not necessary with modern intraoral scanners. However, due to the optical nature of the scanners, wet and shiny surfaces may still be challenging to capture but are often managed inside the mouth, for example, with adequate moisture control [34]. When taking conventional impressions, the quality of models depends on multiple factors within the workflow of an office, a laboratory, and anyone involved in processing the models. Intraoral scanners give you complete control and sole responsibility for the quality of your impressions.

5.3.3 Scanning of Edentulous Jaws

Edentulous jaws are notoriously difficult to scan due to the lack of teeth and the presence of moving soft tissues [35]. This results in fewer reference points for the scanner to stitch the captured data correctly. Although relatively more challenging, it can be achieved with modern intraoral scanners. The scanning protocol varies depending on the thickness, shape, and size of the resid-

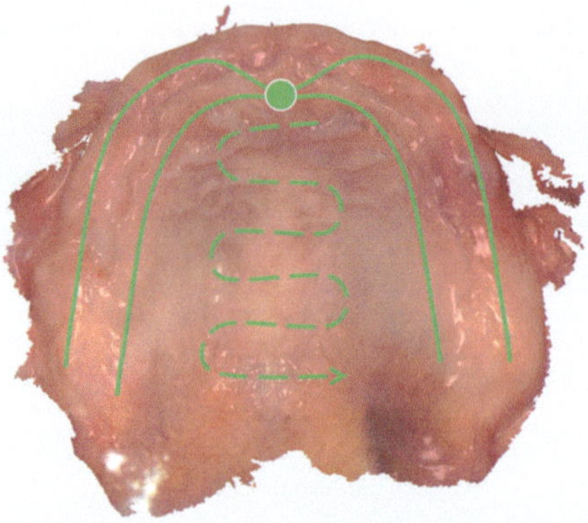

Fig. 5.7 An illustration of the edentulous maxillary scanning strategy

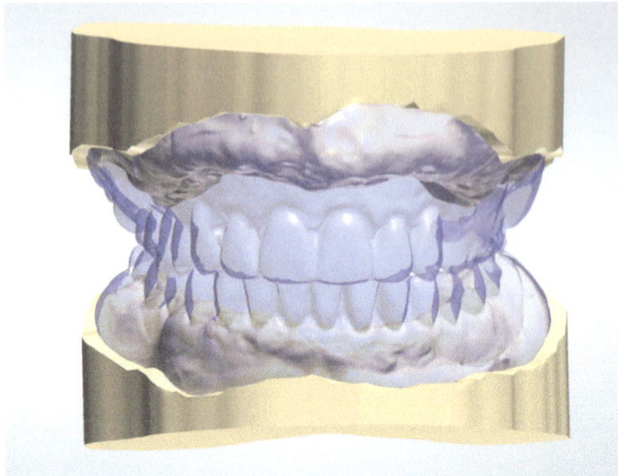

Fig. 5.8 Models aligned using patient's occluding dentures (transparent). (Source: Institute of Digital Dentistry Laboratory, Wellington, New Zealand)

ual ridge. It is crucial to follow scanning protocols carefully when taking a digital impression of an edentulous jaw to minimize any inaccuracies [36, 37] (Fig. 5.7).

The bite of an **edentulous** patient cannot be registered using the same techniques as in the case of patients with dentate arches. If your patient has pre-existing dentures, scanning their bite will not work either due to the lack of reference points shared with the jaw scans—the dentures cover the alveolar ridges and a palate and only a small part of the movable soft tissue stays exposed which provides insufficient data [38]. For this reason, multiple techniques were developed to achieve a bite scan for an edentulous patient. These techniques can involve surgical guides or CT scans for an entirely digitized workflow, or alternatively scanning the fitting surface of dentures. Other more common methods still require conventional impressions, e.g., bite registration with a facebow or a patient's occluding dentures [38, 39]. Stone models can be mounted using the facebow information or the dentures and then scanned using laboratory desktop scanners. The base of the stone models should provide enough reference points to align jaw scans with the bite scan. Some scanners also allow for the mounted models to be scanned with the entire articulator. Alternatively, traditional bite rims/blocks can also be used [40]. When using a conventional impression technique, a custom impression tray is made to fit the patient's anatomy precisely. The tray's rim should be long enough to capture the entire length of the vestibule, and before the material is fully set, the patient is encouraged to move their tongue and cheeks, so the full range of mucosa movement is recorded. For these reasons, a completely impression-free workflow for final full dentures is still not perfect due to the static nature of IOS images [41]. They fail to capture the functional movements of the mouth which are typically captured during border molding in the traditional impression stage. Thus, fully digital dentures are either relined in the mouth or are fabricated with a combination of traditional and digital impressions [42] (Fig. 5.8).

5.4 Computer-Aided Designing (CAD) for Dental Applications

All laboratory scanners come with integrated CAD design software, but not every chairside IOS. Scanners from 3Shape, Planmeca, and CEREC all have their own dedicated chairside and laboratory CAD software available and are purchased seperately. Multiple computer-aided design softwares are available in the market, the most notable being DentalCAD by exocad, Dental System by 3Shape, or CEREC software by Dentsply Sirona. These programs are either integrated into a CAD/CAM system purchased from the supplier or bought separately. Some programs can be either purchased with a one-off payment, while others are subscription-based. This specialized software can be used to design all forms of dental prosthetics such as veneers, crowns, inlays, onlays, short or long-span bridges, both reduced (to make room for porcelain) or full-contour (monolithic), and tooth-borne or implant-supported. Most CAD software also has modules for removable prostheses such as full or partial dentures, including metal frameworks or a bar design. Other indications can include clear aligners, indirect bonding trays, surgical guides, splints, or custom impression trays. CAD software can accomplish almost everything that was achieved with traditional techniques [43]. Always check with your provider whether your particular software package comes with modules (capabilities) you require—just because the software can theoretically do it, it does not mean that your CAD/CAM provider or license enables you to.

If the system is open, you can upload various tooth libraries or, more importantly, **implant libraries**, which allow you to design restorations fitting on titanium bases of your choice. Even if you are using a titanium base compatible with another brand's implant connection (e.g., Medentika Ti-base connecting to an original Nobel implant), you need to make sure that you are using a library compatible with that particular Ti-base (using the same example, Medentika library). Although the connective part is the same, the portion of the Ti-base that is glued to the restoration can differ from the original, authentic Ti-base. Some implant libraries come with **dummy connections**. This dummy library provides your design with fake, unfunctional design and effectively makes it useless for either in-house or a third-party milling center. Restorations designed using these libraries must be sent to a designated milling center where the dummy connection is replaced with genuine, functional geometry and processed afterward.

In some clinical applications, such as digital wax-ups, you can use **3D modeling software** that was not developed specifically for dentistry, e.g., by Pixologic or Meshmixer by Autodesk. Programs like these tend to be cheaper, but they also tend to be harder to use and are not as streamlined for dental applications [44]. Programs that were developed for dentistry guide you through the workflow to achieve a functional result. They focus on the crucial dental aspects, such as finding the path of insertion, blocking out the undercuts to make seating possible, providing tools for effective margination, or helping you respect a specific material's minimum thickness. It is also important to realize that a non-specialized program does not offer tooth templates, meaning that you must upload a template or design the teeth from scratch. On the other hand, dental CAD software comes with multiple templates, usually in the form of a catalog that offers teeth of various morphologies, which will make the designing process more accessible. In the following sections, we will explore in detail the most popular CAD software on the market.

5.4.1 Exocad

Exocad GmbH (stylized as exocad) is one of the leading CAD/CAM companies on the market, with its headquarters in Darmstadt, Germany. The company focuses solely on developing CAD/CAM software that is hardware neutral, completely open, and can be integrated with the CAD/CAM hardware from any manufacturer. Nowadays, exocad has been integrated with systems by Amann Girrbach, Zirkonzahn, KaVo, and others. Exocad allows third-party integrators to customize the software to various extents—it can be as simple as a color change of the graphic user interface or alter the system's capabilities. When you purchase the software through a third-party CAD/CAM provider, you need to handle any exocad issues through them exclusively as exocad itself does not provide such customer service directly. Additionally, if the system is purchased with an entire CAD/CAM system through a provider, it can come with a perpetual license or an annual fee, depending on the arrangement. If you choose to purchase the system separately, you have an option of either a perpetual license or a subscription model (Fig. 5.9).

Exocad's flagship software is the **dental designing software**, often referred to as "exocad," although its proper name is DentalCAD. This program allows for fabricating multiple types of

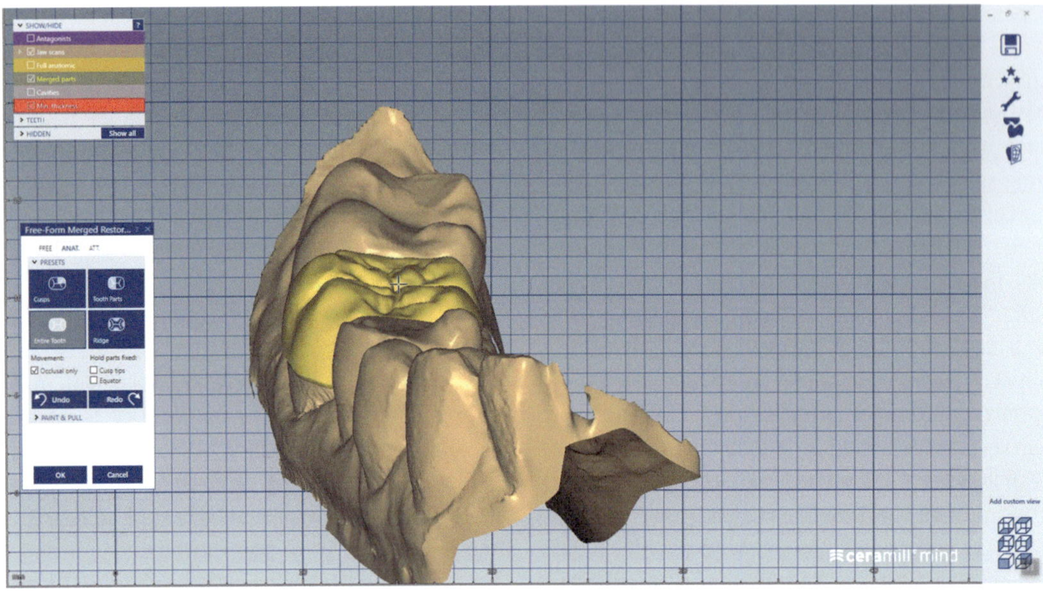

Fig. 5.9 "Free-Form Merged Restoration" tools in the Expert Mode, DentalCAD by exocad integrated with Ceramill CAD/CAM system by Amann Girrbach. (Source: Institute of Digital Dentistry Laboratory, Wellington, New Zealand)

fixed or removable prosthetics, both tooth- and implant-supported. It consists of various modules and add-ons that can be purchased separately. These modules include the following:

- **PartialCAD**: a module add-on for designing metal partial removable frameworks.
- **Smile Creator**: an extension to DentalCAD for digital smile design. It enables a 2D design to be created over patient's photographs and later adapted to 3D scans.
- **Exoplan**: for guided implant surgery and to design surgical guides using CT or CBCT data, intraoral and model scans, and digital smile design.
- **Exocad Ortho Archiver**: for creating and archiving orthodontic models. Exocad has announced its plans to allow tooth setups and bracket placement soon.

DentalCAD has been developed primarily for use in dental laboratories, while ChairsideCAD is a version of the software adapted for clinicians, focusing on chairside dentistry. **Dentalshare** allows data transfer between dental offices, dental laboratories, or milling centers. Both DentalCAD and ChairsideCAD offer a **guided workflow**, the Wizard Mode. It is a restoration design process separated into a step-by-step workflow which is especially useful for beginners. The exocad guided workflow is rigid, with many capabilities and options of the software being hidden. To reveal and enable all the capabilities of the software, you need to switch to the Expert Mode. Besides the dental design software and its modules, exocad has also developed a **project manager** to create orders for any laboratory work. This is essentially a laboratory prescription form and is the part of the software where you input case information, select teeth, type of restoration, and material. From here, you can either scan a case with a laboratory scanner (not necessarily using exoscan by exocad) or upload the scans from any intraoral scanner.

The company has also developed a **CAM manager** for milling, called exocam. This software adapts the program to each milling machine's requirements and capabilities. It allows for material management, placement of the designs within disks or blocks, calculation of tool paths, and review of the milling simulation before the actual production. Furthermore, for 3D printers, exoprint software is also available. However, not all CAD/CAM systems that use DentalCAD by exocad must use exocam or exoprint for production. Some have their CAM managers that can be used in combination with exocad. This software once again is designed to be completely open. STL, OBJ, and PLY files can be freely imported from any desktop or intraoral scanner and exported to be manufactured by any milling machine or 3D printer, although this feature can be limited to a certain extent by the CAD/CAM reseller.

Exocad was previously a privately held company until its recent acquisition by Align Technology in March 2020. The original co-founders (Tillmann Steinbrecher and Maik Gerth) remain as the heads of the company, and it is theorized that this new ownership may eventually lead to some changes in the digital dentistry market, such as complete integration of exocad with iTero scanners or perhaps closing the Chairside CAD to iTero's competitors.

5.4.2 3Shape

3Shape is another company with a primary focus on the CAD aspect of the CAD/CAM workflow. 3Shape was established in Denmark in the year 2000. It started in the hearing aid industry by developing a digital method to produce hearing aid shells and take ear impressions. Their success in this field disrupted the industry. In 2004, 3Shape shifted its focus onto the dental industry after many companies showed interest in their technology. From the beginning, their goal was to achieve a complete switch from analog to digital techniques in dentistry, similar to their previous successes in the field of audiology. 3Shape introduced its first 3D dental laboratory scanner and CAD/CAM software at the International Dental Show (IDS) in 2005. After its success in the dental laboratory sector, 3Shape moved to dental clinics, intending to create a system with all the advantages and none of the drawbacks of existing systems. They have since introduced multiple laboratory desktop **scanners** (E series, D series), intraoral scanners (TRIOS series), and a CBCT scanner (X1). They have become a globally recognized dental CAD/CAM company with headquarters in Denmark and offices worldwide (Fig. 5.10).

The 3Shape Dental System software is designed for laboratory use and can design any fixed or removable restoration or framework, tooth- or implant-supported. Similar to exocad, it also offers multiple add-on modules, some notable examples being:

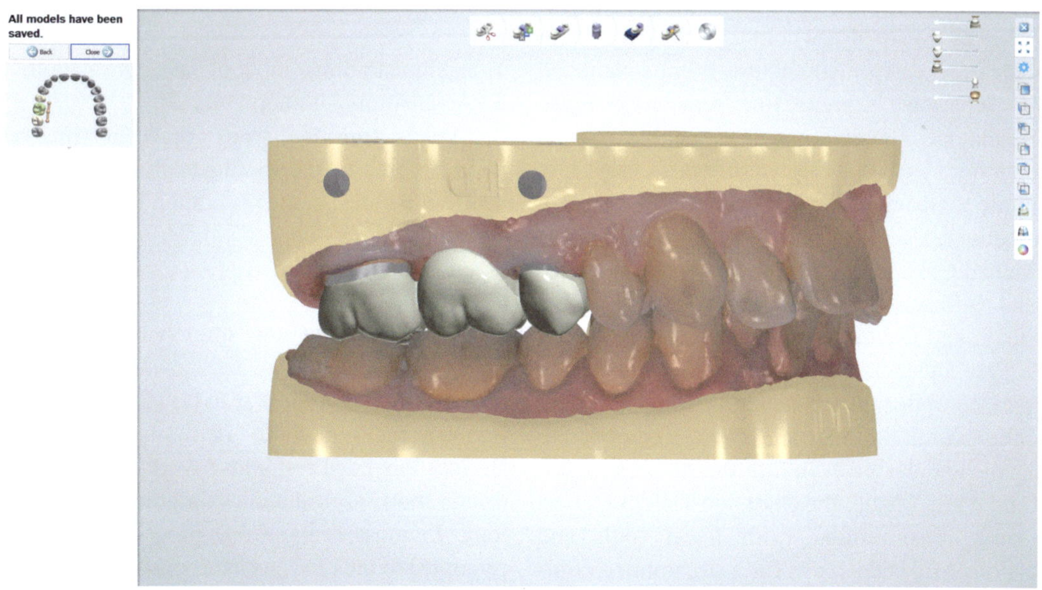

Fig. 5.10 A bridge design in the Dental System by 3Shape. (Source: Institute of Digital Dentistry Laboratory, Wellington, New Zealand)

- **Splint Studio**: design of dental splints using a virtual articulator.
- **Implant Studio**: to plan implant treatment using intraoral or extraoral scans, CBCT scan data, digital restoration design/mock-up, and to design a surgical guide for guided surgery.
- **Smile Design**: software for digital smile design using photographs of the patient.
- **Orthodontic Planner**: treatment planning software where you can design digital study models and assess them using various analyzing tools, with an option of uploading patient's CBCT scans or photographs.
- **Indirect Bonding Studio**: plan and design trays for indirect bonding of brackets.
- **Clear Aligner Studio**: treatment planning with clear aligners and their in-house production.

Although they have developed scanners, their software is also entirely open, compatible with scans from any company. As they do not manufacture milling machines or 3D printers, their output data are also compatible with various CAM systems. 3Shape users can even customize the interface, designing process, design parameters, and material requirements to a great extent. 3Shape software can be purchased either as a part of a whole CAD/CAM system with third-party manufacturing units included or separately (e.g., for remote design). Mandatory subscription fees apply. 3Shape chairside and laboratory software vary too. While their laboratory software, Ortho System Premium, does not differ as much from the clinical Clear Aligner Studio, there is a significant difference between the TRIOS Design Studio and Dental System. The latter is mainly for use in dental laboratories. It is generally more complicated and has more tools and flexibility within the software, and it is equipped to handle and design even full-arch implant cases. On the other hand, the chairside TRIOS Design Studio is simplified, with fewer tools and options, and is designed for a streamlined clinical experience. Similar to exocad, CAD software by 3Shape uses a guided workflow that is separated into multiple steps, each including different tools and functionalities.

5.4.3 CEREC Software and inLab CAD Software

Sirona started developing its first CEREC system in the early 1980s, and in 1985, the first CAD/CAM crown was seated. The crown was milled chairside from Vita Mark 1 (Vita Zahnfabrik), a feldspathic ceramic, using the CEREC 1 prototype [5]. This achievement makes Sirona the longest operating provider of commercial chairside CAD/CAM systems. In 2016, Dentsply International and Sirona Dental Systems merged into Dentsply Sirona. CEREC is one of two CAD/CAM companies that offer an end-to-end CAD/CAM workflow—acquisition, design, and manufacturing without using any third-party software or hardware. The other is Planmeca. The part that makes the CEREC stand out is their chairside milling machines and ceramic furnaces. Once again, all these units vary depending on their use chairside or in a laboratory. CEREC Chairside Software is a CAD program that can be purchased only with the CEREC intraoral scanners, with or without a milling machine. Its workflow is divided into five steps:

1. Administration—input of case information.
2. Acquisition—intraoral scanning.
3. Model—trim the scans, draw margin, set insertion axis, preparation analysis.
4. Design—design the restoration itself.
5. Manufacture—choose a material, nest it, and commence milling.

You can design any single-unit restoration (veneers, inlays, onlays, and crowns, full or partial coverage) or a short-span bridge, both monolithic or reduced, tooth- or implant-supported. Besides this fixed restorative work, you can also design short surgical guides for guided implantology. In most dental CAD programs, you are prompted to pick the most appropriate tooth morphology from the catalog, but CEREC software uses what the company calls "**biogeneric soft-**

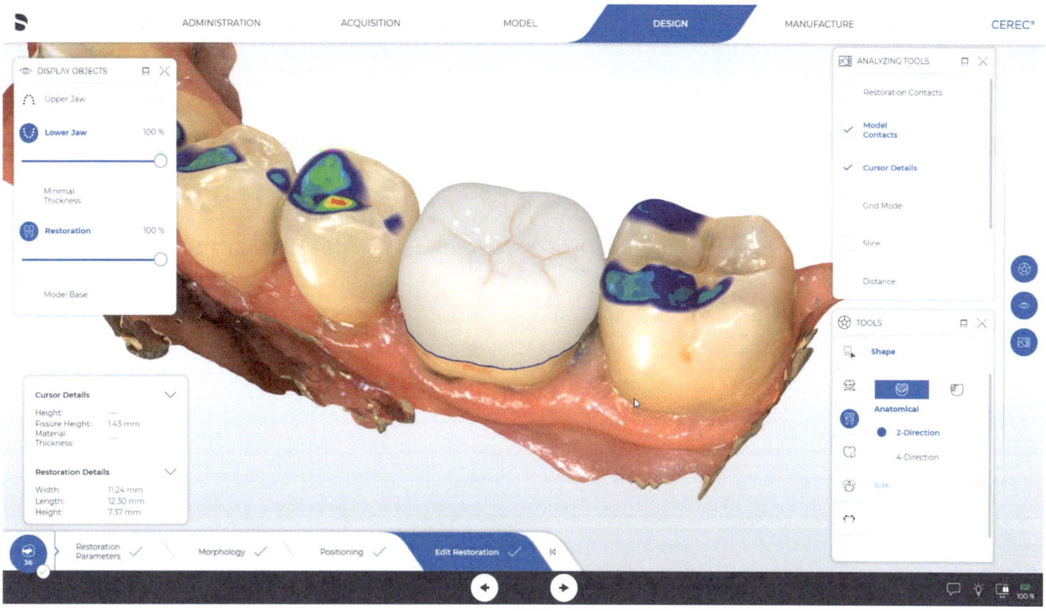

Fig. 5.11 Designing a crown in a design stage, CEREC Chairside software on CEREC Primescan. (Source: Dr. Ahmad Al-Hassiny, Wellington, New Zealand)

ware." At the beginning of the design phase, the software assesses the morphology of the adjacent teeth and generates an initial proposal based on this data [45]. Before customizing the tooth further, you can adjust the natural variation of this proposal using the "biogeneric variation" functionality. Alternatively, you can choose the morphology from a tooth catalog just as you would with any other dental CAD software (Fig. 5.11).

After designing, you proceed to nest the restoration within the same program and in the same workflow. The chairside milling machine (e.g., CEREC MC X, MC XL, or Primemill) processes single unit and bridge blocks exclusively, either polymethyl methacrylate (PMMA) or various ceramic materials for monolithic restorations (e.g., lithium disilicate and resin-based hybrid ceramics). Since 2016, the chairside CEREC system also started to mill zirconia and the release of the induction furnace, the SpeedFire, enabled same-day zirconia restorations. Although the chairside software enables the export of STL files, it is an otherwise closed system and does not allow importing any non-CEREC data.

The **laboratory version** of this software is called **CEREC inLab**. The laboratory version of CEREC comes with a desktop scanner (inEos X5), a milling machine (inLab MC X5, inLab MC XL), and an open system that can be integrated with any third-party hardware, unlike the somewhat closed chairside system. Laboratory CEREC inLab software offers more indications than the CEREC's chairside CAD Software, and the software is divided into more modules. With the basic module, you can design any partial or full-coverage crown, short- or long-span bridges, reduced or full-contour, and models for 3D printing. If you want to design implant-supported restorations and surgical guides, you need to purchase the implantology module. The removable module allows you to design partial and full removable dentures, frameworks, attachment prostheses, bars but also bite splints and custom impression trays. Given this broader indication list, the CEREC series laboratory milling machine (inLab MC[1] X5) also processes material disks, as

[1]MC is an abbreviation for "Milling center" although Dentsply Sirona refers to their milling machines as "milling and grinding" units nowadays.

opposed to chairside blocks exclusively like the chairside mills (MC X, MC XL, and Primemill).

5.4.4 Implant Planning Software

Implant surgical planning, surgical guide design and fabrication is carried out using specialized programs or add-on modules, such as exoplan by exocad, Implant Studio by 3Shape, or Blue Sky Plan by Blue Sky Bio. These programs allow you to align computed tomography (CT) or cone-beam computed tomography (CBCT) scans with surface scans (model, IOS, denture) and/or restoration mock-ups for surgical planning purposes (Fig. 5.12) [46]. The most important part is the DICOM data from your CT/CBCT. It allows you to properly examine the quality of the bone and space for surgery. For instance, in any mandibular surgery avoiding the inferior alveolar nerve is vital [47]. Most implant planning software provides you with tools to manually highlight the nerve to make sure you keep an adequate distance from it (Fig. 5.13). Additionally, CAD software enables you to digitally remove existing teeth, plan treatment, and design a temporary crown for immediate placement within the same program or workflow. The purpose of implant planning software is to place a digital implant in the ideal position for the patient to aid in improving surgical outcomes.

Planning software generally comes with an implant catalog from multiple manufacturers of dental implants and various types and sizes, so you can choose a size that fits you and your patient's needs. Visualization of the bone type in contact with the implant (Fig. 5.14) is also a commonly seen feature in implant software to help you assess the treatment better [46].

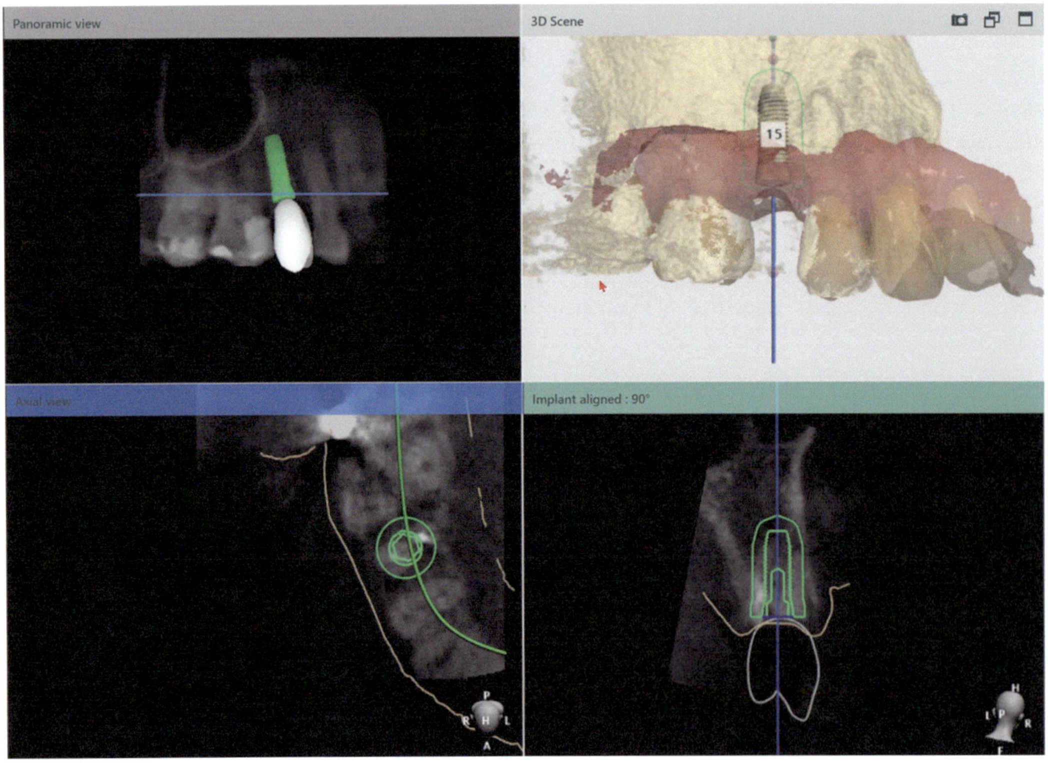

Fig. 5.12 DICOM data as visualized in Implant Studio by 3Shape: panoramic view with an implant and a mock-up crown (top left), CBCT, the implant and an intraoral surface scan (top right), axial view of DICOM data, and the implant (bottom left), alignment visualization of CBCT data, the mock-up, the implant, and the surface scan. (Source: Dr. Ahmad Al-Hassiny, Naenae Dental Clinic, Wellington, New Zealand)

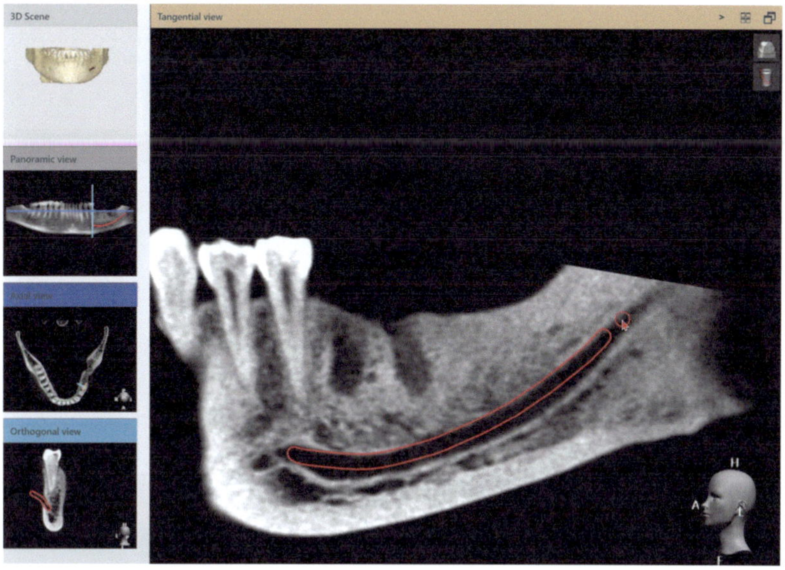

Fig. 5.13 Highlighting of inferior alveolar nerve as observed in "nerve definition" step of Implant Studio (3Shape) workflow. (Source: Dr. Ahmad Al-Hassiny, Naenae Dental Clinic, Wellington, New Zealand)

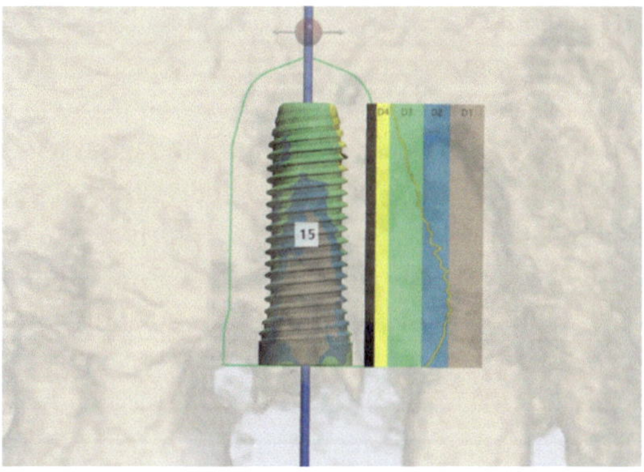

Fig. 5.14 A tool visualizing what type of bone the implant is in contact with (Implant Studio by 3Shape). (Source: Dr. Ahmad Al-Hassiny, Naenae Dental Clinic, Wellington, New Zealand)

These programs also enable you to design a surgical guide corresponding with the implant chosen. The design can be milled, or 3D printed, ideally from a transparent material that can autoclaved to minimize the risk of infection during the surgery. A metal ring purchased from the implant company is then typically inserted in the guide hole post-manufacturing to reinforce the guiding hole's strength and guide the osteotomy. There are also digital libraries that do not require the placement of such metal rings. The benefit of surgical planning using digital technology is the ability to determine the desired prosthetic outcome and surgical requirements before any surgical intervention has taken place [48]. The use of digital technologies has undoubtedly improved the precision and outcomes of surgical placement [47].

5.4.5 Virtual Articulator

With conventional impressions (or 3D printed models), models are mounted on an articulator, either non-adjustable, semi-adjustable, or fully adjustable. The articulator simulating the jaw movements helps to eliminate any occlusal interferences during excursive movements and ensures anterior guidance when necessary, making restorations as functional as possible [49]. Designing a single posterior crown in a centric occlusion is often sufficient. In contrast, more significant cases, especially those requiring anterior guidance, should always be designed using an articulator to protect the temporomandibular joint and ensure the longevity of the restorations and remaining dentition since imbalanced dynamic occlusion can have harmful consequences [49]. Dental CAD software often features virtual articulators as an alternative to mechanical articulators, replicating the design and functionality of various mechanical articular

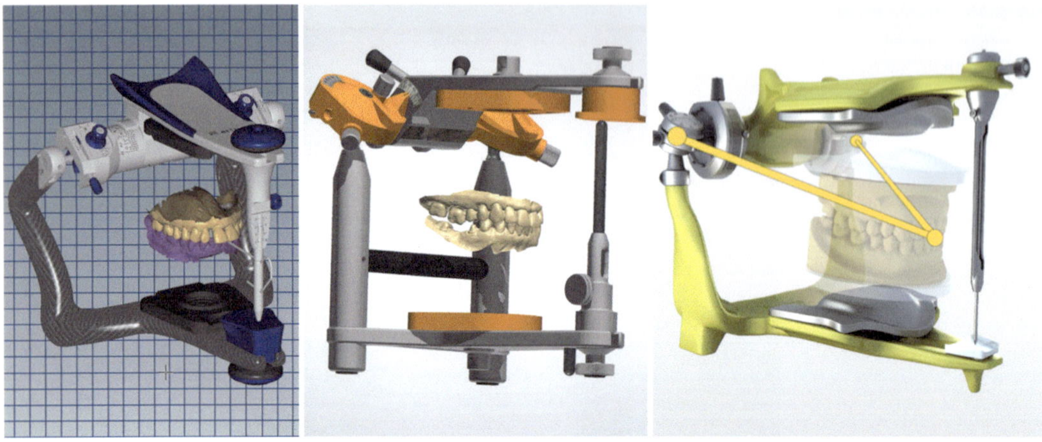

Fig. 5.15 Virtual articulators (left to right): DentalCAD by exocad, Dental system by 3Shape, CEREC Primescan by Dentsply Sirona

models (e.g., Artex CR by Amann Girrbach) [50]. While some software imitates the movements as performed by a mechanical articulator (maxillary scan performing the motion), other programs take advantage of the virtual reality and perform the movements naturally—with the mandible scan appearing in motion [32]. Virtual articulators tend to be fully adjustable, giving you a choice to customize the settings based on your experience/preference or data you collected from a patient. Some CAD software (exocad, CEREC software, inLab) allows importing digitally obtained jaw motion tracking data, usually in XML or CSV format. This can be used to enhance the virtual articulator further and make it as accurate as possible. The main application of a virtual articulator is individualized diagnostics and treatment planning [50] (Fig. 5.15).

5.4.6 Jaw Motion Tracking

Jaw motion tracking (JMT) is an umbrella term for various methods of capturing a patient's mandibular and temporomandibular joint movements. These movements can be recorded using tracking sensors attached to a plate or a framework mounted to the patient's teeth [51]. This tracking device is then connected to dedicated software that records the movements in real-time. These movements combined with DICOM data become a powerful diagnostic tool for multiple disciplines, such as prosthodontics, orthodontics, and periodontics. Additionally, TMJ specialists and maxillofacial surgeons may benefit from this information. The tracking and visualizing systems, such as SICAT Function (SICAT) or Romexis with Planmeca 4D Jaw Motion (Planmeca), additionally allow alignment with intraoral or extraoral surface scans. Systems such as ModJaw Tech in Motion (ModJaw) focus on jaw movement tracking without CBCT data using only intraoral or extraoral scans, as its main focus is restorative dentistry [51]. Programs that record JMT are rarely integrated with CAD software. If you intend to design a prosthetic case using patient's individual jaw movements, the most efficient way is to either export the data (XML, CSV formats) to your dental CAD software, given it is an open software and you have all the necessary modules. In case you cannot export/import the data, you can read the measurements (e.g., sagittal angle, Bennett angle) in the TMJ software and set up the parameters of the virtual articulator manually [52].

5.4.7 Digital Smile Design

Dr. Christian Coachman developed the concept of digital smile design (DSD) in the 2000s [53] and, ever since then, has revolutionized dental treatment planning, especially for cosmetic cases

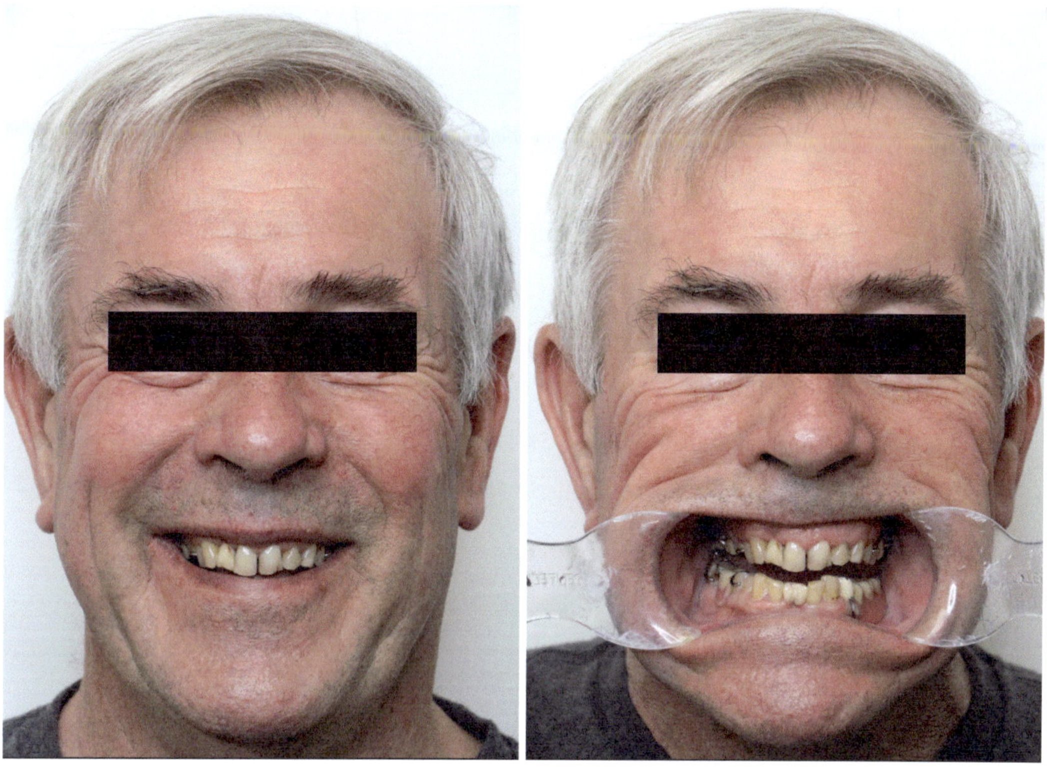

Fig. 5.16 Photographs necessary for digital smile design: smiling (left) and retracted (right). (Source: Dr. Ahmad Al-Hassiny, Naenae Dental Clinic, Wellington, New Zealand)

[54]. DSD is a tool that is used to design the smile of patients digitally. This digital mock-up is presented to the patient before treatment physically starts. The DSD concept helps dental professionals improve communication between clinician and patient, allowing the patient to understand the possible solutions, educating and motivating them about the benefits of the treatment, and increasing the case acceptance [55]. Since then, other companies have developed software using similar methodology, e.g., Romexis by Planmeca, Smile Designer Pro, and CEREC Smile Design [56].

The most basic DSD is strictly 2D, and it uses at least two pictures of the patient: two frontal shots, one with a wide smile and one retracted (Fig. 5.16) [53]. In the software, these two pictures are overlaid over each other, the inner lip line is established, so the graphic simulation stays within the natural smile line. Other pictures such as the noon, smiling profile shot, and occlusal pictures are optional and are not used strictly for motivational purposes but rather for treatment planning [57]. Next, the software is used to establish several points or lines of the patient's facial features (Fig. 5.17). The plane of occlusion is either set up by straightening the picture or marking the middle of the patient's pupils and therefore establishing the bipupilar line. The dentition midline is usually aligned with columella or philtrum and may be deviated from the facial midline. Another guideline is the nasal wings or the inner corners of the eyes (medial canthus)—these points often vertically align with canines [54].

With this information established, the software generates a proposal, which generally starts as an outline that respects the golden ratio. The golden ratio or golden proportion states that if the lateral incisors are given a value of 1, central incisors should be 1.6, and canines 0.8—this is how much of each tooth's proportion should be observable from a direct frontal point of view. Additionally, the perfect width of anterior teeth

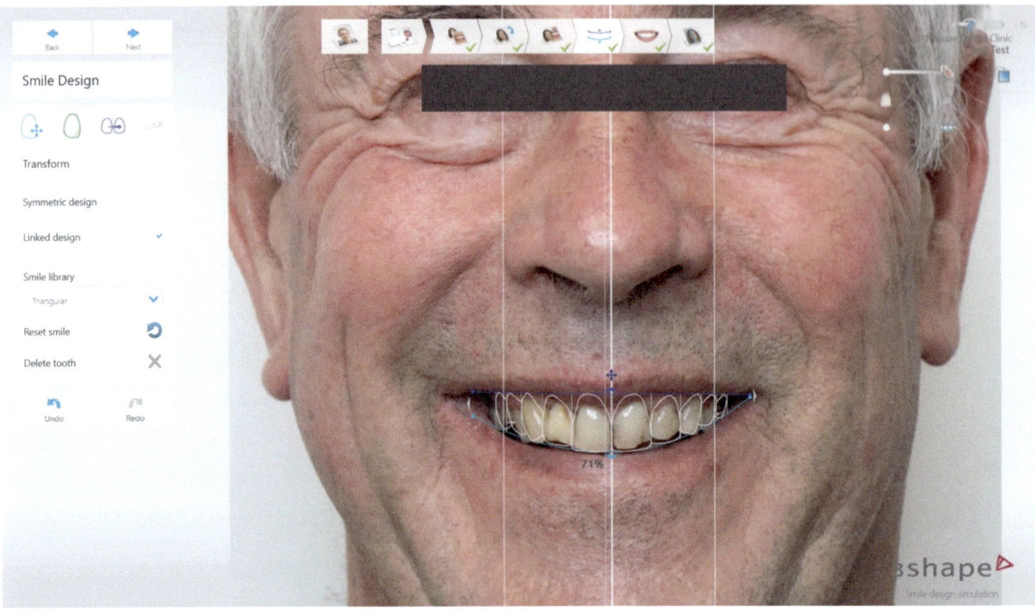

Fig. 5.17 An initial proposal of a digital smile design using facial features to establish guidelines (Smile Design by 3Shape). (Source: Dr. Ahmad Al-Hassiny, Naenae Dental Clinic, Wellington, New Zealand)

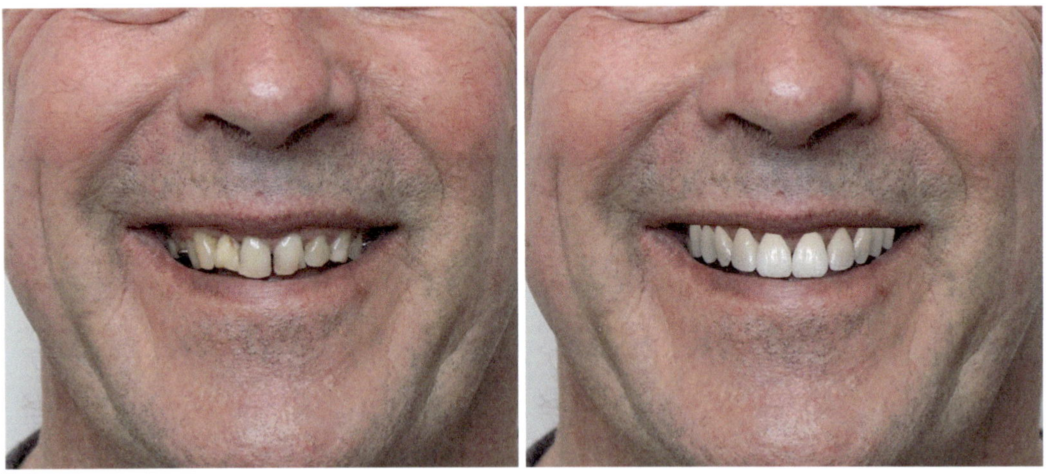

Fig. 5.18 Patient's pre-op smile (left) and digital smile design overlaid (right) over the same photograph. (Source: Dr. Ahmad Al-Hassiny, Naenae Dental Clinic, Wellington, New Zealand)

should be about 70% of their height [53, 54]. A golden ratio is a theoretical tool that will help you establish the perfect smile proposal, but you still need to respect the patient's dentition and make the required adjustments.

Digital smile design is a relatively inexpensive and efficient way of showing the patient what can be achieved (Fig. 5.18). It is also a lot less skill-demanding than traditional wax-ups. DSD opens a discussion about the patient's expectations which are later forwarded to the laboratory. It greatly improves case acceptance as most people struggle to imagine themselves with a different smile [54].

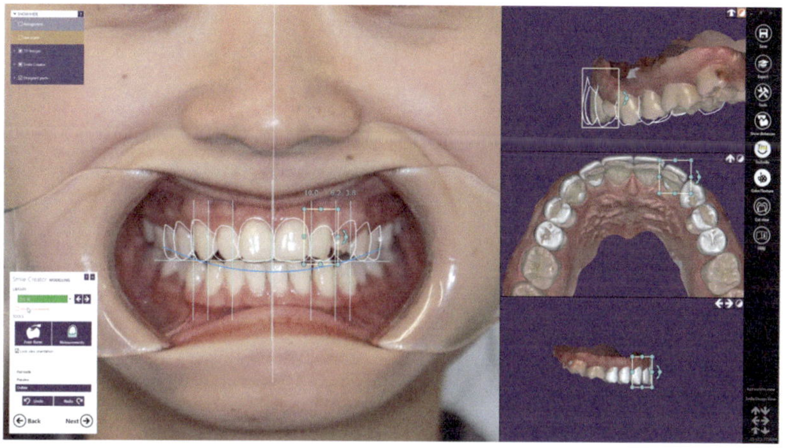

Fig. 5.19 Digital smile design using Smile Creator (exocad). (Source: Dr. Ahmad Al-Hassiny, Naenae Dental Clinic, Wellington, New Zealand)

Digital smile design is typically carried out using photographs, but some software (e.g., DSD by Christian Coachman, Smile Creator by exocad, or SmileFy by Smilefy Inc) integrates the 2D smile design with 3D CAD software. The 2D proposal generates a 3D mock-up that can be printed and transferred into the patient's mouth with a putty stent and temporary material (Fig. 5.19).

5.4.8 Digital Orthodontics

Today digital technology has impacted every aspect of orthodontic treatment. CAD/CAM has applications in **orthodontic treatment**, especially in clear aligners, one of the fastest growing fields within the dental industry [58]. Nowadays, there are many different manufacturers of clear aligners that clinicians have access to, e.g., ClearCorrect (Straumann Group), SureSmile (Dentsply), and most notably Invisalign through their planning software ClinCheck (Align Technology). These services require sending multiple orthodontic pictures of the patient's face, dentition, and scans or models of their teeth and bite. Based on the prescription and directions you provide, a proposal of a treatment plan is provided which you can adjust to meet the patient's expectations and needs. Since 3D printing has become much more affordable, planning orthodontic treatment with clear aligners and manufacturing these aligners in-house has become more popular [59]. Alternatively, instead of using one of the aligner services mentioned above, you can instead purchase treatment planning software (e.g., 3Shape's Clear Aligner Studio, eXceed by Exceed Services, Maestro Ortho Studio Software) and plan treatment yourself and design the clear aligners for in-house production [59, 60]. Planning and fabrication of in-house aligners have been shown in the Fig. 5.20. CAD/CAM can also be useful for conventional orthodontics, thanks to **indirect bonding**. This allows the clinician to bond all the brackets simultaneously in a more reliable, accurate, and faster way that is also more comfortable for the patient. Traditionally, a laboratory process, but with the help of specialized software (e.g., Indirect Bonding Studio by 3Shape, Ortho Studio by Maestro3D) you can plan the placement of the brackets and manufacture the transfer tray in-house using a 3D printer [61]. Additionally, some software has been developed to help with the conventional orthodontic treatments in general, e.g., SureSmile by OraMetrix and OrthoCAD iQ by Align Technology. By using this, the clinician can convert intraoral scans into study models, archive them, and help with patient monitoring. You can also upload photographs, OPG, and CBCT data to assess a patient's situation better and plan the treatment more predictably.

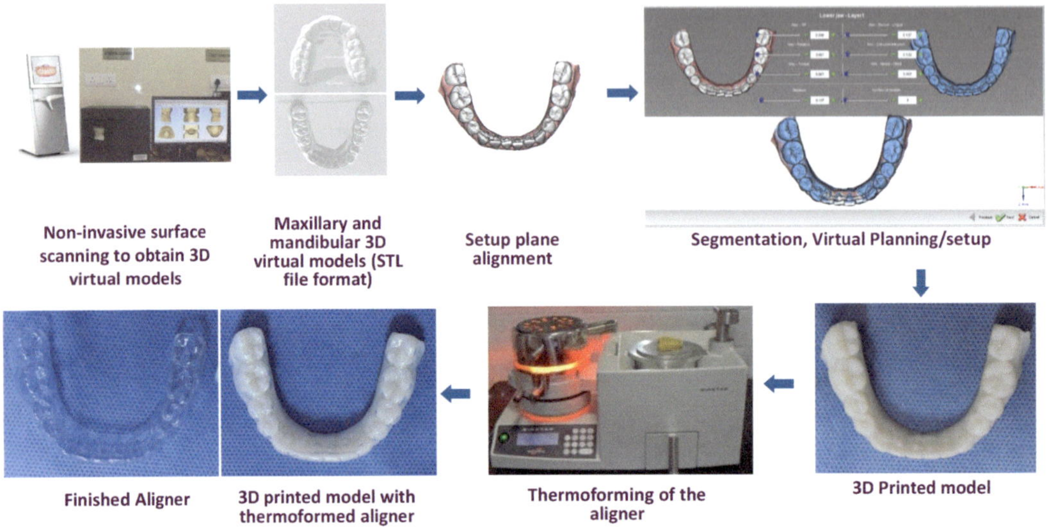

Fig. 5.20 Planning and fabrication of in-house aligners using the Maestro Orthostudio Software. (Source: Dr. Prabhat Kumar Chaudhari, Division of Orthodontics and Dentofacial Deformities, Centre for Dental Education and Research, All India Institute of Medical Sciences, New Delhi, India)

5.4.9 Digital Denture Design

As explained in the previous section, intraoral scanning faces certain limitations when used to fabricate dentures. However, there are constant improvements to this workflow. The complete denture CAD/CAM workflow is a very relevant concept, and many companies are investing heavily into the development of software, hardware, and materials. All laboratory CAD programs offer a complete denture designing module, usually in their laboratory-based varieties. There are now multiple ways to design digital dentures:

- Design both the base and teeth and fabricate them separately (milling or 3D printing or a combination of both).
- Design and fabricate a base only and use pre-made acrylic teeth.
- Design a base and teeth as a one-piece design and mill from a special, multicolored disk.

CAD programs enable you to design a denture as separate elements (base and teeth) or as a one-piece, depending on what kind of approach you choose to fabricate the prosthetic and what software you are using. The most significant advantages of **manufacturing both elements separately** come from the high level of customization. You can design unique teeth and even copy a patient's previous denture. The most significant disadvantage of 3D printed teeth is the relatively poorer aesthetics compared to conventional acrylics [62]. The denture teeth are manufactured from one, relatively opaque material that does not perfectly mimic natural tooth structures and their various levels of translucency. PolyJet printers allow printing from multiple materials at the same time, but this technology has not been widely popular in dentistry as it is relatively slow and prone to technical issues. Milling teeth from multilayer PMMA disks offers a more natural look, although milling tends to take longer than printing and involves a more significant investment in the appropriate milling machine. Both milled or 3D printed materials can be manually **polished**. In theory, you can do this with a handpiece and polishing kit, similar to polishing a filling, for example, but a bigger, desktop polisher is typically used. The denture can be further customized with light-cured **stain and glaze**, e.g., OPTIGLAZE Color by GC. Staining allows you to add characterization and imitate incisal translucency, etc. Glazing (without staining) is generally much faster to do than manual polishing, but it tends to wear off over time or even chip, which eventually leads to staining in the area [63]. Glazing is generally only recommended for temporary dentures, which will eventually be replaced.

If you choose to design only the base and use stock, **premade acrylic teeth,** you do not need much polishing or glazing [64]. In your CAD software's tooth catalog, you pick the teeth you want to use and place them within the base as you wish, independent of each other. You cannot change the teeth' shape in any way as these digital teeth are just a placeholder to create room in the denture base for the final teeth to be put in these spots after the fully customized denture base is fabricated. It is essential to make sure that the type of teeth you order is the same as those planned in the CAD software as otherwise they might not fit (Fig. 5.21).

The third option is to design the **base and teeth as one piece** [65], eliminating the issue of teeth debonding from the denture base. This allows you to fully customize the design, both base and the teeth, and either mill or print the design. Commonly used 3D printers (SLA and DLP) use only one material at a time, so when printing a one-piece denture, it is preferable to do so from a material that matches the desired tooth shade. The same applies if you choose to mill the teeth from a PMMA disk (choose dentine shade, not gingival shade). The base would, of course, look very unnatural if left with a dentine shade. While the teeth are designed full-contour, the base's buccal aspect can be reduced to make room for the light-cured composite in a more natural, gingiva-like color. Some disks are **multilayered** and imitate incisal translucency and the pink base (e.g., Full Denture PMMA Disks by Polident) [65]. The CAM manager indicates where the pink layer ends, and the dentine layer starts to set up the denture accordingly—you aim to place all teeth in the tooth-colored half. In 2020, a new kind of multilayered disk was introduced by Ivoclar Vivadent called Ivotion. This material system comes with PMMA disks with a unique "**shell geometry**" that divides the base and dentine colors into two precise layers that imitate the gingival scalloping [66]. In the CAD program (Dental System by 3Shape exclusively), you can customize both base and teeth, the mor-

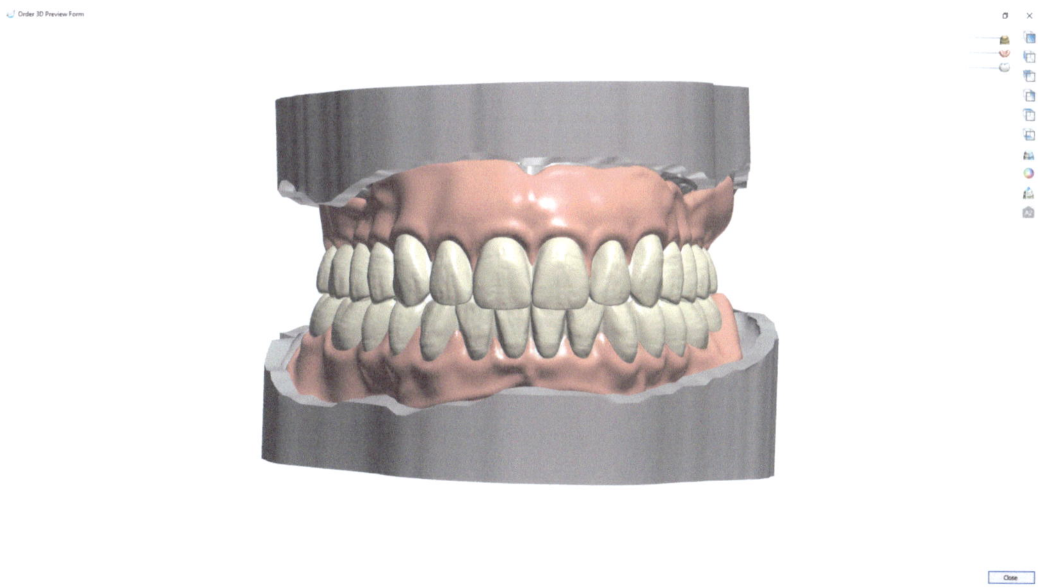

Fig. 5.21 Designing a denture in Dental System by 3Shape. In this case, the base (pink) is fully customized and ready to be 3D printed, while the digital teeth are a placeholder to make room for the premade acrylic teeth that will be bonded to the base. (Source: Institute of Digital Dentistry Laboratory, Wellington, New Zealand)

phology of the teeth, and the width and curvature of the arch. The CAD program limits the position of the teeth to some extent to ensure precise scalloping between the denture base and teeth colors is not breached [62, 64]. Due to this internal geometry of the Ivotion disk, you do not have an option of adjusting the denture's position within the disk in the CAM manager. One disk can only fit one complete denture, and it still requires post-processing (often polishing). At the time of this publication, Ivotion is only available for milling using PrograMill System or Zenotec Select milling machines. This arrangement can change or expand or for a similar system to be developed by Ivoclar's competitors in the near future.

5.5 Computer-Aided Manufacturing for Dental Applications

Manufacturing, or the CAM part of the workflow, is facilitated using either a milling machine or a 3D printer. They both use vastly different technologies with varying accuracy, detail, and speed. Below, we will cover chairside and laboratory CAM systems.

5.5.1 Chairside vs. Laboratory CAM

As we have established previously, most CAD/CAM systems components differ depending on whether they are meant for chairside use or in a dental laboratory. This is again true in the case of milling machines and 3D printers. Milling machines, in particular, can vary wildly in their size or capabilities. The most prominent example of a chairside milling system would be CEREC MC XL by Dentsply Sirona. Still, there are many other chairside milling units available, such as DWX-42W by Roland or PlanMill 40 S by Planmeca. Chairside milling machines are expected to be utilized chiefly for same-day restorative dentistry, which usually means a single monolithic restoration fabricated from ceramic materials or PMMA. Chairside milling machines process blocks that fit either a single restoration or larger blocks for short-span bridges, while laboratory milling machines process mostly large disks that can fit about 20 units [67] (Fig. 5.22).

Chairside milling machines also seem to favor efficiency over detail [68]. Milling can only produce concave details as fine as the smallest bur, and since the smallest bur in chairside milling machines is often about 1 mm in

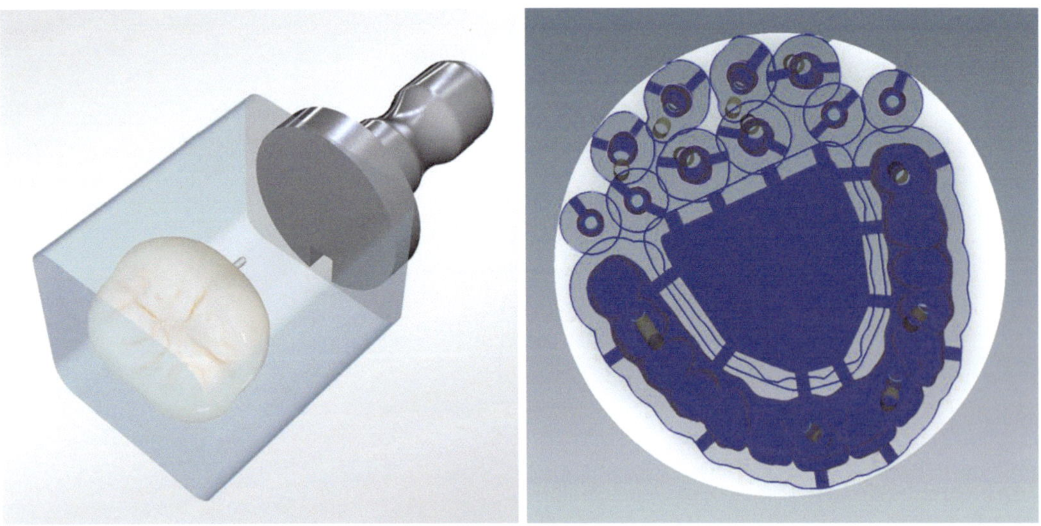

Fig. 5.22 Nesting into a block using a chairside CAM (CEREC Primescan by Dentsply Sirona, right), nesting into a disk using a laboratory CAM (exoCAM by exocad, integrated with an Amann Girrbach CAD/CAM system, left). (Source: Naenae Dental Clinic, Wellington, New Zealand)

diameter, it simply cannot compare with the level of detail achieved with a 0.3 mm bur as commonly used in laboratory milling machines [67]. The CAD software anticipates these limits and designs the intaglio surface accordingly. The narrow spots of intaglio are purposefully designed larger to fit the smallest bur and mill the inside of the crown in a way that makes the crown possible to seat down. Milling machines have some limitations during production. Although CAD software allows you to design shapes that are difficult to mill, the areas that a milling machine cannot produce will stay filled in with the material and might require manual enhancement post-milling using a handpiece and an appropriate set of burs. In the case of single-unit crown restorations, the most commonly undermilled areas are the fissures. Alternatively, in the intaglio surfaces, **overmilling** may occur around any sharp edges of the preparation to enable seating of the restoration. In regards to milling larger designs such as bridges, all-on-x prostheses, or complete dentures, you will often find issues achieving the same level of detail as what is designed in the CAD software. Typical areas that may cause problems are the gumline, interproximal areas in bridges, screw channels in implant restorations, and Ti-base cavities. In the case of these larger designs, most of the undermilled areas are a result of an insufficient bur size and the machine's physical limitations, as the spindle might be unable to reach these areas (Fig. 5.23).

Milling machines are capable of milling in three, four, or five axes. Three-axis milling operates on *X*-, *Y*-, and *Z*-axes, and it has been the standard across the dental industry for decades [67]. Recently, it has been shown to be insufficient when it comes to manufacturing narrow and deep spaces, and for this reason, up to 5-axis milling machines have been developed [69]. In addition to the *X*-, *Y*-, and *Z*-axes, these machines operate in *A* and *B* axes around which the tool or the holder rotates, effectively making hard-to-mill areas accessible from two more sides. Most laboratory CAM software comes with toolpath simulation, which can reveal the areas that cannot be milled. Some parts such as horizontal concave

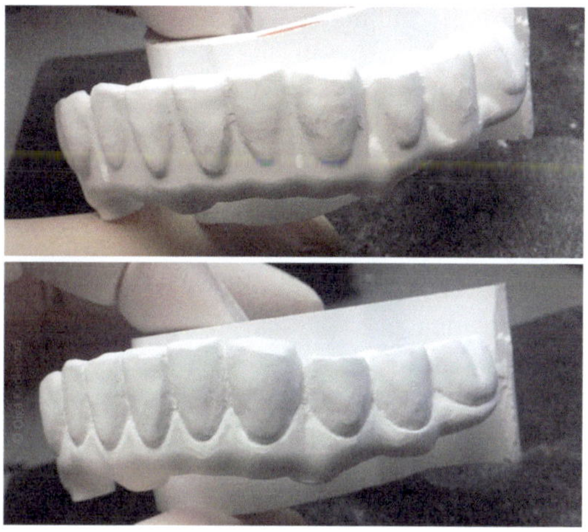

Fig. 5.23 An example of unmilled areas around the gingival margin and interproximal (top) and these areas are manually refined by a dental technician before sintering (bottom picture). (Source: Olga Kadlecová, dental CAD/CAM technician)

spots (gumline, perikymata) can be made accessible for the milling tools by tilting the design within the disk accordingly. When dealing with larger designs, like an all-on-x restoration, tilting the object in the nesting phase is not always possible due to limited space or the machine's limitations. If your machine is capable of 5-axis milling, try to simulate with that strategy instead of the 3-axis one. 3D printers do not face these issues, as additive manufacturing creates the object layer by layer, without tools reaching into cavities [70].

5.5.2 Dry and Wet Milling

Milling processes can be either **dry or wet**, which involves water with coolant. The choice of the milling system depends on the material; e.g., zirconia is preferably milled dry, while lithium disilicate materials require cooling to prevent overheating the ceramic. Some materials can be milled both wet or dry. The choice of milling strategy for materials like that seems to depend on a manufacturer's preference, a machine's abilities, and preset milling strategies. For example, PMMA can be milled both, wet or dry, depending on the setup: If the cooling is sufficient and the speed of the milling tools is low enough, the PMMA should not melt due to the temperature increase during milling, but if you want faster milling, then you might need to choose a wet milling system. When switching between wet and dry milling, you should prevent cross-contamination, especially if you are switching to zirconia, which can be prone to contamination due to residual glass particles from other materials. This involves making sure the milling chamber is clean of any residual particles and having separate "wet" and "dry" water tanks.

5.5.3 Material Processing

After fabrication all materials need some form of post-processing. This may involve polishing, characterisation with stains and glazes and/or processing in a ceramic furnace. Many **materials** can be milled, and 3D printed, but 3D printing dental ceramics (e.g., zirconia or lithium disilicate) has proven very difficult and is currently in a developmental/experimental stage [71]. If you want to manufacture restorations from zirconia or the majority of dental ceramics available, you must opt for a milling machine. The most popular 3D printers in dentistry (stereolithography and digital light processing) work best with photosensitive resins. These materials come in a wide range of colors, strengths, opacities, and translucencies. Some materials can even be sterilized in an autoclave, making them fit for use in oral surgery as surgical guides and prosthodontics for temporary crowns [72]. Metal alloys can also be processed using multiple CAM systems, both by milling and various 3D printers, although the most reliable processing is milling. Milled **metal** frameworks work well as they do not go through any change of state (from liquid to solid), but manufacturing them in this way is straining on the milling equipment and industrial hardware is required, which is outside the scope of most, if not all, dental practices. If you do not intend to mill metal frameworks on a daily basis, consider having them sent to a milling center rather than investing in the expensive machinery. During 3D printing, the metal liquid material is cured and set with a light source, effectively hardening it. The change of material's state during manufacturing is when the object is most prone to warping [73]. It is at this point before the material is fully hardened, war page can occur either due to shrinkage or due to soft material's collapse under its weight. The collapse can be prevented by adding enough supports in vulnerable spots (e.g., wide, flat areas). In dentistry, 3D printing already has a diverse range of applications, and we will only see this grow with time. 3D printing has the potential to enable many new and exciting treatments and approaches to manufacturing dental restorations.

5.6 Summary

Digital dentistry and CAD/CAM technologies are significantly impacting all aspects of dentistry. The history of computer-aided design and manufacturing in dentistry started decades ago, and even though the initial concept was conceived in the 1970s, it has fundamentally remained unchanged. These days, CAD/CAM is increasing in market penetration and is becoming more widely popular in the dental office and in laboratories. Compared to the adoption of innovative technology in any other industry, digital and especially chairside dentistry took a relatively long time to gain the profession's trust. Perhaps it was limited by the technology of its time and only the recent technological breakthroughs that allowed faster and broader market penetration. Nowadays, CAD/CAM has established itself as a powerful tool for dental technicians and dentists of any specialty alike. A modern chairside system can help keep the restorative work in-house or make communication with a laboratory more efficient. Dental surgeons, implant, and maxillofacial specialists can use CAD/CAM to make surgical outcomes more predictable by implementing digital treatment planning and guided surgery into their workflow. CAD/CAM manufacturing has improved all aspects of restorative dentistry making it faster, cleaner, more efficient, and predictable than conventional. There are only a few disciplines where current digital methods have not entirely replaced traditional methods, such as scanning edentulous jaws to manufacture complete dentures.

Competition in the market is ever-growing, driving innovation forward and prices down, making it easy for anyone to find a CAD/CAM system to fit their needs and budget. Switching from conventional techniques does not need to be abrupt or in full right away, but instead can be a gradual transition. Either way, it is crucial to realize that CAD/CAM and digital dentistry is not the future. It is happening right now and it is the inevitable evolution of our procession. Therefore, it is vital to embrace it and implement the digital workflow in daily practice.

Acknowledgments The author wishes to thank Dr. Hamid Al-Hassiny, and Dr. Khilood Ali, for their endless support and mentorship. Also, thanks to Olga Kadlecova for her involvement in this chapter and help with all things related to the Institute of Digital Dentistry.

References

1. Baroudi K, Ibraheem SN. Assessment of chair-side computer-aided design and computer-aided manufacturing restorations: a review of the literature. J Int Oral Health. 2015;7(4):96–104.
2. Richert R, Goujat A, Venet L, Viguie G, Viennot S, Robinson P, et al. Intraoral scanner technologies: a review to make a successful impression. J Healthc Eng. 2017;2017:8427595.
3. Beuer F, Schweiger J, Edelhoff D. Digital dentistry: an overview of recent developments for CAD/CAM generated restorations. Br Dent J. 2008;204(9):505–11.
4. Duret F, Blouin JL, Duret B. CAD-CAM in dentistry. J Am Dent Assoc. 1988;117(6):715–20.
5. Mormann WH. The evolution of the CEREC system. J Am Dent Assoc. 2006;137(Suppl):7S–13S.
6. Davidowitz G, Kotick PG. The use of CAD/CAM in dentistry. Dent Clin N Am. 2011;55(3):559–70, ix.
7. Zarone F, Ferrari M, Mangano FG, Leone R, Sorrentino R. "Digitally oriented materials": focus on lithium disilicate ceramics. Int J Dent. 2016;2016:9840594.
8. Dawood A, Marti Marti B, Sauret-Jackson V, Darwood A. 3D printing in dentistry. Br Dent J. 2015;219(11):521–9.
9. Erozan C, Ozan O. Evaluation of the precision of different intraoral scanner-computer aided design (CAD) software combinations in digital dentistry. Med Sci Monit. 2020;26:e918529.
10. Alghazzawi TF. Advancements in CAD/CAM technology: options for practical implementation. J Prosthodont Res. 2016;60(2):72–84.
11. Suese K. Progress in digital dentistry: the practical use of intraoral scanners. Dent Mater J. 2020;39(1):52–6.
12. Jahangiri L, Akiva G, Lakhia S, Turkyilmaz I. Understanding the complexities of digital dentistry integration in high-volume dental institutions. Br Dent J. 2020;229(3):166–8.
13. Ahn JJ, Bae EB, Lee WS, Cho WT, Huh JB. Comparison of clinical fit of three-unit zirconia fixed prostheses fabricated using chairside and labside CAD/CAM systems. J Dent Sci. 2020;15(2):239–42.
14. Kollmuss M, Kist S, Goeke JE, Hickel R, Huth KC. Comparison of chairside and laboratory CAD/CAM to conventional produced all-ceramic crowns regarding morphology, occlusion, and aesthetics. Clin Oral Investig. 2016;20(4):791–7.
15. Ender A, Mehl A. In-vitro evaluation of the accuracy of conventional and digital methods of obtaining full-arch dental impressions. Quintessence Int. 2015;46(1):9–17.
16. Alikhasi M, Siadat H, Nasirpour A, Hasanzade M. Three-dimensional accuracy of digital impression versus conventional method: effect of implant angulation and connection type. Int J Dent. 2018;2018:3761750.
17. Sadid-Zadeh R, Katsavochristou A, Squires T, Simon M. Accuracy of marginal fit and axial wall contour for lithium disilicate crowns fabricated using three digital workflows. J Prosthet Dent. 2020;123(1):121–7.
18. Khaledi AA, Farzin M, Akhlaghian M, Pardis S, Mir N. Evaluation of the marginal fit of metal copings fabricated by using 3 different CAD-CAM techniques: milling, stereolithography, and 3D wax printer. J Prosthet Dent. 2020;124(1):81–6.
19. Janeva NM, Kovacevska G, Elencevski S, Panchevska S, Mijoska A, Lazarevska B. Advantages of CAD/CAM versus conventional complete dentures - a review. Open Access Maced J Med Sci. 2018;6(8):1498–502.
20. Di Fiore A, Meneghello R, Graiff L, Savio G, Vigolo P, Monaco C, et al. Full arch digital scanning systems performances for implant-supported fixed dental prostheses: a comparative study of 8 intraoral scanners. J Prosthodont Res. 2019;63(4):396–403.
21. Mangano FG, Admakin O, Bonacina M, Lerner H, Rutkunas V, Mangano C. Trueness of 12 intraoral scanners in the full-arch implant impression: a comparative in vitro study. BMC Oral Health. 2020;20(1):263.
22. Christopoulou I, Kappaaklamanos EG, Makrygiannakis MA, Bitsanis I, Tsolakis AI. Patient-reported experiences and preferences with intraoral scanners: a systematic review. Eur J Orthod. 2022;44(1):56–65.
23. Lee SJ, Kim SW, Lee JJ, Cheong CW. Comparison of intraoral and extraoral digital scanners: evaluation of surface topography and precision. Dent J (Basel). 2020;8(2):52.
24. Mandelli F, Gherlone E, Gastaldi G, Ferrari M. Evaluation of the accuracy of extraoral laboratory scanners with a single-tooth abutment model: a 3D analysis. J Prosthodont Res. 2017;61(4):363–70.
25. Víctor Henarejos-Domingo SNM, Roig M, Blasi Á. The digital alveolar cast: a revised approach to an old concept. J Prosthet Dent. 2022;127(2):213–8.
26. Glisic O, Hoejbjerre L, Sonnesen L. A comparison of patient experience, chair-side time, accuracy of dental

arch measurements and costs of acquisition of dental models. Angle Orthod. 2019;89(6):868–75.
27. Papi P, Di Murro B, Penna D, Pompa G. Digital prosthetic workflow during COVID-19 pandemic to limit infection risk in dental practice. Oral Dis. 2021;27(Suppl 3):723–6.
28. Mascitti M, Campisi G. Dental public health landscape: challenges, technological innovation and opportunities in the 21st century and COVID-19 pandemic. Int J Environ Res Public Health. 2020;17(10):3636.
29. Kathree BA, Khan SB, Ahmed R, Maart R, Layloo N, Asia-Michaels W. COVID-19 and its impact in the dental setting: a scoping review. PLoS One. 2020;15(12):e0244352.
30. Mangano F, Gandolfi A, Luongo G, Logozzo S. Intraoral scanners in dentistry: a review of the current literature. BMC Oral Health. 2017;17(1):149.
31. Zimmermann M, Mehl A, Mormann WH, Reich S. Intraoral scanning systems - a current overview. Int J Comput Dent. 2015;18(2):101–29.
32. Lepidi L, Galli M, Mastrangelo F, Venezia P, Joda T, Wang HL, et al. Virtual articulators and virtual mounting procedures: where do we stand? J Prosthodont. 2021;30(1):24–35.
33. Park JS, Lim YJ, Kim B, Kim MJ, Kwon HB. Clinical evaluation of time efficiency and fit accuracy of lithium disilicate single crowns between conventional and digital impression. Materials (Basel). 2020;13(23):5467.
34. Kihara H, Hatakeyama W, Komine F, Takafuji K, Takahashi T, Yokota J, et al. Accuracy and practicality of intraoral scanner in dentistry: a literature review. J Prosthodont Res. 2020;64(2):109–13.
35. Fang JH, An X, Jeong SM, Choi BH. Digital intraoral scanning technique for edentulous jaws. J Prosthet Dent. 2018;119(5):733–5.
36. Lee JH. Improved digital impressions of edentulous areas. J Prosthet Dent. 2017;117(3):448–9.
37. Lo Russo L, Caradonna G, Troiano G, Salamini A, Guida L, Ciavarella D. Three-dimensional differences between intraoral scans and conventional impressions of edentulous jaws: a clinical study. J Prosthet Dent. 2020;123(2):264–8.
38. Lo Russo L, Troiano G, Salamini A, Guida L. Intraoral scans alignment in single edentulous arch cases. J Prosthodont. 2020;29(9):826–8.
39. Fang Y, Fang JH, Jeong SM, Choi BH. A technique for digital impression and bite registration for a single edentulous arch. J Prosthodont. 2019;28(2):e519–e23.
40. Lo Russo L, Ciavarella D, Salamini A, Guida L. Alignment of intraoral scans and registration of maxillo-mandibular relationships for the edentulous maxillary arch. J Prosthet Dent. 2019;121(5):737–40.
41. Goodacre BJ, Goodacre CJ, Baba NZ. Using intraoral scanning to capture complete denture impressions, tooth positions, and centric relation records. Int J Prosthodont. 2018;31(4):377–81.
42. Schmidt MB, Rauch A, Schwarzer M, Lethaus B, Hahnel S. Combination of digital and conventional workflows in the CAD/CAM-fabrication of an implant-supported overdenture. Materials (Basel). 2020;13(17):3688.
43. Al Mortadi N, Eggbeer D, Lewis J, Williams RJ. CAD/CAM/AM applications in the manufacture of dental appliances. Am J Orthod Dentofac Orthop. 2012;142(5):727–33.
44. Buzayan MM, Seong L, Elkezza A, Abidin Z, Yunus N, Sivakumar I. Digital workflow for articulating maxillary and mandibular 3D arch models using an open source 3D modeling software program. Quintessence Int. 2020;51(9):776–9.
45. Arslan Y, Karakoca Nemli S, Bankoglu Gungor M, Tamam E, Yilmaz H. Evaluation of biogeneric design techniques with CEREC CAD/CAM system. J Adv Prosthodont. 2015;7(6):431–6.
46. Mora MA, Chenin DL, Arce RM. Software tools and surgical guides in dental-implant-guided surgery. Dent Clin N Am. 2014;58(3):597–626.
47. Patel N. Integrating three-dimensional digital technologies for comprehensive implant dentistry. J Am Dent Assoc. 2010;141(Suppl 2):20S–4S.
48. Rosenfeld AL, Mandelaris GA, Tardieu PB. Prosthetically directed implant placement using computer software to ensure precise placement and predictable prosthetic outcomes. Part 1: Diagnostics, imaging, and collaborative accountability. Int J Periodontics Restorative Dent. 2006;26(3):215–21.
49. Abduo J. Safety of increasing vertical dimension of occlusion: a systematic review. Quintessence Int. 2012;43(5):369–80.
50. Koralakunte PR, Aljanakh M. The role of virtual articulator in prosthetic and restorative dentistry. J Clin Diagn Res. 2014;8(7):ZE25–8.
51. Solaberrieta E, Barrenetxea L, Minguez R, Iturrate M, De Prado I. Registration of mandibular movement for dental diagnosis, planning and treatment. Int J Interact Des Manuf. 2018;12(3):1027–38.
52. Park JH, Kim JE, Shim JS. Digital workflow for a dental prosthesis that considers lateral mandibular relation. J Prosthet Dent. 2017;117(3):340–4.
53. Coachman C, Calamita M. Digital smile design: a tool for treatment planning and communication in esthetic dentistry. QDT 2012 Quintessence. 2012;35:1–9.
54. Jafri Z, Ahmad N, Sawai M, Sultan N, Bhardwaj A. Digital smile design-an innovative tool in aesthetic dentistry. J Oral Biol Craniofac Res. 2020;10(2):194–8.
55. Omar D, Duarte C. The application of parameters for comprehensive smile esthetics by digital smile design programs: a review of literature. Saudi Dent J. 2018;30(1):7–12.
56. Kurbad A, Kurbad S. Cerec Smile Design—a software tool for the enhancement of restorations in the esthetic zone. Int J Comput Dent. 2013;16(3):255–69.
57. Agnini A, Apponi R, Maffei S, Agnini A. Digital dental workflow for a smile makeover restoration. Int J Esthet Dent. 2020;15(4):374–89.

58. Tamer I, Oztas E, Marsan G. Orthodontic treatment with clear aligners and the scientific reality behind their marketing: a literature review. Turk J Orthod. 2019;32(4):241–6.
59. Tarraf NE, Ali DM. Present and the future of digital orthodontics☆. Semin Orthod. 2018;24(4):376–85.
60. Fudalej P, Bollen AM, Hujoel IA. Relapse of mandibular incisor alignment is not associated with the total posttreatment mandibular rotation. Am J Orthod Dentofac Orthop. 2010;138(4):392.e1–7.
61. Kim J, Chun Y-S, Kim M. Accuracy of bracket positions with a CAD/CAM indirect bonding system in posterior teeth with different cusp heights. Am J Orthod Dentofac Orthop. 2018;153(2):298–307.
62. Bidra AS, Taylor TD, Agar JR. Computer-aided technology for fabricating complete dentures: systematic review of historical background, current status, and future perspectives. J Prosthet Dent. 2013;109(6):361–6.
63. Srinivasan M, Kamnoedboon P, McKenna G, Angst L, Schimmel M, Özcan M, et al. CAD-CAM removable complete dentures: a systematic review and meta-analysis of trueness of fit, biocompatibility, mechanical properties, surface characteristics, color stability, time-cost analysis, clinical and patient-reported outcomes. J Dent. 2021;2021:103777.
64. Goodacre CJ, Garbacea A, Naylor WP, Daher T, Marchack CB, Lowry J. CAD/CAM fabricated complete dentures: concepts and clinical methods of obtaining required morphological data. J Prosthet Dent. 2012;107(1):34–46.
65. Baba NZ, Goodacre BJ, Goodacre CJ, Muller F, Wagner S. CAD/CAM complete denture systems and physical properties: a review of the literature. J Prosthodont. 2021;30(S2):113–24.
66. Lee J-H. Digital workflow for establishing the posterior palatal seal on a digital complete denture. J Prosthodont. 2021;30(9):817–21.
67. Lebon N, Tapie L, Duret F, Attal JP. Understanding dental CAD/CAM for restorations - dental milling machines from a mechanical engineering viewpoint. Part A: Chairside milling machines. Int J Comput Dent. 2016;19(1):45–62.
68. Blatz MB, Conejo J. The current state of chairside digital dentistry and materials. Dent Clin N Am. 2019;63(2):175–97.
69. Abduo J, Lyons K, Bennamoun M. Trends in computer-aided manufacturing in prosthodontics: a review of the available streams. Int J Dent. 2014;2014:783948.
70. Jeong YG, Lee WS, Lee KB. Accuracy evaluation of dental models manufactured by CAD/CAM milling method and 3D printing method. J Adv Prosthodont. 2018;10(3):245–51.
71. Chen Z, Li Z, Li J, Liu C, Lao C, Fu Y, et al. 3D printing of ceramics: a review. J Eur Ceram Soc. 2019;39(4):661–87.
72. Prasad T, Sujatha G, Jayanandan M, Patil S, Raj AT. Three-dimensional printing in reconstructive oral and maxillofacial surgery. J Contemp Dent Pract. 2018;19:1–2.
73. Revilla-León M, Özcan M. Additive manufacturing technologies used for 3D metal printing in dentistry. Curr Oral Health Rep. 2017;4(3):201–8.

3D Printing and Its Applications in Maxillofacial Surgery

6

Ajoy Roychoudhury and Bhaskar Agarwal

Contents

6.1 Introduction

Advancements in computer science and allied technologies have revolutionised every sphere of our lives. Field of Medicine and Dentistry is not left untouched by these advancements. With the advent of sophisticated and high-resolution imaging modalities like Magnetic resonance imaging, Computed Tomography and Cone-Beam Computed Tomography, it has become possible to evaluate the diseased and deformed part in three dimensions. These have given us a possibility to look deeper into the disease process. One such revolutionary technology is 3-dimensional printing, also known as rapid prototyping. The advancements in Imaging modalities, Computer-processing powers, 3D printers and material sciences have opened new doors to deliver more personalised treatment to the patient in a safe and predictable manner.

Three-dimensional (3D) printing, also known as additive manufacturing or rapid prototyping, refers to a process of creating a physical object from a 3D digital model, typically by laying down or solidifying a material layer by layer in succession [1]. Standard terminology in the field of Medical 3D-printing has been an important goal and comprehensive analysis of literature has led to acceptance of '3D printing' as an

A. Roychoudhury (✉)
Department of Oral and Maxillofacial Surgery, AIIMS, New Delhi, India

B. Agarwal
Department of Oral and Maxillofacial Surgery, AIIMS, New Delhi, India

Columbia-Asia Hospital, Ghaziabad, India

© The Author(s), under exclusive license to Springer Nature Switzerland AG 2022

P. K. Chaudhari et al. (eds.), *3D Printing in Oral Health Science*,
https://doi.org/10.1007/978-3-031-07369-4_6

appropriate term [1]. Radiological society of North America has brought the 3D-printing terms to RadLex project, a radiological ontology for use in reporting, decision support, data mining, education and research [2]. There are variety of materials that can be used in 3D printing process. These materials can be plastic, nylon or metals like titanium and cobalt chromium. One can print any 3-dimensional object based on its application and physical requirements, only your imagination is the limit!

3D-printed medical devices have recently caught the attention of healthcare industry. The use of 3D-printed bronchial splint saved a child's life, a 3D-printed cranial plate replaced part of patient's skull, a patient-specific artificial knee transplant restores the patient's mobility and all of these were unimaginable previously. All of these are 3D-printed medical devices having a profound effect on the patient's health. 3D printing finds applications in multiple surgical domains. Orthopedics has the largest share in the number of publications in this field, followed by maxillofacial surgery. This is followed by Cranial surgery and Spinal surgery [3]. Here we have presented our experience gained through the various cases performed with the aid of virtual surgical planning and 3D printing in our unit.

Any discussion on applications of 3D printing in maxillofacial surgery is absolutely incomplete without discussion of Virtual Surgical Planning techniques. Virtual Surgical Planning is also known as Computer-Aided Surgical Simulation (CASS), Digital Surgical Planning and Digital Surgical Simulation. Virtual surgical planning is the first step towards generating a file compatible for 3D printing.

6.2 Virtual Surgical Planning (VSP)

Virtual Surgical Planning is a computer-based application to analyse the anatomy or deformity, perform the surgery and simulate the outcome so as to weigh the planned procedure in terms of safety and predictability and to fabricate the necessary tool and patient-specific armamentarium to safely execute the procedure. VSP provides surgeons with clear 3D visualisation of a patient's anatomy to develop a surgical plan prior to entering the operating room.

Virtual surgical planning requires an image data set obtained through MRI/CTscan/CBCT in the form of DICOM (Digital Imaging and Communications in Medicine) images. The image data set is than fed into surgical planning software where the various operations like segmentation and thresholding are performed to obtain a patient-specific anatomic reconstruction model. This patient-specific anatomic reconstruction can be exported as an STL file for 3D printing or exported to Computer-Aided Designing (CAD) software for designing of cutting guides or patient-specific implants. These cutting guides and patient-specific implants can also be exported as STL file for 3D printing. A typical virtual surgical planning software would include the following components [2]:

1. Data acquisition
2. Medical image analysis
3. 3D anthropometric analysis
4. Surgical simulation
5. Implant/template design via CAD software
6. Implant/template fabrication
7. On-line communication tool
8. Management system

There are numerous virtual surgical planning software packages available. Few of them are FDA approved. Few are open source and freely available whereas others are expensive. Available softwares and their salient features are listed in Table 6.1.

It is not necessary that every time a virtual surgical planning should lead to 3D printing of models or guides. Every case should be planned properly and if a virtual surgical planning software is available, then it should be utilised to get familiar with the anatomy and be prepared for the possible intra-operative problems. Following are the possible ways of how only virtual surgical planning was useful in certain cases:

1. For making measurements that can be transferred intra-operatively without needing a surgical guide. The CBCT/CT scan of the patient

Table 6.1 List of available softwares for virtual surgical planning

S. No	Name of software package	Salient features
1	Mimics Innovation Suite	• FDA approved • Has both Dicom to STl file conversion + CAD software • Very expensive
2	3D systems VSP	• FDA approved • Has both Dicom to STl file conversion + CAD software • Expensive
3	Rhino3D Medical	• Not FDA approved at present • Has both Dicom to STl file conversion + CAD software • Expensive
4	inVesalius	• Not FDA approved at present • Has only Dicom to STl file conversion • Open source and free
5	3D slicer	• Not FDA approved • Has only DICOM to STL file conversion • Open source and free • Powerful software with lot of functionality • Long learning curve is a drawback
6	ITK snap	• Not FDA approved • Has only DICOM to STL file conversion • Open source and free • Comparatively easier to use than 3D slicer
7	Dolphin 3D-imaging Software	• FDA approved • Only dedicated for orthognathic surgical planning and simulation • Easy graphic user interface • Expensive
8	Blue Sky Plan Bio	• FDA approved • Only dedicated for digital implant planning and simulation and fabrication of surgical guides for dental implant placement • Easy to use • Economical
9	Osirix	• FDA approved • Has only DICOM to STL File Conversion • Free to use. Available only for Macbook users
10	Meshmixer/Blender	• Not FDA approved • Have only computer-aided designing features • Have to be used in combination with DICOM to STL file converting softwares • Open source and free

can be fed into the virtual planning software to make certain measurements pertinent for the surgical approach. This way, surgeon is better acquainted with the patient's anatomy and prepared to avoid critical structures and preventing any complications resulting from the injury to these structures. It is well elucidated through Fig. 6.1.

2. For studying the anatomical deformity and the possible measures needed to prevent complications (Fig. 6.2).
3. For planning incision location and length (Fig. 6.3). Various CAD tools in the virtual planning software can be used to mark a virtual incision. The length of the incision can be measured.
4. For planning the osteotomy or odentectomy planes. The possible margins of the tumour for resection can be marked and measurements can be made from anatomical landmarks to reproduce the same osteotomy in the operating room. In case of impacted maxillary

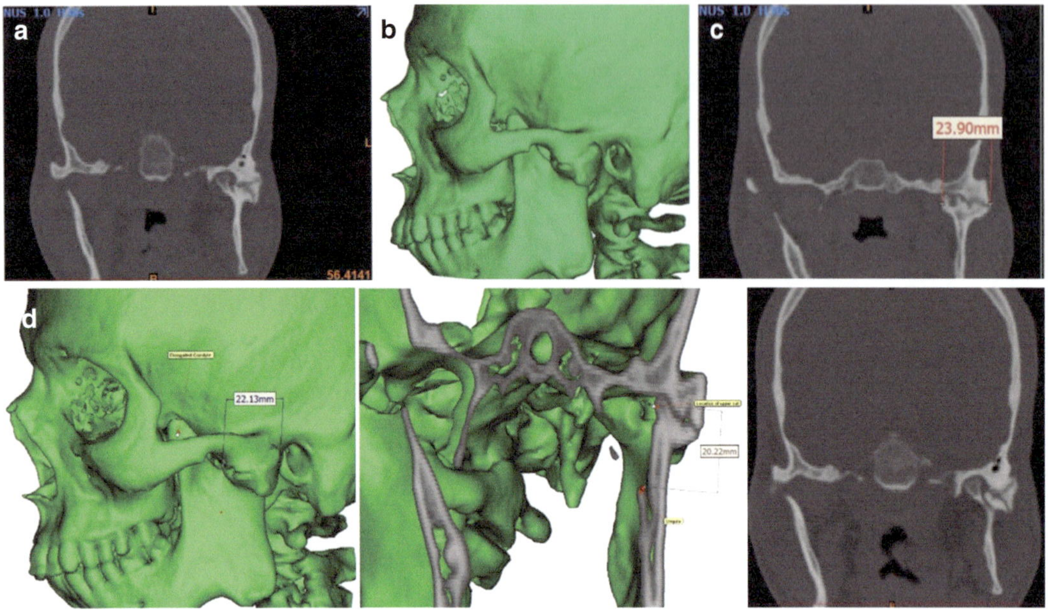

Fig. 6.1 (**a**) CT face showing ankylosis of left temporomandibular joint. (**b**) The elongated coronoid process projecting beyond the zygomatic arch is visible. (**c**) The mesiodistal dimensions of the ankylotic bone that would need to be removed are measured and are about 24 mm. (**d**) Measurement of the ankylotic bone mass in anteroposterior dimension is about 22 mm

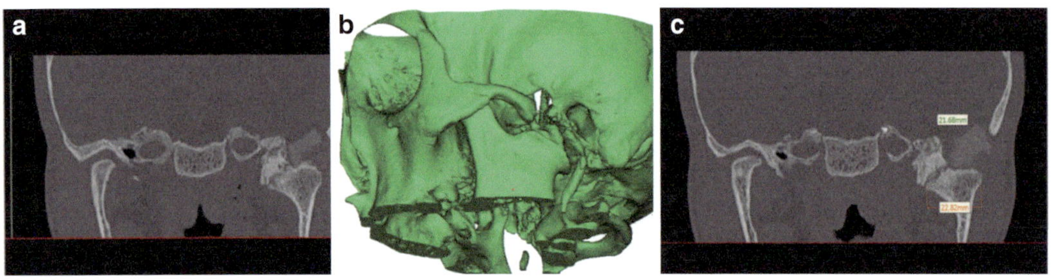

Fig. 6.2 (**a**) CT face showing ankylosis of left temporomandibular joint with a defect in the temporal bone near the skull base and a foreign body projecting into the brain. (**b**) The defect in the skull base is well appreciable with a vertically oriented silastic block projecting into the brain through the skull base defect. (**c**) The mesiodistal dimension of the skull base defect is about 21.60 mm

canines, the sectioning of the tooth for its subsequent removal can be planned (Fig. 6.4). This also facilitates the patient education and resident training.

5. For accurate transfer of the virtual surgical plan to the operating room, patient-specific surgical guides and cutting jigs can be 3D printed.

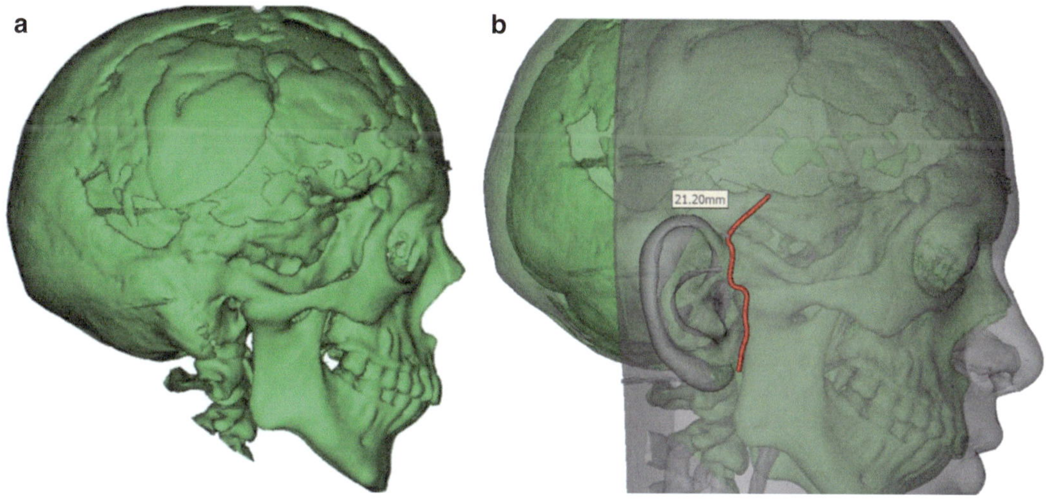

Fig. 6.3 (**a**) 3D reconstruction of a computed tomogram of the face showing multiple defects in parietal and temporal bone following a decompressive craniectomy and repair. There is ankylosis of right temporomandibular joint also. (**b**) The possible incision is marked with the available CAD tools in the software taking care of the temporal extension of the incision ends on sound bone. The length of this extension is measured to be 21.20 mm

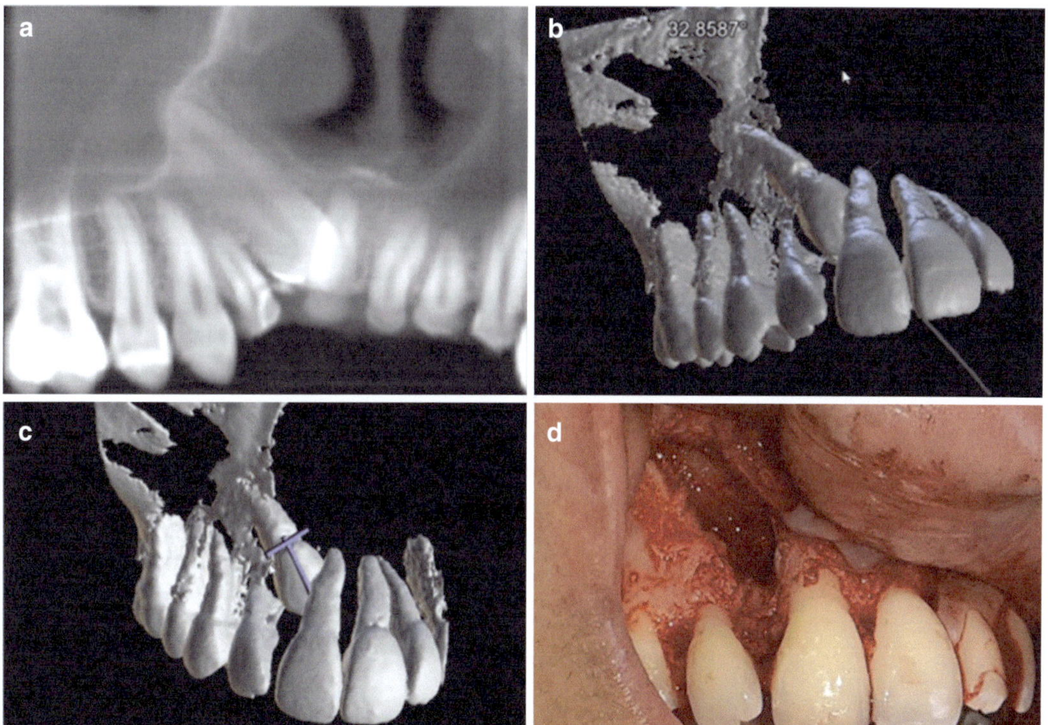

Fig. 6.4 (**a**) Panoramic reconstruction showing an impacted maxillary canine. (**b**) 3D reconstruction of the CBCT shows angulation of the impacted canine and the crown located palatal to the root of central incisor. (**c**) Virtual surgical planning done using CAD tools in the software to decide the tooth sectioning plan. (**d**) Post-operative view. A buccal approach was taken for tooth removal

6.3 3 Dimensional Printing (3D Printing)

3D printing is a rapid process of manufacturing an object with less wastage of the raw material. Opposite to 3D printing is subtractive manufacturing wherein an object is made out of carving from a bigger piece of raw material, similar to how a sculpture is carved out of a rock. Subtractive manufacturing is costly as a lot of raw material is wasted.

There are various 3D-printing technologies available based on the applications.

Four technologies that we have found useful in oral and maxillofacial surgery with the salient features are listed in Table 6.2.

Once the virtual surgical planning is completed and the desired surgical guides and patients-specific implants have been designed in the CAD software, the patient's anatomic model, surgical guides and implants are exported as STL files for 3D printing. These STL files are fed into a slicing software which converts it into a file readable by the 3D printer.

Virtual Surgical Planning was followed by 3D printing for transferring the plan to the patient in the following ways:

Table 6.2 List of 3D-printing technologies useful in oral and maxillofacial surgery

S. No.	Technology	Common materials	Accuracy	Cost	Advantages	Disadvantages	Possible uses
1	Fused deposition modelling	PLA ABS plastics	+	$	Low cost Strong, durable materials	Lower accuracy Model surfaces have prominent stair stepping ridges Not biocompatible Requires post-processing	Study models fabrication Models for pre-operative plate bending
2	Stereolithography apparatus	Epoxy and acrylic-based polymers	+++	$$	Good accuracy Biocompatible (short-term) materials available	Brittle, moderate strength Models are single material Limited colour options Requires post-processing	Surgical guides fabrication Dental implant guides fabrication
3	Selective laser sintering	Plastics, synthetic polymers like nylon, metals	++	$$$	Diverse mechanical properties Variety of materials Material strength sufficient for functional parts Supports free printing. Usually does not require any post-processing	Models are single material based High cost	Surgical guides fabrication
4	Selective laser melting	Metals	++	$$$	Long-term implantable biocompatible materials available	High cost Requires post-processing	Patient-specific implants and metallic surgical guides fabrication

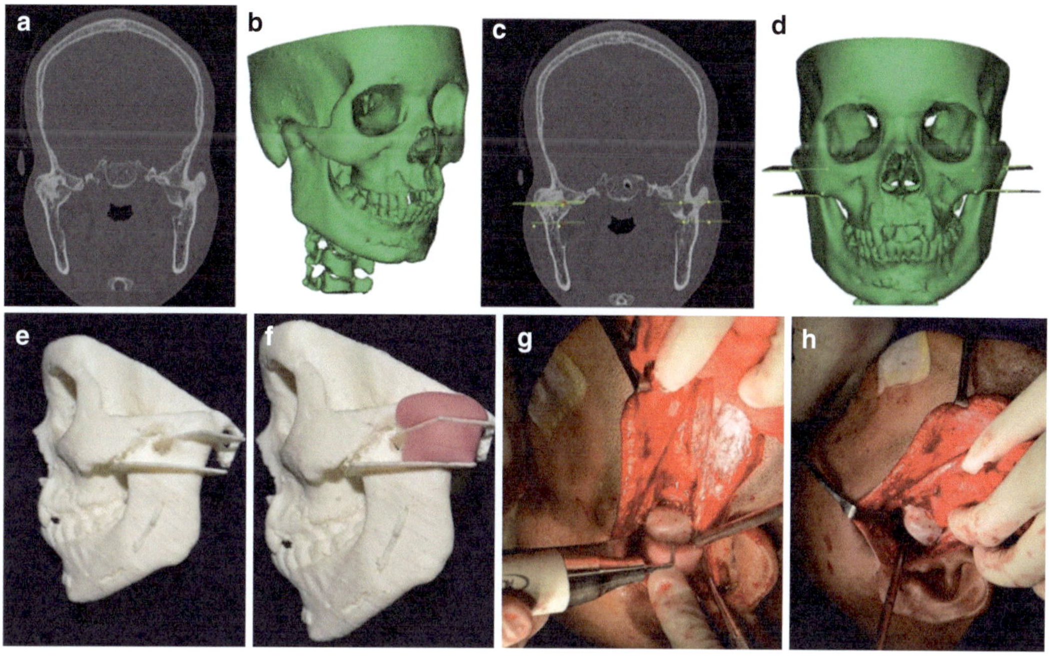

Fig. 6.5 (**a**) CT face of a patient with bilateral temporomandibular joint ankylosis. (**b**) 3D reconstruction of the computed tomogram. (**c**) The green lines showing the location of osteotomy planes on both sides. Note that the superior osteotomy plane is well away from skull base and inferior osteotomy plane is missing the inferior alveolar neurovascular bundles. (**d**) 3D reconstruction of the computed tomogram with the osteotomy planes jutting out. This model along with the osteotomy planes was exported as STL file and 3D printed. (**e**) 3D-printed model fabricated from ABS plastic using FDM technology. (**f**) The surgical guides (pink colour) made from self-curing polymethyl methacrylate resin and were later sterilised. (**g**) Intra-operative view of the guide in position and the osteotomy being performed using a piezo-surgical bone saw. (**h**) A gap of 1.5 cm was made following osteoarthrectomy. Same procedure was followed on the other side using similar surgical guide

1. For manual method of surgical guide fabrication for temporomandibular joint ankylosis release (Fig. 6.5).
2. For Patient-specific total TMJ replacement prosthesis in a case of bilateral TMJ ankylosis (Figs. 6.6 and 6.7).
3. For bending of reconstruction plates for reconstruction of mandible following tumour resection (Fig. 6.8).

6.4 Types of 3D-Printed Models

Based on the various applications as shown above through the examples, the 3D-printed models in the oral and maxillofacial surgery speciality can be classified into four types [4]:

1. Training models
2. Planning models
3. Simulation models
4. Patient-specific surgical guides

Each of these models have a specific property requirement that should dictate the use of material and technology to be used in fabricating them.

6.4.1 Training Models

A purpose of training model is to enhance the quality of teaching. The training model should accurately reproduce the anatomy, should have similar haptic feedback as that of natural bone and should be economical [4]. Haptic feedback relates to the sense of touch. The training model should feel similar to the tissue when it is being cut or drilled. Cadaver or animal models for training are often difficult to obtain, lack specific pathologic features and usually have a high cost. Several training mod-

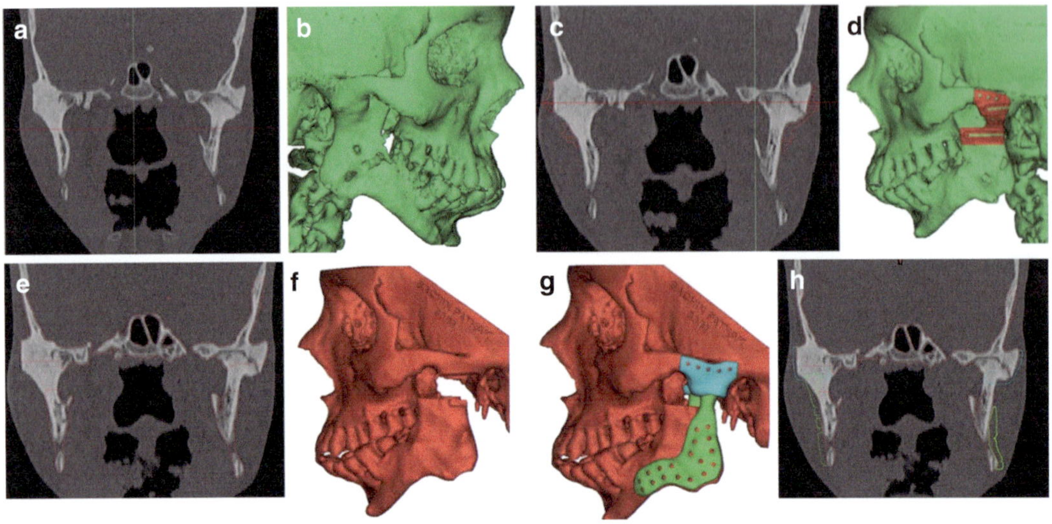

Fig. 6.6 (**a**) CT face showing bilateral TMJ ankylosis. (**b**) The 3D reconstruction of the computed tomogram. (**c**) The superimposition of osteotomy guides (red outline) on to the CT data to verify the fit of the surgical guides. (**d**) The cutting guides superimposed over the bilateral TM joint region on 3D reconstruction. The cutting guides are designed using CAD tools in the designing software. (**e**) The outlines of the mandible and maxilla (red outline) after the removal of ankylosed bony mass, superimposed over the CT images to confirm the accuracy of superimposition. (**f**) The 3D reconstruction of the mandible and maxilla will be used for designing of the patient-specific TMJ replacement prosthesis. (**g**) The superimposition of the ramus component (green) and the fossa component (blue) over the mandible and skull base. (**h**) The fit of the ramus component (green outline) and fossa component (blue outline) on to the bone surface (red outline) is evaluated

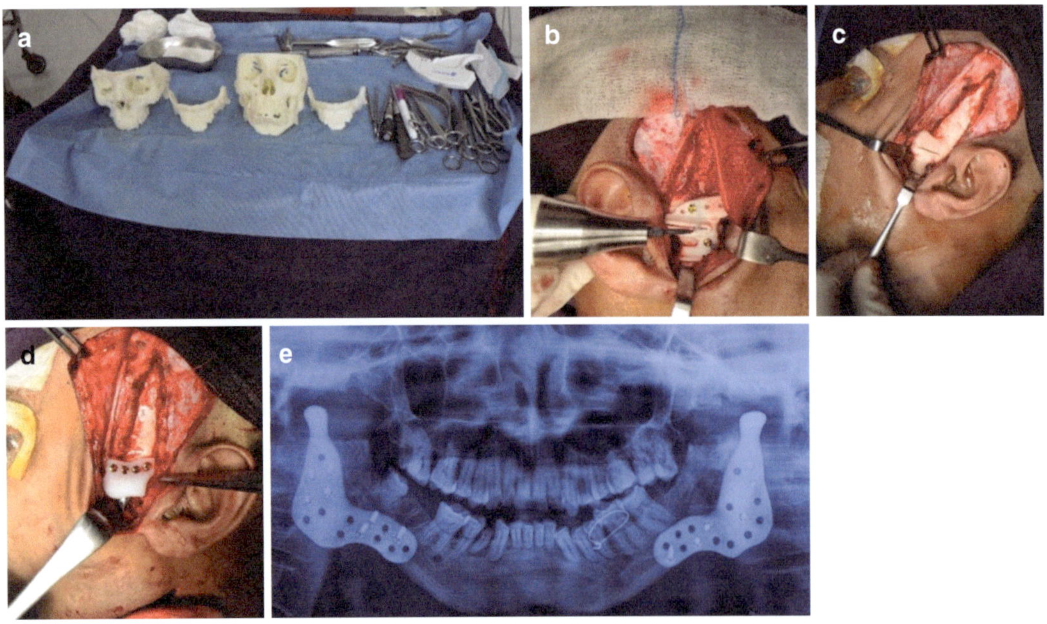

Fig. 6.7 (**a**) Sterilised study models for intra-operative referencing. These models can be plasma sterilised. (**b**) The surgical guides are in place and secured to the bones with screws. (**c**) Osteotomy planes are accurately transferred to the patient with the use of surgical guides. Now the osteoarthrectomy is limited in between these two cuts. (**d**) The ramus and fossa components are secured to the bone with screws at the planned positions. (**e**) Post-operative orthopantomogram showing the position of the ramus components

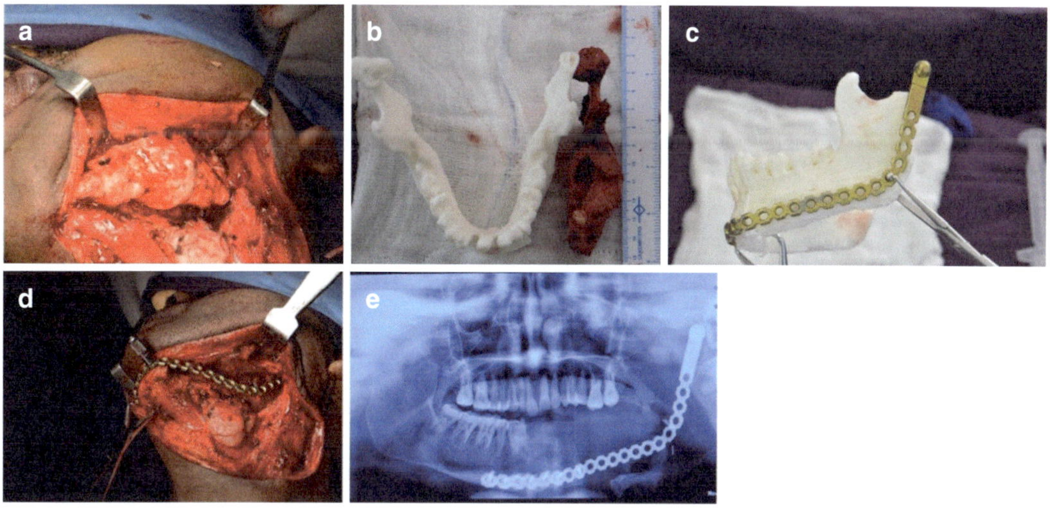

Fig. 6.8 (**a**) The tumour is exposed through the submandibular approach. (**b**) The tumour is resected with disarticulation of left TMJ. The 3D-printed model is modified in way that the 2.4-mm reconstruction plate remains undercontoured so as to prevent any exteriorisation of plate. (**c**) The 2.4-mm reconstruction plate is pre-bent on the 3D-printed model to achieve good facial symmetry at the same time keeping the plate undercontoured to prevent exteriorisation. (**d**) The pre-contoured plate is being secured to the mandible. (**e**) Orthopantomogram showing the reconstruction plate in place with complete removal of the tumour mass

els for temporomandibular joint replacement made in ABS (acrylonitrile butadiene styrene) from fused deposition modelling had been used at our Centre during various total temporomandibular joint replacement workshops (Fig. 6.9). Similar training models for the orthognathic surgery training had also been designed and produced at our institute (Fig. 6.10). ABS has been found to have the haptic feedback closest to natural bone when model was cut with a piezo-surgical unit. The disadvantage of these 3D-printed models is that the material tends to stick to the drills while drilling. These models have a near natural feel when the fixation screws are screwed into the model. Additionally these training models can be customised to facilitate easy cutting of the models. Similar training models have also been used for surgical correction of cleft lip and palate training and rhinoplasty training. Though these models can accurately reproduce the anatomy, but the haptic feedback is always below the expectations. In future with availability of more innovative materials, these models may accurately reproduce the natural haptic feedback.

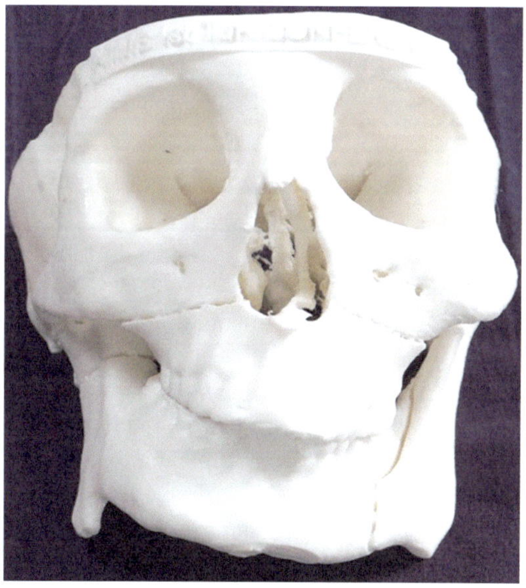

Fig. 6.9 3D-printed training model for concomitant orthognathic surgery with total temporomandibular joint replacement. The model is printed in acrylonitrile butadiene styrene (ABS) plastic. The model has been weakened at the LeFort 1 osteotomy level and TMJ ankylosis level to facilitate easy cutting of the models

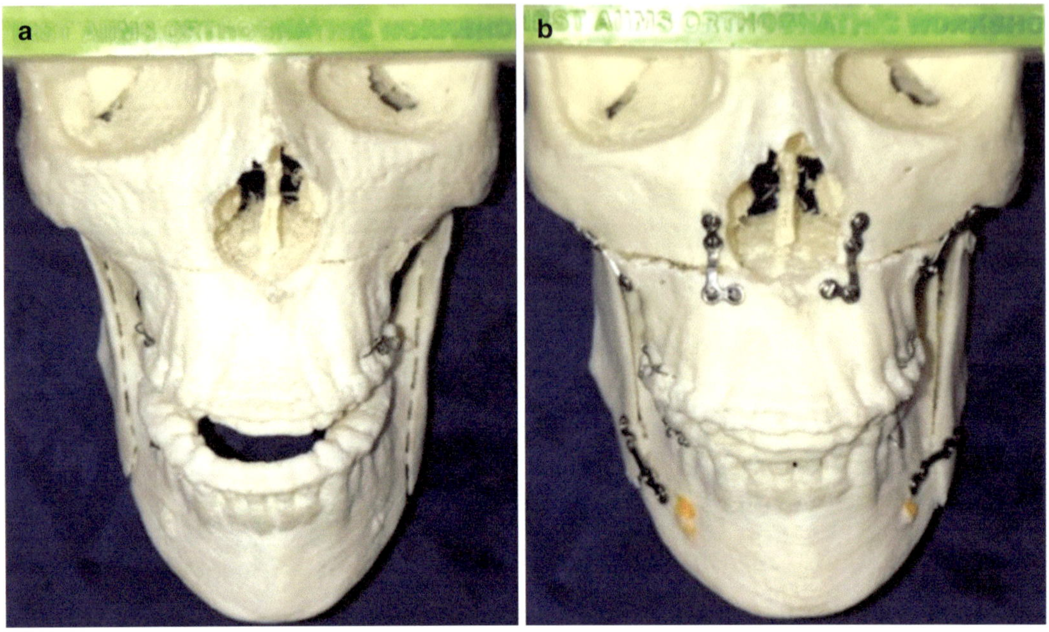

Fig. 6.10 (**a**) 3D-printed training model for orthognathic surgery training. The model is printed in acrylonitrile butadiene styrene (ABS) plastic. The model has been weakened at the Lefort 1 osteotomy level and bilateral sagittal split osteotomy in mandible to facilitate easy cutting of the models. (**b**) 3D-printed training model for orthognathic surgery training. The maxilla and mandible are repositioned and fixed with plates and screws in their final desired position

6.4.2 Planning Models

The purpose of a planning model is to completely understand the patient's anatomy in a complex surgery. These models are also referred to as anatomical models [3], biomodels [5]. Biomodels can be translucent or coloured to help in identification of margins in case of jaw tumours, for implant trial implantation and pre-operative sizing. Currently the 3D printing is most widely used for making biomodels. Cost-benefit analysis of biomodels shows that biomodels can potentially reduce the cost to healthcare providers if operating room time is reduced by 14 min or more [5]. Some authors have found saving of an average of 0.4 h by pre-operative bending of reconstruction plates on a 3D-printed model [6]. The time required for bending of the reconstruction plate also depends on the location and dimension of the surgical defect. These models can also be used for orthognathic surgeries and reconstructive surgeries. Various measurements can be made on these models. A planning model should be highly accurate. Sometimes a virtual surgical planning software is enough to understand the anatomy but a sterilised 3D-printed model is useful in the operating room where the surgical planning software might not be accessible or impractical to handle. The technology and material used for these models should result in an accurate and sterilisable product. It should not leave any frills or residues on the gloves when it is handled. Selective laser sintering and stereolithography are highly accurate, cost-effective and residue-free technologies which can be used for manufacturing planning models. These models have been used for total alloplastic TMJ replacement cases for planning and intra-operative referencing (Fig. 6.11).

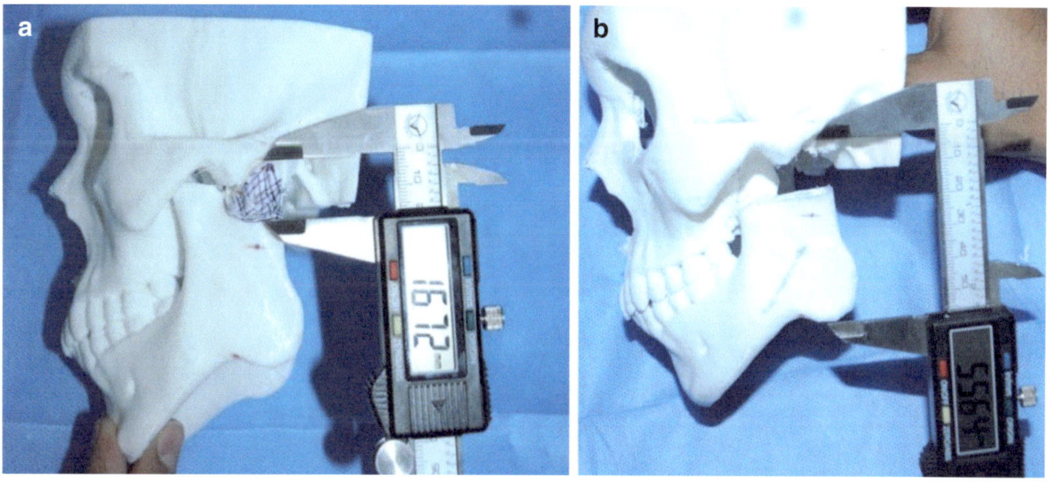

Fig. 6.11 (**a**) This is 3D-printed model of a patient with TMJ ankylosis. Note the measurement being made of the ankylotic bony mass. (**b**) The available ramal length being measured after the ankylotic bone mass has been removed. This measurement is important to decide the size of condylar component for total TMJ replacement

6.4.3 Simulation Models

A simulation model is used to perform a mock surgery before it is done on the patient. It helps to plan the procedure, foresee the hurdles and plan out techniques to circumvent them, also to evaluate the safety and predictability. These models should be accurate with a patient like haptic feedback. In our unit, simulating models are used in complex total temporomandibular joint replacement (Fig. 6.12) cases and for correction of post-traumatic residual deformity cases (Fig. 6.13). Such models are also useful in cases where Lefort III osteotomy is planned in syndromic craniosynostosis patients for treatment of midface retrusion. Sometimes these simulation models may also be carried to operation theatre. In such cases, the simulation models should be made from a sterilisable material and should not leave any residue on the gloves while handling.

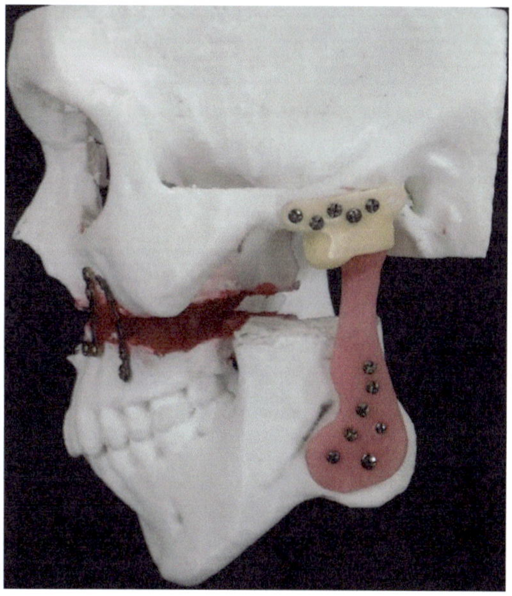

Fig. 6.12 A 3D-printed model used for simulating concomitant orthognathic surgery with alloplastic total temporomandibular joint replacement

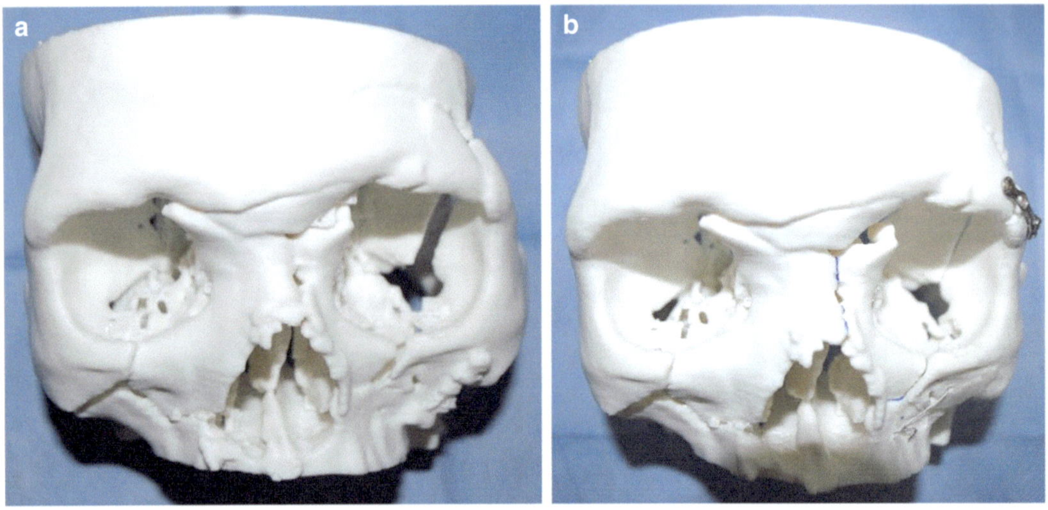

Fig. 6.13 (**a**) 3D-printed model of a patient with increased bizygomatic width due to malunited zygomaticomaxillary complex. You can appreciate the gap between the spheno-zygomatic suture. (**b**) 3D-printed model after repositioning the malunited zygomatic bone. Note the closure of spheno-zygomatic suture

The haptic feedback of these models depends on the mechanical properties of the material used for making the model. The elastic modulus and stiffness of material are few of the parameters which effect the haptic feedback.

6.4.4 Patient-Specific Surgical Guides

Surgical guides are the most popular medical application of 3D printing. Earlier anatomical models/biomodels were the most common 3D-printed models but now the growing importance of 3D-printed guides is noticeable. Patient-specific surgical guides are now gaining momentum as these are now the most commonly reported type of 3D-printed application [1]. These are patient-specific instruments used for making accurate osteotomies or to allow accurate placement of a device. These guides are made from a biocompatible, autoclavable material with a high accuracy. These guides should accurately conform to the structure to which they will be anchored. These can be made from nylon or titanium depending on the site of placement. In cases where a wide surgical exposure is possible and there is no major anatomical structure a nylon guide can be used, like in case of a free fibula osteomyocutaneous graft harvesting procedure. The nylon is weak in thin sections and so need bulk to be sufficiently rigid. In cases of temporomandibular joint replacements, the surgical access available is small and also have trunk of facial nerve in vicinity, this situation demands a surgical guide which is thin and small yet sufficiently rigid to allow its accurate placement. The surgical guides should not leave any residue on coming in contact with the bone saw or drills nor should it deform due to heat generated while drilling. The patient-specific surgical guides are widely used in guided dental implantology cases. These guides can also be used for bone graft harvesting from ramus of mandible (Fig. 6.14).

It is a good idea to keep two sets of these surgical guides ready for surgery in case a guide breaks or deforms on coming in contact with the saw or drills.

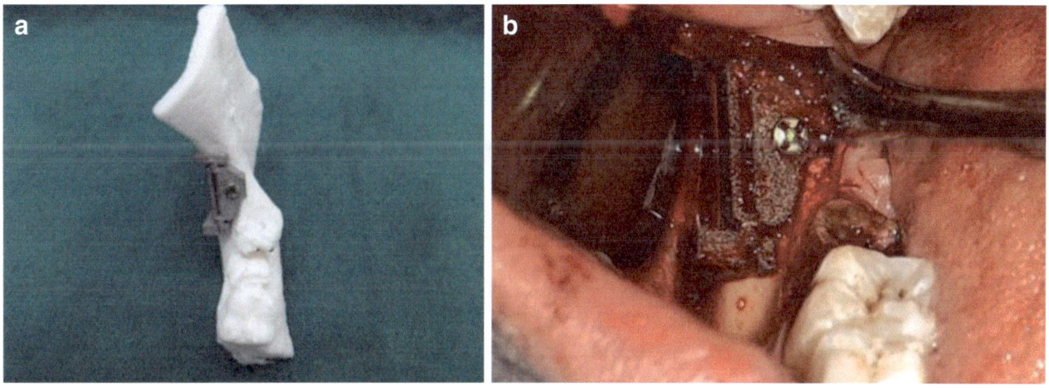

Fig. 6.14 (**a**) Patient-specific surgical guide (black coloured) to harvest cortical bone graft from ramus of mandible fixed on the planning model (white coloured). (**b**) Intra-operative view of the surgical guide secured to the ramus with a single titanium screw

6.5 Patient-Specific Implants (Custom Implants)

Reconstruction of the craniofacial skeleton can be extremely challenging for the most experienced and astute surgeon. Some factors that make such reconstructive procedures challenging are complex anatomy, vital structures in the vicinity, size of defect and chances of infection. Restoration of aesthetics and function is always the primary goal of any reconstructive procedure performed in the craniofacial region. Post-traumatic defects, congenital malformations, defects created due to tumour ablative surgery need precise pre-surgical planning for accurate reconstruction. Auto grafts are usually considered the gold standard for such reconstructions. Their use is limited due to limited availability of donor sites for, donor site morbidity, additional patient discomfort and additional surgical procedure and time. This has led to search for alloplastic reconstructive options. Virtual surgical planning, 3D printing and CAD/CAM coupled with newer biocompatible materials have led to increased use of patient-specific implants in the craniofacial region. The success of these implants depend on material characteristics, design of implant and surgeon's skill. Patient-specific implants have gained popularity over the stock counterparts due to their better adaptation to the region of implantation, reduced surgical time, faster recovery and better cosmesis. Selective Laser Sintering (SLS), Selective Laser Melting (SLM) and Electron Beam Melting (EBM) have enabled the direct manufacturing of the custom implants from biocompatible implantable materials like titanium, Ti6Al4V, chrome cobalt and polyetheretherketone (PEEK).

6.6 Limitations and Areas of Research

The process of conversion of Dicom Data to a 3D model is a manual, time-consuming process vulnerable to human error. The incorporation of AI-based segmentation of the DICOM data to extract the meaningful, desired and accurate 3D model is the area of current research in the field of virtual surgical planning. In the field of 3D printing, the research is focussed on developing cost-effective biocompatible materials and 3D printers for small-scale printing. The focus is also on reducing the printing time so that the emergency cases can benefit from this technology. The 3D printing of patient-specific implants and the incorporation of its complete workflow in an inhouse 3D-printing facility is another exciting field of research.

6.7 Future Perspectives and Summary

The automation and AI script-based workflow will reduce the human error and the time to print. The print quality can be optimised to be biomechanically acceptable yet requiring the minimum raw material for 3D printing. The automatic generation of patient-specific implants based on the patient's normal side or from the bank of pooled data will help to reduce the time to production of patient-specific implant.

3D printing is a new and exciting field. Its use in oral and maxillofacial surgery has been recently introduced. Virtual surgical planning and 3D printing have opened vast areas of possibilities and have allowed surgeon to perform more complex procedure with a higher degree of predictability and safety. It is a great tool for training and education of residents. However, it needs to be understood that the virtual surgical software does not suggest a treatment plan, but only shows the possible outcome of the planned treatment to some extent. Not everything that is planned on the software is practically possible on the patient due to the biological limits of the tissues. Virtual surgical planning and 3D printing are just an adjunct and not a substitute to sound biological and surgical principles. A high degree of surgical experience is required to foresee the possible problems and incorporate the necessary changes in the virtual planning level to circumvent this problem.

With further improvements in the materials and technology, the dream to make life like 3D-printed models could be realised. Till then we need to understand the limitations of this technology and use it judiciously for patient care and resident training.

References

1. Tack P, Victor J, Gemmel P, Annemans L. 3D-printing techniques in a medical setting: a systematic literature review. Biomed Eng Online. 2016;15(1):115.
2. Zhao L, Patel PK, Cohen M. Application of virtual surgical planning with computer assisted design and manufacturing technology to cranio-maxillofacial surgery. Arch Plast Surg. 2012;39(4):309–16.
3. Alexander AE, Wake N, Chepelev L, Brantner P, Ryan J, Wang KC. A guideline for 3D printing terminology in biomedical research utilizing ISO/ASTM standards. 3D Print Med. 2021;7(1):8.
4. Meglioli M, Naveau A, Macaluso GM, Catros S. 3D printed bone models in oral and cranio-maxillofacial surgery: a systematic review. 3D Print Med. 2020;6(1):30.
5. Parr WCH, Burnard JL, Wilson PJ, Mobbs RJ. 3D printed anatomical (bio)models in spine surgery: clinical benefits and value to healthcare providers. J Spine Surg. 2019;5(4):549–60.
6. Lethaus B, Poort L, Böckmann R, Smeets R, Tolba R, Kessler P. Additive manufacturing for microvascular reconstruction of the mandible in 20 patients. J Craniomaxillofac Surg. 2012;40(1):43–6.

Applications of 3D Printing in Periodontal Tissue Regeneration

7

Kunaal Dhingra, Kanika Makker, Flora Verma, and Bhumika Gumber

Contents

7.1 Introduction

The periodontium is a highly organized, three-dimensional, complex structure composed of four specialized tissues viz., alveolar bone, cementum, periodontal ligament, and gingiva that function synergistically to support the teeth, transmit mechanical forces, and prevent microbial invasion. Periodontal disease causes irreversible loss of these supporting tissues and affects 20–50% of the global population [1]. It is a leading cause of tooth loss, which in turn causes compromised function and esthetics, and poor quality of life. The ultimate aim of periodontal treatment is to predictably restore the prediseased state of anatomically and functionally intact periodontium. This necessitates accurate spatiotemporal orchestration of healing in the different compartments and guided multiple tissue formations within 500-μm interfaces [2]. Conventional periodontal therapies have neither precisely nor predictably achieved this objective. The development of newer technology and the advent of 3D printing offer a possible solution to this persistent problem.

3D printing has enabled the creation of biocompatible scaffolds that facilitate the regeneration of the missing tissue. It also includes the printing of three-dimensional functional tissues consisting of living cells and their supporting elements. The latter process, known as "bioprinting", utilizes "bioink" which is a biocompatible hydrogel with high water content, living cells, and modifiable chemical and mechanical

K. Dhingra · K. Makker (✉) · B. Gumber
Periodontics Division, Centre for Dental Education and Research, All India Institute of Medical Sciences, New Delhi, India

F. Verma
Viral Research and Diagnostic Laboratory, Department of Microbiology, Atal Bihari Vajpayee Institute of Medical Sciences and Dr. Ram Manohar Lohia Hospital, New Delhi, India

© The Author(s), under exclusive license to Springer Nature Switzerland AG 2022
P. K. Chaudhari et al. (eds.), *3D Printing in Oral Health Science*,
https://doi.org/10.1007/978-3-031-07369-4_7

properties [3–5]. This enables the 3D-printing technology to customize and fit the tissue defect morphology and architecture [6, 7].

Hixon et al. [8] have utilized 3D-printed molds for customized bone regeneration in cleft-craniofacial defects in children. Furthermore, in a pilot study, Xu et al. [9] fabricated customized polyglycolic acid (PGA)/polylactic acid mandibular condylar scaffolds for cartilage and bone regeneration. The use of these biomaterials not only allows customization of design but also manipulation of porosity and surface texture [10]. It can address variability in thickness, mechanics, and function associated with different areas of the oral cavity.

Unlike the above studies where single tissue regeneration was achieved, regeneration of the periodontium requires simultaneous regeneration of hard and soft tissues. Novel approaches like the use of multiphasic biomaterials represent a significant leap in this regard. Park et al. [11] demonstrated the design and fabrication of composite scaffolds for in-vivo regeneration of human tooth dentin-ligament-bone complexes. These tissue complexes showed obliquely and parallel-oriented fibers within the polycaprolactone (PCL)-PGA-designed constructs forming periodontal ligament, cementum-like tissue, and bone structures.

The use of 3D printing for periodontal regeneration is still in its infancy. Thorough knowledge of the biological mechanisms and 3D-printing systems and materials is necessary to effectively translate this technology into treatment. This systematic review summarizes evidence from published animal and human studies that have assessed the application of 3D printing for periodontal regeneration.

7.2 Methodology

7.2.1 Protocol

The present systematic review was conducted according to the Cochrane Handbook for Systematic Reviews of Interventions Version 6.2 [12] published in the year 2021 and followed the PRISMA (Preferred Reporting Items for Systematic Reviews and Meta-Analyses) guidelines [13].

7.2.2 Literature Search Strategy

The literature search was conducted in the PubMed database from inception till 30th April 2021. References of systematic reviews on 3D printing in periodontal regeneration were searched. Hand searching through relevant journals and the web was also undertaken. Combinations of the following keywords were used to search the PubMed database: three-dimensional printing, 3D printing, 3-D printing, 3D printed, rapid prototyping, selective laser sintering, additive manufacturing, fused deposition modeling, bioprinted, bioprinting, scaffold, periodontal repair, periodontal regeneration, alveolar bone, cementum, periodontal ligament, and gingiva. Articles were included without restriction on the language, year of publication, or status of the publication.

7.2.3 Selection Criteria for Inclusion/Exclusion of Studies

The studies were chosen to be eligible based on a priori participant-intervention-comparison-outcome-study (PICOS) criteria (Table 7.1). Initially, studies were checked for duplicates, followed by the title, and abstract screening. Finally, full texts were retrieved and checked independently by the authors.

7.2.4 Data Extraction

Data extraction was performed independently by the authors. The following data were extracted:

Study characteristics such as author, publication year, and study design

1. Type of intervention
2. Study outcomes

Table 7.1 Participant-intervention-comparison-outcome-study (PICOS) criteria

Category	Inclusion criteria	Exclusion criteria
Participant	Studies on adult human participants, irrespective of gender, with periodontal disease Animal interventional studies with orthotopic or ectopically created defects	–
Intervention	Periodontal regeneration using 3D-printed scaffold with/without biomolecules and cell cultures	3D printing used only for preoperative analysis 3D printing used only for fabrication of simulation model
Comparison	Studies assessing regeneration after using 3D-printed biomaterials With/without control group	–
Outcome	Evaluation of regeneration of any/all periodontal tissue(s) by histological or radiographic methods Evaluation of intraoperative or postoperative complications	–
Study design	Experimental (randomized/nonrandomized clinical trials) and descriptive studies (case reports and case series)	Unsupported expert opinions Letters to editor Analytical studies Reviews Systematic reviews

In animal studies, the following data were extracted:

1. Type of animal
2. Number of subjects
3. Objective
4. Printing technology used
5. Scaffold
6. Biomolecules or cell seeding (if used)
7. Observation period
8. Outcome
9. Complications (if any)

In human studies, the following characteristics were recorded:

1. Sample size
2. Objective
3. Printing technology
4. Scaffold
5. Biomolecules (if any)
6. Follow-up period
7. Complications (if any)

7.2.5 Quality Assessment of Animal Studies

The SYRCLE's risk of bias tool containing ten entries was used to assess the risk of bias in the included animal studies. Six types of bias, i.e., detection bias, selection bias, performance bias, attrition bias, reporting bias, and other biases were assessed and classified as:

1. Low
2. Unclear
3. High

7.2.6 Outcomes Evaluation

The outcome assessment was based on the selection criteria for the included studies. Periodontal regeneration was assessed by radiographic or histological evaluation. Presence of treatment-related complications that may affect the success of the intervention were also considered.

7.3 Results

A total of 164 articles were obtained using the above search strategy, out of which 160 were from the databases and 4 from other sources. Amongst these, duplicate articles were eliminated and the remaining abstracts were scanned for inclusion. Full text of 22 articles was retrieved. Twelve articles fulfilled the inclusion criteria and were, therefore, included in the present system-

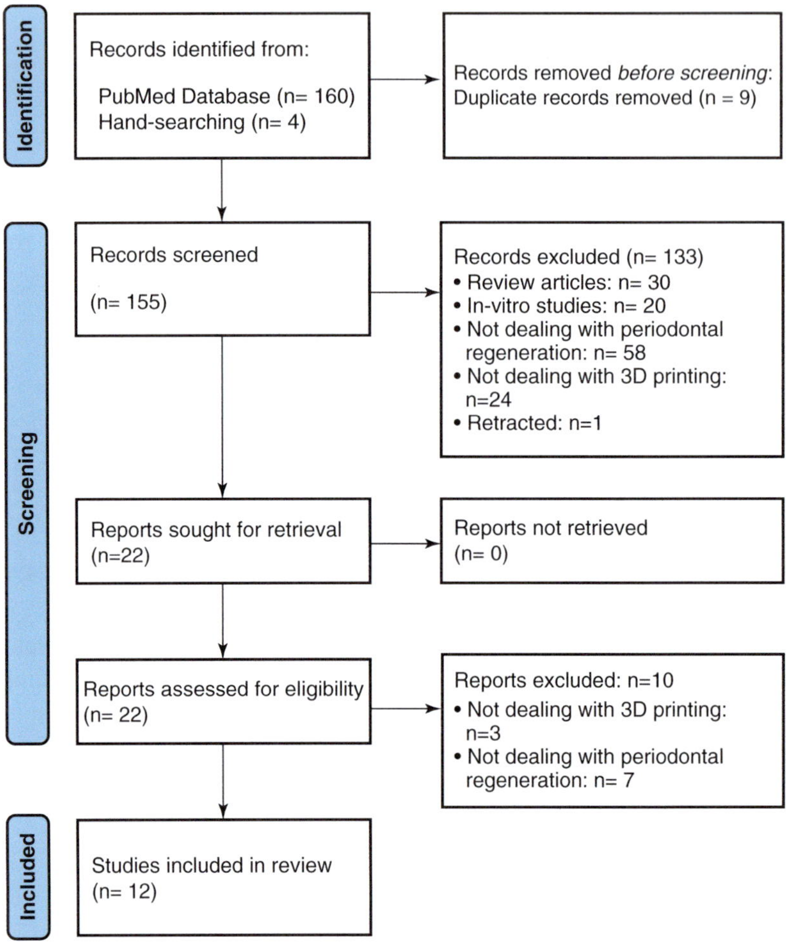

Fig. 7.1 PRISMA flow diagram for the literature search

atic review. These included 11 animal studies and 1 human study. The PRISMA flow diagram for the literature search is presented in Fig. 7.1.

7.3.1 Animal Studies

7.3.1.1 Study Selection

Data of the 11 animal studies published between 2010 and 2021 are presented in Table 7.2.

7.3.1.2 Study Characteristics

In all, 233 animal subjects were used in the 11 studies. The number of animals used in each study ranged from 3 to 64. One study, however, did not report the number of animals used [23]. It was observed that rats were most often used [14–18, 22] followed by mice [11, 19, 20]. There was only one study each on rabbits [23] and dogs [21].

The vast majority were ectopic defects [11, 16, 18–20, 22, 23] followed by orthotopic defects [15, 17, 21]. One study included both surgically created ectopic and orthotopic defects [14].

Fused deposition modeling (FDM) was the most used printing technology [2, 16, 18, 19] followed by microextrusion [14, 21, 23]. All studies used PCL scaffolds, some in combination with PLGA [11] and some with HA. Enhancement of the scaffold using biomolecules/cell cultures was also done in all but one study [21]. Observation time ranged from 3 to 12 weeks.

7.3.1.3 Risk of Bias

The risk of bias in animal studies was evaluated using the SYRCLE's tool (Table 7.3). Only one

Table 7.2 Data extraction of the 11 animal studies published between 2010 and 2021

S. No.	Author	Year	Animal	Number	Objective	Printing technology	Scaffold	Biomolecules used (if any)	Observation period (weeks)	Outcome	Complication (if any)
1	Kim et al. [14]	2010	Sprague-Dawley rats	22	Multitissue periodontal regeneration	Pneumatic extrusion using rapid prototyping machine	PCL-HA	SDF-1 and BMP-7	9	1. Integration of the scaffold and tissue ingrowth 2. Regeneration of putative PDL and bone at scaffold interface	Not specified
2	Park et al. [11]	2010	Immunodeficient mice	8–9/group/time point for four groups at two time points	Multitissue periodontal regeneration	Fused deposition modeling	Bone compartment: PLGA PDL compartment: PCL	Bone compartment: hGF transduced with BMP-7 PDL compartment: Human PDLC	3–6	1. Parallel and obliquely oriented ligamentous fibers 2. Formation of mineralized tissue 3. At 6 weeks, deposition of cementum-like tissue observed	Not specified
3	Park et al. [15]	2012	Athymic rats	48	Multitissue periodontal regeneration	Rapid prototyping	PCL (fiber-guiding)	Human PDLC transduced with adenoviral BMP-7	3–6	1. Good geometric fit to the periodontal defect 2. Compartmentalized multitissue neo-genesis with ligament directionality 3. Cementum-like tissue deposited on dentin surface	Not specified

(continued)

Table 7.2 (continued)

S. No.	Author	Year	Animal	Number	Objective	Printing technology	Scaffold	Biomolecules used (if any)	Observation period (weeks)	Outcome	Complication (if any)
4	Vaquette et al. [16]	2012	Male athymic rats	5	Multitissue periodontal regeneration	Bone compartment: Fused deposition modeling Periodontal compartment: Electrospinning	Biphasic PCL and TCP	OB in FDM PDL cell sheet	8	1. Good tissue integration 2. Higher osteoinduction and bone density in OB-seeded scaffolds 3. Cellular organization, PDL fiber attachment, cementoblast precursors and cementum-like tissue seen	Fraction of dentin block exposed at day 7 in two cases
5	Park et al. [17]	2014	Athymic rats	48	Multitissue periodontal regeneration	Rapid prototyping	PCL (fiber-guiding)	Human PDLC	3–6	1. Good geometric fit 2. Significant regeneration of mineralized tissue at 6 weeks 3. Perpendicular arrangement of type I collagen fibers	Not specified
6	Costa et al. [18]	2014	Male athymic rats	5	Multitissue periodontal regeneration	Bone compartment: Fused deposition modeling Periodontal compartment: Electrospinning	Biphasic PCL	OB	8	1. Good tissue integration 2. Significant new bone at bone/PDL interface 3. Blood vessel penetration in bone and PDL compartments 4. Some level of tissue organization (60° to dentin)	None

7	Lee et al. [19]	2014	Immunodeficient mice	10, 10	Multitissue periodontal regeneration	Fused deposition modeling	PCL/HA	DPSCs, PDLSCs, or ABSCs with human amelogenin, CTGF, and BMP2	6	Formation of bone, PDL, cementum/dentin-like tissue and aligned PDL-like fibers	Not specified
8	Pilipchuk et al. [20]	2016	Immunodeficient mice	6	Aligned bone-ligament-cementum complex	Selective laser sintering for bone compartment	PCL	Human gingival fibroblasts and PDL cells transduced with adenoviral BMP-7	3–6	1. High cellular alignment, nuclear elongation, and bone formation in micropatterned scaffold 2. At 6 weeks, cementum-like tissue seen on dentin surface	None
9	Shim et al. [21]	2017	Beagles	3	Guided bone regeneration	Extrusion using multihead deposition system	PCL and PCL/beta-TCP	None	8	1. Shape and position of membrane were retained 2. Bone formation evident. Significantly higher new bone percentage and significantly lower soft tissue percentage	None

(continued)

Table 7.2 (continued)

S. No.	Author	Year	Animal	Number	Objective	Printing technology	Scaffold	Biomolecules used (if any)	Observation period (weeks)	Outcome	Complication (if any)
10	Dubey et al. [22]	2020	Male Fischer 344 rats	6/time point for two time points	Guided bone regeneration	Melt electro-writing using highly translational multihead bioprinting platform	PCL	MSC	4, 8	1. Delayed degradation of the membrane preventing soft tissue invasion 2. New bone formation from defect margins at 4 and 8 weeks	Not specified
11	Wang et al. [23]	2021	New Zealand white rabbits	Not specified	Guided tissue regeneration	Syringe-based extrusion bioprinting	Bilayer: 1. PCL strontium-doped calcium silicate 2.hGF laden col. hydrogel (bioink)	hGF	12	1. Scaffold structure maintained 2. New bone formation with high trabecular thickness evident	Not specified

Table 7.3 Results of risk of bias in animal studies using the SYRCLE's tool

	Selection bias			Performance bias		Detection bias		Attrition bias	Reporting bias	Other
Author/year	Sequence generation	Baseline characteristics	Allocation concealment	Random housing	Blinding	Random outcome assessment	Blinding			
Kim et al. (2010)	Unclear	Low	Unclear	Unclear	Unclear	Unclear	Low	Low	Low	None
Park et al. (2010)	Unclear	Low	Unclear	Unclear	Unclear	Unclear	Unclear	Unclear	Low	None
Park et al. (2012)	Unclear	Low	Unclear	Unclear	Unclear	Unclear	Unclear	Unclear	Unclear	None
Vaquette et al. (2012)	Unclear	Low	Unclear	Unclear	Unclear	Unclear	Unclear	Low	Low	None
Park et al. (2014)	Unclear	Low	Unclear	Unclear	Unclear	Unclear	Unclear	Unclear	Low	None
Costa et al. (2014)	Unclear	Low	Unclear	Unclear	Unclear	Unclear	Unclear	Unclear	Low	None
Lee et al. (2014)	Unclear	Low	Unclear	Unclear	Unclear	Unclear	Unclear	Unclear	Low	None
Pilipchuk et al. (2016)	Unclear	Low	Unclear	Unclear	Unclear	Unclear	Unclear	Unclear	Low	None
Shim et al. (2017)	Unclear	Unclear	Unclear	Unclear	Unclear	Unclear	Unclear	Unclear	Low	None
Dubey et al. (2020)	Unclear	Low	Unclear	Unclear	Unclear	Unclear	Unclear	Unclear	Low	None
Wang et al. (2021)	Unclear	Low	Unclear	Unclear	Unclear	Unclear	Unclear	Unclear	Low	None

study by Kim et al. [14] reported a low risk of bias in investigator blinding and attrition bias parameters. All studies except Shim et al. [21] reported maintaining similar baseline characteristics for the animals used in the study. The remaining parameters were the same across the included studies.

7.3.2 Human Studies

One human study fulfilled the inclusion criteria and was included in this systematic review. In this study, a large periodontal osseous defect was treated with a 3D-printed polymer scaffold with good short-term but poor long-term results (Table 7.4).

7.4 Discussion

7.4.1 Overview of Reported Studies

3D printing is an ever-growing and ever-evolving field where new innovations are first tested using in-vitro studies. Owing to their speed, accuracy, and cost-effectiveness, in-vitro studies are a powerful tool that drives new developments. However, better insights regarding the clinical application and long-term effects of 3D-printed scaffolds for periodontal regeneration can be deduced from preclinical (animal) and clinical (human) research. Thus, the present systematic review included both preclinical (animal) and clinical (human) to better understand the developments that have cleared preliminary testing and advanced closer to actual clinical applications for periodontal regeneration.

The 11 animal studies reported favorable clinical outcomes and limited serious complications. None of the studies used random sequence generation, allocation concealment for animals, random housing of animals, or random selection of animals for outcome assessment. Most studies had a low risk of bias in terms of maintaining homogeneity in baseline characteristics and the selective reporting parameter (Fig. 7.2). The only human study reported unsuccessful long-term results. Nonetheless, these studies provide important data for the efficacy of 3D-printing technology and lay the foundation for further innovation and investigation.

7.4.2 Scaffold Materials and Design

The constituent materials and design of a scaffold influence its biological and mechanical properties. Numerous types of scaffold materials are available. They can be classified into natural and synthetic polymers, and ceramics. Synthetic polymers are the most commonly used as they can be produced in large quantities at low costs. Further, PCL, a synthetic polymer, is the best known and the most widely used material for 3D printing, particularly for craniofacial biostructures [24]. All studies included in this review used PCL, alone or in combination, for scaffold fabrication. PCL has been approved by Food and Drug Administration (FDA) for clinical use and shows good biocompatibility, miscibility, and mechanical strength [25]. Its thermal stability and a low melting point of ~60 °C make it a good candidate for temperature-based printing techniques [26].

Park et al. [15], Pilipchuk et al. [20], and Dubey et al. [22] used PCL alone. However, Pilipchuk et al. [20] integrated 250-μm high pillars at varying depths into the design of the scaffold to function as microscale cues for guiding tissue alignment. The remaining studies used PCL in combination to overcome its shortcomings of low biodegradability (3–4 years) and low cell adhesion [25, 27]. Vaquette et al. [16] fabricated a biphasic scaffold with PCL and TCP (high rate of resorption), which was also used by Costa et al. [18] with a modification of calcium phosphate coating to increase bone-related gene expression. Park et al. [11] used PCL with PLGA, as this improves not only osteoconductivity but also, printing resolution [28, 29]. A combination of PCL and HA was used by Kim et al. [14], Lee et al. [19], and Rasperini et al. [30] as HA improves osteoblast cell adhesion and proliferation. However, since HA is also slowly resorbed,

Table 7.4 Data extraction of human study ($n = 1$)

Author	Year	Study design	Sample size	Objective	Printing technology	Scaffold	Biomolecules used (if any)	Follow-up period	Outcome	Complication (if any)
Rasperini et al.	2015	Case report	1	Multitissue periodontal regeneration in aggressive periodontitis	Selective laser sintering	PCL-HA	rhPDGF-BB	13 months	1. Scaffold remained covered for 12 months with 3 mm gain in CAL and partial root coverage 2. Primarily CT healing. Minimal bone formation	Scaffold exposure at 13 months followed by graft exposure, dehiscence, and wound failure Scaffold, then removed

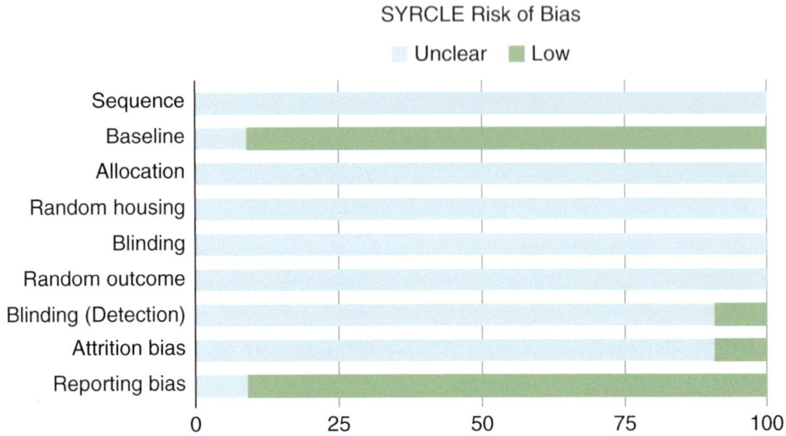

Fig. 7.2 Assessment of risk of bias in animal studies using the SYRCLE's tool

the scaffold used by Rasperini et al. [30] was exposed at 13 months with subsequent microbial contamination of the graft.

7.4.3 Biomolecules

A majority of studies reported the use of cells, growth factors, or both. Cell seeding into the scaffold may be static or dynamic. In the former, acellular scaffolds are printed first and cells are incorporated using a chemical binder. In the latter, cells are printed along with the scaffold material. While this leads to greater uniformity in cell distribution, it decreases the mechanical strength of the scaffold [31]. The most commonly used cell for seeding was hPDLC (Park et al. [15], Pilipchuk et al. [20], and Park et al. [11]). It was seeded statically using fibronectin or phosphoric acid. hPDLCs express Cementum Protein-1, which is a specific marker for cementoblasts and their progenitors. Thus, cell seeding with hPDLC initiates cementum formation and promotes periodontal regeneration [32]. Vaquette et al. [16] and Costa et al. [18] used OB while Lee et al. [19] used a combination of DPSC, PDLSC, and ABSC. DPSC can differentiate into odontoblasts, a property that makes their use favorable even in pulpal regeneration. PDLSCs are multipotent cells and may differentiate into cementoblasts, fibroblasts, and osteoblasts. They can be used in combination with a scaffold or even alone for periodontal regeneration [27].

Cell-free constructs, based on scaffolds loaded with growth factors only, eliminate the need for cell procurement and culture. Kim et al. [14] used SDF-1 and BMP-7 in their study and compared growth-factor-laden scaffolds with growth-factor-free scaffolds. Greater angiogenesis and more cell recruitment to the site of periodontal regeneration were seen in the scaffolds with growth factors. BMP-7 is an FDA-approved biomolecule that demonstrates rapid chondrogenesis, followed by osteogenesis and cementogenesis. Thus, a synergistic interaction exists between the biomolecules and the scaffold wherein bone formation is stimulated by both, osteoconductivity of the scaffold and osteoinductivity of the seeding biomolecules.

7.4.4 Printing Technologies

Extrusion-based printing technology including FDM was most commonly reported in animal studies. It utilizes a mechanical extruder and stage with three-directional movement. In FDM, the material is extruded as filament at high temperatures and deposited on the substrate [33]. Multiphasic scaffold can be constructed using this technology [11, 16, 18]; however, the high temperature used in FDM eliminates the possibility of direct cell printing (bioprinting).

The only reported human clinical study used selective laser sintering (SLS) for scaffold fabrication in which a high-powered laser is used to "sinter" powdered material. It is superior to FDM as it is self-supporting thereby allowing greater design freedom. However, SLS-printed scaffolds are porous and brittle and show considerable cooling shrinkage [34].

7.4.5 Periodontal Tissue Regeneration

The regeneration of the periodontal complex requires the re-establishment of its typical "sandwich structure". Costa et al. [18] modified the biphasic scaffold used by Vaquette et al. [16] and demonstrated increased mineralization and bone formation. Better penetration of blood vessels and some degree of tissue orientation were also evident. Next, Lee et al. [19] developed a highly compartmentalized and seamless triphasic scaffold that achieved the formation of all three supporting structures. These studies recapitulated the "sandwich structure" but achieving precise PDL orientation remained elusive.

Attempts to achieve proper collagen bundle formation and alignment have been made by Park et al. [11, 15, 17]. In the first study, parallel and obliquely oriented fibers were formed by using a composite scaffold. In the latest, the regenerated collagen bundles were perpendicularly oriented to the root surface.

To the best of our knowledge, 3D printing for regeneration of gingival tissue in vivo has not been reported yet. Rasperini et al. [30] reported partial root coverage and 3 mm gain in clinical attachment level; however, these results were short-lived.

7.4.6 Merits and Demerits

The merits of this systematic review include an extensive literature search that was conducted without limitations of language, publication status, or year to ensure maximal data collection. Secondly, SYRCLE's risk of bias tool, an adaptation of the Cochrane risk of bias tool was used for animal studies unlike arbitrary, self-generated tools used in many systematic reviews pertaining to animal studies. With these merits, there are also some limitations of the study. Due to the novelty of this technology and therefore, scarcity of randomized trials, case reports were also included. Another limitation is the number and type of animals used, and the observation period varied greatly between the studies giving rise to heterogeneity in these spheres.

7.4.7 Summary and Future Prospects

Challenges to greater acceptability and more extensive use of 3D printing include optimization of the printing process and development of suitable biomaterials that allow good resolution printing at high speed and low cost. These challenges call for strengthening the collaboration between experts in the fields of medicine, engineering, and materials science. Also, clinical studies applying new developments should be undertaken to assess their benefit over conventional methods of treatment.

References

1. Nazir M, Al-Ansari A, Al-Khalifa K, Alhareky M, Gaffar B, Almas K. Global prevalence of periodontal disease and lack of its surveillance. Sci World J. 2020;2020:2146160. https://doi.org/10.1155/2020/2146160.
2. Park CH. Biomaterial-based approaches for regeneration of periodontal ligament and cementum using 3D platforms. Int J Mol Sci. 2019;20(18):4364. https://doi.org/10.3390/ijms20184364.
3. Chimene D, Kaunas R, Gaharwar AK. Hydrogel bioink reinforcement for additive manufacturing: a focused review of emerging strategies. Adv Mater. 2020;32:e1902026.
4. Guillotin B, Guillemot F. Cell patterning technologies for organotypic tissue fabrication. Trends Biotechnol. 2011;29:183–90. https://doi.org/10.1016/j.tibtech.2010.12.008.
5. Morgan FL, Moroni L, Baker MB. Dynamic bioinks to advance bioprinting. Adv Healthc Mater. 2020;2020:e1901798. https://doi.org/10.1002/adhm.201901798.
6. Matai II, Kaur GG, Seyedsalehi AA, McClinton AA, Laurencin CTC. Progress in 3D bioprinting technology for tissue/organ regenerative engineer-

ing. Biomaterials. 2020;226:119536. https://doi.org/10.1016/j.biomaterials.2019.119536.
7. Sun WW, Starly BB, Daly ACA, Burdick JAJ, Groll JJR, Skeldon GG, Shu WW, Sakai YY, Shinohara MM, Nishikawa MM, et al. The bioprinting roadmap. Biofabrication. 2020;12:022002. https://doi.org/10.1088/1758-5090/ab5158.
8. Hixon KR, Melvin AM, Lin AY, Hall AF, Sell SA. Cryogel scaffolds from patient-specific 3D-printed molds for personalized tissue-engineered bone regeneration in pediatric cleft-craniofacial defects. J Biomater Appl. 2017;32(5):598–611. https://doi.org/10.1177/0885328217734824. Epub 2017 Oct 5
9. Xu H, Han D, Dong JS, Shen GX, Chai G, Yu ZY, Lang WJ, Ai ST. Rapid prototyped PGA/PLA scaffolds in the reconstruction of mandibular condyle bone defects. Int J Med Robot. 2010;6(1):66–72. https://doi.org/10.1002/rcs.290.
10. Oberoi G, Nitsch S, Edelmayer M, Janjić K, Müller AS, Agis H. 3D printing-encompassing the facets of dentistry. Front Bioeng Biotechnol. 2018;6:172. https://doi.org/10.3389/fbioe.2018.00172.
11. Park CH, Rios HF, Jin Q, Bland ME, Flanagan CL, Hollister SJ, Giannobile WV. Biomimetic hybrid scaffolds for engineering human tooth-ligament interfaces. Biomaterials. 2010;31(23):5945–52. https://doi.org/10.1016/j.biomaterials.2010.04.027.
12. Higgins JPT, Thomas J, Chandler J, Cumpston M, Li T, Page MJ, Welch VA (editors). Cochrane handbook for systematic reviews of interventions version 6.2 (updated February 2021). Cochrane. 2021. Available from: www.training.cochrane.org/handbook.
13. Page MJ, McKenzie JE, Bossuyt PM, Boutron I, Hoffmann TC, Mulrow CD, et al. The PRISMA 2020 statement: an updated guideline for reporting systematic reviews. PLoS Med. 2021;18(3):e1003583. https://doi.org/10.1371/journal.pmed.1003583.
14. Kim K, Lee CH, Kim BK, Mao JJ. Anatomically shaped tooth and periodontal regeneration by cell homing. J Dent Res. 2010;89(8):842–7. https://doi.org/10.1177/0022034510370803.
15. Park CH, Rios HF, Jin Q, Sugai JV, Padial-Molina M, Taut AD, Flanagan CL, Hollister SJ, Giannobile WV. Tissue engineering bone-ligament complexes using fiber-guiding scaffolds. Biomaterials. 2012;33(1):137–45. https://doi.org/10.1016/j.biomaterials.2011.09.057.
16. Vaquette C, Fan W, Xiao Y, Hamlet S, Hutmacher DW, Ivanovski S. A biphasic scaffold design combined with cell sheet technology for simultaneous regeneration of alveolar bone/periodontal ligament complex. Biomaterials. 2012;33(22):5560–73. https://doi.org/10.1016/j.biomaterials.2012.04.038.
17. Park CH, Rios HF, Taut AD, Padial-Molina M, Flanagan CL, Pilipchuk SP, Hollister SJ, Giannobile WV. Image-based, fiber guiding scaffolds: a platform for regenerating tissue interfaces. Tissue Eng Part C Methods. 2014;20(7):533–42. https://doi.org/10.1089/ten.TEC.2013.0619.
18. Costa PF, Vaquette C, Zhang Q, Reis RL, Ivanovski S, Hutmacher DW. Advanced tissue engineering scaffold design for regeneration of the complex hierarchical periodontal structure. J Clin Periodontol. 2014;41(3):283–94. https://doi.org/10.1111/jcpe.12214.
19. Lee CH, Hajibandeh J, Suzuki T, Fan A, Shang P, Mao JJ. Three-dimensional printed multiphase scaffolds for regeneration of periodontium complex. Tissue Eng Part A. 2014;20(7–8):1342–51. https://doi.org/10.1089/ten.TEA.2013.0386.
20. Pilipchuk SP, Monje A, Jiao Y, Hao J, Kruger L, Flanagan CL, Hollister SJ, Giannobile WV. Integration of 3D printed and micropatterned polycaprolactone scaffolds for guidance of oriented collagenous tissue formation in vivo. Adv Healthc Mater. 2016;5(6):676–87. https://doi.org/10.1002/adhm.201500758.
21. Shim J, Won J, Park J, Bae J, Ahn G, Kim C, Lim D, Cho D, Yun W, Bae E, Jeong C, Huh J. Effects of 3D-printed polycaprolactone/β-tricalcium phosphate membranes on guided bone regeneration. Int J Mol Sci. 2017;18(5):899.
22. Dubey N, Ferreira JA, Daghrery A, Aytac Z, Malda J, Bhaduri SB, Bottino MC. Highly tunable bioactive fiber-reinforced hydrogel for guided bone regeneration. Acta Biomater. 2020;113:164–76. https://doi.org/10.1016/j.actbio.2020.06.011.
23. Wang CY, Chiu YC, Lee AK, Lin YA, Lin PY, Shie MY. Biofabrication of gingival fibroblast cell-laden collagen/strontium-doped calcium silicate 3D-printed bi-layered scaffold for osteoporotic periodontal regeneration. Biomedicine. 2021;9(4):431. https://doi.org/10.3390/biomedicines9040431.
24. Raveau S, Jordana F. Tissue engineering and three-dimensional printing in periodontal regeneration: a literature review. J Clin Med. 2020;9(12):4008. https://doi.org/10.3390/jcm9124008.
25. Woodruff MA, Hutmacher DW. The return of a forgotten polymer—polycaprolactone in the 21st century. Prog Polym Sci. 2010;35(10):1217–56.
26. Azimi B, Nourpanah P, Rabiee M, Arbab S. Poly (ϵ-caprolactone) fiber: an overview. J Eng Fibers Fabrics. 2014;9(3):155892501400900.
27. Ma Y, Xie L, Yang B, Tian W. Three-dimensional printing biotechnology for the regeneration of the tooth and tooth-supporting tissues. Biotechnol Bioeng. 2019;116(2):452–68. https://doi.org/10.1002/bit.26882.
28. Fielding GA, Bandyopadhyay A, Bose S. Effects of silica and zinc oxide doping on mechanical and biological properties of 3D-printed tricalcium phosphate tissue engineering scaffolds. Dent Mater. 2012;28(2):113–22.
29. Grynpas MD, Pilliar RM, Kandel RA, Renlund R, Filiaggi M, Dumitriu M. Porous calcium polyphosphate scaffolds for bone substitute applications in vivo studies. Biomaterials. 2002;23(9):2063–70.
30. Rasperini G, Pilipchuk SP, Flanagan CL, Park CH, Pagni G, Hollister SJ, Giannobile WV. 3D-printed bioresorbable scaffold for periodontal repair. J

Dent Res. 2015;94(9 Suppl):153S–7S. https://doi.org/10.1177/0022034515588303.

31. Kačarević ŽP, Rider PM, Alkildani S, et al. An introduction to 3D bioprinting: possibilities, challenges and future aspects. Materials (Basel). 2018;11(11):2199. https://doi.org/10.3390/ma11112199.
32. Kadokura H, Yamazaki T, Masuda Y, et al. Establishment of a primary culture system of human periodontal ligament cells that differentiate into cementum protein 1-expressing cementoblast-like cells. In Vivo. 2019;33(2):349–52. https://doi.org/10.21873/invivo.11480.
33. Zein I, Hutmacher DW, Tan KC, Teoh SH. Fused deposition modeling of novel scaffold architectures for tissue engineering applications. Biomaterials. 2002;23(4):1169–85.
34. Williams JM, Adewunmi A, Schek RM, Flanagan CL, Krebsbach PH, Feinberg SE, Hollister SJ, Das S. Bone tissue engineering using polycaprolactone scaffolds fabricated via selective laser sintering. Biomaterials. 2005;26(23):4817–27. https://doi.org/10.1016/j.biomaterials.2004.11.057.

Contemporary Applications of 3D Printing in Prosthodontics

8

Aditi Nanda, Shankar Iyer, Mathew T. Kattadiyil, Veena Jain, Harsimran Kaur, and Dheeraj Koli

Contents

A. Nanda · V. Jain · D. Koli (✉)
Division of Prosthodontics, Centre for Dental Education and Research, All India Institute of Medical Sciences (AIIMS), New Delhi, India

S. Iyer
Malo Smile USA, Elizabeth, NJ, USA

Department of Periodontics and Restorative Dentistry, Rutgers School of Dental Medicine, Newark, NJ, USA

M. T. Kattadiyil
Advanced Education Program in Prosthodontics, School of Dentistry, Loma Linda University, Loma Linda, CA, USA
e-mail: mkattadiyil@llu.edu

H. Kaur
Manav Rachna Dental College and Hospital, Faridabad, Haryana, India

8.1 3D Printing Applications in Prosthodontics

Computer engineered prosthesis has been revolutionizing dentistry at a rapid pace. Despite this transformation, we are living in an age where additive manufacturing (AM) or 3D printing methods of dental prosthesis fabrication coexist and differentiate from the prosthesis fabricated by digitally driven subtractive manufacturing (milling) and traditional manual methods. However, among the three viable solutions of prosthesis fabrication, 3D printing technologies are pushing the boundaries of prosthodontics to distinct, new frontiers rapidly.

3D printing helps to fabricate a tangible prosthesis of numerous shapes by successive addition

© The Author(s), under exclusive license to Springer Nature Switzerland AG 2022

P. K. Chaudhari et al. (eds.), *3D Printing in Oral Health Science*,
https://doi.org/10.1007/978-3-031-07369-4_8

Table 8.1 Sequence of events in AM manufacturing

Step 1	Preparation of a 3D CAD model (computer-aided designing) (from a digital scan or after virtual designing of object to be printed)
Step 2	Creation of .STL (standard tessellation)/.OBJ (object) file
Step 3	Slicing software to break up the CAD model into layers
Step 4	Transfer of files and determination of the path of the tool to print the object
Step 5	AM process/3D Printing
Step 6	Post-processing of 3D printed object

Table 8.2 Factors to consider while selecting AM services [3, 4]

Type of prosthesis to be printed
Accuracy (trueness and precision of printed object)
Speed of printing
Value outcome in terms of cost of investment vs. output return
Characteristics (mechanical, surface, esthetic) and quality of prosthesis to be printed
Compliance of materials used for each method with regulatory requirements of the region of service delivery
Patient satisfaction

Table 8.3 Applications of AM in prosthodontics

Direct application	Partial applications
Surgical guides	Complete denture
Splints and overlay prosthesis	Removable partial denture
Molds for MFP prosthesis	Metal frameworks for fixed or implant prosthesis
Wax patterns	Ceramic printing
Provisional fixed dental prosthesis	
Study cast/dies	

of layers of the material as determined by the computer design. The creation of this tangible object is made possible through different technologies including stereolithography (SLA), selective laser sintering (SLS), fused deposition modeling (FDM), and digital light processing (DLP) [1, 2]. Regardless of the technology implemented, there are a series of steps involved in the fabrication of an additionally manufactured prosthesis and the same have been enumerated in Table 8.1.

In such mosaic availability of technologies, the impending selection of a method of fabrication of prosthesis can befuddle the dentist. Factors to consider when utilizing the 3D printing are numerous and have been shown in Table 8.2 [3, 4].

The versatility of 3D printing in its applications in the field of prosthodontics is overarching which is evident in its widespread utility in all the domains. However, while a definitive 3D printing solution is provided in some conditions, in other applications the solution may only be partial (as seen in Table 8.3). This distinction is largely dependent on the post-processing/post-printing treatment required, which in turn is dependent on the material used for printing. Minimal post-processing steps are involved in direct application, while elaborate steps of post-processing/post-printing are involved in partial solutions.

The broad range of services offered by 3D printing in fabrication of a prosthesis can be accomplished as in office or with assistance from a dental laboratory. For example, smaller restorations with minimal throughput such as splints, provisional restorations, and surgical guides can be efficiently accomplished by using in-office printers. Restorations with an elaborate process, requiring multiaction procedures, and elaborate post-processing are better handled by laboratory printers. Professionals (dentists and laboratory personnel) should clearly know their needs and expected outcomes before they embark to purchase printers (either bench-top/office-based printers which are used preferably in clinics or laboratory-based printers). General considerations include what needs to be printed (purpose of printing), how many prints are expected per day, the upfront cost and recurring budget (per unit material cost, service cost, and maintenance cost) that can be allocated for the printer and the associated products, size of printer, and service availability. Some printers may be needed for education and training purpose only. These printers can belong to any resin printer category, unrelated to biocompatibility. However, if a biocompatible 3D printed object is desired, one has to ensure agreeability of printer infrastructure with the available biomaterials (polymer-based, metals, or ceramics). A range of brands offer multiple solutions, with a wide spectrum of choice in printer size and application. Logically speaking, a printer having chairside operation should preferentially be lightweight, compact, and work with minimum noise production, besides having enhanced speed of operation. Laboratory printers on the other hand should be able to fulfill high printing volumes requirements in order to support the large scale service to a number of dental clinics.

While a discussion on direct comparison among these options is beyond the scope of this textbook, nevertheless this chapter shall lay down the premise of 3D printing in prosthodontic applications enumerating the technical specifications and details for each application.

8.2 3D Printing for Complete Denture Fabrication

Complete dentures have an expanded role in prosthodontic service. Traditional methods, however, involve numerous time-consuming and arduous, technique-sensitive procedures, involving numerous clinical visits and long haul laboratory support. It has been estimated (in the year 2019) that nearly 50 million dentures are produced every year, globally [2, 5–7].

Retaining its status of chosen form of treatment especially in middle- and lower-income countries like India, and the depreciating number of skilled removable denture laboratory technicians world over, the struggle to make ends meet by conventional, time drawn procedures of complete denture fabrication due to lack of labor and manual skills is evident. 3D printed dentures can be a game changer in times to come by simplifying the procedures of technical laboratory aspects, which are often the rate-limiting steps in complete denture fabrication. The philosophy, principles, and clinical steps of conventional complete dentures are same in digital denture fabrication. The difference predominantly exists in the execution of laboratory procedures, which are imparted a digital spin. The utilization of the keen interest of the younger generation in digital tools, computer, and graphics can be put to benefit for the purpose of fabrication of dentures by AM [8, 9]. This section shall review the technical details that a clinician and technical staff should familiarize before embarking on digital denture practice (as seen in Fig. 8.1).

8.2.1 Procedural Details of 3D Printing in Removable Complete Dental Prosthodontics

8.2.1.1 Clinical Visit I: Impression Making (Two-Staged or Single-Staged) and Bite Registration

After a thorough patient examination, the impression making procedure can be conventional two-staged procedures (primary and final impression) or a single-staged procedure incorporating a wash impression in the same visit. The discretion

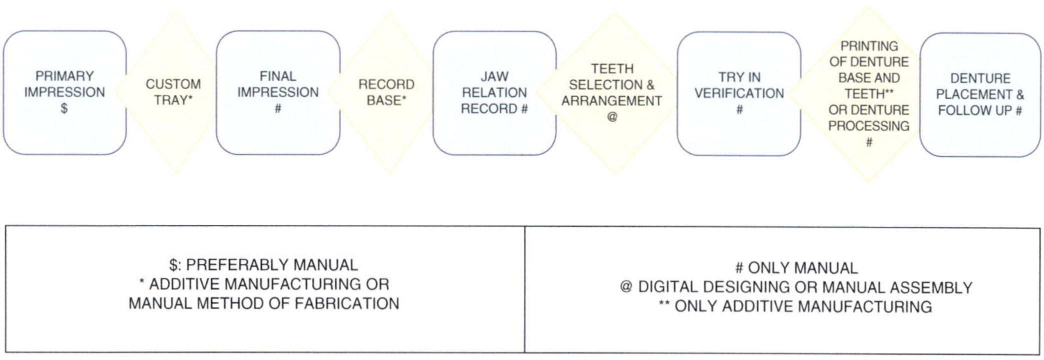

Fig. 8.1 Throughput of dentures (manual and additive manufacturing)

Table 8.4 Options of intraoral scanners. The operators are encouraged to verify the compatibility of output of the scanner with the designing software

Manufacturer and name of intraoral scanner	Light source used
TRIOS from 3Shape, Denmark	Confocal microscopy
CEREC Primescan/ OMNI scan	Structured light (optical triangulation and confocal microscopy)
Carestream	Structured light
Dental Wings, Montreal, Quebec, Canada	Blue laser—Multiscan Imaging™ technology
Emerald®Planmeca, Helsinki, Finland	Red, green, and blue lasers—Projected Pattern Triangulation™
iTero Align Technologies	Confocal laser scanning using a laser beam (red)

of the operator has a role to play in the success of the outcome. Recent reports have also suggested use of stereolithography (by using UV sensitive liquid resin) to print custom trays for the second stage of impression [1, 10, 11]. Although initial strides have been made toward intraoral scanning of the edentulous arches [1, 12] (Table 8.4) and the comfort associated with the procedure of scanning is beneficial, the outcome of such attempts has been questionable primarily due to lack of possibility to attain border seal and difficulty to stitch the images due to lack of clear anatomic landmarks [1, 10–14]. Additionally, the presence of fluids in the oral cavity may impose difficulties in attaining the scan [15].

For bite/occlusal registration record, anatomical devices such as central bearing device [16], an anatomic measuring device [17], or nobelium's intraoral tracers can be used [18]. Conventional bite/occlusal record in wax on a physiologically extended record base on a physical model can also be used. Relevant details in the bite/occlusal registration record must be registered such as lip line, smile line, midline, canine line, and occlusal plane. Recent literature also suggests use of 3D facial scanning in natural rest position with Frankfurt's plane parallel to floor and in exaggerated smile position [19]. This has been documented to help create a virtual patient analog for 3D evaluation of trial in subsequent steps. It must be borne in mind that the facial scanning is not mandatory and that the intraoral scanners and facial scanners are not a part of the 3D printing system and need to be assembled separately from a third party.

Laboratory Procedure I

Step 1: Preparation of Virtual Study Models

This can be accomplished by laboratory scanning of the stone model prepared from the impression and transfer of the digital data. Alternatively, laboratory scanning of the impression for digitization of data can also be done (as seen in Fig. 8.2). It is also highlighted that the laboratory scanners are also not a part of the 3D printing system and need to be acquired separately from the third party (Table 8.5).

The laboratory scanner captures the point cloud data of the cast model or impression by using a two or more camera system and movement of the platform on which the cast or impression rests along a specific number of axes [20]. The data, as collected by the scanner, are stored in a 3D file format that can be read and modified by a 3D processing software. The information in the 3D file format can be either in binary text for instance in STL (tessellation format) or plain text such as OBJ (Wavefront OBJect) or ASC (ASCII: American Standard Code for Information Interchange), depending on the digitizing system. The STL format is simpler and more popular. It is also the native file format to printing, and it encodes the surface of the object into a triangular mesh. The smaller the size of the triangle is, the greater the resolution is. The main limitation of

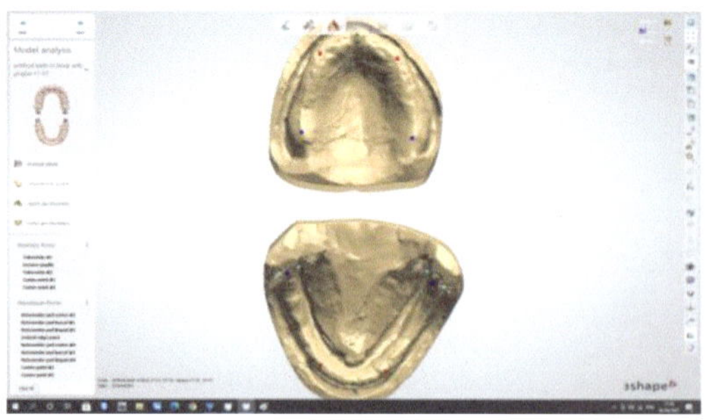

Fig. 8.2 Preparation of virtual maxillary and mandibular models and landmark identification

Table 8.5 Options of laboratory scanners. The operators are encouraged to verify the compatibility of output of the scanner with the designing software

Manufacturer and name of laboratory scanner	Light source used
Generation RED E Scanner from 3Shape Copenhagen, Denmark	Redline laser
Ceramill Amann Girrbach, Germany	Blue light technology
Everest Scan from KaVo Dental, Germany	White light fringe projection
Lava Scan ST, 3M ESPE	White light fringe projection
Zeno Scan S100 from Wieland, Germany	Red laser
Zfx™ Evolution plus by Zimmer Biomet	LED GreenLight Technology
Medit T710	LED blue light with phase-shifting triangulation
inEos X5 by Dentsply Sirona	Digital stripe light projection with blue light
7 Series Dental Wings scanner	Blue laser illumination

the STL file format is that it contains information of only the geometry (size and shape), and the details of color and texture are missing. The OBJ file format is more universal format but more complex. It creates the encoding by using polygons such as hexagons or quadrilaterals, thereby creating a smoother mesh, which is more precise, and contains information about the geometry as well as texture.

Step 2: Designing of Denture

The digital file of the scanned edentulous arches is then sent to CAD software (Table 8.6) in order to proceed with the denture designing [1, 2, 11, 19–21].

In a nutshell, designing of complete denture includes the following procedures:

1. **Identification of Landmarks in the Maxillary and Mandibular Arches**. For the maxillary arch, this includes identification of incisive papilla, corner of the arch related to canine eminence, and prominent part of tuberosity (Fig. 8.2). In the mandibular arch, the retromolar pad (prominent part, buccal extent, and lingual extent), center of arch (to generate the midline plane), and canine eminence or cornerstone of the arch are located (Fig. 8.2).
2. **Elimination of Gross Undercuts**. Gross undercuts that need to be relieved are eliminated or blocked out by the virtual block-out tool based on the path of insertion (Figs. 8.3 and 8.4). The software allows for control of extent and amount of the block out. The need to block out should be verified by the operator based on clinical assessment of the undercut. It is not required in all the cases, and caution must be borne that excess block out can predispose to loss of intimate adaptation of denture base to the tissue, potentially resulting in loss of retention.

Table 8.6 Options of CAD software for digital prosthesis

Full Denture Module by Exocad	inLab SW 20.0 by Dentsply Sirona
Pala Design Studio by Kulzer	DWOS by Dental Wings
Full Denture Design by 3Shape	The denture module by Blue Sky Plan
BDCreator® PLUS by Zfx (Zimmer Biomet)	

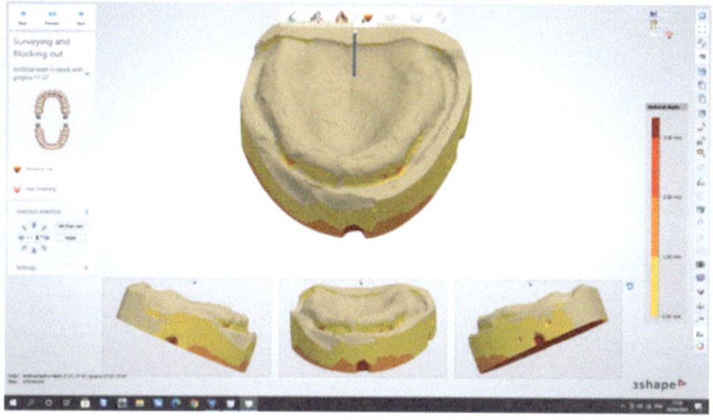

Fig. 8.3 Undercuts relieved in maxillary virtual model

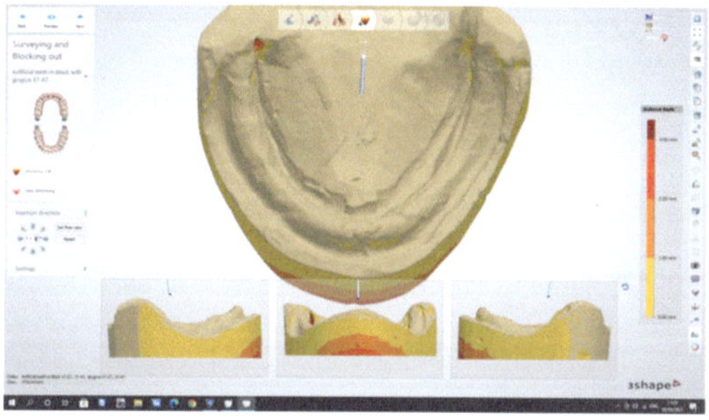

Fig. 8.4 Undercuts relieved in mandibular virtual model

3. **Delineation of Denture Base Area**. Outline of denture bearing area is marked on the maxillary and mandibular cast (Figs. 8.5 and 8.6). The principle of physiological coverage of denture bearing area must be followed. The thickness of denture base is also determined, and for non-load-bearing regions, such as the facial surface of denture dictated by esthetics, the thickness can be reduced. The ease of computer graphics allows for ease of adjustment of the border extensions (Fig. 8.7).

   Simultaneously, sockets for printed teeth are created in the denture base (Fig. 8.8). Since unlike a traditional denture, long root of the ridge lap region of artificial teeth does not guarantee a stronger bond, excessive depth of the recess of teeth socket should be avoided. Also, with excessive root length, the translucency of the tooth can make it appear unnatural. Hence, conventionally a recess is cut at 1 mm depth and with 15°–30° bevel to accommodate the teeth.

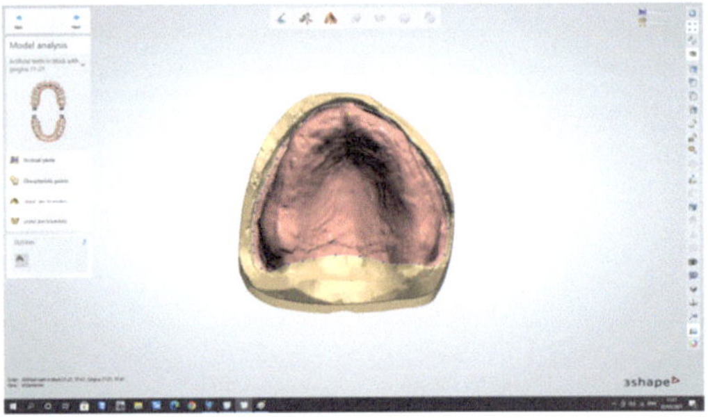

Fig. 8.5 Outline of maxillary denture bearing area

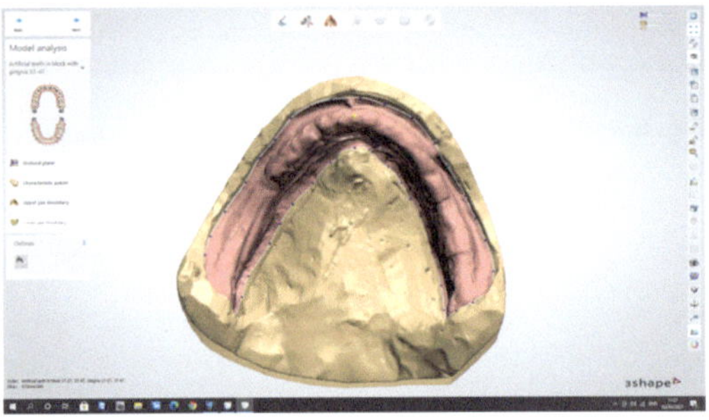

Fig. 8.6 Outline of mandibular denture bearing area

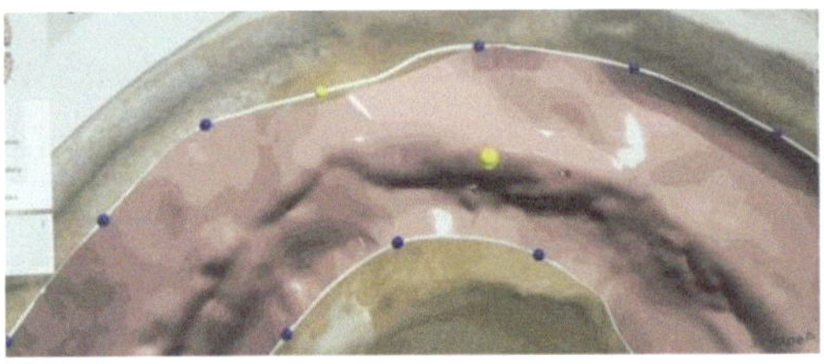

Fig. 8.7 Easy delineation of boundary of denture bearing area

4. **Virtual Articulation**. The bite/occlusal registration record is scanned in the laboratory scanner and superimposed on the virtual models (Fig. 8.9a, b). For the same, the wax bite (occlusal rim) can be sprayed with a matting spray to reduce the reflection of the light and enable complete registration of the details. Commercial options of the spray include CEREC Optispray; Programil by Ivoclar; Scanspray by Renfert; high-resolution scanning spray by Midmark if suggested by the scanner manufacturer.
5. **Establishing the Occlusal Plane**. The position of the occlusal plane is established by using three landmarks on the bite record (midline and anterior one-third of retromolar pad). The accuracy of bite/occlusal registration record in this aspect cannot be more overemphasized. Midline, as registered in the occlu-

Fig. 8.8 Customizing the ridge lap area of the teeth to be printed

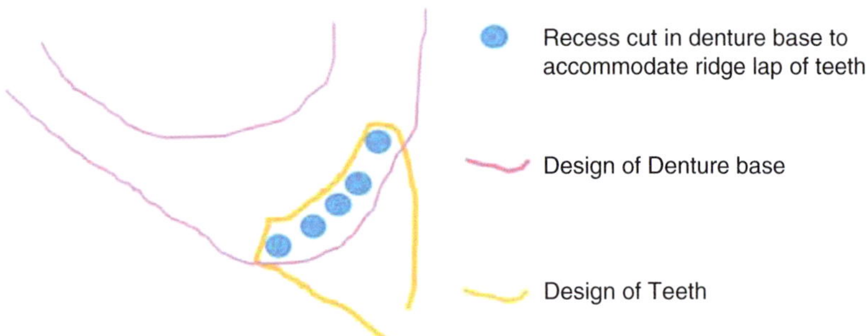

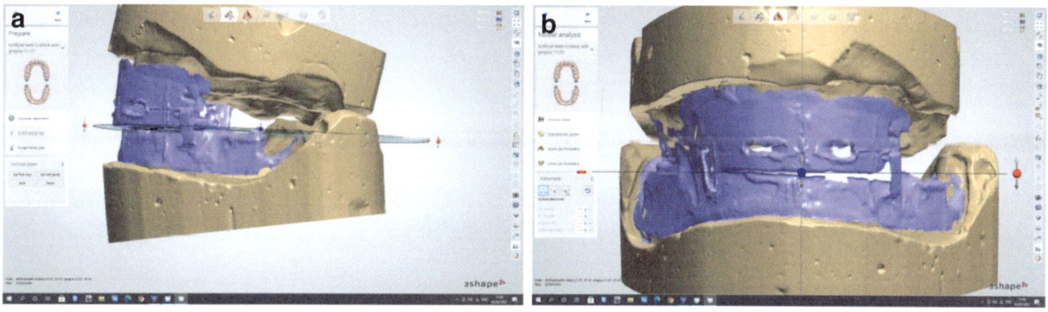

Fig. 8.9 Scanned bite/occlusal registration is superimposed on virtual models. (**a**) Occlusal plane is established (lateral view). (**b**) Occlusal plane is established (frontal view)

sal record, is made to coincide with the *y*-axis of the virtual plane analyzer. The occlusal plane can be rotated to establish a favorable plane of orientation.

6. **Teeth Selection**. Denture teeth can be selected from the tooth library [2, 11, 20, 21]. Interarch distance can be measured to further help in teeth selection (Fig. 8.10). It is pertinent to point out that denture teeth library supported by CAD software may vary. An operator must check the compatibility of the software used for designing with the library available. For instance, 3Shape CAD software supports the teeth library provided by Ivoclar Vivadent, Candulor, Kulzer, Vertex, and Vita. Additionally, the output supported by the CAD software in terms of assembly type may also vary. The output may be either a monoblock assembly type (usually available with the entire digital teeth library as seen with all the earlier mentioned commercial options, except Vita), or stand-alone assembly of teeth may also be possible as enabled by Kulzer, Vertex Dental, and Ivoclar digital libraries. It is also emphasized that operators must consistently verify the digital library support as it can evolve with time.
7. **Teeth Arrangement**. Teeth are aligned on the virtual models according to the occlusal plane (Fig. 8.11a, b). Adjustments to the ridge lap, proximal surface, axial surface, and occlusal surface can be done by using the specific tools of designing software (Fig. 8.12a–c).

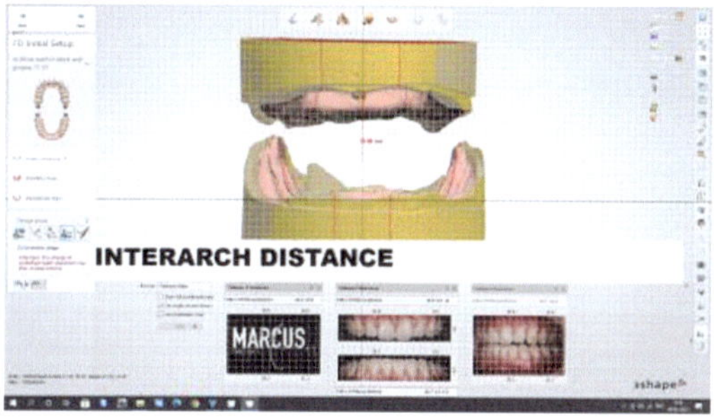

Fig. 8.10 Selection of teeth from the library and use of interarch distance

8. **Establishing the Cameo Surface**. Soft tissue can then be simulated by using the virtual tool as shown in Fig. 8.13a. Minor additions and deletions can also enhance the final outcome by using the virtual tools as shown in Fig. 8.13b. The entire teeth arrangement is then completed as shown in Fig. 8.14a–d.

Step 3: Nesting and Incorporation of Supports

The files created by the denture design software are exported from design software and imported into the software that nests the files for printing in the printer. The mesh of designed denture base and teeth is then interpreted by the slicing software, thereby preparing the object (division of the virtual digital structure into individual, horizontal print layers) for the 3D printer by using

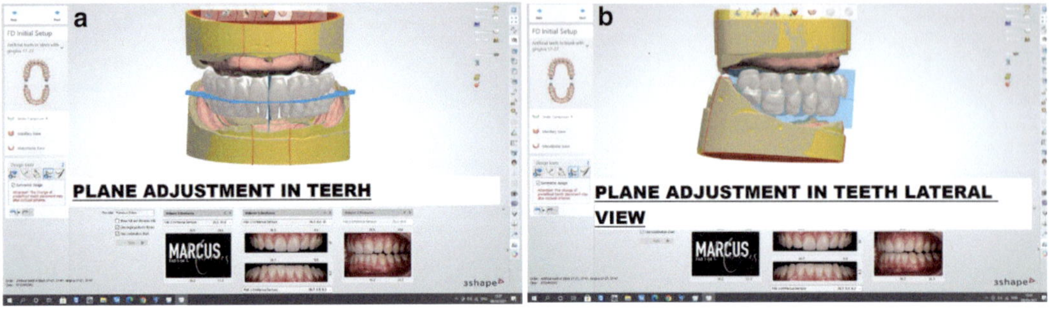

Fig. 8.11 (**a**) Adjustment of teeth according to occlusal plane (frontal view). (**b**) Adjustment of teeth according to occlusal plane (lateral view)

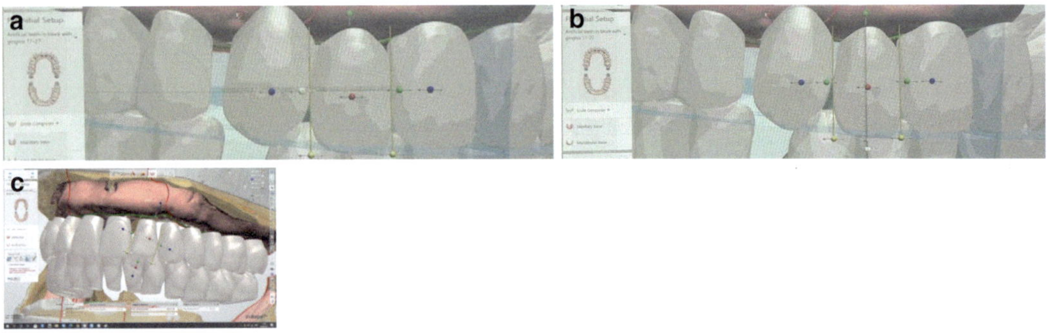

Fig. 8.12 (**a**) Adjustment of width of individual tooth. (**b**) Adjustment of length of individual tooth. (**c**) Adjustment of angulation of individual tooth

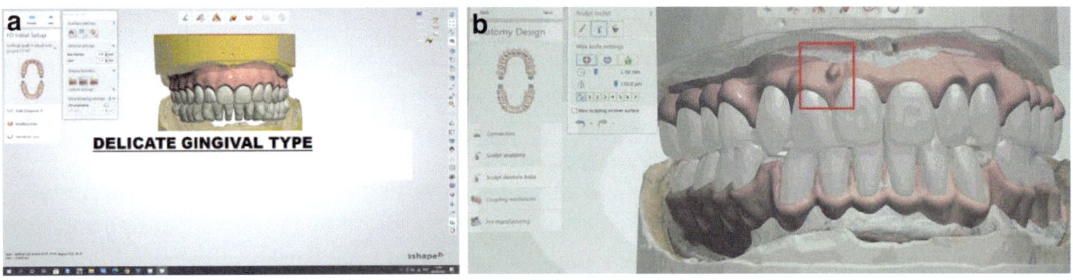

Fig. 8.13 (**a**) Selection and sculpting of gingival wax up. (**b**) Individual areas can be sculpted by use of virtual tools

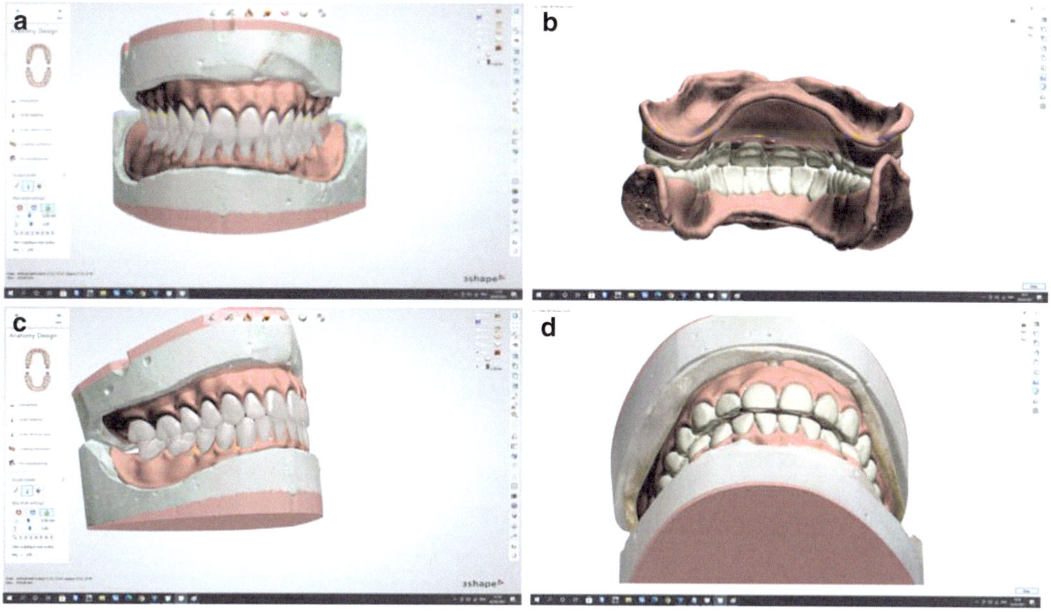

Fig. 8.14 (**a**) Completion of arrangement (frontal view). (**b**) Completion of arrangement (lingual view). (**c**) Completion of arrangement (lateral view). (**d**) Completion of arrangement

the numerical control (CNC) programming language (such as *.gcode files). Thus, the slicing software (Table 8.7) commands the 3D printer on the specific action to perform, including direction of movement, speed of use, and temperature settings.

It is important to discuss a few details of the printer in this context. It is obvious that the first layer of the print is extremely important. The first layer of the print is built upon the build plate or build platform, and it needs to adhere correctly without any movement. In case of movement of the first layer, the subsequent layers do not align as per plan and the print does not conform to the recommended geometric details required. The build plate or platform characteristics are important to affect the crucial first step of buildup. The orientation or arrangement of the denture base to be printed (nesting) and the placement of the supports are also crucial in this aspect (Figs. 8.15 and 8.16). Nesting is the layout strategy that helps to arrange the 3D files by maximizing the space in 3D printer without compromising the accuracy [22].

Due to its contour, numerous overhangs and undercuts are present in prosthesis. Consequently, during printing, there may not be any layer present directly beneath the printed portion. The purpose of the support is to provide a temporary hold to this overhang portion of the prosthesis during the printing, thereby holding the prosthesis in the selected alignment throughout the printing process. Supports in turn consist of three components, namely the rafts, scaffolding, and touch points. The raft forms a base (attachment layer) that adheres to the build platform. Scaffolding is the emerging stems from the raft that have the purpose of securing the object as it prints. Touch points are areas of contact where the scaffolding meets the prosthesis. Supports are planned both for the denture base and for the teeth, individually, unless it is for a monolithic trial placement. The supports touch points must be avoided at the area of junction of teeth and the denture base and must be planned on the arch, cusp tips, and incisal edges of teeth facing the build platform. Supports have to be printed in a low density of the selected material so as to enable easy removal after fabrication [5].

Table 8.7 Printers available for digital printing of dentures with corresponding software and light source. CAD libraries interpreted by the printers and the material for printing may be specific. Consumers are encouraged to check the throughput process of integration of printer with 3D design software and certified material that the printer can print. Direct light processing uses visible light, and SLA uses laser for printing purpose. Accuracy in *z*-axis refers to minimum height of each printed layer

Printer	Software used	Light source for printing	Technical specifications Accuracy (μm)/ resolution (pixel)	Material used	Resource
SprintRay MoonRay S100 3D printers	RayWare	Digital light processing (405 nm light-emitting diode)	Accuracy in *x*-axis and *y*-axis: 75–100 μm *z*-axis: 20 μm	Open system including SprintRay materials (for monolithic try-in) as well as certified materials like NextDent Denture Base; DENTCA Denture Teeth; DENTCA Denture Base	https://sprintray.com/moonray-desktop-3d-printer/technical-specifications/
Carbon printer (M series)	Browser-based software interface	Carbon Digital Light Synthesis™ (Carbon DLS™)[a]	Accuracy *x*-axis and *y*-axis: 75 μm	Closed system Used by 1. DENTCA Denture Base and Denture Teeth 2. Lucitone Digital Print™ 3D Denture resin	https://www.carbon3d.com/products/m2-3d-printer/
Rapid shape (D20 II)	3DPrinterOS NetFabb	385-nm UV LED	Accuracy in *x*-axis and *y*-axis: 34 μm	Open system	https://www.rapidshape.de/images/dental/datasheet/RASH_Datasheet_D20.pdf
Dental Wings D30 and D40printer (in partnership with Rapid Shape)	DWOS Dental Wings Inc., Canada (Straumann group brand)	385-nm UV LED	Resolution for D40: dual HD 1920 × 1080 px Resolution for D30 and D20: HD 1920 × 1080 px	Open system	https://pdf.medicalexpo.com/pdf/rapidshape-gmbh/d30-d40/102944-151175.html
Asiga Digital Dental Solutions	Asiga Composer	High-power UV 385-nm LED	Accuracy in *x*-axis and *y*-axis: 62 μm	DentaBASE and DentaTOOTH (by Asiga dental materials) Open material compatibility with Next Dent, Dentca, FotoDent (by Dreve, GmBh) and numerous other materials	https://www.asiga.com/products/printers/max_series/max_lcd/
NextDent 5100 printer from 3D System printers	3D Sprint, 3D Systems' advanced software	405-nm noncontact membrane digital light printing (DLP)[b]	Resolution: 1920 × 1080 pixels Accuracy in *x*-axis and *y*-axis:65 μm	Closed system: NextDent Denture 3D+	https://www.3dsystems.com/3d-printers/nextdent-5100

Kulzer Cara Print	Kulzer Cara CAM software	Digital light projection (DLP)	Accuracy in x-axis and y-axis: 54 μm Accuracy in z-axis: 30–100 μm	Closed system dima® Print Materials (denture base in pink, teeth, and try-in)	https://www.kulzer.com/en/en/products/cara-print-4-0.html
Envision One	The Envision One Rapid Prototype (RP) software	385-nm (UV) LED[c]	Accuracy x-axis 82 μm and y-axis 105 μm Accuracy in z-axis: 50–150 μm	Closed system: NextDent Denture 3D+ E1-Dent 400 MFH (microfilled hybrid crown and bridge material)	https://envisiontec.com/wp-content/uploads/2019/02/2018-Envision-One-Dental.pdf
Miicraft 125	Utility software (MiiUtility)	Digital light projection (DLP) 385/405 nm wavelength	Accuracy in x-axis and y-axis is 65 μm	Open system Any material with Class I and II a certification	https://miicraft.com/product/
Formlabs	PreForm software	Stereolithography (by using laser)	Accuracy in x-axis and y-axis: 100–140 μm Accuracy in z-axis: 25 μm	Closed system than can run in open mode Open mode disables the wiper and heater which can cause trouble when printing non-Formlabs resins	https://formlabs.com/
Roland DWP-80S	Ver 1.1	Digital light projection (DLP)	Accuracy x-axis, y-axis, and z-axis: data not available	Closed system, indicated for custom trays, base plates, and frameworks for lost wax technique	https://www.rolanddga.com/support/products/dental-cad-cam/dwp-80s-dental-3d-printer

nm nanometers, *μm* micrometers

[a]Includes CLIP (continuous liquid interface production) technology and thermal processing

[b]Figure 8.4 Technology (an air-based adhesion solution to overcome the sticking of newly solidified resin to tank bottom)

[c]Continuous digital light manufacturing (cDLM): Continuous motion of the build plate in the Z direction eliminates any rocking of the platform that can allow for accuracy and eliminates center to edge inaccuracy

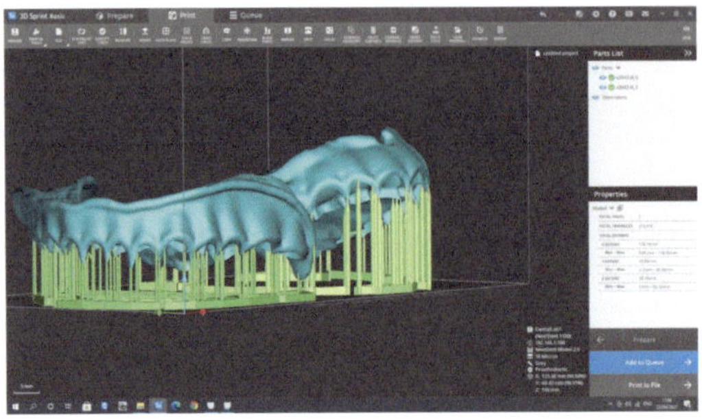

Fig. 8.15 Nesting and addition of supports (for denture base)

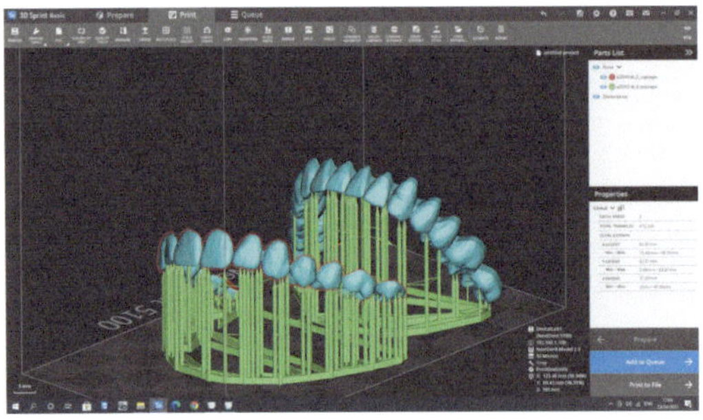

Fig. 8.16 Nesting and addition of supports (for teeth)

Step 4: Printing

The printing process is initiated after inserting the build platform or build plate, resin cartridge, and resin tank or tray. It is important to shake the resin cartridge thoroughly, and for this, a special equipment such as a roller/tilting stirring device for mixing 3D printing materials may be recommended by some manufacturers before pouring in the resin tray of the printer by the specific printer manufacturer.

Step 5: Post-processing

The printed denture base and teeth are removed from the build platform (by a specially recommended tool). The printed objects have to be washed in isopropyl alcohol (96% for about 10 min) to dissolve any uncured resin. An ultrasonic bath can be used for the same. Compressed air can be used to remove alcohol from the objects. The supports have to be removed in a stepwise manner. The raft is removed first followed by the entire scaffolding. The area after removal of the support has to be smoothened out. If the teeth have been printed together, the embrasures may have to be widened in order to create a more esthetic and natural appearance. The denture base and teeth should be washed in isopropyl alcohol one last time to ensure complete removal of debris.

Step 6: Assembly

The printed denture teeth (Figs. 8.17 and 8.18) are inserted into the denture base (Figs. 8.19 and 8.20) to verify the fit. The unpolymerized denture base resin is painted in the sockets for the teeth in the denture base. The teeth are inserted in the socket and held in position by applying firm pressure. Any excess resin must be removed by a fine brush. Light cure unit can be used to cure the junction of the teeth and denture base. Any deficiencies can be corrected by re-addition of resin and subsequent repolymerizing.

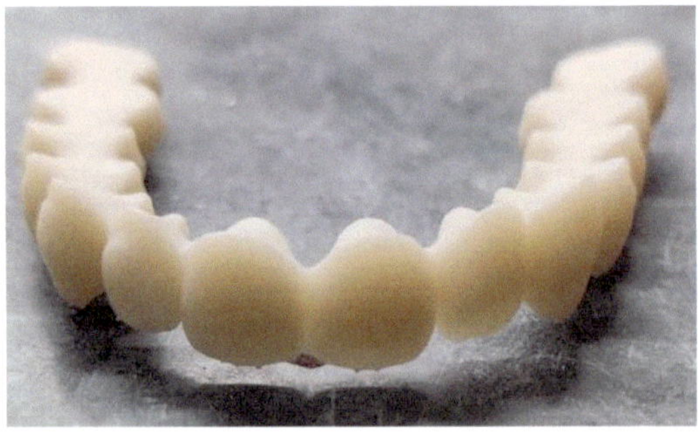

Fig. 8.17 3D printed teeth (by DLP)

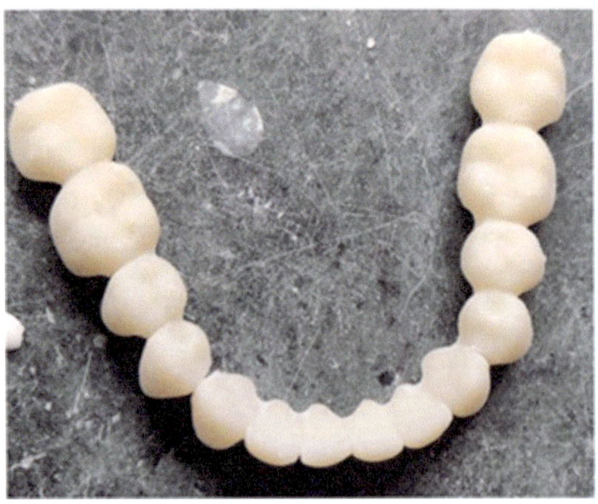

Fig. 8.18 3D printed teeth (by DLP)

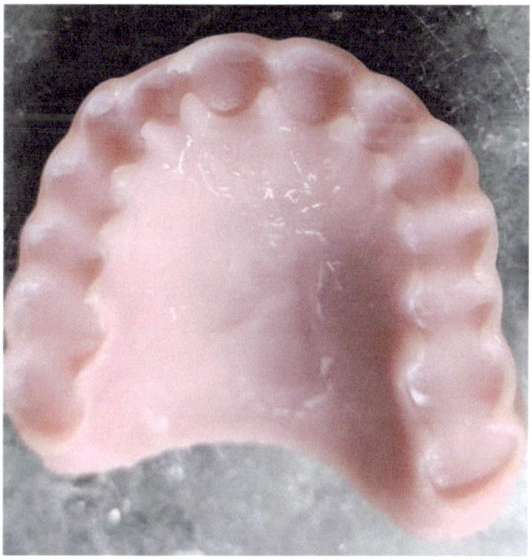

Fig. 8.19 3D printed denture base (DLP)

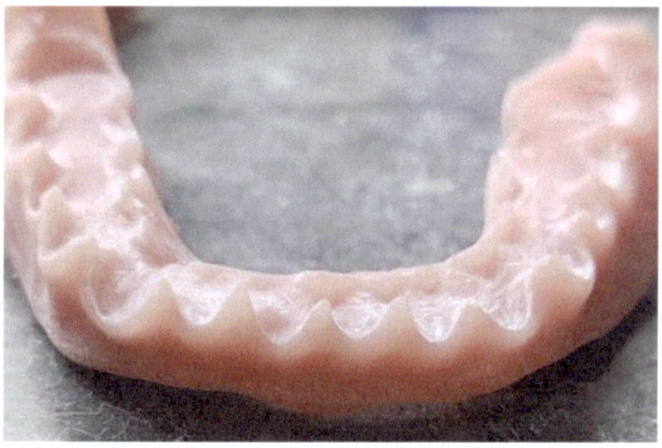

Fig. 8.20 3D printed denture base (DLP)

Step 9: Post-cure

Finally, immersion of the denture is done in glycerin at about 80 °C for a final cure in a dedicated post-cure unit for about 30 min. The unit has a temperature control, with a full light spectrum (usually within the range of 300–550 nm) for completion of polymerization and must ensure illumination from all sides. The denture should be flipped to complete polymerization of opposite side. Traditional polishing can be completed by using a wool buff and pumice.

8.2.1.2 Clinical Visit II

Most of the commercial systems supporting denture printing allow the printing of a try-in base in monolithic resin (Fig. 8.21). This step holds tremendous value just like it does in conventional or

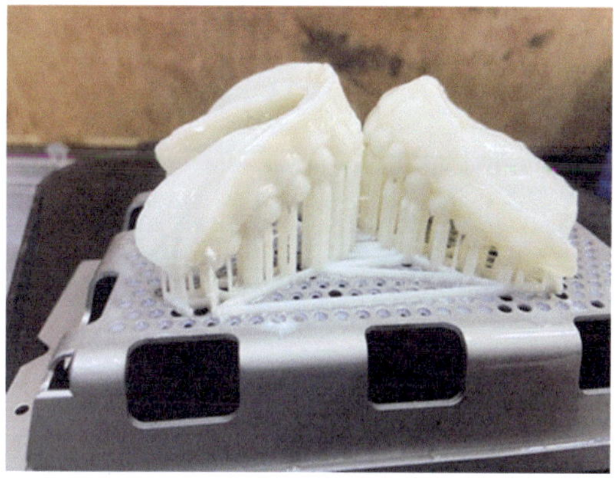

Fig. 8.21 3D printed monolithic try-in base before removal of supports and post-curing

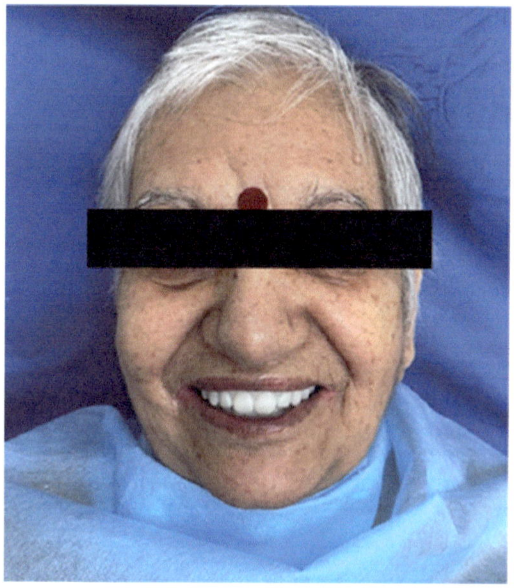

Fig. 8.22 Verification of esthetics, vertical dimension

traditional denture fabrication procedure. The trial base can be verified for fit, extension, and adaptation of the denture base on denture bearing area. Verification of jaw relation (Fig. 8.22), visibility of teeth, esthetics, and general alignment of teeth is also possible and is strongly recommended. However, in contrast to conventional dentures, only minor alterations in tooth length adjustment (Fig. 8.23) or tooth shape (Fig. 8.24) is possible with the monolithic trial. The corrections are then scanned by laboratory scanner to improvise on the existing design. Some authors have also suggested a completely virtual try-in by using a face scan,

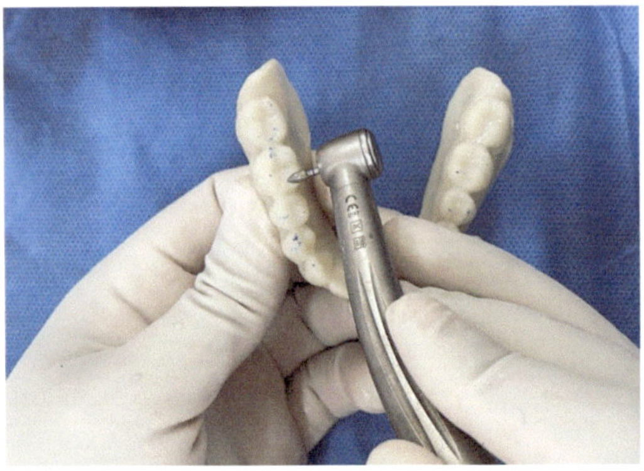

Fig. 8.23 Minor adjustments in trial base can be accomplished and scanned before printing

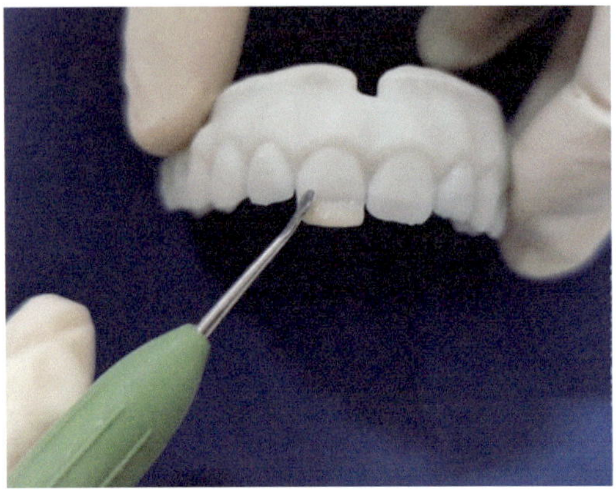

Fig. 8.24 Alteration of length of teeth by addition of composite

intraoral scan, and the designed denture data, although the success of the same has been documented to be limited [21, 23]. Certain other authors have quantitatively evaluated tissue adaptation of a CAD and printed denture by using wax and found it comparable to compression-molded, manual denture base fabrication technique. However, a successful method of try-in is still not established in an infallible manner with AM dentures [24].

Laboratory Procedure II

In case, modifications have been done to the monolithic trial denture, the denture is scanned, and design is rectified. For final denture base to be printed nesting, addition of supports, printing process per se, post-processing wash, removal of supports, assembly, and post-cure are performed as discussed earlier.

8.2.1.3 Clinical Visit III

The completed denture (Fig. 8.25a–c) can be placed in the patient following guidelines in conventional denture placement.

Take-Home Points Regarding 3D Printed Removable Complete Denture

When compared with traditional or digitally milled dentures, 3D printed dentures provide a plethora of features that are bound to change the

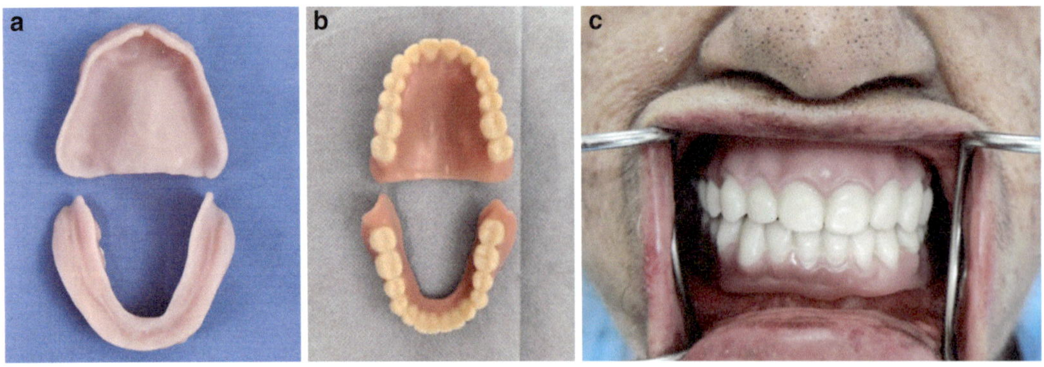

Fig. 8.25 (**a**) 3D printed dentures (intaglio surface). (**b**) 3D printed dentures (polished and occlusal surface). (**c**) 3D printed denture (intraoral)

prosthodontic landscape in future [22–37]. The throughput takes less time when compared to conventional denture fabrication protocol (Fig. 8.1). The cost of inventory and material wastage is less when compared to milled complete dentures. The resin substrate that is unused in 3D printing can be recycled for the subsequent prosthesis. This is a more sustainable form of prosthesis fabrication that helps in conservation of resources, adds to economic benefits in long run, and has minimum environmental concern [25, 26]. Patient satisfaction has been reported to be higher when compared to conventional dentures in a few studies, probably due to reduced chairside time and reduced number of patient visits [8, 27, 28]. Substantially improved fit has been observed in digitally printed bases when compared with denture bases fabricated by milling or traditional compression molding [29]. 3D printed dentures allow for use of truly customized teeth when compared with traditional dentures where pre-manufactured denture teeth have to be adjusted to fit within the arch specifics of the patient. Regarding the properties of the printed teeth such as resistance to chipping, tensile fracture resistance, and wear resistance, reports suggest an outcome comparable to that of the prefabricated denture teeth [30, 31]. Additionally, accurate, consistent, and high-quality results with intuitive adaptable tools are a feature that assures a strong foothold of digitally printed dentures in times to come [5, 25]. Archiving of electronic records, creation of patient repositories, and drastically reduced generation of biohazard waste is a win–win with 3D printed dentures. The applicability of digital dentures is widespread, encompassing maxillary and mandibular removable complete dentures, single arch dentures, denture duplication, immediate dentures, and implant overdentures [1, 2, 11].

However, 3D printing for dentures still has a few miles to cover mainly due to the associated shortcomings. Much skepticism has been associated with the bond strength of the printed teeth adhered to the printed base [1, 32]. Contradictory findings have been reported for fit accuracy and dimensional accuracy of 3D printed dentures when compared to milled denture base in a few studies, leading to a lack of robust unanimous consensus [25, 33]. Divergent observations have also been made for mechanical properties of the materials used for 3D printing especially in relation to flexural properties and surface hardness through the in vitro studies. Although these values have been found to be within clinical limits, much needs to be achieved in terms of upgrading the materials for 3D printed denture [34–36]. Much has been also been discussed in the literature regarding the effect of build direction or orientation of the layers during printing on the mechanical properties, particularly compressive strength. Many studies favor printing in a direction such that the direction of load is perpendicular to direction of printing (vertical direction) [37]. High initial cost of equipment and the need to train staff and develop a strong collaboration between the dentist, technician, and research and development industry for a successful competitive advantage are also strongly urged to enable a deeper percolation of 3D printed denture fabrication in prosthodontic treatment options [8, 38]. The inherent risk of microbial colonization, particularly *Candida* species, is still associated with the material used for 3D printing of dentures. Attempts to overcome the same have been made by incorporation of titanium oxide and its corresponding nanoparticles in the literature [39–42]. The inability to acquire aesthetic gradations and hence a vital tone in the teeth due to their monochromatic appearance in AM is evident. This is in contrast to the cross-linked, layered teeth used for conventional methods of denture fabrication or the teeth milled out of layered pucks for CAD CAM dentures. Observing the pros and cons of AM manufactured dentures, one can also consider a hybrid approach of digital denture fabrication by using milled resin teeth with printed base for attaining the desired outcome. Also, despite the attempts to improvise the materials, water absorption and color changes of the resins have also been observed. Conclusively, 3D printing of complete dentures is a work in progress, having a growth curve of its own. Leveraging on numerous advantages, it can prove to have a monumental role in transforming the operations and strategies of dentists, especially in sub-developed regions of the world.

8.3 Fixed Prosthodontics

Reconstitution of integrity of dental arch, lost either due to congenital or acquired causes, by fixed dental prosthesis (FDP) has profound improvements on the quality of life of the patients [43]. Much of the recent practice has been focused on optimization of technology to provide contemporary rehabilitation methods and facilitate the otherwise technically intensive manual procedures [1]. Among these methods, AM has been considered to impart superior-quality, circumvent laborious procedures, eliminate the possibility of human error, and reduce the cost and overall treatment time [44–47]. The current section shall discuss the available AM technologies as applied in assembly line of FDP fabrication.

8.3.1 3D Printed Casts and Dies

There are three common ways of fabricating tooth-borne FDPs by additive manufacturing technique. The first is complete, anatomic digital designing and printing of the prosthesis (in metal or ceramic). The second is to digitally design and print the metal framework. On the printed framework, ceramic is veneered to create external surface form and establish proximal and occlusal contact points. The third method is to digitally design and print a castable pattern (in 3D printed wax or resin) and use this pattern for traditional method of prosthesis fabrication. Complete virtual designing is possible for the first option. However, for the second and third options, fabrication of the prosthesis requires creation of an accurate analog. The traditional gypsum-based casts are prone to fracture and abrasion [48]. The 3D printed casts or dies are resistant to deformation as they are materialized in nylon or styrene (Fig. 8.26) after obtaining digital impressions or digitization of the physical impression [49]. They overcome the shortcomings of gypsum cast and also provide the possibility of fabrication of multiple replicas simultaneously [48].

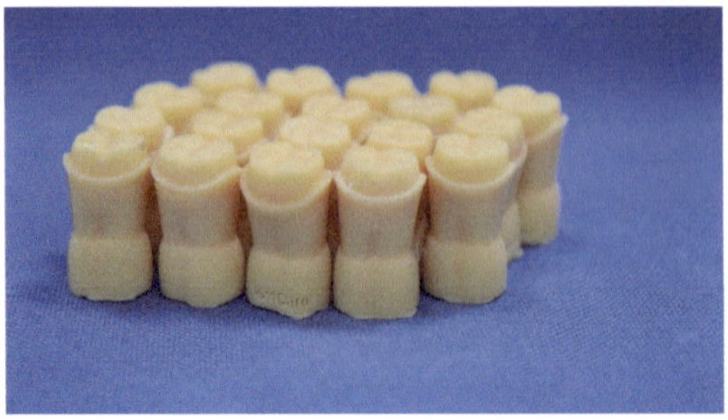

Fig. 8.26 Resin-based dies made by DLP

Consequently, for the fabrication of the definitive restoration, a realistic analog or a cast is important if the framework is printed. 3D printing for dental casts or dies can be done by using material jetting (MJ) or by digital light processing (DLP) printers. The material jet printers deposit tiny drops of supporting as well as building materials. These printers provide good accuracy to the die or casts. They are, however, unable to reproduce detail and have inferior mechanical properties. The DLP (UV or light-emitting diodes) as in complete dentures fabrication photopolymerizes resin, providing fast production of dies, although the choice of material used is restricted [49–53]. It has been supported that the volumetric changes in 3D printed casts are well within clinical acceptable limits for diagnosis as well as prosthesis fabrication. Among the options of 3D printers, DLP using UV light shows the least volumetric change of the casts and dies [49, 54, 55].

8.3.2 Metal Frameworks of Tooth-Supported and Implant-Borne FDPs

Metal frameworks can either be fabricated by direct metal printing or by burnout and casting of the 3D printed wax/plastic pattern. While the 3D printed pattern used for burnout and casting is prepared by SLA/DLP (Table 8.8), the metal framework is fabricated by employing powder bed fusion (PBF) technology, which includes

Table 8.8 Printers commonly used for fixed dental prosthesis (laboratory-based models)

Commercial company	System	Technology	Resolution/layer thickness	Build materials	Resource
BEGO	Varseo XS	DLP (405 nm) followed by light cure	Resolution in *x*-axis and *y*-axis: 50 μm	VarseoWax CAD	https://www.bego.com/3d-printing/3d-printer/3d-printer-varseo-xs/
3D systems	ProJet 1200	Micro-stereolithography for materials that can be casted	Layer thickness: 0.03 mm. Resolution in *x*-axis and *y*-axis: 56 μm	VisiJet FTX Green (plastic material), FTX Cast (UV-curable plastic with wax), FTX Gray (UV-curable plastic)	https://www.3dsystems.com/images/projet-1200-3d-printer-parts-3
	DMP Flex 100	DMLS (direct metal laser sintering). Laser power: 100 W Type: fiber laser. Wavelength: 1070 nm	10–100 μm	Laser form CoCr (C and B)	https://www.3dsystems.com/3d-printers/dmp-flex-100
	ProX® DMP 200 (rebranded Phenix system)	DMLS (direct metal laser sintering). Laser power: 300 W Type: fiber laser. Wavelength: 1070 nm	30 μm	Laser form CoCr	https://3dprintedparts.com/3d-printers/prox-dmp-200/

(continued)

Table 8.8 (continued)

Commercial company	System	Technology	Resolution/layer thickness	Build materials	Resource
Concept Laser, GE Additive	Concept Laser Mlab	DMLS (direct metal laser sintering). Laser power: 100 W Type: fiber laser. Focus diameter of 50 μm	15–30 μm	Remanium star® CL (CoCrW) by Dentaurum	https://www.ge.com/additive/additive-manufacturing/machines/dmlm-machines/mlab-cusing
	Concept Laser Mlab R	DMLS (direct metal laser sintering). Laser power: 100 W Type: fiber laser. Focus diameter of 50 μm	15–30 μm	Titanium Ti6Al4V ELI Grade 23, Titanium CPTi Grade, Remanium star® CL (CoCrW), rematitan® by Dentaurum	https://www.ge.com/additive/additive-manufacturing/machines/dmlm-machines/mlab-r
	Concept Laser Mlab 200R	DMLS (direct metal laser sintering). Laser power: 200 W Type: fiber laser. Focus diameter of 75 μm	15–30 μm	Titanium Ti6Al4V ELI Grade 23, Titanium CPTi Grade, remanium star® CL (CoCrW), rematitan® by Dentaurum	https://www.ge.com/additive/additive-manufacturing/machines/dmlm-machines/mlab-200r
	M2 cusing Multilaser	Laser system: Dual × 200 W with variable focus diameter (50–500 μm)	20–80 μm	Titanium alloy (TiAl64V ELI) Pure titanium Grade 2 Nickel-based alloy Cobalt chromium alloy (Remanium star by Dentaurum)	https://www.ge.com/additive/sites/default/files/2018-02/M2%20cusing%20Multilaser_EN.pdf
Renishaw, UK	Renishaw AM250 AM	Laser power: 200 W type: fiber laser Focal diameter: 70 μm	20–100 μm	Cobalt chrome Titanium stainless steel,, aluminum, and other inconel-based alloys	https://resources.renishaw.com/gen/details/am250%2D%2D73743
EOSINT	EOS M 100 (Electro Optical system)	Laser type: ytterbium (Yb) fiber laser; with power of 200 W Focus diameter 40 μm	40 μm	Open system BEGO (Co—Cr—Mo) SP2 powder BEGO Wirobond, Bremen, Germany EOS Cobalt Chrome SP2 EOS Titanium Ti64	https://www.eos.info/en/3d-printing-examples-applications/people-health/medical-3d-printing/dental

EKZO, SLM Solutions GmbH	SLM 125HL	Fiber laser with 70–100 μm of focus diameter	20–75 μm	Closed system for Inconel (nickel, chromium, and iron alloy) and Hastelloy X (nickel-based alloys), chrome cobolt alloys, stainless steel, aluminum, and titanium Ti6Al4V Pure titanium	http://www.3dprintekzo.be/products-services/3d-printers/slm-125hl/
	SLM 280HL	Fiber laser with 80–115 μm of focus diameter	20–75 μm		http://www.3dprintekzo.be/products-services/3d-printers/slm-hl280/
	SLM 500HL	Fiber laser with 80–115 μm of focus diameter	20–75 μm		http://www.3dprintekzo.be/products-services/3d-printers/slm-500hl/
Arcam EBM, GE Additive manufacturing	Arcam EBM Spectra L	Electron beam with power 4500 W	–	Closed system Ti and Ti alloys	https://www.ge.com/additive/additive-manufacturing/machines/arcam-ebm-spectra-l
	Arcam EBM Spectra H	Electron beam with single-crystalline cathode of power 6000 W	–	Closed system Ti and Ti alloys	https://www.ge.com/additive/additive-manufacturing/machines/ebm-machines/arcam-ebm-spectra-h
	Arcam EBM A2X	Electron beam with tungsten filament cathode Power 3000 W 250 μm beam diameter	–	Closed system Ti and Ti alloys	https://www.ge.com/additive/additive-manufacturing/machines/ebm-machines/arcam-ebm-a2x
	Arcam EBM Q10 plus and Arcam EBM Q20 plus	Electron beam with tungsten filament cathode Power 3000 W 140 μm beam diameter	–	Closed system Ti, Ti alloys Chrome cobalt alloys	https://www.ge.com/additive/sites/default/files/2020-04/EBM_Q10plus_Bro_4_US_EN_v1_0.pdf

direct metal laser sintering (DMLS), selective laser melting (SLM), and electron beam melting (EBM) (Table 8.8). In a generic way, the impression is scanned using the laboratory scanner or an intraoral impression of the abutment teeth is made (Tables 8.4 and 8.5). The prosthesis is designed by using the CAD software (Table 8.6 and Fig. 8.27a–c). The nesting or orientation of the prosthesis to be printed is done (Fig. 8.27d), and supports are virtually added (Fig. 8.27e). The data are then transferred in the universal language (for instance STL) to the printer where the slicing software is used and 3D printing is completed (Fig. 8.28a, b) by DMLS or EBM.

The method of fusing the metal is different in all the three techniques. Post-print processing, such as hardening, heat treatment, or hot isostatic pressing (HIP), is performed so as to make the components' properties suited to meet specific requirements. The final outcome of the printed and processed metal is largely dependent on the technology used, the grain size of metal powder (3–14 μm), the orientation of build platform, use of vacuum/argon/nitrogen/helium atmosphere, speed and size of beam for printing, energy power (100–500 W to 7 kW), and energy source (fiber lasers, Nd:YAG, or electron beam), and layer thickness (10–50 mm) [56, 57].

It is noteworthy that the current AM technologies use either chrome cobalt (SLS) or titanium alloys (EBM) for metal framework fabrication. Nickel-based alloys have not gained popularity due to possibility of allergic reactions. As both titanium and cobalt are highly reactive with oxygen, strict control of non-oxidative atmosphere is ensured when the metal undergoes AM. Besides

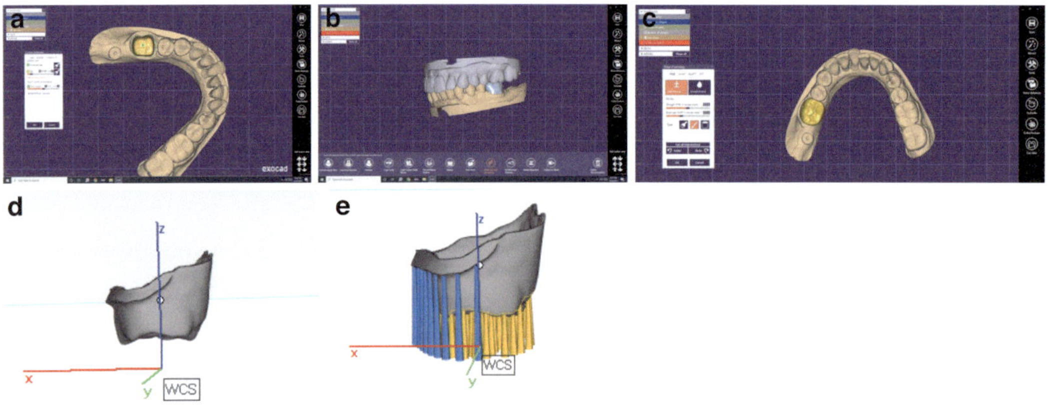

Fig. 8.27 (**a**) Determination of margin and spacer. (**b**) Designing of restoration. (**c**) Designing of restoration. (**d**) Nesting of the virtual prosthesis. (**e**) Addition of supports

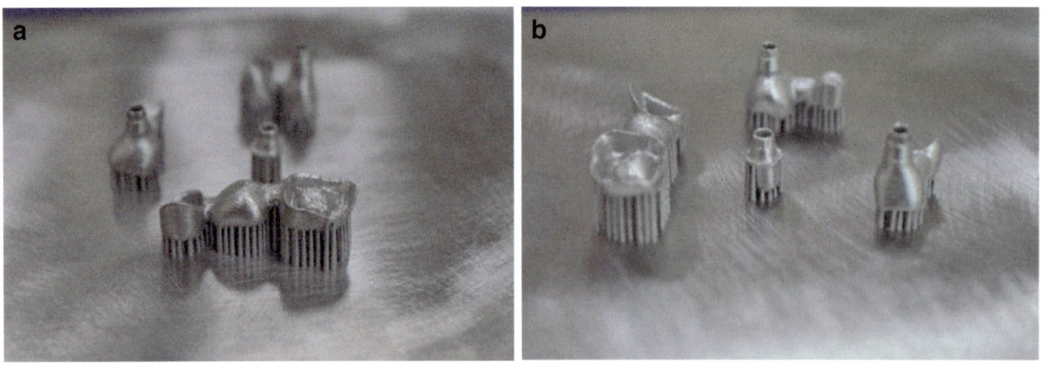

Fig. 8.28 (**a**, **b**) 3D printed frameworks prior to post-print processing

the principal elements, the minor elements in the metal powders to be printed have defined roles. Molybdenum in metal powder reduces susceptibility to pitting corrosion by influencing the grain size. Tungsten causes solid solution strengthening. Silicone and manganese enhance the alloy fluidity. Niobium influences intermetallic phase formation and solution strengthening [58–61]. Current studies suggest mechanical properties of metal frameworks fabricated using AM procedures are comparable to those that are milled and better than those that are fabricated through conventional casting techniques [59, 62–65].

3D printing of metals offers reduced fabrication cost and time and enhanced product density, overcomes the casting defects, and minimizes human intervention. It has, however, been pointed out that a rougher surface microstructure (3–4 μm) is observed with metal 3D printing when compared to conventional casting (2–3 μm) which may be clinically relevant. An upside associated with AM of metal prosthesis is that the bond strength between veneering ceramic and 3D printed alloys is comparable to cast metals, although failure types need further analysis [58, 66]. Literature search also reveals favorable and clinically acceptable marginal gap and fit of prostheses (single and multiunit) fabricated by using 3D printing, which is smaller when compared to the prosthesis fabricated by burnout and casting procedures [4, 58, 60].

8.3.2.1 Direct Metal Laser Sintering (DMLS)

DMLS is a manufacturing process for producing complex 3D components directly from 3D CAD data by using laser without using any machining (Table 8.8). DMLS completely melts the powder at a higher temperature, rather than heating and then using pressure to agglomerate the powder, as seen in SLS. The complete melting of metal ensures that the metal is liquefied long enough to coalesce powder grains into a homogenous mass ensuring minimum porosity, thereby controlling the grain structure and minimizing the risk of failure. A laser beam is focused onto a layer of powdered metal, these areas fuse into thin solid layer, and another layer of powder is then laid down over this. The slices of framework are produced and fused with the previous ones until framework or coping is finished. The metal powders used in conjunction with AM technologies are a mixture of particles with sizes ranging between 3 and 14 μm [59, 67, 68]. The chrome cobalt (CoCr) alloy used in DMLS technique has comparatively less molybdenum, while nickel and beryllium are not present in the composition anymore when compared to the alloy used for casting. The alloys are provided by numerous manufacturers with minor variation in compositions (EOS SP2 Cobalt Chrome, GmbH, Munich, Germany; 3D systems Dent Wise, Rock Hill South Carolina; Concept Laser, Germany, BEGO Wirobond C+, Germany; SLM Solutions MediDent).

The technique of producing copings by DMLS technique has many advantages, such as complete control over the framework and coping designing, highly reduced manufacturing time and costs, improved product density, and elimination of casting defects as seen with traditional casting procedures [60, 69]. DMLS has been documented to allow high precision (approximately 25 μm) and good mechanical strength [67]. The challenges associated with casting (such as shrinkage) and milling (such as high hardness of chrome cobalt alloy) are circumvented by using direct metal laser sintering (DMLS) that assures uniform quality [70].

8.3.2.2 Electron Beam Melting

This is similar to DMLS in the concept, except that the working beam that is used to melt the metal is an electron beam of over 3000 W (Table 8.8). The metal melting is conducted in vacuum in both the procedures. Its use had been first developed by Arcam (taken over by GE in the year 2016) for designing aircrafts, in Sweden, in 1997. The advantages of EBM is that it has significantly high manufacturing speed process as it can separately heat the powder at several places simultaneously, as against the laser which must scan the surface point by point. Additionally, the higher temperature of heating and preheating the powder limits the deformations due to minimal residual stresses and creation of a martensiticfree microstructure. However, with EBM, the beam is wider when compared to the laser beam, thus reducing the accuracy. The minimum z-axis resolution or layer thickness achievable is 50–70 μm with EBM, which is higher than in DMLS (25 μm). It is more commonly employed for use with titanium family of powders encompasses the Ti6Al4V, Ti64ELI, and TiCP alloys. Both Ti6Al4V and Ti6AL4VELI are light alloys (have high strength to weight ratio), characterized by excellent mechanical properties and corrosion resistance combined with low specific weight and biocompatibility. Due to the high oxidation potential of titanium, vacuum is used to minimize contamination due to oxidation and maintain the chemical composition of printed metal. Vacuum also helps to minimize air collisions of electrons, which can otherwise result in loss of energy [66–70].

8.3.3 Prosthesis Fabricated in Ceramics and Zirconia

With outstanding characteristics such as corrosion resistance, esthetics, and biocompatibility, ceramics have been versatile for numerous prosthesis including indirect restorations and implant components. The additive manufacturing techniques used for fabrication of ceramic structures are still in the phase of nascency. One potential technique is the lithography-based ceramic manufacturing (LCM) technology or vat photopolymerization, also known as stereolithography, based on building the structure layer by layer, by using selective light to cure photosensitive ceramic suspension (Table 8.9). The ceramic suspension contains photocurable resin (an organic binder based on acrylate and methacrylate chemistry) and homogenously dispersed ceramic particles of inorganic nonmetallic compounds [71–75].

In this technology, after the CAD information is digitally transferred to the 3D printer, the ceramic-loaded liquid or slurry is dispensed in the rotating vat. The movable build platform is dipped into the material, which is then selectively exposed to visible light through light-emitting diodes (LED) and a digital mirror device (DMD) to generate a 3D green part or green bodies in a layer by layer manner. Depending on the geometry of the green part to be printed, the slurry systems can vary in filler content and crosslink density of the organic matrix so that the green part attains sufficient mechanical strength. The next step is debinding where the organic binder is burned out. Following this, the green parts are cleaned to remove the non-cured suspension with compressed air and cleaning fluid for about 5 min. An excessive cleaning can introduce defects. Lastly, sintering (thermal treatment) is done, resulting in densification of ceramic parts in various predesigned geometries with improved mechanical properties. Thus, this is an indirect printing process where in the shaping is done by the printer, but the final properties of the material are obtained in the sintering furnace (Fig. 8.29).

The LCM technology is particularly useful in fabrication of thin occlusal veneers when compared to subtractive milling process. LCM technology imparts high accuracy and greater internal adaptation, especially in regions of thin borders and feather edges, which can get chipped during manual or milling process that can adversely lead to a discrepancy between the design and the fabricated part. This shortcoming is circumvented in AM [76–79].

Through an in vitro study, it has been reported that the microstructure of lithography-based sintered alumina is homogenous and dense. Nevertheless, due to the micro-porosities present, the mechanical properties such as strength and fracture toughness have been reported to be less as compared to CAD/CAM zirconia. Although a favorable outcome with lithography-based sintered alumina has been observed, the further improvements have been simultaneously recommended [78, 80]. Successful outcome has also been attained for stereolithography-based AM of lithium disilicate glass-ceramic with high precision and dimensional accuracy [81, 82]. Favorable outcomes have been reported in relation to strength, especially when compared to milled zirconia crowns [83–85]. Conflicting outcomes have been reported regarding internal and marginal adaptation for 3D printing zirconia crowns

Table 8.9 Lithography-based ceramic manufacturing (laboratory-based model)

Commercial support	Printer	Technology	Minimum layer thickness	Accuracy in *x*-axis and *y*-axis	Resolution	Materials used	Resource
Lithoz, Austria	CeraFab 7500	DLP (lithography-based ceramic manufacturing (LCM)) Blue light source: 465 nm[a]	0.03 mm	0.04 mm	635 dpi (high-resolution imparts good surface quality)	1. Aluminum oxide: LithaLox HP500 2. Hydroxy apatite: LithaBone HA 400 3. Tricalcium phosphate: LithaBone TCP 300 4. Zirconium oxide: LithaCon 3Y 230 5. Lithium disilicate, ATZ, ZTA and Bioglass	https://www.lithoz.com/

[a]An upside-down setup causing non-cured suspension to run back into the reservoir for the production of next layer utilizing only a small volume of suspension and easier cleaning process of nonprinted material from the printed part

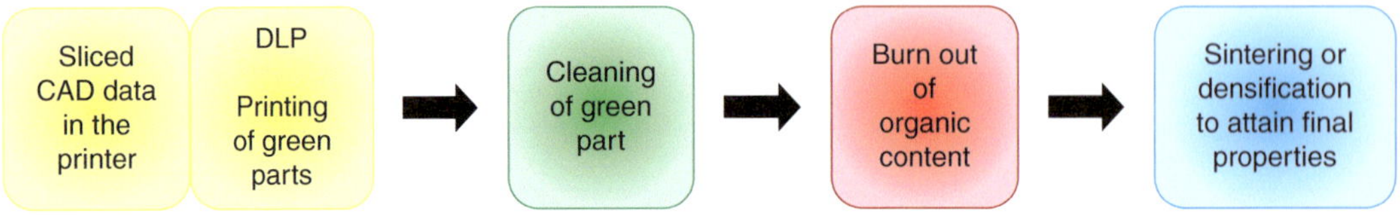

Fig. 8.29 Sequence of events for LCM

with some studies suggesting less than ideal internal and marginal adaptation for clinical application [83]. Contrary outcomes have been made for trueness of zirconia crowns by some other authors [86]. Thus, it has been concluded that the reliability of 3D printed zirconia depends largely on the volume fraction of zirconia in the slurry used for printing. As the volume fraction increases, the flowability of zirconia-based suspensions decreases resulting in less reliable 3D print [87]. Conclusively, numerous studies are skeptical about 3D printing in ceramics from reaching the full potential primarily due to persistent technical barriers.

8.3.4 Provisional Crowns/Bridges Fabrication by 3D Printing

Interim restorations or provisional prosthesis is integral for eventual success of the prosthodontic treatment. They are required for a shorter period until the definitive restoration has been fabricated or for a longer duration in patients undergoing full mouth rehabilitation and multiple endodontic and orthodontic treatments. Besides protecting the pulp and promoting periodontal health, these restorations also have esthetic utility in rehabilitation of anterior teeth. They must also permit easy plaque removal and must possess suitable mechanical properties like flexural strength and fracture toughness. Consequently, interim restorations have an important purpose in the success of the eventual outcome of the prosthodontic treatment. DLP and SLA are two commonly employed 3D printing methods by which provisional restorations are fabricated in resin.

Clinically acceptable flexural strength of 3D printed micro-hybrid filled light-cured resin has been reported in the literature, although the strength reported is less than that of the milled provisional material [88, 89]. The hardness of 3D printed provisional restorations has been found to be acceptable and higher than the milled provisional restorations [88]. The dimensional accuracy and surface roughness of 3D printed provisional restorations have been found to have clinically satisfactory values, although printer settings, polymer properties, nesting, and build direction immensely influence the outcome [90]. Similar or better marginal fit has been documented in 3D printed polymer crowns when compared to CAD CAM milled crowns [4, 91–93]. In fact, the satisfactory outcome as observed with the polymer printed provisional restorations has encouraged its use in 4 unit provisional and full mouth provisional fabrication [89, 90, 94]. An important property of provisional materials is obtaining and maintaining color stability to impart satisfactory esthetic outcome. Results of studies, however, reveal that color matching with AM provisional materials is difficult as the materials used are different from the ones used conventional manually fabricated interim restorations. The authors also suggest use of custom shade guides in order to obtain desired shades in interim restorations with 3D printed materials predictably [60]. An analysis of the documented literature in relation to 3D printed interim prosthesis reveals that most of the studies pertain to laboratory studies and there is dearth of data from clinical outcomes. It may then be concluded that although 3D printed interim materials are strongly making a foothold in prosthodontics, yet full potential of these restorations can be released only with sustained improvement in materials science per se.

8.4 3D Printing in Implant Dentistry

The impact of 3D printing in implant dentistry is monumental. The use of additive manufacturing in dental implant planning has improved the treatment results with minimal complications by offering tailor-made solutions to numerous clinically challenging problems. Models for surgical planning, single piece surgical guides, stackable surgical guides, dental implants, implant abutments, and osteotomy cutting guides are the numerous applications in which 3D printing has successfully made inroads in implant dentistry. Despite the versatility of 3D printing in terms of materials, machines, and eventual outcome, the application can be reliably embraced only after understanding the technology including the engineering and material science.

8.4.1 Surgical Guide/Template and Models

The digital data workflow includes deriving the DICOM (digital information for communications in medicine data) data from the CT scans and translating these data into the prosthetically driven dental implant placement by using surgical guides and models. 3D printed mock surgical models (Fig. 8.30) and guides (Figs. 8.31, 8.32, 8.33, and 8.34) enhance the outcome of the treatment in terms of implant placement, biomechanical loading, esthetics, and longevity.

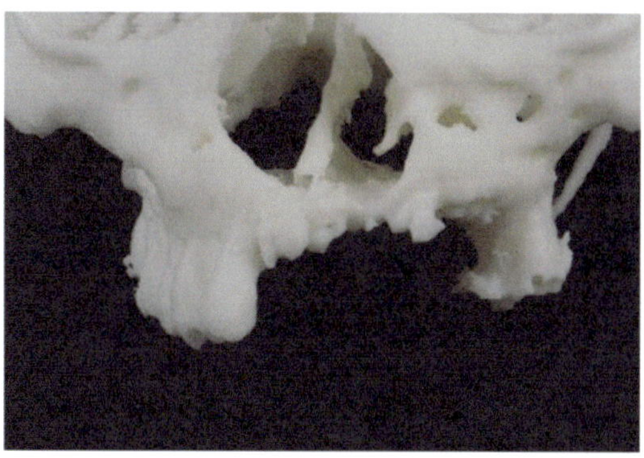

Fig. 8.30 Surgical model for treatment planning fabricated by AM

Surgical guide and models can be prepared by stereolithography (SLA), fused deposition model (FDM), binder jetting (BJ) with the use of polymer-like poly(methyl methacrylate) (PMMA), thermoplastic polymers such as poly(lactic acid) (PLA), acrylonitrile butadiene styrene (ABS), PEEK (poly ethyl ether ketone), and DMLS by using titanium alloy, cobalt chrome. Many authors have suggested the superiority of the 3D printed method over the milling or subtractive method [95, 96]. Among the different options, PolyJet technique fabricated surgical guides have been documented to be more accurate and reproducible even after 1 month when compared to DMLS printed guides [97]. The accuracy of fit of a surgical guide will directly influence the accuracy of implant surgery. With the use of 3D planning software (like Blue Sky Bio, 3Shape, Simplant virtual implant placement, coDiagnostiX™, Dental Wings, Simplant Pro™; Smop™; NobelClinician™; Implant Studio), the location, optimal dimension, and angulations of implant placement can be decided (Fig. 8.31a–d) [98]. In conditions requiring bone reduction prior to implant placement, fabrication of stackable guides (aligned multilayered guides) by using digital planning and AM is proving to impart efficient and predictable outcomes (Fig. 8.34a–e).

The digital imaging and communications in medicine (DICOM) data created from the cone beam computed tomography (CT) scan and the data from the laboratory scan of the dental cast are overlapped and used to determine the most ideal implant positions. The surgical guide is virtually designed, and the STL file created is sent to the printer for fabrication of the 3D printed surgical guide (Figs. 8.32 and 8.33). Sleeves when incorporated in surgical guides further add to the purpose of positioning of implants. These sleeves may be prefabricated or stock metal sleeves. Alternately, their information is incorporated in the designing software and printed as metal sleeve-free guide [99–103]. It has been stated in the literature that SLA-produced surgical guides provide substantial accuracy in terms of angula-

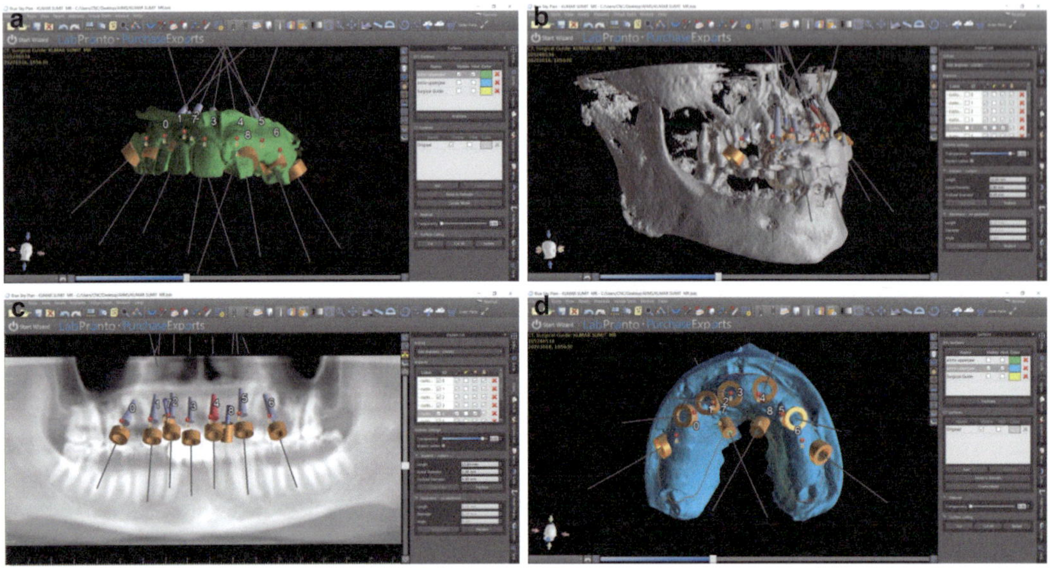

Fig. 8.31 (**a–c**) Prosthodontically driven planning for immediate implant placement on a virtual model. (**d**) Designing of prosthodontically driven implant surgical guide and creation of STL files for 3D printing of guide

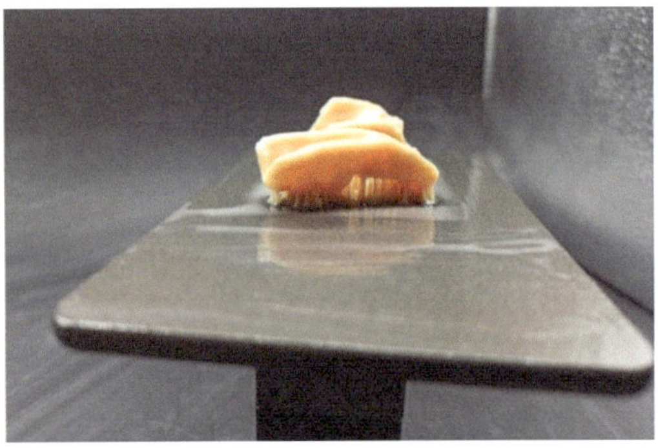

Fig. 8.32 3D printed guide in resin

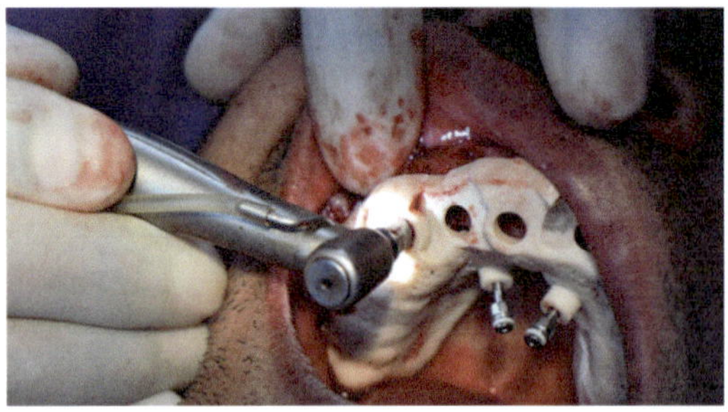

Fig. 8.33 Surgical placement of implant by using the 3D printed guide

tion and apical positioning of dental implants [104, 105]. This particularly holds true when compared to manual thermoplastic guides. Significant accuracy has been noted in implant head and apex position when 3D printed guides are used [106]. The same holds true even for mucosa-based guides, where the resiliency of mucosa may otherwise be thought to have a negative effect [107].

Among the various printers, PolyJet and SLA 3D printers have been reported to meet the required accuracy for clinical applications in dentistry [108, 109]. FDM printed surgical guides made of lignin and PLA have been documented as proof of concept and recommended to be a cost-effective, easy to sterilize, and environmental-friendly alternative to SLA printed surgical guides [103], although the accuracy of surgical guides fabricated by FDM is questionable [110]. Regarding the sleeves, angular deviation of the implants has been more with metal sleeve-incorporated 3D printed surgical guides when compared to metal sleeve-free 3D printed guides [100, 111]. In fact, luting of metal sleeves has been reported to increase production time and error due to variability of luting of sleeve in the slot. The same can thus induce errors in implant positioning [103, 112]. When compared to milled guides, 3D printed guides have

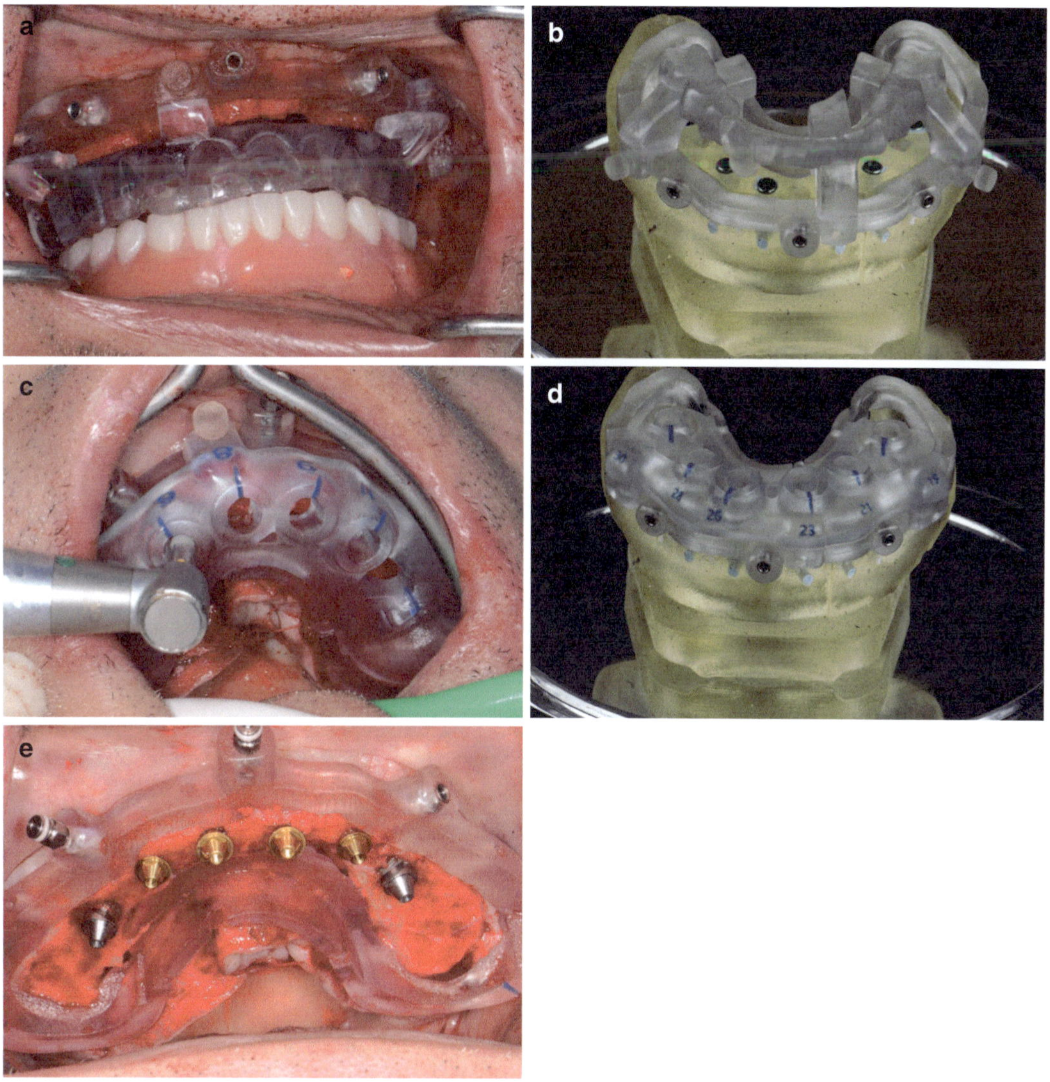

Fig. 8.34 (**a**, **b**) Foundation guide (first part of stackable guide system) for the purpose of guiding the bone reduction secured in position before fixation. The position of the guide is verified by approximation of occlusal splint (second part of the guide system) with opposing arch. The occlusal splint is aligned and connected to the foundation guide by a locking mechanism that permits easy assembly and removal. (**c**, **d**) Implant placement guide (third part of the stackable guide system) aligned and connected to the fixed foundation guide by locking mechanism after bone reduction. (**e**) After implant placement and connection of multiunit abutments, the foundation guide is retained to orient the prefabricated prosthesis

been found to be similar in accuracy. Considering the ease of fabrication, high accuracy, reduced wastage of material resources, and laboratory time, 3D printed guides have been stated to be more effective in execution of predictable implant surgery [113]. It is, however, emphasized that steam sterilization is strongly discouraged for the surgical guides due to linear expansion of the photo-polymeric resin material used in the fabrication of these guides resulting in increase in outer dimensions and a decrease in inner dimensions that can cause inaccurate implant placement [114].

A preliminary model can also be printed by using the DICOM data (Fig. 8.30). This model can be used for mock surgical steps in order to plan the surgery on patient more precisely [115]. Literature, however, suggests that the 3D printed models, although deemed clinically acceptable for orthodontic purposes (with an acceptable range of error of <100 to 500 μm) may not necessarily be acceptable for prosthodontic workflow where high accuracy of 59–150 μm is required. This is even more relevant in context of binder jetting printers where the discrepancy of more than 500 μm has been reported. One can conclusively recommend that the choice of printer largely depends on the workflow for a specific application [55, 115–121]. In a generic way, SLA and DLP models have been the most suitable in terms of accuracy for full arch models. Caution is warranted for printers using binder jetting and those using thermal technologies (like CLIP by Carbon M2). It has also been suggested that optimal accuracy and reduced printing time can be attained if the layer thickness (selection of z-axis) is restricted to 100 μm for both SLA and DLP. A lesser layer thickness may improve accuracy but will take excessive amount of time to complete printing. Additionally, if the filling pattern of the model is hollow instead of solid, one can minimize material wastage, time of build, and cost as well [118]. Regarding the design of the base of the model, horseshoe base, although conserves a large amount of material, has been associated with changes in transverse direction that render the models clinically unacceptable [122]. Lastly, the post-processing and storage of the 3D printed models can also affect the dimensional accuracy of the models created, and one must keep that into consideration.

8.4.2 Implant Abutments and Subperiosteal Implants

Implant abutments have an important role in predictably affecting the clinical outcome including longevity and esthetics of the treatment [123, 124]. In some clinically challenging conditions, the abutments provided by manufacturers may not be able to restore in an optimal manner due to limitations of available prosthetic space and angulations. A prosthesis fabricated on an inappropriately selected abutment can jeopardize the overall success by deflecting the force vectors unfavorably, compromising hygiene, or esthetics. In such conditions, customizing the abutments is the solution to meet the clinical demand and bring about clinical success. The biggest challenge in customizing the abutments is to attain accuracy, passivity of fit, a biologically sound, and mechanically equilibrated implant-abutment junction [125, 126]. To that effect, 3D printing has been a game changer in providing options of customized healing abutments, provisional abutments, or definitive abutments and bringing about esthetic and functional benefits to patients [127–129].

Among the wide choice of materials, metals (chrome cobalt and titanium alloy) and polyether ether ketone (PEEK) have gained immense popularity for the fabrication of customized abutment by using 3D printing. Intraoral implant position can be recorded directly from the oral cavity by implant scan body or indirectly from the working models. With the use of implant prosthesis design software, the size and shape together with the contours of the emergence can be determined for the abutment. Additive manufacturing of chrome cobalt and titanium alloy with DMLS results in passive prosthesis without wastage of material which is otherwise witnessed in milling technique. Additionally, complex implant prosthesis superstructures are easily fabricated by metal printing technology as compared to conventional lost wax technique. DMLS enables fabrication of framework and superstructures with marginal fit and accuracy comparable to milled (CAM) technology. In a study comparing 3-unit implant-supported screw-retained chrome cobalt frameworks fabricated by conventional casting, CAD CAM milling, and DMLS, least marginal gap has been reported by selective laser printing (35 μm) [130]. However, certain other studies have reported higher vertical marginal discrepancies (2.5, 11.3, 11.83 μm) with 3D laser-sintered chrome cobalt abutments at implant-abutment junction [131–133]. The chief problems encountered with laser sintering according to literature is roughness of mating surfaces inducing the microgap and hindering passive fit [13, 126, 132–

134]. Nevertheless, the misfit value of laser-sintered chrome cobalt abutments is within clinically acceptable limits. Additionally, the possibility of manufacturing the abutment economically can be beneficial, especially in low-income group of countries [60, 131, 133].

Another aspect that needs deliberation is deformation associated in 3D printed metal framework due to ceramic veneering [135]. Literature reports suggest that in cases where one piece abutment-crown is considered, post-ceramic veneering of 3D printed chrome cobalt is associated with higher linear and angular discrepancies (38.9 ± 16.6 μm) than in the milled conditions (36.9 ± 15.6 μm) although the discrepancy is clinically acceptable [136]. Similar considerations have been raised by other studies including those comparing 3D printed titanium and chrome cobalt with milled frameworks. The possible reason for the same is contraction during the heating and cooling firing cycle or differences between the thermal coefficients of the ceramic and the alloy. These changes are reported to be greatest during the oxidizing and glaze cycles. Additionally, these disparities may also be attributed to different metal alloy composition used to print the frameworks [137]. Regarding 3D printed titanium frameworks for complete arch implant-supported prosthesis, both SLM and EBM have shown clinically acceptable discrepancy in *x*-, *y*-, and *z*-axis [138].

Another material that has been making inroads in implant dentistry is polyether ether ketone (PEEK). FDM technology is conventionally used for fabrication of prosthesis by using PEEK material [139–141]. PEEK can be used for the fabrication of customized implants as the material is durable, is lightweight, and has excellent biocompatibility. As a high-performance polymer with high strength to weight ratio, the versatility of PEEK has been attempted widely as 3D printed implant components and 3D printed implants as well.

Fabrication of subperiosteal implants to replace missing teeth is an alternative option to circumvent advanced regenerative surgical procedures in numerous anatomically challenging clinical conditions with severely reduced bone volume. Amalgamation of the traditional subperiosteal implant methods with contemporary 3D printing techniques now provides tailor-made solutions with good accuracy. The main advantage of using 3D printing for subperiosteal implants is the single surgical intervention for insertion of implants. DMLS or EBM have proven to be clinically viable methods for fabricating accurate subperiosteal implants, with satisfactory clinical outcomes [142, 143]. For nonmetal subperiosteal implant fabrication, PEEK is a viable option, and to that effect research to improve its elastic modulus and bioactivity at nanoscale is underway [144, 145].

8.5 Maxillofacial Prosthesis Fabricated by 3D Printing

Maxillofacial prosthodontics is a specialty of dentistry that deals with the rehabilitation of patients with acquired and congenital defects of the head and neck region. The conventional method is laborious and time-consuming, associated with a large deal of unpredictability. The introduction of digital technology with additive manufacturing for intraoral and extraoral defects has made the fabrication of maxillofacial prosthesis more predictable requiring reduced time and thus helps improve quality of life with enhanced post-treatment outcomes. Geomagic and Mimics are the two most commonly used software to predictably design the maxillofacial prosthesis [146, 147].

8.5.1 Extraoral Defects (Facial Prosthesis and Cranial Implants)

Rehabilitation of missing facial structure ear, nose, eye, and cranial defects caused due to congenital malformation, disease, surgery, or trauma is accomplished by the use of an extraoral prosthesis. The steps of digital workflow in extraoral prosthesis require facial scanning with the use of a structured light facial scanner such as Artec 3D. The scanned image records the extent of the defect, undercuts, and associated anatomical details. The scanned data are then imported to CAD software like MIMICS (Materialise interactive medical image control system from Materialise) or ZBrush (pixologic) software. The missing structures are designed by using virtual anatomical sculpting tool, and the edges of the prosthesis are modified according to soft tissue adjacent to the defects. Contralateral non-aberrant organ (like ear or orbit) or preoperative photograph (for midfacial defect) can be used to accomplish the anatomic virtual sculpting [148].

A prototype can be printed in wax by material jetting (for instance, VisiJet M3 Hi-Cast wax by using a printer specific for the purpose such as ProJet 3510 CPXPlus from 3D systems). This prototype can be tried on the patients to preview the gross morphology as well as finer details such as marginal adaptability. Minor alterations can be done to this wax prototype, and the same can be directly invested for fabrication of definitive prosthesis [148].

An alternative method is to follow a complete digital route for investing by creating a negative mold of the prosthesis virtually. This will require designing the negative mold around the previously designed virtual prototype. This mold can be prepared in durable polyamide (nylon) material by using SLS compatible printer. Once the mold is printed, silicone can be packed traditionally, within the mold, and fabrication of the prosthesis is proceeded in the conventional manner [148, 149].

It is noteworthy that exclusive digital printing of prosthesis is sparsely documented in the literature [150]. Although 3D printed polylactic acid (printed by FDM), polyamide (printed by SLS), and polycarbonate (printed by stereolithography) have been documented to have good dimensional detail, shortcomings have been noted in inadequate skin texture reproduction. Among the three options, however, FDM has been identified as the most efficient in reproducing skin details [151].

The reports on direct printing of silicone prosthesis are recent and document restricted clinical application [152–155]. Unkovskiy et al. presented a complete digital throughput for the manufacture of a nasal prosthesis from a pure, solvent-free silicone (ACEO Silicone General Purpose; Wacker Chemie AG) printed by ejection technique. In its basic formulation, the 3D printable silicone (ACEO Silicone General Purpose; Wacker Chemie AG) has been reported to be free from acrylates or urethanes and is reinforced with a filler, cross-linker, and addition-cure platinum catalyst (activated by UV light). The Si-H groups of the cross-linking agent react with the vinyl groups of the polymer to form a 3D network. The high viscosity and low surface energy of the silicone are a challenge to consistently enable ejection of the silicone through the printer. To overcome this, the compatible 3D printer (ACEO IMAGINE printer) has a dosing valve to formulate the droplets according to the STL mesh by shearing the silicone material and ejecting at high frequency (drop on demand technology). UV curing of each layer helps to attain a layer thickness (resolution in z-axis) of 0.3–0.4 mm. The support material is easily washed off with water, and the post-cure treatment includes heating at 200 °C for 4 h [155, 156].

Despite advancements, 3D printing of silicone is still associated with limitations due to staircase effect (which need to be eliminated by manual finishing) and the necessity to extrinsically apply colorants to enable esthetic shade match. The authors have also reported compromised margin adaptation probably due to a high z-axis or layer thickness of 0.3–0.4 mm when compared to conventional process which yields a thickness of 0.1 mm only. Lastly, it has been emphasized that the documented silicone used by the authors is not medical grade. Conclusively, with much scope for improvement, our wait for a definite 3D printed silicone prosthesis is far from over, even now [155, 156].

3D printing has also been used in the fabrication of implant surgical guide for the extraoral implant placement for ear, nose, or orbital prosthesis. The DICOM file from the computer tomography scans is used to design the implant surgical guide within the confinement of prosthetic design as determined by using contralateral mirror images or by freehand sculpting. The virtual designing software is used to design the guide which can be printed in the aforementioned

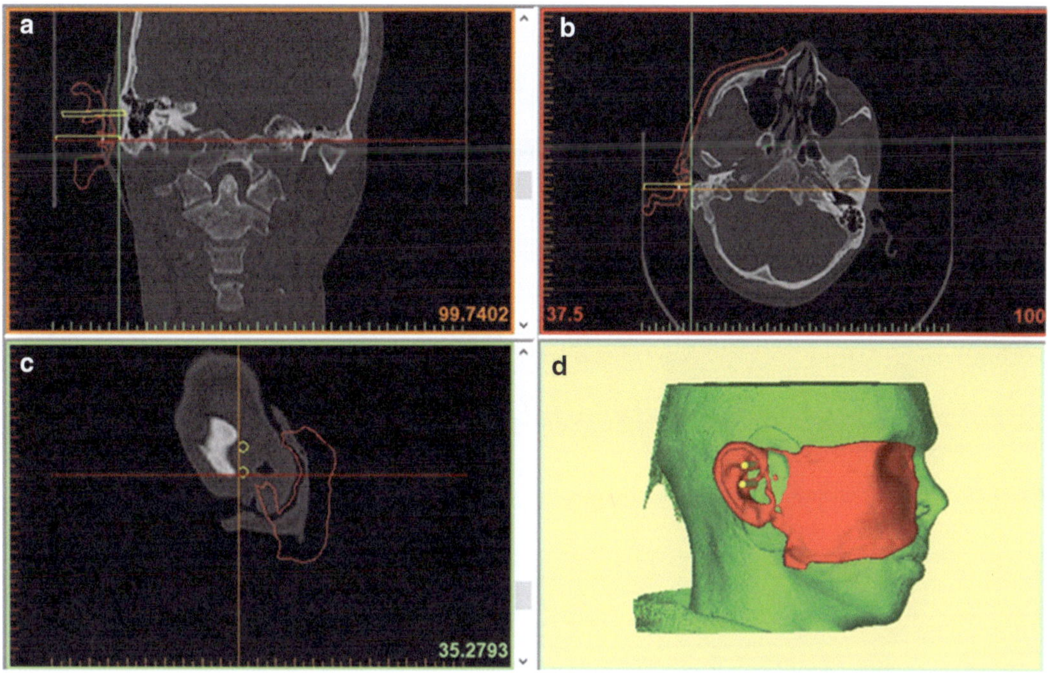

Fig. 8.35 (**a**) Integration of the STL file of the planned prosthesis (mirror image of contralateral ear) with the Digital Imaging and Communications in Medicine (DICOM) file (obtained from Computer tomography scan) in the planning software (for instance, Blue Sky Plan, Blue Sky Bio, LLC) (Frontal view); (**b**) transverse view; (**c**): sagittal view; (**d**) complete design of surgical guide

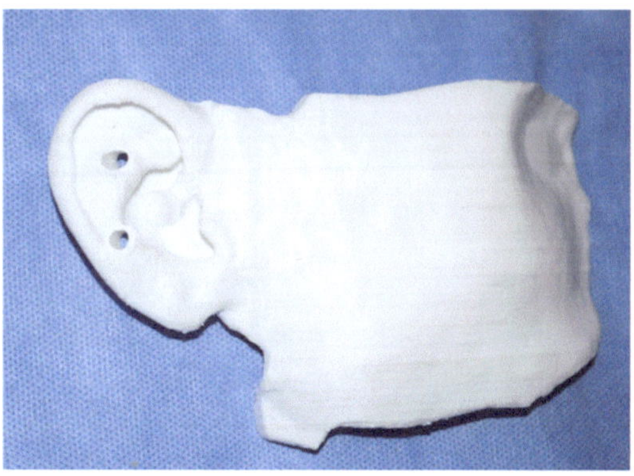

Fig. 8.36 3D printed surgical guide for auricular implant insertion

polymers (FDM, SLA, DMP) (Figs. 8.35a–d and 8.36) [157, 158].

Similarly, for rehabilitation of cranial defects by cranioplast, the data are imported from the DICOM computer tomography file of the patients and virtual planning can be done in designing software (Fig. 8.37a). An alternate method involves printing the skull analog (Fig. 8.37b) and use of that analog for manual method of prosthesis fabrication.

8.5.2 Intraoral Defects (Mandibular and Mandibular)

Patients planned for mandibulectomy or maxillectomy can be greatly benefitted by AM. The use of virtually preplanned and 3D printed osteotomy guides for harvesting microvascular (fibula) graft of the resected arch enables precise transfer and accurate position of the graft in the dental arch for the purpose of restoring the continuity of the arch and subsequent rehabilitation within the realms of prosthodontic considerations. This promises optimal functional and esthetic outcome [159–162]. While some authors recommend simultaneous implant placement in the grafted bone, subsequent implant insertion is still a popular approach for many clinical condi-

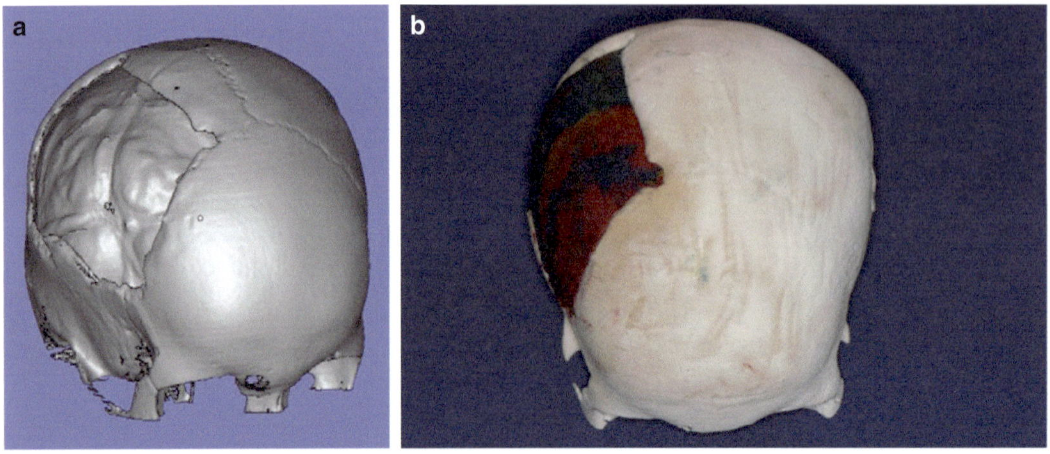

Fig. 8.37 (**a**) Virtual analog of cranial defect. (**b**) 3D printed analog

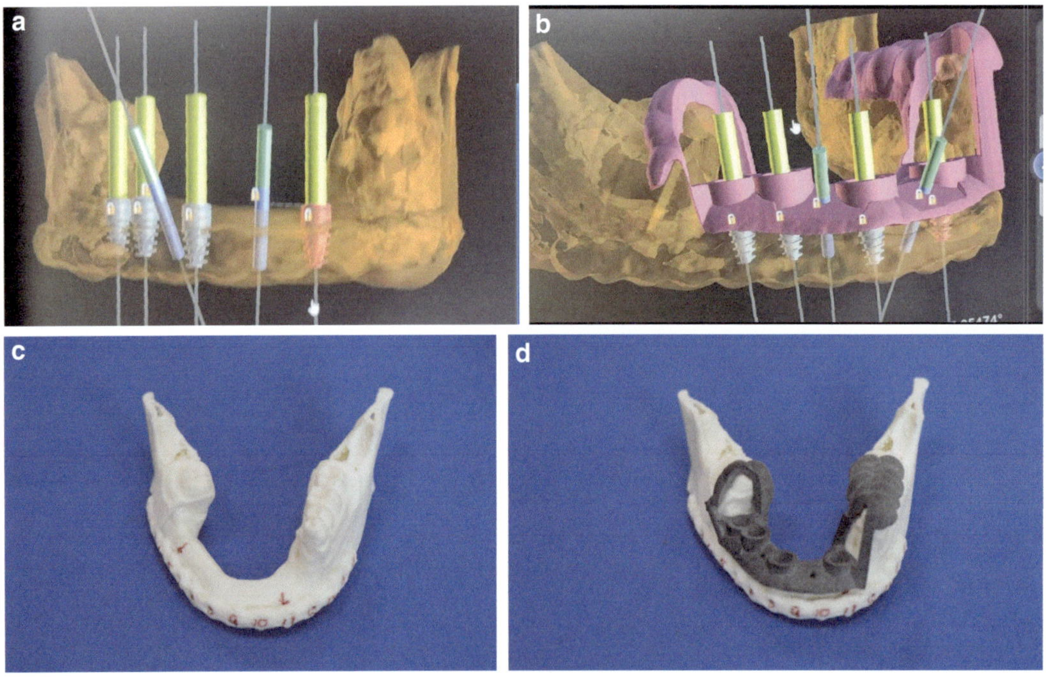

Fig. 8.38 (**a**) Use of DICOM data for implant planning. (**b**) Designing of the surgical guide. (**c**) AM model fabricated in nylon by using SLS. (**d**) Tooth-supported surgical guide fabricated by using SLS on the surgical model

tions. In case implant rehabilitation is planned after grafting, the CBCT data of the reconstructed mandible with radiopaque marker stent prosthesis are used to obtain DICOM data. These data are then used in implant planning software (such as Blue Sky Bio, Simplant) (Fig. 8.38a). Virtual planning of implant location corresponding to prospective occlusion is designed. The surgical guide is then designed (as shown in Fig. 8.38b). These data are then sent to the 3D printer. Surgical guides and surgical models (Fig. 8.38c, d) can be printed in polymer by using additive manufacturing techniques. The surgical guide helps dental implant placement with precision in the grafted bone, abridging the length of the surgery. This helps to attain the functional occlusion promising longevity of treatment.

After the osseointegration of implants, the digital technology can again be used for implant rehabilitation as has been discussed in the section of fixed prosthesis or by implant-supported overdenture (the overdenture fabrication can be done by using 3D printing as discussed in section of complete denture).

In maxillectomy patients, the surgical guide fabrication for placement of zygoma implants or conventional intraoral implants can be done similarly as has been explained for mandibulectomy. The fabrication of obturators by using the CT data of the patients can be done by following indirect or direct techniques. In the indirect technique, the 3D printed model of the maxillary defect is obtained in polymer by using SLA or FDM techniques. The obturator prosthesis can be fabricated on the 3D printed models [163, 164]. In direct technique, the digital planning of obturator plate prosthesis can be done followed by printing in biocompatible PMMA resin, in wax printing, or in PEEK. The wax-printed plate can be converted into conventional obturator prosthesis by using resin and silicones. The advantage of this technique is tremendous in fabrication of surgical obturator or interim obturator where one can proceed directly with the CT scan data of the patient before surgical resection [165–168].

Role of 3D printed appliances is also observed in patient-specific brachytherapy appliance which is used to curtail the amount of radiation dose to the patient particularly in the IMRT (intensity-modulated radiotherapy technique) [169]. The use of 3D printing is tremendous especially in anatomically challenging regions complicated by irregular tissue contours such as the orbit [170, 171] or in deep vaults of palatal regions [172]. One must, however, be wary that most 3D printing materials have not been well characterized, in terms of their radiologic tissue equivalence. There is belief that different printers and print materials may, in fact, have different densities, ushering a variability of Hounsfield-to-density relationship. This puts a question on the consistency and radiologic suitability of 3D printed materials. Conclusively, there is a strong recommendation to standardize the materials for brachytherapy appliance in terms of the radiographic densities in order to use 3D printing safely and without loss of benefits of radiotherapy [173].

8.6 Metal-Based Partial Dentures

A shift toward rising proportion of partially edentulous patients has been witnessed all across the globe. Concurrent to the option of implant rehabilitation, an alternate option includes well designed removable partial dentures in such conditions, particularly in cases where economic challenges are co-existent. In light of the prevailing scenario, the impending need to advance the technologies associated with removable partial denture fabrication procedure has been emphasized in the literature [174].

Traditional lost wax technique and investment casting have now been annexed by contemporary methods such as computer-aided designing and manufacturing (subtractive or additive). The computer-aided subtractive manufacturing has a few shortcomings that have been circumvented by the additive manufacturing process. This includes reduced wastage of the raw materials, possibility to print varied shapes, and multiple frameworks simultaneously [175]. Additionally, in the fabrication of milled frameworks, the accuracy of final product is adversely affected by progressive blunting of cutting tool (due to use), which is not an issue of concern in printed frameworks.

When AM is used for partial denture framework fabrication, two options of fabrication are possible. Either a RPD framework can be designed, printed in resin, and used for casting (indirect printing or hybrid method) or a direct method is used where RPD framework is printed in metal [176–178]. The printed resin framework provides the advantage of clinical trial and modification before casting, a possibility not available for metal printed frameworks [179]. In both the options, virtual planning and designing

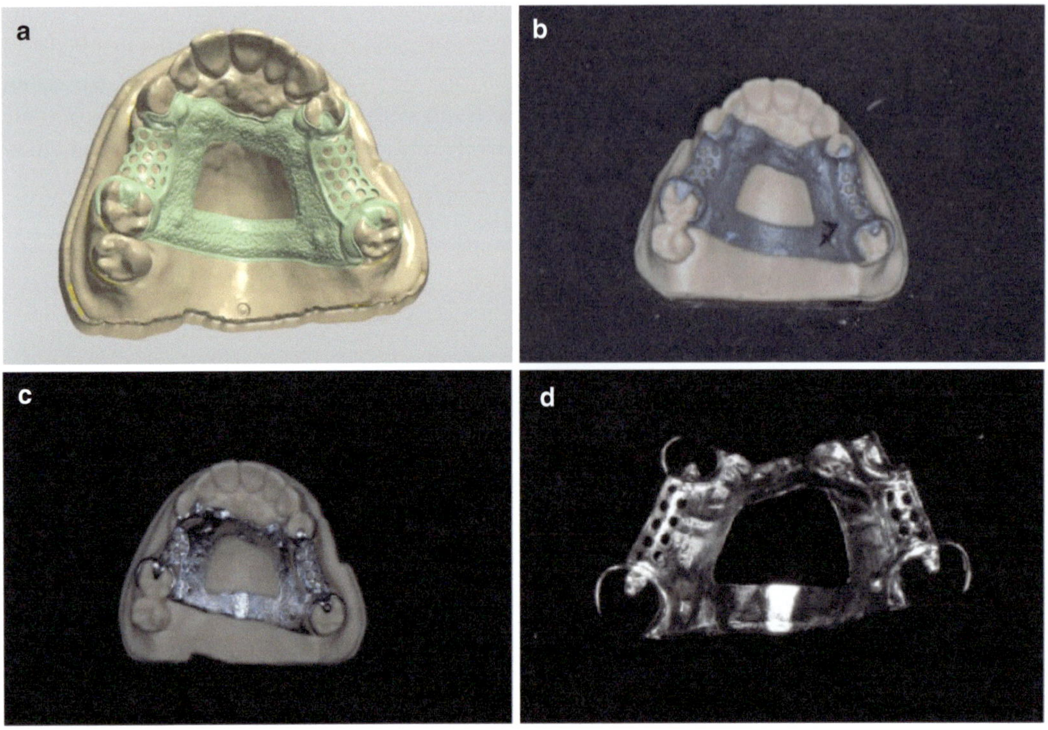

Fig. 8.39 (**a**) Virtual planning of partial denture; (**b**) printed partial denture framework before sintering; (**c**) printed partial denture framework after sintering; (**d**) printed partial denture framework. (Images curtsey: Dr. Majed Altoman: graduate student, School of Dentistry, Loma Linda University, California, USA)

can be done for geometric analysis and placement of components of partial denture framework with high level of accuracy (Fig. 8.39a–d) [177, 179, 180].

Stereolithography or DLP is used for printing the resin framework of partial dentures. Powder bed fusion (PBF) including selective laser sintering (SLS), selective laser melting (SLM), or electron beam melting (EBM) is used for printing the metal-based frameworks. SLS uses carbon dioxide laser energy for sintering and consolidation of the metal particles layer by layer. The drawback with SLS is that melting of particles is not achieved in the process, which is critical for achieving the desired mechanical properties. SLM and EBM completely accomplish the procedure of melting either by using ytterbium high-power laser or electron beam in an atmosphere of argon, nitrogen, or helium [9, 10, 24, 175, 179, 181–183].

The features that make SLM more favorable than conventional casting are fewer steps of fabrication, conservation of labor and time, superior homogeneity in microstructure amounting to improved mechanical properties, clinically acceptable accuracy, increased resistance to distortion resulting in favorable distribution of forces to remaining teeth and supporting structures, and possibility to save digital data for future use [174, 178, 181–191]. Although preliminary data indicate a favorable outcome in terms of fit and esthetic improvement when compared with the conventional technique, the literature on 3D printed removable prosthesis frameworks is still varied [181].

8.6.1 Accuracy of Metal Printed Partial Dentures

Fit of a prosthesis refers to the degree of discrepancy evident between any part of the prosthesis and oral tissues, manifesting either as a gap or as an indentation into the tissues due to excessive tissue contact [192]. It is an important parameter for longevity, as fit is directly related to health of abutment teeth and the residual bone [193]. Fit of a prosthesis has been documented to be the most commonly reported source of dissatisfaction (76%) contributing to discomfort, movement, or damage to the associated tissues tantamounting to lack of function, plaque retention, and poor aesthetics. In fact, improper fit has been suggested to be the primary cause patients avoid wearing the RPD [176, 192, 194]. Reports for fitness accuracy of printed frameworks vary greatly because of the variability of the methods used to evaluate fit. While some methods may be too subjective such as clinical visual assessment or assessment on a definitive cast (analog methods), other methods may be more objective as they use digital assessment and color mapping [176]. Clinically acceptable accuracy of fit (a gap of 50–311 μm) has been observed with conventional SLS printed frameworks [181]. The accuracy of fit of CAD printed RPD (discrepancy of 0.2 mm) has been documented to be less when compared with conventional cast partial denture (discrepancy of 0.04 mm) [181]. Distinct better fitness accuracy of conventional framework over printed RPD framework has been documented in other literature reports as well [174, 179, 194–196]. In a FEA study followed by in vitro tests, it was deduced that printed RPD frameworks achieve a peak pressure on the tissues lower than clinical pressure pain thresholds (PPT) and ensure a more evenly distributed contact pressure. This can promise less pressure-induced mucosa lesions, minimized pain and discomfort, and potentially reduced long-term residual ridge resorption by using printed RPD frameworks [197].

It is recommended to consider the accuracy of fit for specific structural components in printed RPD framework [196]. In a printed RPD framework, maximum discrepancy in fit has been noted in the anterior strap region probably due to inaccuracy during scanning of complete arch and the concavity in the palate. For the same reasons, in the region of the lingual bar, discrepancy has been observed at the center of the lingual bar of SLS framework. Additionally, in general, the documented internal discrepancy in printed RPD frameworks has been reported to be less at the periphery than at the center [198]. Contradictory findings of fit have been reported for adaptation of rests on the intended surface [196]. In a clinical study, it was concluded that nearly 76% of rests of AM printed cast partial frameworks did not contact the intended surface and thus did not fulfill the intended function [194]. Conversely, in a separate study, it has been suggested that among the components of the RPD, best fit has been observed in the region of rests and reciprocal plates (50 μm) [179, 196]. Lastly, although the accuracy of claps is satisfactory for SLS framework, the time of RPD placement [196] concerns about loss of adaptation of claps to the abutment tooth, with a nearly 60% misfit between clasps and abutment teeth has been documented after nearly 8 years of use for printed RPD [199].

When the accuracy is compared between printed resin used for casting of RPD framework and printed metal RPD framework (SLS), a noticeable difference is observed. The overall discrepancy is smaller for SLS framework promising a superior accuracy and reproducibility of fabrication of metal RPD framework printed with SLS [196].

It has also been suggested that the method of preparation of analog to fabricate the RPD can tremendously affect the fit of the printed denture. Printable resin casts for RPD fabrication are less accurate than the stone casts probably due to errors in resin printing process [179]. Completely digital fabrication (intraoral digital scan and SLM) had significantly better fitness than did the traditional analog (physical impression and cast) [200].

8.6.2 Retentive Force of AM Printed Clasps

The retentive force of AM clasps has been documented to be less than the cast clasps [201, 202]. The documented mean retentive force value of cast clap is 13.6 N, while for the AM claps is 15.7 N. Additionally, a significant decrease in retentive force has been observed with cast clasps when compared with AM printed clasps. The consistency of retentive force in the clasps may have an impact on the longevity of the treatment [202].

8.6.3 Internal Porosity and Surface Roughness

Size, shape, and melting temperature of particles largely determine the surface quality of the printed RPD. Preheating the particles close to melting temperature enables curtailing the energy required for AM and thus promises a smooth surface texture [203]. It has been documented that although both SLM and EBM exhibit favorable results and excellent properties, EBM structures have a significantly rougher finish, lesser strength, and greater hardness as compared to SLM structures [204].

8.6.4 Patient Satisfaction

Laser-sintered RPDs have been associated with better outcome than conventional cast RPD in regards to general satisfaction, ability to speak, ability to clean, comfort, ability to masticate, masticatory efficiency, and maintenance of oral health status [177, 205]. The possible reason quoted by some authors for this is that laser-sintered cobalt chromium alloy is harder and denser and has better microstructural organization and higher strength than cast alloys, thus resulting in better stability and retention of the prosthesis [59, 206]. In a study comparing three different methods of RPD fabrication, including the analog method (physical impression and cast), analog and digital (laboratory scanner to scan stone cast and SLM), and digital (intraoral digital scan and SLM) by using a dichotomous questionnaire, the completely digital method of fabrication of RPD was found to yield maximum perception of fitness [200].

It would be safe to conclude that though printed partial denture frameworks are making inroads into prosthodontic practice, literature providing a robust support to the same is scarce. Clinical studies and in vitro research on designing and 3D printed removable prosthesis frameworks are largely inconclusive. However, preliminary data do not dismiss the 3D printing technique of removable partial framework fabrication as reports are confined well in clinically acceptable limits.

8.7 Occlusal Splints and Miscellaneous Appliances

The use of AM to fabricate splints or splint like appliances for several applications such as management of temporomandibular joint disorder, obstructive sleep apnea, interim prosthesis in full mouth rehabilitation procedures (Fig. 8.40a, b), and appliance for brachytherapy has been gaining popularity [70, 207–217].

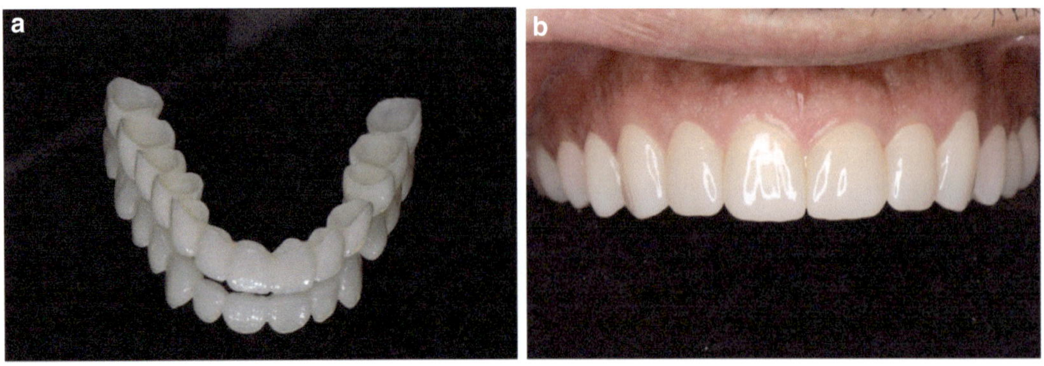

Fig. 8.40 (**a**, **b**) Full-coverage interim prosthesis fabricated by AM for evaluation during full mouth rehabilitation

Stereolithography using laser or digital light processing is most commonly employed for solidifying the polymer used in fabrication of these appliances [212]. Offering the advantage of consistent quality control and speed, simultaneously circumventing the limitations imposed by manual methods and shortcomings of materials like gypsum, wax, and resin for compression molding, the utility of 3D printing in fabrication of these appliances is tremendous. The process of fabrication of such appliances by 3D printing can be either completely digital or a hybrid (amalgamation of digital and analog approach). A purely digital approach uses intraoral data (digital impression of arches and bite/occlusal registration record) obtained from intraoral scanning for designing (Table 8.4). Hybrid approach uses conventional impression making and bite recording techniques in order to procure a model which is scanned by a desktop scanner (Table 8.5). The protocol subsequent to designing stays same for both the approaches. Designing of the appliance as per requirements is done by using the specific software (Table 8.6).

8.7.1 Fabrication of Splints, Overlays, and Interim Appliances by AM

The virtual analog of the patients is mounted on the virtual articulator by using the bite record obtained. The condylar guidance and incisal guide table are set on the virtual articulator as per the records obtained from the patient in order to attain the occlusal scheme desired. For designing, it is recommended to determine path of insertion followed by delineating retentive undercuts and blocking out the interferences. The boundary of the appliance is marked. A shell of about 1.5–2 mm thickness is designed on the demarcated area of the appliance. The occlusal surface of the shell is modified in order to establish the occlusal contact points as per the occlusal scheme desired. The data are exported to a compatible printer. The slicer software prepares the data for printing. Printing is eventually accomplished by digital light processing/SLA of resin.

8.7.2 Appraisal of Properties of Appliances Fabricated by AM

The overall outcome of the appliance fabricated by AM depends on the accuracy of fabrication procedure together with quality, long-term stability, and biocompatibility of material. There is lack of availability of long-term clinical outcome with AM splints [212, 218]. Regarding accuracy, while a high degree of precision has been reported in printed splints when compared to milled splints [219], the trueness of 3D printed splints has been found to be inferior to the CAD CAM milled occlusal splint appliance. When compared to conventional heat polymerized appliances, better internal fit and equivalent number of occlusal adjustments have been observed in 3D printed splints [218].

Regarding the mechanical properties of the printed appliances, it had been suggested that these properties depend more on the material than technique of fabrication [36]. For instance, material considerations are extremely important in context of the property of wear. Contradictory findings have been reported for wear of printed splints. A higher material wear and less favorable material properties have been documented with printed occlusal splints in some studies [219, 220–222], and lower values have been documented in others [223]. Lower surface hardness, but higher flexural strength in printed occlusal splint materials when compared to other polymethyl methacrylate acrylic (PMMA) resins, has been documented in the literature [36, 224]. When compared to milled resins, 3D printed resins have been documented to have lower flexural strength and hardness values [221].

It is again highlighted that when evaluating the outcome of properties of printed objects, the anisotropic behavior of printed specimens with regard to build orientation and positioning has been observed [225]. The working angle on the build platform and, hence, the direction of the layers seem to be of particular importance [212]. In this context, 90° specimens with layer orientation parallel to the axial load show the superior flexural strength and flexural modulus [225,

226]. It has also been suggested that highest hardness, elastic modulus, and lowest surface roughness have been found when printing is done at 45° [227]. Also, objects printed on the edges of build platform are more prone to inaccuracies than those in the center [225]. Among the technical aspects determining the outcome, post-print processing has an important role. The time of post-processing is critical as a high degree of conversion enhances mechanical properties. However, caution must be taken not to exceed the recommended time of post-processing, as more than the recommended time can cause a reduction in flexural strength [228–230].

3D printed resins have high water sorption and solubility when compared to pressed and milled resins [221, 226]. This property is of considerable importance as the method of storage in water can have a considerable effect on mechanical properties, especially flexural properties. In fact, water storage has been documented to have a greater effect on flexural properties than different printing directions [226, 230].

The effect of printing layer thickness has also been found to determine the mechanical properties of printed splints largely. Literature reports suggest that reduced print surface layer thickness enhances the flexural strength and hardness of the 3D printed splint [186, 190, 195].

Microbial adhesion to oral splints can be a concern for medically compromised patients due to occurrence of *Candida*, periodontitis, or caries. It has been observed that AM of oral splints has been associated with increased adhesion of *Candida albicans* when compared to conventionally fabricated splints. This may be attributed to higher water sorption and solubility when compared to pressed and milled resins [231].

Advantages of using 3D printing for miscellaneous appliance fabrication include reproducibility, accuracy by elimination of errors associated with conventional steps, abridged fabrication time, the possibility to archive data, and reprint additional appliances for the patient as the need arises.

8.8 Summary

Sustainable development goals carry a big promise for all living beings on the planet. 3D printing is a part of the massive change that can help achieve this goal. Dentistry in general and prosthodontics in particular have adopted this change positively, as are witnessed in the diverse applications of 3D printing. The reduced overhead costs, impressive speed of fabrication of a prosthesis resulting in faster turnaround time, conservation of resources and man-hours, archiving of the data enabling replication, and generation of complex architectures with greater interior detailing accounting for greater accuracy and fit are among the many credible factors bound appeal even those who are currently lagging in embracing the of this technology. Maturing the prosthodontic practice with tailor-made AM solutions requires a thorough understanding of the related mechanics and materials aspects related to 3D printing. In order to excel in applications of AM, one also needs to be very proficient in conventional prosthodontic protocols. It requires a sound understanding of basic foundation principles of prosthodontics. This technology is a tool or an adjunct that can enhance the outcome and is a work in progress. With expansion of technological advancements we can deliver better products in times to come.

Along those lines, it is emphasized that operators must ally with manufactures specializing in dental printers and choose among their services diligently. Investing in any type of printer requires integration of numerous events such as acquiring the hardware, software, ensuring a steady supply of consumables materials, discerning the techniques, understanding the material properties, and overall process. This requires exclusive training and support services dedicated toward the same. A compromise at any level in the chain of fabrication can jeopardize the overall outcome. For instance, while designing, operators are encouraged to choose a program that befits their skills and yet allows a technically sound design. In this context, it will be hard to miss the necessity to have a thorough understanding of the post-processing procedures and execute them in the prescribed manner in order to ensure a high-quality product.

Specific considerations when comparing different brands include size of printer, accuracy, resolution, speed, quality of materials, and rage of use of materials. For instance, razor-thin margins, a requirement of direct restorative and implant prosthesis, require high resolution of the printer. Speed of printing is undoubtedly an important consideration especially when the purpose is for early clinical utilization of the product (such as surgical guides) or where high volume of laboratory printing is desired.

Consumers are also encouraged to be aware of range of materials that can be printed and whether the printer offers use of open system (open parameters) or proprietary materials only. An open system printer supporting numerous materials from varied manufacturer globally provides maximum flexibility and productivity. Operators must, however, ensure that when using third-party materials, clinically acceptable quality and accuracy are achievable.

AM is one of the most dynamic development of science requiring consistent progressive outlook to move ahead. The overall performance gains and potential efficiencies of a well-orchestrated AM are redefining prosthodontics today and can be foreseen to evolve more in the future!

Acknowledgements are appropiate

1. Precision Ceramic Artz, Mohali, Mr. Sudhir Bansal, Mr. Manish Gupta.
2. INNOVATIVE 3D PRINTING SOLUTIONS LLP, Mr. Mitesh Patil, Mr. Ganesh Padwal, Mr. Kalpesh Jadhav, Mr. Mohd. Anwar Khan, Mr. Mohd Rizwan Khan for their technical assistance and data collection in relation to Fixed dental prosthesis section.
3. Dr. Amit Sharma and Dr. Arun Kumar: Senior resident, Division of Prosthodontics, CDER, AIIMS, New Delhi.
4. Dr. Majed Altoman: Graduate student, School of Dentistry, Loma Linda University.

References

1. Pillai S, Upadhyay A, Khayambashi P, Farooq I, Sabri H, Tarar M, et al. Dental 3D-printing: transferring art from the laboratories to the clinics. Polymers. 2021;13(1):157.
2. Anadioti E, Musharbash L, Blatz MB, Papavasiliou G, Kamposiora P. 3D printed complete removable dental prostheses: a narrative review. BMC Oral Health. 2020;20(1):343.
3. Cristache C, Totu E, Grosu A, Ene O, Beuran I, Burlibasa M. Nanocomposite for rapid prototyped complete denture eighteen months follow-up on clinical performance. Rev Chim. 2019;70(2):387–92.
4. Alharbi N, Wismeijer D, Osman RB. Additive manufacturing techniques in prosthodontics: where do we currently stand? A critical review. Int J Prosthodont. 2017;30(5):474–84.
5. Formlabs introduces 3-D printed dentures [Internet]. Dentistryiq.com. 2019 [cited 25 October 2021]. Available from: https://www.dentistryiq.com/dentistry/products/restorative-products/article/16363613/formlabs-introduces-3d-printed-dentures.
6. Felton DA. Complete edentulism and comorbid diseases: an update. J Prosthodont. 2016;25(1):5–20.
7. Bidra AS, Taylor TD, Agar JR. Computer-aided technology for fabricating complete dentures: systematic review of historical background, current status, and future perspectives. J Prosthet Dent. 2013;109(6):361–6.
8. Kattadiyil MT, Jekki R, Goodacre CJ, Baba NZ. Comparison of treatment outcomes in digital and conventional complete removable dental prosthesis fabrications in a predoctoral setting. J Prosthet Dent. 2015;114(6):818–25.
9. Fernandez MA, Nimmo A, Behar-Horenstein LS. Digital denture fabrication in pre- and postdoctoral education: a survey of U.S. dental schools. J Prosthodont. 2016;25(1):83–90.
10. Wu J, Cheng Y, Gao B, Yu H. A novel digital altered cast impression technique for fabricating a removable partial denture with a distal extension. J Am Dent Assoc. 2020;151(4):297–302.
11. Maragliano-Muniz P, Kukucka ED. Incorporating digital dentures into clinical practice: flexible workflows and improved clinical outcomes. J Prosthodont. 2021;30(S2):125–32.
12. Lee SJ, Betensky RA, Gianneschi GE, Gallucci GO. Accuracy of digital versus conventional implant impressions. Clin Oral Implants Res. 2015;26(6):715–9.
13. Wilk BL. Intraoral digital impressioning for dental implant restorations versus traditional implant impression techniques. Compend Contin Educ Dent. 2015;36(7):529–30, 532–3.
14. Patzelt SB, Vonau S, Stampf S, Att W. Assessing the feasibility and accuracy of digitizing edentulous jaws. J Am Dent Assoc. 2013;144(8):914–20.

15. Flügge TV, Schlager S, Nelson K, Nahles S, Metzger MC. Precision of intraoral digital dental impressions with iTero and extraoral digitization with the iTero and a model scanner. Am J Orthod Dentofac Orthop. 2013;144(3):471–8.
16. Massad JJ, Connelly ME, Rudd KD, Cagna DR. Occlusal device for diagnostic evaluation of maxillomandibular relationships in edentulous patients: a clinical technique. J Prosthet Dent. 2004;91(6):586–90.
17. Alqarni H, AlHelal A, Kattadiyil MT. Computer-engineered complete denture fabrication with conventional clinical steps: a technique to overcome protocol limitations. J Prosthet Dent. 2019;122(5):430–4.
18. Akerly WB. Recording jaw relationships in edentulous patients. Dent Clin N Am. 1996;40(1):53–70.
19. Hassan B, Greven M, Wismeijer D. Integrating 3D facial scanning in a digital workflow to CAD/CAM design and fabricate complete dentures for immediate total mouth rehabilitation. J Adv Prosthodont. 2017;9(5):381–6.
20. Clark WA, Duqum I, Kowalski BJ. The digitally replicated denture technique: a case report. J Esthet Restor Dent. 2019;31(1):20–5.
21. Schweiger J, Stumbaum J, Edelhoff D, Güth JF. Systematics and concepts for the digital production of complete dentures: risks and opportunities. Int J Comput Dent. 2018;21(1):41–56.
22. Kraemer Fernandez P, Unkovskiy A, Benkendorff V, Klink A, Spintzyk S. Surface characteristics of milled and 3D printed denture base materials following polishing and coating: an in-vitro study. Materials (Basel). 2020;13(15):3305.
23. Schweiger J, Güth JF, Edelhoff D, Stumbaum J. Virtual evaluation for CAD-CAM-fabricated complete dentures. J Prosthet Dent. 2017;117(1):28–33.
24. Chen H, Wang H, Lv P, Wang Y, Sun Y. Quantitative evaluation of tissue surface adaption of CAD-designed and 3D printed wax pattern of maxillary complete denture. Biomed Res Int. 2015;2015:453968.
25. Kalberer N, Mehl A, Schimmel M, Müller F, Srinivasan M. CAD-CAM milled versus rapidly prototyped (3D-printed) complete dentures: an in vitro evaluation of trueness. J Prosthet Dent. 2019;121(4):637–43.
26. Presotto AGC, Barão VAR, Bhering CLB, Mesquita MF. Dimensional precision of implant-supported frameworks fabricated by 3D printing. J Prosthet Dent. 2019;122(1):38–45.
27. Saponaro PC, Yilmaz B, Johnston W, Heshmati RH, McGlumphy EA. Evaluation of patient experience and satisfaction with CAD-CAM-fabricated complete dentures: a retrospective survey study. J Prosthet Dent. 2016;116(4):524–8.
28. AlHelal A, Goodacre BJ, Kattadiyil MT, Swamidass R. Errors associated with digital preview of computer-engineered complete dentures and guidelines for reducing them: a technique article. J Prosthet Dent. 2018;119(1):17–25.
29. McLaughlin JB, Ramos V Jr, Dickinson DP. Comparison of fit of dentures fabricated by traditional techniques versus CAD/CAM technology. J Prosthodont. 2019;28(4):428–35.
30. Choi JJE, Uy CE, Plaksina P, Ramani RS, Ganjigatti R, Waddell JN. Bond strength of denture teeth to heat-cured, CAD/CAM and 3D printed denture acrylics. J Prosthodont. 2020;29(5):415–21.
31. Cha HS, Park JM, Kim TH, Lee JH. Wear resistance of 3D-printed denture tooth resin opposing zirconia and metal antagonists. J Prosthet Dent. 2020;124(3):387–94.
32. Park C, Kee W, Lim HP, Park SW. Combining 3D-printed metal and resin for digitally fabricated dentures: a dental technique. J Prosthet Dent. 2020;123(3):389–92.
33. Lee S, Hong SJ, Paek J, Pae A, Kwon KR, Noh K. Comparing accuracy of denture bases fabricated by injection molding, CAD/CAM milling, and rapid prototyping method. J Adv Prosthodont. 2019;11(1):55–64.
34. Arslan M, Murat S, Alp G, Zaimoglu A. Evaluation of flexural strength and surface properties of prepolymerized CAD/CAM PMMA-based polymers used for digital 3D complete dentures. Int J Comput Dent. 2018;21(1):31–40.
35. Aguirre BC, Chen JH, Kontogiorgos ED, Murchison DF, Nagy WW. Flexural strength of denture base acrylic resins processed by conventional and CAD-CAM methods. J Prosthet Dent. 2020;123(4):641–6.
36. PrpiĆ V, Schauperl Z, ĆatiĆ A, DulčiĆ N, ČimiĆ S. Comparison of mechanical properties of 3D-printed, CAD/CAM, and conventional denture base materials. J Prosthodont. 2020;29(6):524–8.
37. Bonnet G, Batisse C, Bessadet M, Nicolas E, Veyrune JL. A new digital denture procedure: a first practitioners appraisal. BMC Oral Health. 2017;17(1):155.
38. Kado D, Sakurai K, Sugiyama T, Ueda T. Evaluation of cleanability of a titanium dioxide (TiO_2)-coated acrylic resin denture base. Prosthodont Res Pract. 2005;4(1):69–76.
39. Arai T, Ueda T, Sugiyama T, Sakurai K. Inhibiting microbial adhesion to denture base acrylic resin by titanium dioxide coating. J Oral Rehabil. 2009;36(12):902–8.
40. Tsuji M, Ueda T, Sawaki K, Kawaguchi M, Sakurai K. Biocompatibility of a titanium dioxide-coating method for denture base acrylic resin. Gerodontology. 2016;33(4):539–44.
41. Totu EE, Nechifor AC, Nechifor G, Aboul-Enein HY, Cristache CM. Poly(methyl methacrylate) with TiO_2 nanoparticles inclusion for stereolithographic complete denture manufacturing—the future in dental care for elderly edentulous patients? J Dent. 2017;59:68–77.
42. Alharbi N, Osman R, Wismeijer D. Effects of build direction on the mechanical properties of 3D-printed

complete coverage interim dental restorations. J Prosthet Dent. 2016;115(6):760–7.
43. Carlsson GE, Omar R. Trends in prosthodontics. Med Princ Pract. 2006;15(3):167–79.
44. Joda T, Zarone F, Ferrari M. The complete digital workflow in fixed prosthodontics: a systematic review. BMC Oral Health. 2017;17(1):124.
45. Beguma Z, Chhedat P. Rapid prototyping—when virtual meets reality. Int J Comput Dent. 2014;17(4):297–306.
46. Berman B. 3-D printing: the new industrial revolution. Bus Horiz. 2012;55(2):155–62.
47. Sun J, Zhang FQ. The application of rapid prototyping in prosthodontics. J Prosthodont. 2012;21(8):641–4.
48. Ellakany P, Al-Harbi F, El Tantawi M, Mohsen C. Evaluation of the accuracy of digital and 3D-printed casts compared with conventional stone casts. J Prosthet Dent. 2020;127(3):438–44.
49. Park ME, Shin SY. Three-dimensional comparative study on the accuracy and reproducibility of dental casts fabricated by 3D printers. J Prosthet Dent. 2018;119(5):861.e1–7.
50. Hazeveld A, Huddleston Slater JJ, Ren Y. Accuracy and reproducibility of dental replica models reconstructed by different rapid prototyping techniques. Am J Orthod Dentofac Orthop. 2014;145:108–15.
51. Liu Q, Leu M, Schmitt S. Rapid prototyping in dentistry: technology and application. Int J Adv Manuf Technol. 2005;29(3–4):317–35.
52. Keating AP, Knox J, Bibb R, Zhurov AI. A comparison of plaster, digital and reconstructed study model accuracy. J Orthod. 2008;35(3):191–201; discussion 175.
53. Xie R, Li D, Chao S. An inexpensive stereolithography technology with high power UV-LED light. Rapid Prototyp J. 2011;17(6):441–50.
54. Aly P, Mohsen C. Comparison of the accuracy of three-dimensional printed casts, digital, and conventional casts: an in vitro study. Eur J Dent. 2020;14(2):189–93.
55. Papaspyridakos P, Chen YW, Alshawaf B, Kang K, Finkelman M, Chronopoulos V, et al. Digital workflow: in vitro accuracy of 3D printed casts generated from complete-arch digital implant scans. J Prosthet Dent. 2020;124(5):589–93.
56. Murr L, Gaytan S, Ramirez D, Martinez E, Hernandez J, Amato K, et al. Metal fabrication by additive manufacturing using laser and electron beam melting technologies. J Mater Sci Technol. 2012;28(1):1–14.
57. Yadroitsev I, Krakhmalev P, Yadroitsava I, Johansson S, Smurov I. Energy input effect on morphology and microstructure of selective laser melting single track from metallic powder. J Mater Process Technol. 2013;213:606–13.
58. Chou WT, Chuang CC, Wang YB, Chiu HC. Comparison of the internal fit of metal crowns fabricated by traditional casting, computer numerical control milling, and three-dimensional printing. PLoS One. 2021;16(9):e0257158.
59. Koutsoukis T, Zinelis S, Eliades G, Al-Wazzan K, Rifaiy MA, Al Jabbari YS. Selective laser melting technique of Co-Cr dental alloys: a review of structure and properties and comparative analysis with other available techniques. J Prosthodont. 2015;24(4):303–12.
60. Revilla-León M, Sadeghpour M, Özcan M. A review of the applications of additive manufacturing technologies used to fabricate metals in implant dentistry. J Prosthodont. 2020;29(7):579–93.
61. Konieczny B, Szczesio-Wlodarczyk A, Sokolowski J, Bociong K. Challenges of Co-Cr alloy additive manufacturing methods in dentistry-the current state of knowledge (systematic review). Materials (Basel). 2020;13(16):3524.
62. Qian B, Saeidi K, Kvetková L, Lofaj F, Xiao C, Shen Z. Defects-tolerant Co-Cr-Mo dental alloys prepared by selective laser melting. Dent Mater. 2015;31(12):1435–44.
63. Parthasarathy J, Starly B, Raman S, Christensen A. Mechanical evaluation of porous titanium (Ti6Al4V) structures with electron beam melting (EBM). J Mech Behav Biomed Mater. 2010;3:249–59.
64. Takaichi A, Suyalatu, Nakamoto T, Joko N, Nomura N, Tsutsumi Y, Migita S, et al. Microstructures and mechanical properties of Co-29Cr-6Mo alloy fabricated by selective laser melting process for dental applications. J Mech Behav Biomed Mater. 2013;21:67–76.
65. Al Jabbari YS, Koutsoukis T, Barmpagadaki X, Zinelis S. Metallurgical and interfacial characterization of PFM Co-Cr dental alloys fabricated via casting, milling or selective laser melting. Dent Mater. 2014;30(4):e79–88.
66. Hesse H, Özcan M. A review on current additive manufacturing technologies and materials used for fabrication of metal-ceramic fixed dental prosthesis. J Adhes Sci Technol. 2021; https://doi.org/10.1080/01694243.2021.1899699.
67. Barucca G, Santecchia E, Majni G, Girardin E, Bassoli E, Denti L, et al. Structural characterization of biomedical Co-Cr-Mo components produced by direct metal laser sintering. Mater Sci Eng C Mater Biol Appl. 2015;48:263–9.
68. Venkatesh KV, Nandini VV. Direct metal laser sintering: a digitised metal casting technology. J Indian Prosthodont Soc. 2013;13(4):389–92.
69. Horn TJ, Harrysson OL. Overview of current additive manufacturing technologies and selected applications. Sci Prog. 2012;95(Pt 3):255–82.
70. Abduo J, Lyons K, Bennamoun M. Trends in computer-aided manufacturing in prosthodontics: a review of the available streams. Int J Dent. 2014;2014:783948.
71. Zocca A, Colombo P, Gomes C, Günster J. Additive manufacturing of ceramics: issues,

potentialities, and opportunities. J Am Ceram Soc. 2015;98(7):1983–2001.

72. Zhao X, Evans JRG, Edirisinghe MJ, Song J. Direct ink-jet printing of vertical walls. J Am Ceram Soc. 2002;85(8):2113–5.
73. Derby B. Additive manufacture of ceramics components by inkjet printing. Engineering. 2015;1(1):113–23.
74. Bourell DL, Marcus HL, Barlow JW, Beaman JJ. Selective laser sintering of metals and ceramics. Int J Powder Metal. 1992;28:369–81.
75. Gonzalez JA, Mireles J, Lin Y, Wicker RB. Characterization of ceramic components fabricated using binder jetting additive manufacturing technology. Ceram Int. 2016;42:10559–64.
76. Schwentenwein M, Schneider P, Homa J. Lithography-based ceramic manufacturing: a novel technique for additive manufacturing of high-performance ceramics. Adv Sci Technol. 2014;88:60–4.
77. Scheithauer U, Schwarzer E, Moritz T, Michaelis A. Additive manufacturing of ceramic heat exchanger: opportunities and limits of the lithography-based ceramic manufacturing (LCM). J Mater Eng Perform. 2017;27(1):14–20.
78. Uçar Y, Aysan Meriç İ, Ekren O. Layered manufacturing of dental ceramics: fracture mechanics, microstructure, and elemental composition of lithography-sintered ceramic. J Prosthodont. 2019;28(1):e310–8.
79. Hartmann M, Pfaffinger M, Stampfl J. The role of solvents in lithography-based ceramic manufacturing of lithium disilicate. Materials (Basel). 2021;14(4):1045.
80. Dehurtevent M, Robberecht L, Hornez JC, Thuault A, Deveaux E, Béhin P. Stereolithography: a new method for processing dental ceramics by additive computer-aided manufacturing. Dent Mater. 2017;33(5):477–85.
81. Baumgartner S, Gmeiner R, Schönherr JA, Stampfl J. Stereolithography-based additive manufacturing of lithium disilicate glass ceramic for dental applications. Mater Sci Eng C Mater Biol Appl. 2020;116:111180.
82. Schönherr JA, Baumgartner S, Hartmann M, Stampfl J. Stereolithographic additive manufacturing of high precision glass ceramic parts. Materials (Basel). 2020;13(7):1492.
83. Li R, Wang Y, Hu M, Wang Y, Xv Y, Liu Y, Sun Y. Strength and adaptation of stereolithography-fabricated zirconia dental crowns: an in vitro study. Int J Prosthodont. 2019;32(5):439–43.
84. Zandinejad A, Methani MM, Schneiderman ED, Revilla-León M, Bds DM. Fracture resistance of additively manufactured zirconia crowns when cemented to implant supported zirconia abutments: an in vitro study. J Prosthodont. 2019;28:893–7.
85. Ioannidis A, Bomze D, Hämmerle CHF, Hüsler J, Birrer O, Mühlemann S. Load-bearing capacity of CAD/CAM 3D-printed zirconia, CAD/CAM milled zirconia, and heat-pressed lithium disilicate ultra-thin occlusal veneers on molars. Dent Mater. 2020;36:e109–e16.
86. Wang W, Yu H, Liu Y, Jiang X, Gao B. Trueness analysis of zirconia crowns fabricated with 3 dimensional printing. J Prosthet Dent. 2019;121:285–91.
87. Jang KJ, Kang JH, Fisher JG, Park SW. Effect of the volume fraction of zirconia suspensions on the microstructure and physical properties of products produced by additive manufacturing. Dent Mater. 2019;35:e97–106.
88. Madhav V, Digholkar S, Palaskar J. Evaluation of the flexural strength and microhardness of provisional crown and bridge materials fabricated by different methods. J Indian Prosthodont Soc. 2016;16(4):328–34.
89. Joo HS, Park SW, Yun KD, Lim HP. Complete-mouth rehabilitation using a 3D printing technique and the CAD/CAM double scanning method: a clinical report. J Prosthet Dent. 2016;116(1):3–7.
90. Dikova TD, Dzhendov DA, Ivanov D, Bliznakova K. Dimensional accuracy and surface roughness of polymeric dental bridges produced by different 3D printing processes. Arch Mater Sci Eng. 2018;2(94):65–75.
91. Lee WS, Lee DH, Lee KB. Evaluation of internal fit of interim crown fabricated with CAD/CAM milling and 3D printing system. J Adv Prosthodont. 2017;9(4):265–70.
92. Peng CC, Chung KH, Yau HT, Ramos V Jr. Assessment of the internal fit and marginal integrity of interim crowns made by different manufacturing methods. J Prosthet Dent. 2020;123(3):514–22.
93. Peng CC, Chung KH, Ramos V Jr. Assessment of the adaptation of interim crowns using different measurement techniques. J Prosthodont. 2020;29(1):87–93.
94. Katreva I, Dikova T, Tonchev T. 3D printing – an alternative of conventional crown fabrication: a case report. J IMAB Annu Proc (Sci Pap). 2018;24(2):2048–54.
95. Tahmaseb A, Wismeijer D, Coucke W, Derksen W. Computer technology applications in surgical implant dentistry: a systematic review. Int J Oral Maxillofac Implants. 2014;29(Suppl):25–42.
96. Cristache CM, Gurbanescu S. Accuracy evaluation of a stereolithographic surgical template for dental implant insertion using 3D superimposition protocol. Int J Dent. 2017;2017:4292081.
97. Chen L, Lin WS, Polido WD, Eckert GJ, Morton D. Accuracy, reproducibility, and dimensional stability of additively manufactured surgical templates. J Prosthet Dent. 2019;122(3):309–14.
98. Kernen F, Kramer J, Wanner L, Wismeijer D, Nelson K, Flügge T. A review of virtual planning software for guided implant surgery - data import and visualization, drill guide design and manufacturing. BMC Oral Health. 2020;20(1):251.
99. Elkomy MM, Khamis MM, El-Sharkawy AM. Clinical and radiographic evaluation of implants

placed with fully guided versus partially guided tissue-supported surgical guides: a split-mouth clinical study. J Prosthet Dent. 2021;126(1):58–66.
100. Oh KC, Shim JS, Park JM. In vitro comparison between metal sleeve-free and metal sleeve-incorporated 3D-printed computer-assisted implant surgical guides. Materials (Basel). 2021;14(3):615.
101. Oh KC, Park JM, Shim JS, Kim JH, Kim JE, Kim JH. Assessment of metal sleeve-free 3D-printed implant surgical guides. Dent Mater. 2019;35(3):468–76.
102. Sarhan MM, Khamis MM, El-Sharkawy AM. Evaluation of the accuracy of implant placement by using fully guided versus partially guided tissue-supported surgical guides with cylindrical versus C-shaped guiding holes: a split-mouth clinical study. J Prosthet Dent. 2021;125(4):620–7.
103. Pieralli S, Spies BC, Hromadnik V, Nicic R, Beuer F, Wesemann C. How accurate is oral implant installation using surgical guides printed from a degradable and steam-sterilized biopolymer? J Clin Med. 2020;9(8):2322.
104. Turbush SK, Turkyilmaz I. Accuracy of three different types of stereolithographic surgical guide in implant placement: an in vitro study. J Prosthet Dent. 2012;108(3):181–8.
105. Bencharit S, Staffen A, Yeung M, Whitley D III, Laskin DM, Deeb GR. In vivo tooth-supported implant surgical guides fabricated with desktop stereolithographic printers: fully guided surgery is more accurate than partially guided surgery. J Oral Maxillofac Surg. 2018;76(7):1431–9.
106. Bell CK, Sahl EF, Kim YJ, Rice DD. Accuracy of implants placed with surgical guides: thermoplastic versus 3D printed. Int J Periodontics Restorative Dent. 2018;38(1):113–9.
107. D'haese R, Vrombaut T, Hommez G, De Bruyn H, Vandeweghe S. Accuracy of guided implant surgery in the edentulous jaw using desktop 3D-printed mucosal supported guides. J Clin Med. 2021;10(3):391.
108. Kim T, Lee S, Kim GB, Hong D, Kwon J, Park JW, et al. Accuracy of a simplified 3D-printed implant surgical guide. J Prosthet Dent. 2020;124(2):195–201.e2.
109. Herschdorfer L, Negreiros WM, Gallucci GO, Hamilton A. Comparison of the accuracy of implants placed with CAD-CAM surgical templates manufactured with various 3D printers: an in vitro study. J Prosthet Dent. 2021;125(6):905–10.
110. Li H, Wang T, Sun J, Yu Z. The effect of process parameters in fused deposition modelling on bonding degree and mechanical properties. Rapid Prototyp J. 2018;24(1):80–92.
111. Sittikornpaiboon P, Arunjaroensuk S, Kaboosaya B, Subbalekha K, Mattheos N, Pimkhaokham A. Comparison of the accuracy of implant placement using different drilling systems for static computer-assisted implant surgery: a simulation-based experimental study. Clin Implant Dent Relat Res. 2021;23(4):635–43.
112. Shah NP, Khanna A, Pai AR, Sheth VH, Raut SR. An evaluation of virtually planned and 3D-printed stereolithographic surgical guides from CBCT and digital scans: an in vitro study. J Prosthet Dent. 2021; https://doi.org/10.1016/j.prosdent.2020.12.035.
113. Henprasert P, Dawson DV, El-Kerdani T, Song X, Couso-Queiruga E, Holloway JA. Comparison of the accuracy of implant position using surgical guides fabricated by additive and subtractive techniques. J Prosthodont. 2020;29(6):534–41.
114. Sharma N, Cao S, Msallem B, Kunz C, Brantner P, Honigmann P, et al. Effects of steam sterilization on 3D printed biocompatible resin materials for surgical guides-an accuracy assessment study. J Clin Med. 2020;9(5):1506.
115. Etemad-Shahidi Y, Qallandar OB, Evenden J, Alifui-Segbaya F, Ahmed KE. Accuracy of 3-dimensionally printed full-arch dental models: a systematic review. J Clin Med. 2020;9(10):3357.
116. Wan Hassan WN, Yusoff Y, Mardi NA. Comparison of reconstructed rapid prototyping models produced by 3-dimensional printing and conventional stone models with different degrees of crowding. Am J Orthod Dentofac Orthop. 2017;151(1):209–18.
117. Jin SJ, Kim DY, Kim JH, Kim WC. Accuracy of dental replica models using photopolymer materials in additive manufacturing: in vitro three-dimensional evaluation. J Prosthodont. 2019;28(2):e557–62.
118. Rungrojwittayakul O, Kan JY, Shiozaki K, Swamidass RS, Goodacre BJ, Goodacre CJ, et al. Accuracy of 3D printed models created by two technologies of printers with different designs of model base. J Prosthodont. 2020;29(2):124–8.
119. Joda T, Matthisson L, Zitzmann NU. Impact of aging on the accuracy of 3D-printed dental models: an in vitro investigation. J Clin Med. 2020;9(5):1436.
120. Emir F, Ceylan G, Ayyildiz S. In vitro accuracies of 3D printed models manufactured by two different printing technologies. Eur Oral Res. 2021;55(2):80–5.
121. Nestler N, Wesemann C, Spies BC, Beuer F, Bumann A. Dimensional accuracy of extrusion- and photopolymerization-based 3D printers: in vitro study comparing printed casts. J Prosthet Dent. 2021;125(1):103–10.
122. Camardella LT, de Vasconcellos Vilella O, Breuning H. Accuracy of printed dental models made with 2 prototype technologies and different designs of model bases. Am J Orthod Dentofac Orthop. 2017;151(6):1178–87.
123. Kalman L. 3D printing of a novel dental implant abutment. J Dent Res Dent Clin Dent Prospect. 2018;12(4):299–303.
124. Magne P, Silva M, Oderich E, Boff LL, Enciso R. Damping behavior of implant-supported restorations. Clin Oral Implants Res. 2013;24(2):143–8.

125. Kotick PG, Blumenkopf B. Abutment selection for implant restorations. Inside Dentistry. 2011;7(7):14–6.
126. Gonzalo E, Vizoso B, Lopez-Suarez C, Diaz P, Pelaez J, Suarez MJ. Evaluation of milled titanium versus laser sintered Co-Cr abutments on the marginal misfit in internal implant-abutment connection. Materials (Basel). 2020;13(21):4873.
127. Hartman MJ. A workflow to design and fabricate a customized healing abutment from a dynamic navigation virtual treatment plan. Compend Contin Educ Dent. 2021;42(2):86–92.
128. Kalman L. In vitro assessment of a novel additive manufactured titanium implant abutment. J Clin Exp Dent. 2021;13(2):e99–e103.
129. Joda T, Bürki A, Bethge S, Brägger U, Zysset P. Stiffness, strength, and failure modes of implant-supported monolithic lithium disilicate crowns: influence of titanium and zirconia abutments. Int J Oral Maxillofac Implants. 2015;30(6):1272–9.
130. Akçin ET, Güncü MB, Aktaş G, Aslan Y. Effect of manufacturing techniques on the marginal and internal fit of cobalt-chromium implant-supported multiunit frameworks. J Prosthet Dent. 2018;120(5):715–20.
131. Alonso-Pérez R, Bartolomé JF, Ferreiroa A, Salido MP, Pradíes G. Evaluation of the mechanical behavior and marginal accuracy of stock and laser-sintered implant abutments. Int J Prosthodont. 2017;30(2):136–8.
132. Fernández M, Delgado L, Molmeneu M, García D, Rodríguez D. Analysis of the misfit of dental implant-supported prostheses made with three manufacturing processes. J Prosthet Dent. 2014;111(2):116–23.
133. Molinero-Mourelle P, Cascos-Sanchez R, Yilmaz B, Lam WYH, Pow EHN, Del Río Highsmith J, Gómez-Polo M. Effect of fabrication technique on the microgap of CAD/CAM cobalt-chrome and zirconia abutments on a conical connection implant: an in vitro study. Materials (Basel). 2021;14(9):2348.
134. Hjalmarsson L, Örtorp A, Smedberg JI, Jemt T. Precision of fit to implants: a comparison of Cresco™ and Procera® implant bridge frameworks. Clin Implant Dent Relat Res. 2010;12(4):271–80.
135. Abduo J, Lyons K, Bennani V, Waddell N, Swain M. Fit of screw-retained fixed implant frameworks fabricated by different methods: a systematic review. Int J Prosthodont. 2011;24(3):207–20.
136. Revilla-León M, Sánchez-Rubio JL, Pérez-López J, Rubenstein J, Özcan M. Discrepancy at the implant abutment-prosthesis interface of complete-arch cobalt-chromium implant frameworks fabricated by additive and subtractive technologies before and after ceramic veneering. J Prosthet Dent. 2021;125(5):795–803.
137. Svanborg P, Eliasson A, Stenport V. Additively manufactured titanium and cobalt-chromium implant frameworks: fit and effect of ceramic veneering. Int J Oral Maxillofac Implants. 2018;33(3):590–6.
138. Revilla-León M, Ceballos L, Martínez-Klemm I, Özcan M. Discrepancy of complete-arch titanium frameworks manufactured using selective laser melting and electron beam melting additive manufacturing technologies. J Prosthet Dent. 2018;120(6):942–7.
139. Zanjanijam AR, Major I, Lyons JG, Lafont U, Devine DM. Fused filament fabrication of PEEK: a review of process-structure-property relationships. Polymers (Basel). 2020;12(8):1665.
140. Najeeb S, Zafar MS, Khurshid Z, Siddiqui F. Applications of polyetheretherketone (PEEK) in oral implantology and prosthodontics. J Prosthodont Res. 2016;60(1):12–9.
141. Han X, Sharma N, Xu Z, Scheideler L, Geis-Gerstorfer J, Rupp F, et al. An in vitro study of osteoblast response on fused-filament fabrication 3D printed PEEK for dental and cranio-maxillofacial implants. J Clin Med. 2019;8(6):771.
142. Mangano C, Bianchi A, Mangano FG, Dana J, Colombo M, Solop I, et al. Custom-made 3D printed subperiosteal titanium implants for the prosthetic restoration of the atrophic posterior mandible of elderly patients: a case series. 3D Print Med. 2020;6(1):1.
143. Surovas A. A digital workflow for modeling of custom dental implants. 3D Print Med. 2019;5(1):9.
144. Honigmann P, Sharma N, Okolo B, Popp U, Msallem B, Thieringer FM. Patient-specific surgical implants made of 3D printed PEEK: material, technology, and scope of surgical application. Biomed Res Int. 2018;2018:4520636.
145. Mounir M, Atef M, Abou-Elfetouh A, Hakam MM. Titanium and polyether ether ketone (PEEK) patient-specific sub-periosteal implants: two novel approaches for rehabilitation of the severely atrophic anterior maxillary ridge. Int J Oral Maxillofac Surg. 2018;47(5):658–64.
146. Jamayet NB, Abdullah YJ, Rajion ZA, Husein A, Alam MK. New approach to 3D printing of facial prostheses using combination of open source software and conventional techniques: a case report. Bull Tokyo Dent Coll. 2017;58(2):117–24.
147. Mussi E, Furferi R, Volpe Y, Facchini F, McGreevy KS, Uccheddu F. Ear reconstruction simulation: from handcrafting to 3D printing. Bioengineering (Basel). 2019;6(1):14.
148. Unkovskiy A, Brom J, Huettig F, Keutel C. Auricular prostheses produced by means of conventional and digital workflows: a clinical report on esthetic outcomes. Int J Prosthodont. 2018;31(31):63–6.
149. Yadav S, Narayan AI, Choudhry A, Balakrishnan D. CAD/CAM-assisted auricular prosthesis fabrication for a quick, precise, and more retentive outcome: a clinical report. J Prosthodont. 2017;26:616–21.
150. Memon AR, Wang E, Hu J, Egger J, Chen X. A review on computer-aided design and manufacturing of patient-specific maxillofacial implants. Expert Rev Med Devices. 2020;17(4):345–56.
151. Unkovskiy A, Spintzyk S, Axmann D, Engel EM, Weber H, Huettig F. Additive manufacturing: a

comparative analysis of dimensional accuracy and skin texture reproduction of auricular prostheses replicas. J Prosthodont. 2019;28(2):e460–8.
152. Jindal SK, Sherriff M, Waters MG, Coward TJ. Development of a 3D printable maxillofacial silicone: Part I. Optimization of polydimethylsiloxane chains and cross-linker concentration. J Prosthet Dent. 2016;116(4):617–22.
153. Jindal SK, Sherriff M, Waters MG, Smay JE, Coward TJ. Development of a 3D printable maxillofacial silicone: Part II. Optimization of moderator and thixotropic agent. J Prosthet Dent. 2018;119(2):299–304.
154. Farook TH, Jamayet NB, Abdullah JY, Rajion ZA, Alam MK. A systematic review of the computerized tools and digital techniques applied to fabricate nasal, auricular, orbital and ocular prostheses for facial defect rehabilitation. J Stomatol Oral Maxillofac Surg. 2020;121(3):268–77.
155. Unkovskiy A, Spintzyk S, Brom J, Huettig F, Keutel C. Direct 3D printing of silicone facial prostheses: a preliminary experience in digital workflow. J Prosthet Dent. 2018;120(2):303–8.
156. ACEO® - Unique Drop on Demand Technology with 70 Years of Silicone Knowhow [Internet]. ACEO. 2021 [cited 26 October 2021]. Available from: https://www.aceo3d.com/technology.
157. Domingue D, Sinada N, White JR. Digital surgical planning and placement of osseointegrated implants to retain an auricular prosthesis using implant software with cone-beam computed tomography and 3D-printed surgical guides: a case report. Clin Case Rep. 2020;9(1):203–9.
158. Domingue D, Glenn NC, Vest A, White JR. Osseointegrated implant-retained auricular prosthesis constructed using cone-beam computed tomography and a prosthetically driven digital workflow: a case report. Clin Case Rep. 2020;9(1):37–45.
159. Kadowaki M, Kubo T, Fujikawa M, Tashima H, Nagayama H, Ishihara O, et al. A two-tiered structure device based on stereolithography for residual mandible repositioning in mandibular reconstruction with fibular flap. Microsurgery. 2017;37(6):509–15.
160. Yuan X, Xuan M, Tian W, Long J. Application of digital surgical guides in mandibular resection and reconstruction with fibula flaps. Int J Oral Maxillofac Surg. 2016;45(11):1406–9.
161. Patel SY, Kim DD, Ghali GE. Maxillofacial reconstruction using vascularized fibula free flaps and endosseous implants. Oral Maxillofac Surg Clin North Am. 2019;31(2):259–84.
162. Rohner D, Bucher P, Hammer B. Prefabricated fibular flaps for reconstruction of defects of the maxillofacial skeleton: planning, technique, and long-term experience. Int J Oral Maxillofac Implants. 2013;28(5):e221–9.
163. Beh YH, Farook TH, Jamayet NB, Dudley J, Rashid F, Barman A, et al. Evaluation of the differences between conventional and digitally developed models used for prosthetic rehabilitation in a case of untreated palatal cleft. Cleft Palate Craniofac J. 2021;58(3):386–90.
164. Murat S, Gürbüz A, Kamburoğlu K. Fabrication of obturator prosthesis by fusing CBCT and digital impression data. Int J Comput Dent. 2018;21(4):335–44.
165. Rodney J, Chicchon I. Digital design and fabrication of surgical obturators based only on preoperative computed tomography data. Int J Prosthodont. 2017;30(2):111–2.
166. Tasopoulos T, Chatziemmanouil D, Karaiskou G, Kouveliotis G, Wang J, Zoidis P. Fabrication of a 3D-printed interim obturator prosthesis: a contemporary approach. J Prosthet Dent. 2019;121(6):960–3.
167. Tasopoulos T, Chatziemmanouil D, Kouveliotis G, Karaiskou G, Wang J, Zoidis P. PEEK maxillary obturator prosthesis fabrication using intraoral scanning, 3D printing, and CAD/CAM. Int J Prosthodont. 2020;33(3):333–40.
168. Xepapadeas AB, Weise C, Frank K, Spintzyk S, Poets CF, Wiechers C, et al. Technical note on introducing a digital workflow for newborns with craniofacial anomalies based on intraoral scans - Part II: 3D printed Tübingen palatal plate prototype for newborns with Robin sequence. BMC Oral Health. 2020;20(1):171.
169. Rooney MK, Rosenberg DM, Braunstein S, Cunha A, Damato AL, Ehler E, et al. Three-dimensional printing in radiation oncology: a systematic review of the literature. J Appl Clin Med Phys. 2020;21(8):15–26.
170. Subashi E, Jacobs C, Hood R, Kirsch DG, Craciunescu O. A design process for a 3D printed patient-specific applicator for HDR brachytherapy of the orbit. 3D Print Med. 2020;6(1):15.
171. Zhao Y, Moran K, Yewondwossen M, Allan J, Clarke S, Rajaraman M, et al. Clinical applications of 3-dimensional printing in radiation therapy. Med Dosim. 2017;42(2):150–5.
172. Lancellotta V, Pagano S, Tagliaferri L, Piergentini M, Ricci A, Montecchiani S, et al. Individual 3-dimensional printed mold for treating hard palate carcinoma with brachytherapy: a clinical report. J Prosthet Dent. 2019;121(4):690–3.
173. Craft DF, Kry SF, Balter P, Salehpour M, Woodward W, Howell RM. Material matters: analysis of density uncertainty in 3D printing and its consequences for radiation oncology. Med Phys. 2018;45(4):1614–21.
174. Arnold C, Hey J, Schweyen R, Setz JM. Accuracy of CAD-CAM-fabricated removable partial dentures. J Prosthet Dent. 2018;119(4):586–92.
175. Suzuki Y, Shimizu S, Waki T, Shimpo H, Ohkubo C. Laboratory efficiency of additive manufacturing for removable denture frameworks: a literature-based review. Dent Mater J. 2021;40(2):265–71.
176. Al Mortadi N, Alzoubi KH, Williams R. A scoping review on the accuracy of fit of removable partial dentures in a developing digital context. Clin Cosmet Investig Dent. 2020;12:551–62.
177. Williams RJ, Bibb R, Eggbeer D, Collis J. Use of CAD/CAM technology to fabricate a removable

partial denture framework. J Prosthet Dent. 2006;96(2):96–9.
178. Bibb R, Eggbeer D, Williams R. Rapid manufacture of removable partial denture frameworks. Rapid Prototyp J. 2006;12(2):95–9.
179. Soltanzadeh P, Suprono MS, Kattadiyil MT, Goodacre C, Gregorius W. An in vitro investigation of accuracy and fit of conventional and CAD/CAM removable partial denture frameworks. J Prosthodont. 2019;28(5):547–55.
180. Eggbeer D, Bibb R, Williams R. The computer-aided design and rapid prototyping fabrication of removable partial denture frameworks. Proc Inst Mech Eng H. 2005;219:195–202.
181. Pordeus MD, Santiago Junior JF, Venante HS, Bringel da Costa RM, Chappuis Chocano AP, Porto VC. Computer-aided technology for fabricating removable partial denture frameworks: a systematic review and meta-analysis. J Prosthet Dent. 2021; https://doi.org/10.1016/j.prosdent.2020.06.006.
182. Wu J, Zhang C, Gao B, Wang X, Zhao X. A study on the fabrication method of removable partial denture framework by computer-aided design and rapid prototyping. Rapid Prototyp J. 2012;18:318–23.
183. Tasaka A, Okano H, Shimizu T, Kato Y, Higuchi S, Yamashita S. Influence of reinforcement bar on accuracy of removable partial denture framework fabricated by casting with a 3D-printed pattern and selective laser sintering. J Prosthodont Res. 2021;65(2):213–8.
184. Ali M, Nairn RI, Sherriff M, Waters NE. The distortion of cast cobalt-chromium alloy partial denture frameworks fitted to a working cast. J Prosthet Dent. 1997;78(4):419–24.
185. Fenlon MR, Juszczyk AS, Hughes RJ, Walter JD, Sherriff M. Accuracy of fit of cobalt-chromium removable partial denture frameworks on master casts. Eur J Prosthodont Restor Dent. 1993;1(3):127–30.
186. Augthun M, Zyfuss M, Spiekermann H. The influence of spruing technique on the development of tension in a cast partial denture framework. Int J Prosthodont. 1994;7(1):72–6.
187. Brudvik JS, Reimers D. The tooth-removable partial denture interface. J Prosthet Dent. 1992;68(6):924–7.
188. Wataha JC. Alloys for prosthodontic restorations. J Prosthet Dent. 2002;87(4):351–63.
189. Arafa KAO. Assessment of the fit of removable partial denture fabricated by computer-aided designing/computer aided manufacturing technology. Saudi Med J. 2018;39:17–22.
190. Lang LA, Tulunoglu I. A critically appraised topic review of computer-aided design/computer-aided machining of removable partial denture frameworks. Dent Clin N Am. 2014;58(1):247–55.
191. Moldovan O, Rudolph H, Luthardt RG. No clear evidence on the clinical performance of different removable prosthetic options in partially edentulous patients. Clin Oral Investig. 2016;20:1435–47.
192. Frank R, Brudvik J, Leroux B, Milgrom P, Hawkins N. Relationship between the standards of removable partial denture construction, clinical acceptability, and patient satisfaction. J Prosthet Dent. 2000;83(5):521–7.
193. Mamoun J. The path of placement of a removable partial denture: a microscope based approach to survey and design. J Adv Prosthodont. 2015;7(1):76–84.
194. Dunham D, Brudvik J, Morris W, Plummer K, Cameron S. A clinical investigation of the fit of removable partial dental prosthesis clasp assemblies. J Prosthet Dent. 2006;95(4):323–6.
195. Ye H, Ning J, Li M, Niu L, Yang J, Sun Y, et al. Preliminary clinical application of removable partial denture frameworks fabricated using computer-aided design and rapid prototyping techniques. Int J Prosthodont. 2017;30(4):348–53.
196. Tasaka A, Shimizu T, Kato Y, Okano H, Ida Y, Higuchi S, et al. Accuracy of removable partial denture framework fabricated by casting with a 3D printed pattern and selective laser sintering. J Prosthodont Res. 2020;64(2):224–30.
197. Chen J, Ahmad R, Suenaga H, Li W, Sasaki K, Swain M, et al. Shape optimization for additive manufacturing of removable partial dentures—a new paradigm for prosthetic CAD/CAM. PLoS One. 2015;10(7):e0132552.
198. Lee J-W, Park J-M, Park E-J, Heo S-J, Koak J-Y, Kim S-K. Accuracy of a digital removable partial denture fabricated by casting a rapid prototyped pattern: a clinical study. J Prosthet Dent. 2017;1:118.
199. Keltjens HM, Mulder J, Kayser AF, Creugers NH. Fit of direct retainers in removable partial dentures after 8 years of use. J Oral Rehabil. 1997;24:138–42.
200. Tregerman I, Renne W, Kelly A, Wilson D. Evaluation of removable partial denture frameworks fabricated using 3 different techniques. J Prosthet Dent. 2019;122:390–5.
201. Takahashi K, Torii M, Nakata T, Kawamura N, Shimpo H, Ohkubo C. Fitness accuracy and retentive forces of additive manufactured titanium clasp. J Prosthodont Res. 2020;13:1883–958.
202. Schweiger J, Güth JF, Erdelt KJ, Edelhoff D, Schubert O. Internal porosities, retentive force, and survival of cobalt-chromium alloy clasps fabricated by selective laser-sintering. J Prosthodont Res. 2020;64(2):210–6.
203. Stansbury JW, Idacavage MJ. 3D printing with polymers: challenges among expanding options and opportunities. Dent Mater. 2016;32(1):54–64.
204. Koike M, Greer P, Owen K, Lilly G, Murr LE, Gaytan SM, et al. Evaluation of titanium alloys fabricated using rapid prototyping technologies-electron beam melting and laser beam melting. Materials (Basel). 2011;4(10):1776–92.
205. Almufleh B, Emami E, Alageel O, de Melo F, Seng F, Caron E. Patient satisfaction with laser-sintered removable partial dentures: a crossover pilot clinical trial. J Prosthet Dent. 2018;119:560–7.

206. Awad MA, Feine JS. Measuring patient satisfaction with mandibular prostheses. Community Dent Oral Epidemiol. 1998;26(6):400–5.
207. Herpel C, Schwindling FS, Held T, Christ L, Lang K, Schwindling M, et al. Individualized 3D-printed tissue retraction devices for head and neck radiotherapy. Front Oncol. 2021;11:628743.
208. Zaid M, Bajaj N, Burrows H, Mathew R, Dai A, Wilke CT, et al. Creating customized oral stents for head and neck radiotherapy using 3D scanning and printing. Radiat Oncol. 2019;14:148.
209. Zaid M, Koay EJ, Bajaj N, Mathew R, Xiao L, Agrawal A, et al. A prospective parallel design study testing non-inferiority of customized oral stents made using 3D printing or manually fabricated methods. Oral Oncol. 2020;106:104665.
210. Şenayli A, Çankaya G, Öztorun CI, Oflaz H, Şenel E. Clinical trials of 3D printing splints to avoid contracture development in burned children. Turk J Med Sci. 2021;51(5):2543–53.
211. Dedem P, Türp JC. Digital Michigan splint - from intraoral scanning to plasterless manufacturing. Int J Comput Dent. 2016;19:63–76.
212. Schweiger J, Edelhoff D, Güth JF. 3D printing in digital prosthetic dentistry: an overview of recent developments in additive manufacturing. J Clin Med. 2021;10:2010.
213. Ahmed MK, Ahsanuddin S, Retrouvey JM, Koka KS, Qureshi H, Bui AH, et al. Fabrication of nasoalveolar molding devices for the treatment of cleft lip and palate, using stereolithography additive manufacturing processes and computer-aided design manipulation software. J Craniofac Surg. 2019;30:2604–8.
214. Revilla-León M, Besné-Torre A, Sánchez-Rubio JL, Fábrega JJ, Özcan M. Digital tools and 3D printing technologies integrated into the workflow of restorative treatment: a clinical report. J Prosthet Dent. 2019;121(1):3–8.
215. Torabi K, Farjood E, Hamedani S. Rapid prototyping technologies and their applications in prosthodontics, a review of literature. J Dent (Shiraz). 2015;16(1):1–9.
216. Oberoi G, Nitsch S, Edelmayer M, JanjiĆ K, Müller AS, Agis H. 3D printing-encompassing the facets of dentistry. Front Bioeng Biotechnol. 2018;6:172.
217. Dawood A, Marti B, Sauret-Jackson V, Darwood A. 3D printing in dentistry. Br Dent J. 2015;219:521–9.
218. Wedekind L, Güth JF, Schweiger J, Kollmuss M, Reichl FX, Edelhoff D, et al. Elution behavior of a 3D-printed, milled and conventional resin-based occlusal splint material. Dent Mater. 2021;37:701–10.
219. Vasques MT, Mori M, Laganá DC. Three-dimensional printing of occlusal devices for temporomandibular disorders by using a free CAD software program: a technical report. J Prosthet Dent. 2020;123:232–5.
220. Lutz AM, Hampe R, Roos M, Lümkemann N, Eichberger M, Stawarczyk B. Fracture resistance and 2-body wear of 3-dimensional-printed occlusal devices. J Prosthet Dent. 2019;121(1):166–72.
221. Berli C, Thieringer FM, Sharma N, Müller JA, Dedem P, Fischer J, et al. Comparing the mechanical properties of pressed, milled, and 3D-printed resins for occlusal devices. J Prosthet Dent. 2020;124:780–6.
222. Marcel R, Reinhard H, Andreas K. Accuracy of CAD/CAM-fabricated bite splints: milling vs 3D printing. Clin Oral Investig. 2020;24:4607–15.
223. Reyes-Sevilla M, Kuijs RH, Werner A, Kleverlaan CJ, Lobbezoo F. Comparison of wear between occlusal splint materials and resin composite materials. J Oral Rehabil. 2018;45(7):539–44.
224. Prpic V, Slacanin I, Schauperl Z, Catic A, Dulcic N, Cimic S. A study of the flexural strength and surface hardness of different materials and technologies for occlusal device fabrication. J Prosthet Dent. 2019;121:955–9.
225. Unkovskiy A, Bui PH, Schille C, Geis-Gerstorfer J, Huettig F, Spintzyk S. Objects build orientation, positioning, and curing influence dimensional accuracy and flexural properties of stereolithographically printed resin. Dent Mater. 2018;34:e324–33.
226. Väyrynen VO, Tanner J, Vallittu PK. The anisotropicity of the flexural properties of an occlusal device material processed by stereolithography. J Prosthet Dent. 2016;116:811–7.
227. Grymak A, Aarts JM, Ma S, Waddell JN, Choi JJE. Comparison of hardness and polishability of various occlusal splint materials. J Mech Behav Biomed Mater. 2021;115:104270.
228. Xu Y, Xepapadeas AB, Koos B, Geis-Gerstorfer J, Li P, Spintzyk S. Effect of post-rinsing time on the mechanical strength and cytotoxicity of a 3D printed orthodontic splint material. Dent Mater. 2021;37:e314–27.
229. Reymus M, Stawarczyk B. Influence of different postpolymerization strategies and artificial aging on hardness of 3D-printed resin materials: an in vitro study. Int J Prosthodont. 2020;33:634–40.
230. Perea-Lowery L, Gibreel M, Vallittu PK, Lassila L. Evaluation of the mechanical properties and degree of conversion of 3D printed splint material. J Mech Behav Biomed Mater. 2021;115:104254.
231. Schubert A, Bürgers R, Baum F, Kurbad O, Wassmann T. Influence of the manufacturing method on the adhesion of Candida albicans and Streptococcus mutans to oral splint resins. Polymers (Basel). 2021;13(10):1534.

9 3D-Printed Surgical Positioning Rib Graft Jig in Combined Orthodontic-Surgical Management of Pruzansky/Kaban Type IIB and Type III Hemifacial Microsomia

Dror Aizenbud, Tal Capucha, Hagai Hazan Molina, Edlira Zere, Dekel Shilo, Omri Emodi, and Adi Rachmiel

Contents

D. Aizenbud (✉) · H. H. Molina
Department of Orthodontics and Craniofacial Anomalies, School of Graduate Dentistry, Rambam Health Care Campus, Technion—Israel Institute of Technology, Haifa, Israel

Oral Biology Research Laboratory, Faculty of Medicine, Technion—Israel Institute of Technology, Haifa, Israel
e-mail: aizenbud@ortho.co.il

T. Capucha · D. Shilo · O. Emodi
Department of Oral and Maxillofacial Surgery, Rambam Health Care Campus, Haifa, Israel

E. Zere
Department of Orthodontics and Craniofacial Anomalies, School of Graduate Dentistry, Rambam Health Care Campus and Faculty of Medicine, Technion—Israel Institute of Technology, Haifa, Israel

A. Rachmiel
Department of Oral and Maxillofacial Surgery, Rambam Health Care Campus, Faculty of Medicine, Technion—Israel Institute of Technology, Haifa, Israel
e-mail: a_rachmiel@rmc.gov.il

© The Author(s), under exclusive license to Springer Nature Switzerland AG 2022
P. K. Chaudhari et al. (eds.), *3D Printing in Oral Health Science*,
https://doi.org/10.1007/978-3-031-07369-4_9

9.1 Introduction

Oculo–auriculo–vertebral spectrum according to Gorlin et al. [1] is related to a variety of developmental anomalies stemming from disorders in the first and second pharyngeal arches, during the first 6 weeks of gestation. The common condition, hemifacial microsomia (HFM), affects primary aural, oral, and mandibular development. The more complex and heterogeneous condition, Goldenhar syndrome, also affects vertebral development. The children affected by this spectrum often present a variety of underdevelopment of facial structures such as the mandible, maxilla, ear, facial soft tissue and muscles, and the facial nerve [2]. They are associated with notable deformities, including facial asymmetry, chin deviation, occlusal abnormalities, and potential compromise of the airway [3]. The resulting functional disorders may include alterations in breathing, feeding, hearing, and speech, which are best treated by a multidisciplinary craniofacial team to provide specialized coordinated treatment [2].

Pruzansky/Kaban classification [4, 5] refers to three types of clinical categories based on the structure and function of the affected mandible and temporomandibular joint (TMJ), which determines the surgical plan accordingly (Fig. 9.1). Type I: mild hypoplasia of glenoid, ramus, and condyle. Type IIA: moderate hypoplasia of glenoid, ramus, and condyle, where good function of TMJ is preserved. Type IIB: mild-to-moderate hypoplasia of glenoid, ramus, and condyle, where TMJ is markedly abnormally placed inferiorly, medially, and anteriorly. Type III: totally absent mandibular ramus, condyle, and TMJ [2].

Young children associated with the spectrum present Pruzansky/Kaban type IIB or type III HFM, that often manifests severely hypoplastic mandible and severe adjunct malfunctions which include mal-breathing and mal-nutrition. Consequently, mandatory early age surgical intervention is indicated. However, insufficient mandibular bone mass in the predevelopmental age prevents the preferred condylar reconstruction procedures of autogenous bone grafts, distraction osteogenesis, and prosthesis implantation [6, 7]. Alternately, a variety of autogenous grafts such as clavicle and sternoclavicular joint [8, 9], fibula [10], iliac bone [11], and metatarsal bone [12] have been suggested for condylar reconstruction in young children. Costochondral grafts (CCG) are considered the best donor site for condylar reconstruction techniques in young children with type IIB and type III HFM [13].

CCG incorporates a cartilage-bone junction, which is considered a growth center that includes considerable growth potential that may improve mandibular function, facial appearance, and normal growth induction of the affected maxilla [14]. The involved growth center cartilage is controlled by intrinsic and extrinsic factors [15]. According to Moss and Rankow's functional matrix theory [16], the downward and forward mandibular growth is a secondary response to the attached organs and tissues. Thus, condylar cartilage growth is secondary, affected by the mandible function and development. According to this functional concept, optimal rib growth of the CCG may eliminate the hypoplastic mandible negative influence on the normal maxillary growth. Therefore, CCG surgical application in young children with type IIB and type III HFM may result in reduced secondary midface defor-

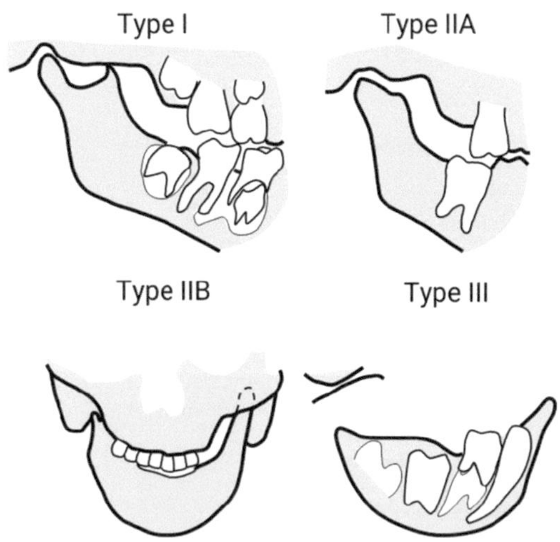

Fig. 9.1 Kaban's modification of the Pruzansky classification system of the mandible in craniofacial microsomia [5]

mation and less skeletal asymmetry [17]. Consequently CCG may improve malocclusion, occlusal cant, and malfunction before primary school age [18].

The ideal surgical outcome and the future growth of the CCG are related to adequate cartilage size (approximately 7 mm), preservation of the periosteal and perichondrial sleeve, blunt dissection and technique, brief immobilization, and early joint loading [18]. Furthermore, precise position and direction of costochondral rib graft might be crucial factors that influence the growth. However, CCG precise surgical positioning is a challenge in patients with Pruzansky/Kaban type IIB and type III mandibular hypoplasia when the glenoid fossa is difficult to identify or in some cases where it might be absent [18]. This chapter describes 3D-printed jig aimed to manage the precise suitable placement position and fixation of the costochondral rib graft in relation to the contralateral unaffected temporomandibular joint and soft tissues. The jig may preserve the rib's cartilage cap and prevent fracture at the costochondral junction and decrease the risk of interference of cartilage growth.

9.2 Three-Dimensional Planning and Printing

3D printers and digital imaging software platforms have leveraged the evolution of the combined orthodontic-surgical management of cranio-maxillofacial disorders. These advanced technologies have culminated the two-dimensional standard of care planning, into 3D-personalized planning of large craniofacial bone defect repairs [19]. The workflow used for 3D virtual planning and printing provides in-house manufactured, sterilizable patient-specific surgical guides to maximize the implant placement position combined with ideal fixation [19–21]. These evolved advancements in orthodontic-surgical planning allow better control over the process of craniofacial reconstruction. The software platform includes computer-assisted design (CAD) software used for bone segmentation (Philips IntelliSpace Portal 11). This software allows the extraction of the anatomical structures of interest from the digital images. In addition, a different CAD software is used for patient-specific surgical guide designs (Geomagic Freeform, 3D systems) [20, 21].

The workflow begins with 2D sections obtained from the computed tomography (CT). Highly accurate image acquisition is required for the process. Next, 3D rendering of the 2D slices is performed using the segmentation software (Philips IntelliSpace Portal 11). In the process, a standard tessellation language (STL) file is created from the DICOM files. The STL file is utilized for the 3D design software (Geomagic Freeform, 3D systems), facilitating advanced computer-assisted design (CAD) for the virtual planning of the operation.

The 3D reconstruction in the segmentation software (Philips IntelliSpace Portal 11) allows the surgeon to perform a complete overview of the recipient site, in the case presented herein of Pruzansky/Kaban type IIB, the facial bones architecture (Fig. 9.2) and the donor site, the rib cage (Fig. 9.3).

This exploration of the anatomical structure is extensive and crucial for proper surgical planning. For example, the patient described in Figs. 9.2 and 9.3 was virtually diagnosed with bicipital rib anomaly. After proper evaluation of the potential donor site for grafted bone and the morphology of the cranio-maxillofacial defects, the recipient site, the planning stage commences.

Pruzansky/Kaban type IIB and type III HFM virtual design of mandibular ramus and condyle reconstruction require several measurements. These are taken by means of the 3D design software. The ramus-condyle length is measured from the gonial angle to the condyle's most superior edge on the unaffected side. This measurement represents the required CCG harvested length.

The 3D models of the donor rib and the facial bones (including the maxillomandibular complex) are printed, and the location of the planned osteotomies is marked, for the specific CCG harvesting procedure (Fig. 9.4).

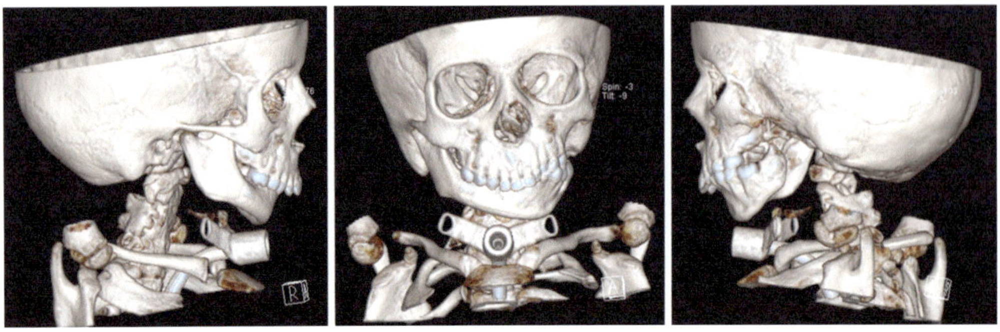

Fig. 9.2 A 4-year-old representative case of Pruzansky/Kaban type III malformation with moderate hypoplasia of glenoid, ramus, and condyle. The child was intubated early due to adjunct severe mal-functions which included mal-breathing

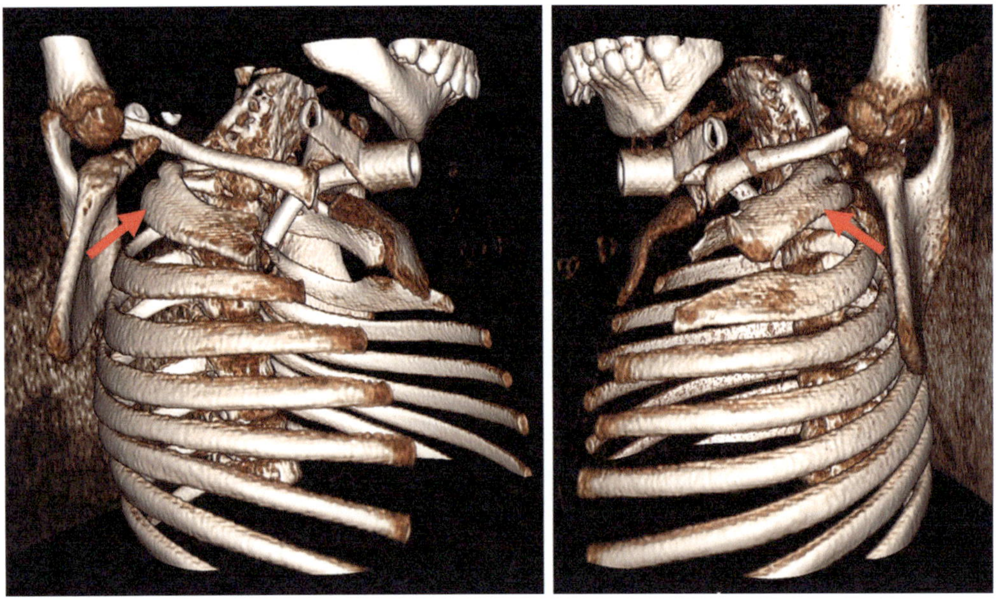

Fig. 9.3 Right and left view of the rib cage (CCG donor site), of the Pruzansky/Kaban type III malformation case. Note the bicipital rib anomaly (red arrow) accidentally revealed

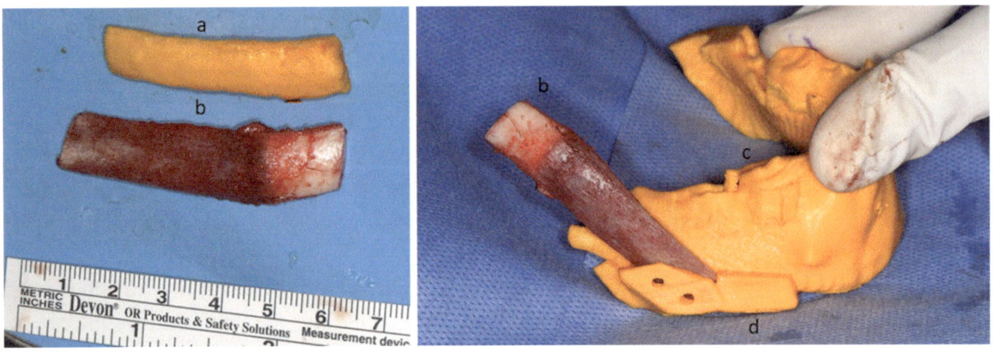

Fig. 9.4 The 3D models of the donor rib and the recipient mandible: (a) Printed model of the donor rib. (b) Real donor rib harvested. (c) Printed model of the Pruzansky/Kaban type III malformed mandible with hypoplasia ramus, and condyle. (d) Printed jig aimed to manage the precise suitable position placement and fixation of the costochondral rib graft in relation to the contralateral unaffected ramus and condyle angulation

The models allow validation of the virtual design and further adjustments prior to the actual surgery. These virtual models allow an advanced level of educating the patient about the surgery, surgical rehearsal for residents, and experienced surgeons' training. During performance of the surgical procedure, the models allow accurate real-time planning while directly visualizing the surgically exposed anatomical structures. At this stage, the optimized position of the grafted rib is considered and dictated. In order to replicate this planned position in the real surgical field during the procedure, a surgical guide (Jig) is designed. The surgical guide carries the donor grafted rib from one side to precisely fit it in the craniofacial recipient location over the hypoplastic defected mandible on the other side.

Thus, the surgical guide serves as an accurate key for the surgeon to properly position the grafted rib in order to maximize the outcome and minimize unwanted side effects. Furthermore, during this stage of virtual planning, the location of the fixation screws of the CCG is planned while taking into consideration the location of the tooth buds.

The mirroring tool of the software is used to maximize accuracy of the virtual surgical design. Because the human skull is considered symmetrical, a mirror image of the unaffected mandibular side is first depicted. Thereafter, the harvested individualized CCG can be virtually positioned over the missing area and the specific vector is ideally selected (Fig. 9.5a, b). Using the transparent tool of the software, the tooth buds' location inside the alveolar ridges can be identified over the transparent bone (Fig. 9.5c). This information allows precise placement of the CCG and accurate planning of the positional screws or the fixation miniplate holes, while protecting the tooth buds (Fig. 9.6).

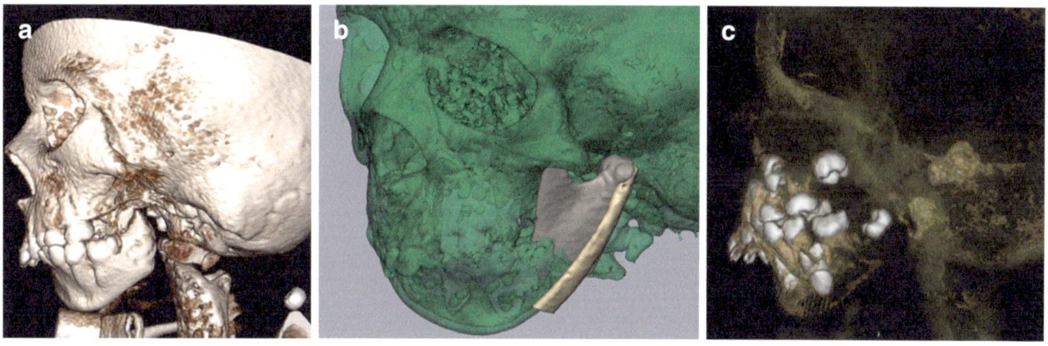

Fig. 9.5 (**a**) 3D CT imaging of left Pruzansky/Kaban type III HFM presenting a malformed left mandible lacking ramus, and condyle. (**b**) 3D CT imaging using the software mirroring tool. The unaffected mandibular right side was superimposed over the defected left side. The harvested CCG can be virtually positioned over the missing left area according to the ideal symmetric position of the mirror picture of the right unaffected side. (**c**) 3D CT imaging using software to identify the location of the tooth buds in the alveolar ridge

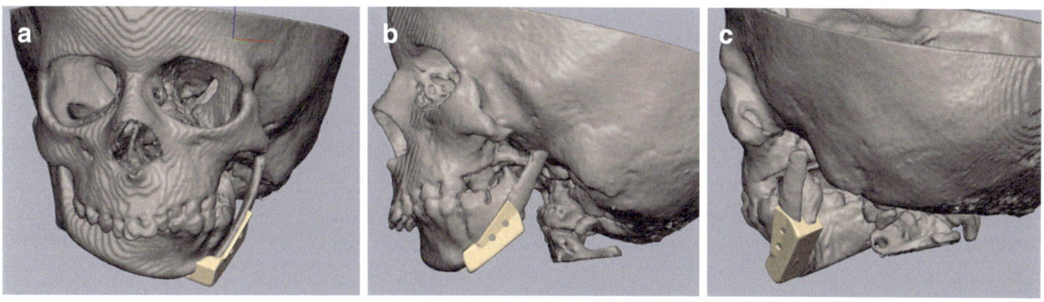

Fig. 9.6 Presurgical accurate planning of the jig (from different views) embracing the defected mandible and the donor rib. The positional screws were precisely planned considering the anatomical landmarks, the CCG, and the tooth buds' location

Collectively, based on the ideal virtual planning of the CCG position and the precise location of the fixation holes, a surgical guide (jig) is planned and printed. This jig allows the individualized ideal positioning of the CCG which: (a) Symmetrically restores the ramus and condyle structure; (b) Surgically precisely guides the condyle into the glenoid fossa through the predetermined accurate CCG length, preventing overloaded bone contacts and unwanted stress inside the joint capsule, which may lead to graft resorption; (c) Accurate graft fixation while properly maintaining tooth buds intact.

9.3 The Surgical Technique

The CCG surgical procedure involves teamwork. One team concentrates on the type IIB/type III HFM anomaly and prepares the recipient site of the face according to a standard sterile method. A 4–5-mm length skin incision is performed according to the Risdon approach [22] namely 1.5 cm below the inferior border of the mandible, exposing the mandibular angle of the affected side. A blunt dissection is then performed by preserving the marginal mandibular branch of the facial nerve. The periosteum is incised along the inferior border of the mandible to expose the margin of the malformed mandible. Decortication of the posterior border and lower border of the mandible is carried out. A pocket is then created in the direction of the glenoid fossa.

Simultaneously, the second team focuses on the CCG harvesting procedure. A local anesthetic of 1% lidocaine with 1:100,000 epinephrine is injected into the subcutaneous layer, through the depth of the rib, along the planned incision. Next, a 4-cm incision in the skin is performed over the fourth or fifth rib, using a #15-blade scalpel. Electrocautery is used to dissect the underlying subcutaneous tissue. The rectus abdominis muscle is identified, running vertically through the surgical field. It is horizontally separated and retracted superiorly and inferiorly. The underlying rib and osseocartilaginous junction are identified and bluntly dissected from the surrounding tissue. Special consideration is given to avoid perforation of the chest wall and subsequent pneumothorax. A horizontal incision is made through the periosteum over the midline of the rib carefully preserving the cartilaginous cap. A standard #9 molt periosteal elevator is used to dissect the subperiosteal plane around the rib to be harvested. Attention is focused on maintaining the periosteum over the osseocartilaginous junction and gently pulling and reconstituting it by proper suturing after removal of the rib, allowing regeneration of the rib. A transection is then performed through the cartilage by leaving 3 mm of the cartilage margin over the rib. Once the accurate length of the CCG segment is reflected based on the virtual planning, the rib segment is cut and delivered as a graft. The chest cavity is filled with saline and a Valsalva maneuver is performed to rule out pleural perforation. Accurate suturing of the periosteum and rectus muscle in layers is carried out by using 3.0 vicryl sutures. Subcutaneous tissues are then approximated using 4.0 vicryl sutures and intracuticular 4.0 biocin is applied to the chest skin.

The CCG is first positioned on the prefabricated 3D-printed jig and fixated by using the first fixation hole (Fig. 9.7a). An orthodontic splint is preliminarily planned, fabricated, and cemented to stabilize the displaced mandible, in order to fit the donor rib. Then the deviated mandible is moved forward and downward toward the unaffected side according to the virtual plan. The mandible position is maintained and stabilized by using the predesigned orthodontic splint creating an ipsilateral open bite (Fig. 9.7b, c). The mandibular cant that appears toward the affected side is leveled and the chin point is aligned with the facial midsagittal plane. Next the jig is placed on the hypoplastic mandibular margins allowing accurate positioning of the rib graft as planned. The graft is then secured to the mandibular angle using the second fixation hole (Fig. 9.7b, c). Finally, the first screw which secures the graft to the positional jig is removed and replaced by a long screw that secures the CCG to the mandible as planned (Fig. 9.7b, c).

The surgical field is then copiously irrigated with normal saline and sutured in a routine fashion by using 3.0 vicryl and 5.0 nylon sutures.

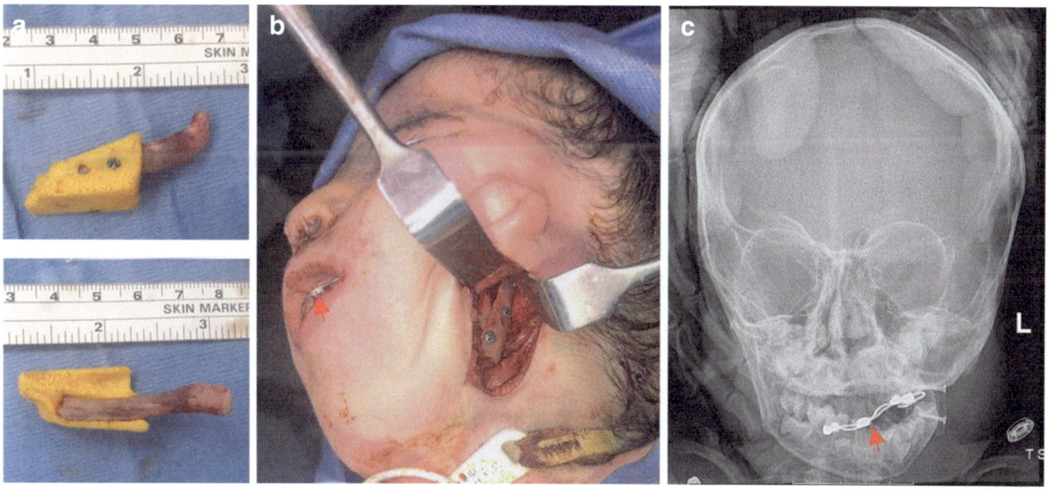

Fig. 9.7 (**a**) Initially, the donor rib graft is positioned over the printed surgical jig using the first fixation hole. (**b**) After cementing the intraoral orthodontic splint (red arrow), the jig is placed on the hypoplastic mandibular margins allowing accurate positioning of the rib graft as planned and it is secured to the mandibular angle using the second fixation hole. Thereafter, the first screw which secures the graft to the surgical jig is replaced by a longer screw that secures the rib graft to the mandibular body. (**c**) Postoperative posteroanterior cephalogram view

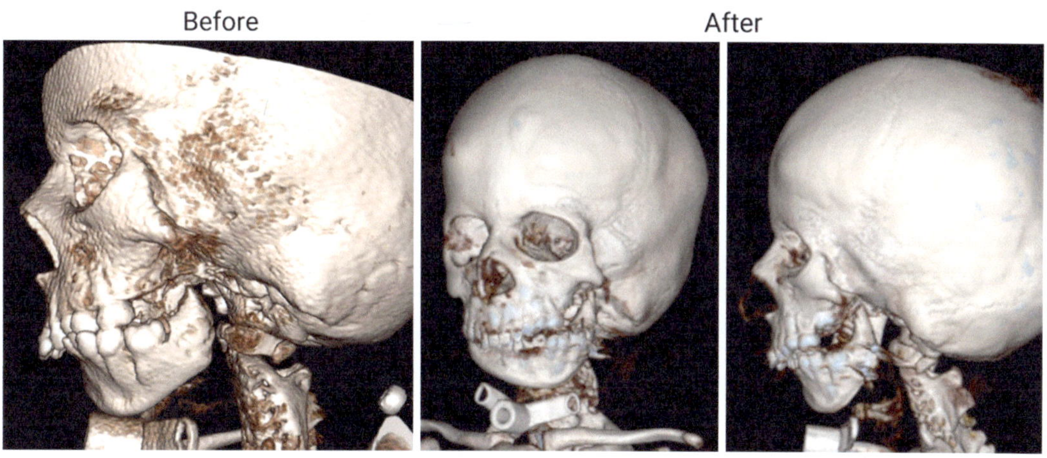

Fig. 9.8 Preoperative and 2-year postoperative 3D CT imaging of CCG implantation in left Pruzansky/Kaban type III HFM

Upon completion, the patient is transferred to the Pediatric Intensive Care Unit for 24-h observation. The child is placed on a soft diet for 2 weeks and a regimen of postoperative physiotherapy to improve mouth opening and lateral jaw movement. A radiographic follow-up is conducted using panoramic radiographs and posteroanterior cephalogram (Fig. 9.7c). 3D CT imaging is performed 2 years postoperatively to reveal optimal rib growth of the CCG, which eliminates the hypoplastic mandible and skeletal asymmetric with reduced secondary midface deformation (Fig. 9.8).

9.4 The Orthodontic Technique

Young children with Pruzansky/Kaban type IIB and type III HFM often present with primary dentition or early mixed dentition. Permanent first molar eruption is delayed or blocked due to hypoplastic mandible and missing posterior permanent molar tooth buds (Figs. 9.2 and 9.5a, c). Mouth opening is limited as a result of the unilateral hypoplastic mandible and partial or complete absence of the TMJ structure in the affected side. The general objectives of the treatment in these cases include correction of the skeletal asymmetry and improvement of the oral malfunctions, which are accomplished by means of the following specific treatment procedures: (1) elimination of the secondary midface deformation and leveling of the maxillary dentition, (2) mandibular advancement through affected TMJ reconstruction, (3) overjet and overbite correction, and (4) soft tissue balancing.

A CCG is a common method used to reconstruct the ramus-condyle unit, because it has growth potential. At this early age of surgical rib graft intervention, in the presence of primary dentition or early mixed dentition phase, a routine presurgical orthodontic preparation stage with a fixed orthodontic appliance is not applicable. However, a routine preoperatively combined orthodontic-surgical evaluation can be performed. This includes multidisciplinary clinical measurements and radiographic analysis using orthopantograms, lateral and P-A cephalograms, and CBCT. In such cases, the defected mandible-deviated position is recorded along with the amount of mandible asymmetry related to the facial midsagittal plane, chin deviation, the canting effect of the occlusal plane, and the reactive skeletal and dentoalveolar maxillary deficiency involvement of the affected side. Accordingly, the downward and forward amount of mandibular repositioning is clinically assessed in reference to the surrounding soft and hard tissue. Thus, the 3D dimensions of the CCG are dictated depending on the required correction of the vertical height of the ramus as reflected in the clinical examination and radiographic analyses, in order to design the optimal mandibular position (Fig. 9.9a). Using these advanced technologies, the positional surgical jig is then collectively planned and printed in advance as described (Figs. 9.5a–c and 9.6).

The occlusal relationship of the deviated mandible is analyzed by means of intraoral scans and the printed dental models or alternately by plaster dental models. The future postsurgical posterior open bite is established on the affected side according to mandibular downward and forward repositioning toward the midsagittal plane, while the 3D CCG implantation is planned. Maxillary and mandibular orthodontic-surgery splints are designed to stabilize the mandible in its new position. The splints are used to level the man-

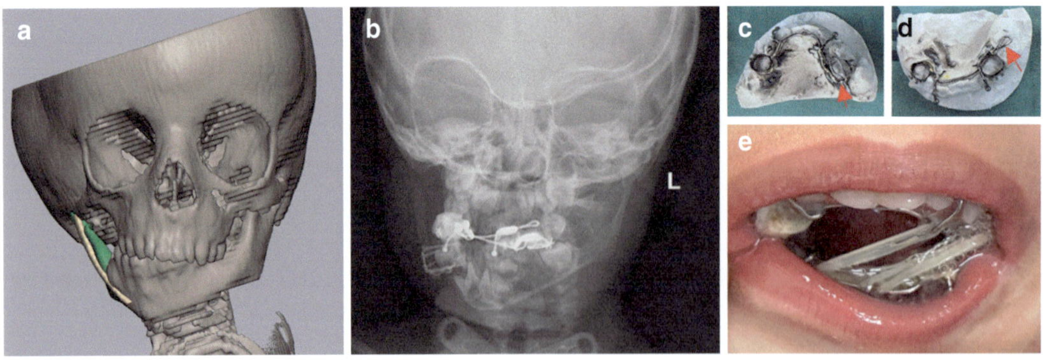

Fig. 9.9 (**a**) 3D imaging of CCG planning. The repositioned mandible is anteriorly and vertically displaced. The implanted donor rib is placed over the affected side (right), according to the mirror image of the unaffected side (left; in green). (**b**) Postoperative posteroanterior cephalogram view. (**c**) Working model of the upper orthodontic-surgery splint with ball clasps for intraoral elastics and a metal wire banded as a platform aimed to incorporate acrylic resins (red arrow). (**d**) Working model of the lower orthodontic-surgery splint with ball clasps for intraoral elastics and a metal wire banded as a platform aimed to incorporate acrylic (red arrow). (**e**) Intraoral elastics connected to upper and lower ball clasps of the splints postoperatively, aimed to stabilize the displaced mandible in order to reduce pressure from the implanted rib

dibular cant and align the chin point with the midsagittal plane. A specific retention is prepared on a banded metal wire platform on the splint (Fig. 9.9b–d—red arrows) to incorporate acrylic resins. The acrylic is built during the surgical rib implantation procedure. The newly positioned mandible creates a posterior open bite and a vertical space. The acrylic resins are incorporated over the splint platform into this vertical space. The acrylic resins help stabilize the displaced mandible by occlusal contacts and teeth indentations. In addition, in order to support the displaced mandible in its new position and to reduce stress on the implanted donor rib created by masticatory muscles, intraoral elastics are used as well. The elastics are connected to ball clasps on the upper and lower splints, which help stabilize the mandible by counteracting the vector and the force of the masticatory muscles.

9.5 Discussion

The optimal treatment approach and the optimal timeline for individuals with HFM vary widely. Each patient requires an individualized treatment plan, specifically tailored to his or her specific needs. These are dependent on the clinical manifestations presented by the individual patient and on the severity of the craniofacial abnormalities. Other considerations may include the motivation of the patient and the family for treatment, the availability of psychosocial support, and the availability of the multidisciplinary team of experts required to coordinate the treatment protocol [2]. However, comprehensive monitoring of weight gain, feeding, growth, and airway patency is essential in this population to indicate whether immediate intervention is required.

Pruzansky/Kaban HFM type I and IIA surgical interventions aim to achieve ramus lengthening through vertical mandibular osteotomy with or without bone graft, typically performed after skeletal maturity. Nonetheless, Pruzansky/Kaban type IIB and III, which include severely hypoplastic mandible, indicate early intervention of condylar reconstruction [6–13]. Condylar reconstruction can be applied by means of autogenous grafts or alternately by distraction osteogenesis and prosthesis implantation [7]. However, young children who present type IIB and type III deformities often manifest insufficient bone for distraction osteogenesis. The prosthesis implantation treatment alternative is not applicable for these growing patients due to the lack of potential growth [18]. Therefore, the classical protocol includes costochondral rib bone graft of the ramus and condyle with reconstruction of the glenoid fossa to create a functioning temporomandibular joint [13]. This surgical intervention is designed to restore the patient's craniofacial form and function and must take into account the expected facial growth pattern [2]. CCG has been suggested as the ideal implant of choice not only because of its growth capabilities, similar to that of the condylar cartilage, but also because it was proved both histologically and physiologically to be well suited for the function of replacement of the mandibular condyle [23]. However, CCG-reported complications may include unpredictable growth, lack of regional soft tissue, and decreased vascularity that contribute to advanced resorption [24–26]. The ideal timing of intervention is when the occlusal cant is initially evidenced, which generally correlates with dental eruption in children aged 2–5 years [5]. Therefore, high success rate and functional improvement of 80% have been reported when CCG is performed in children aged 3–9 years. Nevertheless, the success rate decreases to 50% for patients older than 14 years [13, 27]. Furthermore, the quality of the rib graft has been reported to be suboptimal in patients older than 5 years [3]. Other advantages that have been reported for early surgery in children with type IIB and type III mandibular hypoplasia include a relatively easy surgical intervention in the sense of the donor and the recipient sites both from the perspective of the surgeon and of the patient, as well as psychologic benefits for these young patients and their families [5, 13].

This chapter describes the 3D-advanced technologies of planning and printing a surgical positioning rib jig for accurate condylar reconstruction using CCG. The digital imaging software platforms enable preplanning of the orthodontic and surgical procedures for accurate CCG positioning. The virtual planning combined with the orthodontic and surgical techniques which are described in detail is aimed to reduce facial deformity and achieve the best facial symmetry. By increasing facial height and shifting the midline of the face to normal, a significant improvement is expected in the mandibular function, facial appearance, and occlusal patterns restoring its altered biomechanics. Because of the early CCG intervention in growing children, reduced facial deformity is expected over time as normal growth potential of the mandible is induced by means of the CCG implant [16]. The accurate position and direction of the costochondral rib graft are considered major factors that influence the growth of CCG [18]. Hence, the printed surgical positioning rib graft jig is essential to dictate absolute CCG positioning to restore mandibular function, facial esthetics, and proper CCG growth. This jig is of exceptional importance in cases where glenoid fossa is difficult to identify. In these cases, the mirror image of the contralateral unaffected side aids in determining the most suitable position and proper orientation of the rib graft placement while taking into consideration the contralateral temporomandibular joint and the soft tissues of the affected side.

Finally, it is noteworthy that further changes may occur upon growth and development of the children given that the growth of CCG is unpredictable. Combined orthodontic and surgical treatments are used based on the situation of the patients until completion of growth and as necessary to attain a long-term effect. In this chapter, CCG is used only to reconstruct condyle. There is no doubt that future investigations concerning the use of CCG will focus on the secondary deformities of the maxilla and the soft tissues as well.

9.6 Summary

In patients with Pruzansky/Kaban type IIB and III HFM, orthodontic-surgical intervention to restore the height of the ramus with CCG leads to an immediate correction of facial asymmetry. The advantages of the 3D-printed surgical positioning rib graft jig technique include preplanning of the orthodontic and surgical procedures for accurate CCG positioning. The virtual planning results in full control of the surgery, accurate osteotomies orientation, and precise rib graft fixation screws. These may contribute to a significant reduction in price, reduction in duration of the surgery, superior performance, and highly accurate results. The limitations include the need to master the CAD programs.

References

1. Gorlin RJ, Jue KL, Jacobsen U, Goldschmidt E. Oculoauriculovertebral dysplasia. J Pediatr. 1963;63:991–9.
2. Birgfeld CB, Heike C. Craniofacial microsomia. Semin Plast Surg. 2012;26:91–104.
3. Tahiri Y, Chang CS, Tuin J, Paliga JT, Lowe KM, Taylor JA, Bartlett SP. Costochondral grafting in craniofacial microsomia. Plast Reconstr Surg. 2015;135:530–41.
4. Pruzansky S. Not all dwarfed mandibles are alike. Birth Defects. 1969;1:120–9.
5. Kaban LB, Moses MH, Mulliken JB. Surgical correction of hemifacial microsomia in the growing child. Plast Reconstr Surg. 1988;82:9–19.
6. Pluijmers BI, Caron CJ, Dunaway DJ, Wolvius EB, Koudstaal MJ. Mandibular reconstruction in the growing patient with unilateral craniofacial microsomia: a systematic review. Int J Oral Maxillofac Surg. 2014;43:286–95.
7. Wolford LM, Bourland TC, Rodrigues D, Perez DE, Limoeiro E. Successful reconstruction of nongrowing hemifacial microsomia patients with unilateral temporomandibular joint total joint prosthesis and orthognathic surgery. J Oral Maxillofac Surg. 2012;70:2835–53.
8. Wolford LM, Cottrell DA, Henry C. Sternoclavicular grafts for temporomandibular joint reconstruction. J Oral Maxillofac Surg. 1994;52:119–28.
9. Singh V, Verma A, Kumar I, Bhagol A. Reconstruction of ankylosed temporomandibular joint: sternoclavicular grafting as an approach to management. Int J Oral Maxillofac Surg. 2011;40:260–5.
10. Guo L, Ferraro NF, Padwa BL, Kaban LB, Upton J. Vascularized fibular graft for pediatric mandibular reconstruction. Plast Reconstr Surg. 2008;121:2095–105.
11. Kummoona R. Chondro-osseous iliac crest graft for one stage reconstruction of the ankylosed TMJ in children. J Maxillofac Surg. 1986;14:215–20.

12. Bond SE, Saeed NR, Cussons PD, Watt-Smith SR. Reconstruction of the temporomandibular joint by the transfer of the free vascularised second metatarsal. Br J Oral Maxillofac Surg. 2004;42:241–5.
13. Ross RB. Costochondral grafts replacing the mandibular condyle. Cleft Palate Craniofac J. 1999;36:334–9.
14. Padwa BL, Mulliken JB, Maghen A, Kaban LB. Midfacial growth after costochondral graft construction of the mandibular ramus in hemifacial microsomia. J Oral Maxillofac Surg. 1998;56:122–8.
15. Ellis ER, Carlson DS. Histologic comparison of the costochondral, sternoclavicular, and temporomandibular joints during growth in Macaca mulatta. J Oral Maxillofac Surg. 1986;44:312–21.
16. Moss ML, Rankow RM. The role of the functional matrix in mandibular growth. Angle Orthod. 1968;38:95–103.
17. Mulliken JB, Ferraro NF, Vento AR. A retrospective analysis of growth of the constructed condyle-ramus in children with hemifacial microsomia. Cleft Palate J. 1989;26:312–7.
18. Wu BZ, Ma L, Li Y, Chen S, Yi B. Costochondral graft in young children with hemifacial microsomia. J Craniofac Surg. 2017;28:129–33.
19. Lopez CD, Witek L, Torroni A, Flores RL, Demissie DB, Young S, Cronstein BN, Coelho PG. The role of 3D printing in treating craniomaxillofacial congenital anomalies. Birth Defects Res. 2018;110:1055–64.
20. Shilo D, Capucha T, Goldstein D, Bereznyak Y, Emodi O, Rachmiel A. Treatment of facial deformities using 3D planning and printing of patient-specific implants. J Vis Exp. 2020;(159). https://doi.org/10.3791/60930
21. Capucha T, Shilo D, Blanc O, Turgeman S, Emodi O, Rachmiel A. 3D planning and printing of patient-specific implants for reconstruction of bony defects. J Vis Exp. 2020;(162).
22. Risdon F. Ankylosis of the temporomaxillary joint. J Am Dent Assoc. 1934;21:1933–7.
23. El-Sayed KM. Temporomandibular joint reconstruction with costochondral graft using modified approach. Int J Oral Maxillofac Surg. 2008;37:897–902.
24. Guyuron B, Lasa CI Jr. Unpredictable growth pattern of costo chondral graft. Plast Reconstr Surg. 1992;90:880–886, discussion 887–889.
25. Figueroa AA, Gans BJ, Pruzansky S. Long-term follow-up of a mandibular costochondral graft. Oral Surg Oral Med Oral Pathol. 1984;58:257–68.
26. Perrott DH, Umeda H, Kaban LB. Costochondral graft construction/reconstruction of the ramus/condyle unit: long-term follow-up. Int J Oral Maxillofac Surg. 1994;23:321–8.
27. Goerke D, Sampson DE, Tibesar RJ, et al. Rib reconstruction of the absent mandibular condyle in children. Otolaryngol Head Neck Surg. 2013;149:372–6.

Bioprinting Applications in Craniofacial Regeneration

10

David T. Wu, Hieu M. Pham, Owen Tao, Kevin Y. Wu, and Simon D. Tran

Contents

D. T. Wu
Department of Oral Medicine, Infection, and Immunity, Harvard School of Dental Medicine, Harvard University, Boston, MA, USA

Harvard John A. Paulson School of Engineering and Applied Sciences, Harvard University, Cambridge, MA, USA

Wyss Institute for Biologically Inspired Engineering, Harvard University, Boston, MA, USA

Craniofacial Stem Cells and Tissue Engineering Laboratory, Faculty of Dental Medicine and Oral Health Sciences, McGill University, Montreal, QC, Canada
e-mail: davidt_wu@hsdm.harvard.edu

H. M. Pham
New York University College of Dentistry, New York University, New York, NY, USA
e-mail: hieu.pham@mail.mcgill.ca

O. Tao
Dr. Gerald Niznick College of Dentistry, Rady Faculty of Health Sciences, University of Manitoba, Winnipeg, MB, Canada
e-mail: owen.tao@mail.mcgill.ca

K. Y. Wu
Faculty of Medicine, University Laval, Quebec, QC, Canada
e-mail: yang.wu3@mail.mcgill.ca

S. D. Tran (✉)
Craniofacial Stem Cells and Tissue Engineering Laboratory, Faculty of Dental Medicine and Oral Health Sciences, McGill University, Montreal, QC, Canada
e-mail: simon.tran@mcgill.ca

© The Author(s), under exclusive license to Springer Nature Switzerland AG 2022
P. K. Chaudhari et al. (eds.), *3D Printing in Oral Health Science*,
https://doi.org/10.1007/978-3-031-07369-4_10

10.1 Introduction

Three-dimensional (3D) bioprinting technology has progressed at a rapid pace since its invention in the 1980s. Charles W. Hull (Chuck Hull) had first described the 3D printing technique under the name of stereolithography. Since then, multiple different techniques and methods have emerged. These techniques all aim for the same objective: to create 3D structures that mimic the external and internal structures of the anatomic sites, to provide scaffolds for cell attachment and migration, and to initiate tissue regeneration. This concept of 3D bioprinting, combined with the advancement of tissue engineering, has been proposed as a promising strategy to reconstruct and replace damaged tissues and diseased organs in many areas of medicine and dentistry, including craniofacial tissue regeneration [1]. In particular, 3D bioprinting technology allows for precise manufacturing of biocompatible scaffolds with complex 3D architectures using cell sources and other biomaterials [2].

Craniofacial tissues have highly complex 3D architectures with sophisticated multicellular interactions. Due to this complexity, complete regeneration of craniofacial structures from congenital malformations, trauma, and resective surgeries is extremely challenging. Despite the advances in the field of craniofacial reconstruction, conventional regenerative strategies still have difficulty mimicking the complex architectures and the biological interactions of this anatomical site [2]. To date, the development and advancement of 3D bioprinting technology are still in its early phase. In fact, 3D bioprinting is mainly used in research settings, and its clinical application has been limited by its ability to mostly fabricate simple homogeneous tissues as opposed to heterogeneous tissues in clinical settings [3].

Currently, reconstruction of extensive or complex craniofacial defects requires local or regional flap, or sometimes microvascular transfer of free flaps as the gold standard treatment [4]. These reconstructive procedures have significant limitations including donor site morbidity as well as size mismatch to the recipient site, leading to compromised aesthetics and function [5]. Therefore, 3D bioprinting technology in combination with tissue engineering strategies presents a promising alternative to the current reconstruction techniques. The aim of this chapter is to provide a comprehensive overview of major concepts in 3D bioprinting including the bioprinting process, armamentarium, types of bioprinters, clinical application in craniofacial regenerative medicine, limitations, and future perspectives.

10.2 3D Bioprinting Process

The basic process of 3D bioprinting in craniofacial regeneration can be classified into three phases including pre-bioprinting phase, bioprinting phase, and post-bioprinting phase (Fig. 10.1).

10.2.1 Pre-bioprinting Phase

The pre-bioprinting phase involves (1) digital imaging and computer-assisted design; (2) biomaterial selection and bioink preparation; and (3) cell selection, isolation, culture, and preparation [6].

First, digital imaging of the defect or structure to be replaced is acquired via cone beam computed tomography (CBCT), computed tomography (CT), or magnetic resonance imaging (MRI). These imaging modalities are the most commonly used for medical and dental application of 3D bioprinting [7]. After imaging, the Digital Imaging and Communications in Medicine (DICOM) files are processed with computer-assisted design (CAD) softwares. In addition, tomographic reconstruction is performed to achieve segmented 2D images for the layer-by-layer 3D bioprinting process. Subsequently, Standard Triangle/Tesselation Language (STL) files are generated and sent to the bioprinter [3].

Biomaterial and bioink selection is another crucial part of the pre-bioprinting phase and is determined by the type of 3D printer used as well as specific mechanical, rheological, and biological requirements of the final tissue construct or organ discussed in Sect. 10.4 [8].

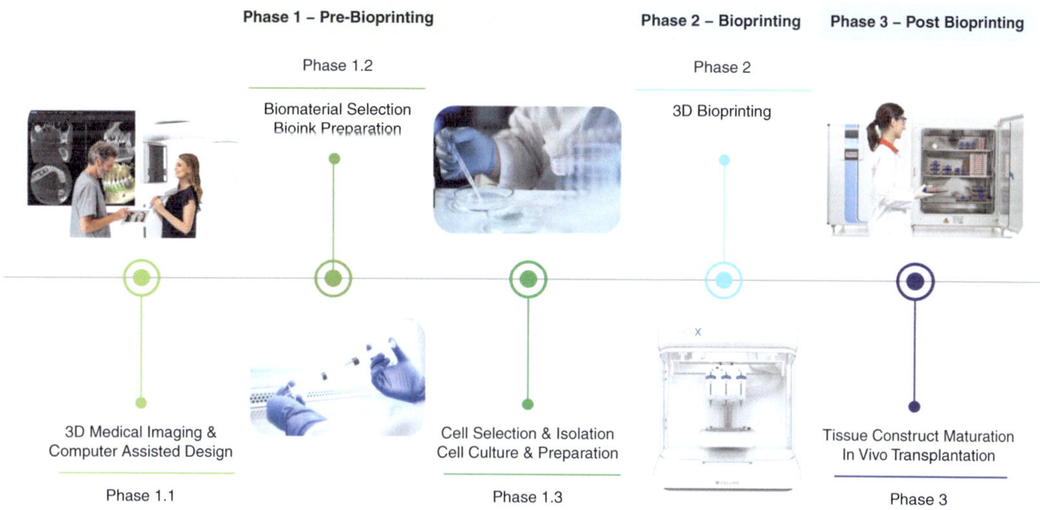

Fig. 10.1 3D bioprinting process: The basic process of 3D bioprinting in craniofacial regeneration can be classified into three phases including (1) pre-bioprinting phase, (2) bioprinting phase, and (3) post-bioprinting phase. The pre-bioprinting phase involves (1) 3D digital imaging and computer-assisted design; (2) biomaterial selection and bioink preparation; (3) cell selection, isolation, culture, and preparation [6]. The post-bioprinting phase involves tissue construct maturation in a bioreactor, and in vivo transplantation [6]

Prior to the bioprinting phase, isolation, expansion, and quality assessment of the desired cells represent another important step. The different types of cells including their sources, characteristics, and advantages are described in Sect. 10.3.1. It is important to ensure that cells have adequate viability, proliferative, differentiation, and extracellular matrix production potential. In addition, cells can be supplemented with biologics and growth factors-enriched culture media to enhance cell viability, proliferation, and differentiation. Currently, only two growth factors are approved by the Food and Drug Administration (FDA) for clinical applications in craniofacial regeneration including human recombinant platelet-derived growth factor-BB (rhPDGF-BB) and human recombinant bone morphogenetic protein-2 (rhBMP-2). Additional biologics include fibroblast growth factor (FGF), insulin-like growth factor (IGF), transforming growth factor beta (TGF-B), and vascular endothelial growth factor (VEGF). Finally, other culture medium supplements can be used to potentiate cell viability and growth including vitamins, hormones, and other macronutrients (Fig. 10.2) [9].

10.2.2 Bioprinting Phase

The bioprinting phase involves the deposition of bioink, cells, and signaling molecules with a bioprinter to form a tissue construct. Bioink and cells are prepared and transferred to their respective cartridges, installed in the printer, and the bioprinting process is initiated to print 3D structures with specific microarchitecture. The main types of bioprinters, their advantages, and respective mechanisms of action are reviewed in Sect. 10.4.

10.2.3 Post-bioprinting Phase

The post-bioprinting phase is key to ensuring the manufacturing of reliable 3D tissue constructs with appropriate structural integrity and biological function [7]. One important component of this

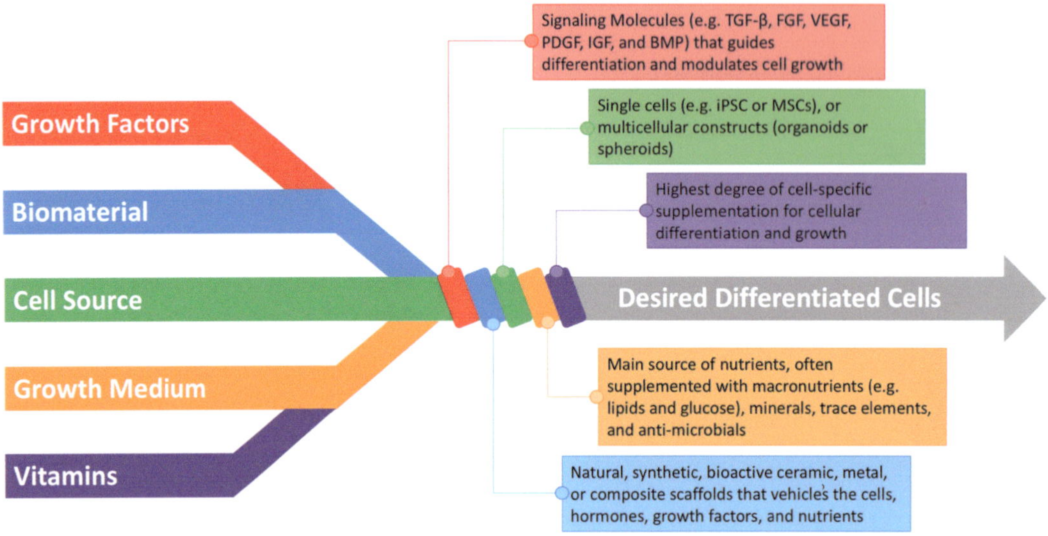

Fig. 10.2 Cell preparation for 3D bioprinting: Key components of cellular preparation for 3D bioprinting include (1) growth factors (2) biomaterials (3) cell source (4) growth medium (5) vitamins. Achieving an optimal combination of the above elements allow desired cells to proliferate, differentiate, and synthesize extracellular matrix to enhance the quality of the 3D bioprinted tissue construct or organ

phase is tissue maturation. The transfer of 3D bioprinted tissue constructs into an incubator or bioreactor allows for enhanced survival, maturation, vascularization, and remodeling prior to in vivo implantation [3]. Recent advances in bioreactor design enable convective nutrient transport, creation of microgravity environment, and compression for dynamic mechanical stimulation [6]. Once the tissue construct or bioprinted organ is ready to be used, a surgical team will perform the surgical implantation or transplantation in animals or patients to address the clinical problem.

10.3 3D Bioprinting Armamentarium

Tissue engineering combines the field of biology and engineering to develop functional substitutes for damaged tissues. The creation of functional tissue engineered constructs requires three main components termed "Tissue Engineering Triad," which includes (1) cells, (2) scaffold, and (3) regulators [10]. 3D bioprinting utilizes the principles of tissue engineering and combines these three key tissue building blocks with spatial precision to enhance tissue structure, architecture, and functionality.

10.3.1 Cellular Component

The first and most important component of the tissue engineering triad is the cells [10]. They are the fundamental building blocks that reconstitute the 3D bioprinted tissue construct and/or organ. Many factors must be considered during cell selection for 3D bioprinting. Ideally, the user should be able to control the proliferative properties of the cells as excessive or insufficient proliferation can lead to complications. This is commonly evident in multicellular constructs that overgrow and develop a necrotic core due to hypoxia. Additionally, researchers should be able to predict or control the timing for cell proliferation [6]. Also, cells must be able to withstand the mechanical and physiological stresses associated with 3D bioprinting as cell viability can be affected by stresses such as sheer forces, changes in temperature and pH, and the presence of chemicals, toxins, and enzymes [6, 11]. Finally, cell viability may also be greatly altered by the bioprinting technique and properties of the scaffold material selected to contain and support the cells.

Due to their complexity and intricacy, more than one type of cells is required to adequately reconstruct the desired tissue and/or organ. For example, in alveolar bone regeneration, osteoblasts, osteoclasts, and osteocytes are required for bone repair and remodeling; epithelial cells and fibroblasts provide structural and barrier functions; and endothelial cells form vasculature to support osteogenesis [6, 12, 13]. In addition, stem cells and progenitor cells are required to provide the tissue construct with self-renewing abilities. Stem cells are undifferentiated cells that have the potential to divide indefinitely and give rise to various cell lineages. In contrast, progenitor cells have limited proliferative capabilities and determined set of cell fates and thus can only differentiate into certain cell types. One particularly important type of stem cells is the mesenchymal stem cells (MSCs). As many craniofacial structures are derived from MSCs, they are an integral component to craniofacial regeneration [14].

Cells may be categorized based on their differentiation potential or source. Firstly, cells may be selected based on their differentiation capabilities: undifferentiated stem cells {totipotent, pluripotent [e.g., embryonic stem cells (ESCs) and induced pluripotent stem cells (iPSCs)], multipotent stem cells (e.g., MSCs, oligopotent, or omnipotent stem cells)}, and differentiated somatic cells [15]. Secondly, cells may also be selected based on their sources: endogenous cells from the donor or exogenous cells from another organism. Exogenous cells may pose immunogenicity challenges [16]. Lastly, in light of 3D bioprinting, cells may also be selected as single cells or larger clusters of cells termed spheroids and organoids. This subsection will explore the use of single cells and multicellular constructs for 3D bioprinting of craniofacial tissue.

10.3.1.1 Single Cells

Single cells are particularly useful in 3D bioprinting to creating vascular channels and capillaries that are composed of a single layer of endothelial cells. In addition, stem cells and progenitor cells from various sources are used in 3D bioprinting. The most common source of stem cells and progenitor cells for craniofacial regeneration come from the bone marrow. Although bone marrow-derived stem and progenitor cells have been extensively used in tissue engineering and regenerative medicine, their harvest is rather invasive and involves bone marrow aspiration from long bones or iliac crests, which may lead to patient discomfort and morbidity.

Since the discovery and characterization of multipotent mesenchymal stem cells (MSCs) from the bone marrow, MSCs from other tissues have been identified and characterized including umbilical cord blood, adipose tissues, and dental tissues. These MSCs are capable of differentiating into cell lineages including osteogenic, chondrogenic, myogenic, and adipogenic [17]. From the early 2000s, significant progress has been made toward identifying different human MSC-like stem/progenitor cells from dental and oral sources. These cells include periodontal ligament stem cells (PDLSCs), dental pulp stem cells (DPSCs), stem cells from exfoliated deciduous teeth (SHED), stem cells from apical papilla (SCAP), dental follicle progenitor cells (DFPCs), and gingiva-derived MSCs (GMSCs) [18–23]. Dental stem cells have the advantage of being easily accessible, thus avoiding the need for invasive harvest in comparison to BM-derived MSCs. Together, this group of cells represents a promising cell source for 3D bioprinting for craniofacial regeneration.

Other cell types are useful in 3D bioprinting of functional craniofacial tissue and structures including salivary glands, nerves, and vasculature; three examples are highlighted in this section. To begin with, exogenous salivary gland stem cells may be used in 3D bioprinting as a building block for functional salivary organoids [24, 25]. In addition, in a preclinical animal study, Zhang et al. demonstrated that human gingiva-derived MSCs (GMSCs) can be differentiated into both neuronal and Schwann-like cells and be used in 3D bioprinting to generate nerve constructs that promoted nerve regeneration and functional recovery in bridge segmental defects in rat facial nerves [26]. Finally, human umbilical vein endothelial cells (HUVECs) may be used in combination with MSCs to induce the formation of pre-vascular networks leading to improved cell viability and proliferation [27].

10.3.1.2 Multicellular Constructs

Fabricating 3D multicellular constructs grown in suspension is an alternative to growing cells in a monolayer fashion. In fact, cells tend to aggregate in clusters and form 3D constructs termed spheroids and organoids. While both spheroids and organoids are well-organized multicellular structures, there are a few defining features that differentiate the two types of constructs. On one hand, spheroids are derived from cell line monoculture, have transient cell organization, and only represent a component of tissue. They are difficult to maintain long term and depend on cell-cell and cell-environment interactions to proliferate and survive. On the other hand, organoids are heterogeneous multilineage constructs derived from stem cells and/or progenitor cells, which possess the ability to differentiate and self-renew. Consequently, organoids can better recapitulate organ physiological parameters and can be maintained in culture for a longer period of time [28, 29]. Ultimately, cells grown in 3D cultures will have increased cell-cell and cell-extracellular matrix (cell-ECM) interactions compared to cells cultured in a monoplane orientation [28]. Thus, 3D cellular models are more physiologically relevant and biologically applicable to 3D bioprinting tissue engineered constructs.

Another advantage of using spheroids and organoids in 3D bioprinting is that they require fewer amount of scaffolding material to support the cells, thus applicable to "scaffold-free printing." More specifically, spheroids and organoids can synthesize and secrete their native ECM; thus, only a minimal amount of scaffold is required their initial formation and subsequent bioprinting. The benefits of multicellular constructs include reduced costs and efforts associated with the fabrication of cell-laden hydrogels, enhanced biocompatibility, and physiological relevance as the cell construct secretes native ECM; thus, the use of exogenous materials is reduced [2, 28, 29].

Craniofacial structures pose challenges in tissue reconstruction due its various multicellular interactions and complex anatomical features [2]. By harnessing the power of spheroids and organoids in 3D bioprinting, researchers have the ability to print homotypic and heterotypic multicellular constructs with higher spatial resolution and density and, thus, may be able to recreate complex tissues such as vascularized bone, cartilage, periodontium, and whole teeth. Specific clinical applications of 3D bioprinting are reviewed in detail in Sect. 10.5 [2].

10.3.2 Biomaterials

The next component to the tissue engineering triad is the biomaterial scaffolds. This component encompasses all natural and synthetic biomaterials, or a combination of both, used to provide structural support and a favorable microenvironment for cells. Biomaterials can be engineered for tunable release of regulators such as growth factors (GF).

Biomaterials used during the bioprinting process to encapsulate cells are termed bioink [3]. Bioinks can serve as cell encapsulation material to provide cells with protection. In addition, bioinks can be printed onto acellular biomaterial ink scaffolds with higher rigidity, which provides the construct with higher structural integrity. Bioinks are typically composed of cell-laden hydrogels consisting of natural or synthetic materials. In contrast, acellular biomaterial ink scaffolds can be composed of a wider selection of materials depending on desired properties and its intended use [30, 31]. Researchers and clinicians may select a biomaterial based on its rheological, mechanical, chemical, and biological properties, which should ultimately reflect the target organ or tissue's native physiological environment. These properties may include pH levels, biocompatibility, immunogenicity, cytotoxicity, degradation rate, inductivity, stiffness, viscoelasticity, and strength. Other properties that may be considered include the material's tunability, reproducibility, cost, availability, printability, and complexity of use [32, 33].

The main advantage of 3D bioprinted scaffolds compared to 3D printed scaffolds is its micron-level precision of cell positioning throughout the scaffold. This characteristic enables researchers and clinicians to create a more desirable and viable scaffold for tissue reconstruction [1]. However, in comparison to the traditional acellular 3D printing method, 3D bioprinting requires additional considerations due to the presence of cells. These considerations include cell positioning, the degree of heat generated, sheering forces, maximum compressive moduli of the biomaterial, and their respective impact on cell viability. Additional printing parameters may affect cell viability and proliferation such as vibrating frequencies, voltage, and mechanical impact during the printing process [1, 31–34]. As a result, some printing techniques and materials may not be suitable for 3D bioprinting of living tissue constructs.

It is suggested that bioinks should have minimal incorporation of synthetic biopolymers to minimize unwanted changes and effects on cells [31]. However, natural biomaterials often have significantly lower mechanical strength compared to synthetic biomaterials and thus cannot be used to create certain craniofacial tissues with high mechanical strength requirements. For example, craniofacial bone has a compressive moduli between 100 MPa and 20 GPa. While alginate is highly biocompatible and fibrin is highly biologically active, both materials have very low compressive moduli (~5 kPa). Comparatively, while the use of synthetic materials such as polyethylene glycol (PEG) with cells may be less favorable, PEG confers higher physical and mechanical strength (~300–350 kPa) needed for harder tissues or areas of high stress such as bone and teeth. These obstacles can be overcome by using natural-synthetic composite bioink and/or simultaneously using 3D printing and bioprinting techniques together [1].

The various biomaterials used in 3D bioprinting have been categorized into the following five categories: natural materials, synthetic materials, bioactive ceramics and cements, metals, and hybrids and composites. It is important to note that there are hundreds of biomaterials being researched, and even more when considering the possible combinations of materials used to create composite gels and scaffolds. Thus, this section will provide an overview of the most commonly used materials and notable composites (Table 10.1).

10.3.2.1 Natural Materials

The main advantage of natural material is its bioactivity and ability to induce cellular activity. For instance, researchers have used protein-based natural biomaterials such as collagen, elastin, laminin, fibrin, fibronectin, and gelatin as scaffolds to mimic the cell's native ECM, which can enhance cell differentiation, proliferation, and migration. Previous studies on the use of ECM-like scaffolds and 3D bioprinting have demonstrated, both in vitro and in vivo, the successful fabrication of various tissue engineered constructs including skin, bone, and cartilage, cardiovascular tissue, hepatic tissue, neuronal tissue, and cornea tissue [46]. Furthermore, there is increasing interest in the use of more complex protein scaffolds containing more than one type of protein substrate such as decellularized ECM (dECM) and decellularized bone matrix (DBM). These materials are natural ECM that have been cleared of native cells, debris, and other immunogenic components leaving intact structure and microarchitectures composed of collagen, adhesive proteins, growth factors, proteoglycans, and glycosaminoglycans (GAGs). Subsequently, dECM can be reseeded with desired cells. In the case of 3D bioprinting, dECM can be further processed into bioinks to be bioprinted [47, 48]. Other protein-based natural materials used in 3D bioprinting include albumin, keratin, and silk fibers.

Natural biomaterials can also be carbohydrate-based (alginate, chitin, chitosan, cellulose, starch, glycosaminoglycans (GAGs), and hyaluronic acid) as shown in Table 10.1 [3, 35, 39, 49]. While some natural carbohydrate-based materials are not found in the human body (e.g., alginate, cellulose, and chitin), their unique properties including biocompatibility, affordability, and accessibility make them excellent biomaterial scaffold candidates for research [50,

Table 10.1 Biomaterial scaffolds used in 3D bioprinting

Biomaterial	Key properties			
	Cost (low-high)	Bioactivity	Degradation rate (low-high)	Unique features
Natural				
Carbohydrate-based E.g., alginate, agarose, dextran, chitin, and chitosan, cellulose, starch, glycosaminoglycans (GAGs), and hyaluronic acid (HA) [35–37]	Low-high • Cost of HA, GAGs, chitin and chitosan are med-high	High bioactivity	Low-high	Antibacterial properties; very low mechanical properties; tunable
Protein-based E.g., keratin, collagen, gelatin, laminin, elastin, fibrin, fibronectin, albumin, silk fibers, decellularized extracellular matrix (dECM), decellularized bone matrix (DBM) [35, 37, 38]	Med • Cost of recombinant human proteins are typically high	High bioactivity	Med-high	High biocompatibility; very low mechanical properties; osteoconductive; osteoinductive; low compressive strength; low reproducibility • dECM and DBM have variable results due to processing method/ technique • Silk fibers have high mechanical property
Synthetics				
Biodegradable E.g., polylactic acid (PLA), polyglycolic acid (PGA), polylactic-co-glycolic acid (PLGA), poly-ε-caprolactone (PCL), polyether urethane (PU), polyvinyl alcohol (PVA), polymerization of methyl methacrylate (PMMA) [35, 37, 39]	Low	Bioinert	Low • PGA and PLA have high degradation rates	Hydrophobic; poor cell adhesion; poor osteoinduction; highly tunable; highly reproducible; acidic degradation byproducts; high printing resolution; porous; moderate mechanical properties • PLA has osteoconductivity • PGA has low compressive strength
Nonbiodegradable E.g., polyethylene glycol (PEG), polyethylene glycol dimethacrylate (PEGDMA), porous polyethylene (PPE), polyetherketoneketone (PEKK)**, polyetheretherketone (PEEK) [38, 39]	Low	Bioinert	Nondegradable	Highly resembles bone; high biocompatibility; durable; risk of bacterial infections; moderate mechanical properties
Bioactive ceramics and cements				
Calcium phosphate-based E.g., calcium phosphate, biphasic calcium phosphate (BCP), β-tricalcium phosphate (β-TCP), hydroxyapatite (HA) [35, 40–42]	Low	High bioactivity	Med • β-TCP has a low degradation rate	Highly resembles bone; osteoinductive; osteoconductive; osteointegrative; high mechanical properties; risk of infections; little injectability in bulk; brittle; reproducible; porous structure

Table 10.1 (continued)

Biomaterial	Key properties Cost (low-high)	Bioactivity	Degradation rate (low-high)	Unique features
Others E.g., calcium carbonate, calcium sulfate, aluminum oxide, calcium silicate, silicon, bioactive glass, zirconia, akermanite, diopside [32, 36, 38, 40–42]	Low	High bioactivity • Aluminum oxide and zirconia are bioinert	Nondegradable • Akermanite has a controllable degradation rate	High hardness; high wear resistance; osteoinductive; brittle • Bioactive glass has low fracture resistance, mechanical strength, brittle, and porous nanostructures which allows for cell adhesion • Silicon has some evidence of angiogenesis and osteogenesis effect • Diopside is capable of thermal expansion
Metals				
Nonbiodegradable Gold, stainless steel, zinc oxide, titanium alloy, cobalt alloy, tantalum [32, 36, 38, 39, 41, 43, 44]	Low-med	Bioinert • Titanium has osteointegrative properties • Tantalum has bioactive properties	Nondegradable	High wear resistance; ductile; high mechanic strength; risk of stress shielding; biomolecules cannot be added into scaffold; thermostability • Titanium has high compatibility and strength-to-weight ratio, similar strength modulus to bone, lower risk of stress shielding
Biodegradable Magnesium, magnesium alloy, iron alloy [38, 43–45]	Low	Med bioactivity	Med-high • Magnesium alloys degrade faster than iron-alloys	Osteoconductive; porous; moderate-high mechanical properties; lower risk of stress shielding than nonbiodegradable metals; ductile; degradation byproduct can cause local acidic environment and form gas pockets; biomolecules cannot be added into scaffold
Composites and hybrids				
Can be any combination of biomaterial previously mentioned	Characteristics will vary based on the composite composition, as desired by the researcher			

51]. However, a major drawback of using naturally occurring materials is the variability in material composition depending on its source. Consequently, this can affect reproducibility and reliability, thus the quality of the research [51].

10.3.2.2 Synthetic Materials

Synthetic materials unlike natural materials are much more reproducible due to its controlled manufacturing conditions. As a result, their use in biomedical research provides more consistent and reliable data [32]. Another advantage of synthetic materials is their superior physical properties such as higher mechanical strength, compressive moduli, and stress-bearing capabilities. Furthermore, many synthetic materials such as polylactic acid (PLA), polyglycolic acid (PGA), and polylactic-co-glycolic acid (PLGA) are thermoplastic and thus can be easily manipulated into desired shapes and microstructures [52]. However, the main concern of synthetic materials is its degradation byproducts. For example, PLA and PGA produce carbon dioxide as they degrade and thus can lead to hypercapnia, an acidic environment, and consequently necrosis of proximal tissue. Another concern with synthetic material stems from its common bioinert property, which can result in rejection of the material in vivo [34].

Other synthetic materials commonly used for 3D bioprinting include poly-ε-caprolactone (PCL), polyethylene glycol (PEG), polyethylene glycol dimethacrylate (PEGDMA), porous polyethylene (PPE), polymerization of methyl methacrylate (PMMA), polyether urethane (PU), polyetherketoneketone (PEKK), and polyetheretherketone (PEEK) [3, 35, 39, 49]. Here, we will further classify synthetic materials into two subgroups, biodegradable and nonbiodegradable (Table 10.1).

10.3.2.3 Bioactive Ceramics and Cements

Bioactive ceramics and cements are great candidates for use in 3D bioprinting due to their chemical properties resembling the mineral components of natural bone. Typically, this group of biomaterials exhibits excellent biocompatibility, high mechanical stiffness, brittleness, low elasticity, and slow degradation rate. However, its most notable advantage is its osteoinductive property, hence its popular use in bone regeneration [32, 34]. The most commonly used bioactive ceramics are those with a mineral phase composed of calcium and phosphate, such as hydroxyapatite (HA), β-tricalcium phosphate (β-TCP), biphasic calcium phosphate (BCP), and calcium phosphate. Other types of ceramics or cements include calcium carbonates, calcium sulfates, calcium silicate, silicon, bioactive glasses, zirconia, and aluminum oxide [3, 35, 53–55]. Some ceramics such as bioactive glass or HA can be further improved with the incorporation of silicone which can promote angiogenesis and bone ingrowth. Other researchers have explored the addition of metallic ions such as copper and/or cobalt into bioactive glass which has been shown to induce angiogenesis [3]. The major drawbacks to using bioactive ceramics are its brittleness and porous property which makes it difficult to sustain high mechanical loading required for bone remodeling [34]. Here we subcategorize bioactive ceramics and cements into two groups: calcium phosphate-based and non-calcium phosphate-based.

10.3.2.4 Metals

Metals are the last group of materials being used in tissue engineering. Metals are generally incorporated into bioinks to increase its stiffness, processability, and printability [56]. Metals used in bioprinting include gold, zinc oxide, iron, stainless steel, titanium alloys, and cobalt alloys. Advantages that metals have to offer are its superior mechanical properties, bioinert, and nondegradable which allow for it to last a long time, even in high-stress areas such as bone and teeth [3, 32, 53–55]. While most metals are nondegradable, there is increasing research on biodegradable metals such as magnesium alloys and iron alloys [43].

It has been suggested that metal-based scaffolds can cause stress shielding due to its higher relatively higher elastic modulus which can result in bone resorption and therefore leaving subjects prone to implant failures [3, 54]. To date, titanium-based constructs are the most widely used metal for craniofacial reconstruction due to its biocompatibility, high strength-to-weight ratio, elastic modulus, nonabsorbable characteristic, and potential for bone ingrowth [36].

10.3.2.5 Hybrids and Composites

Due to each type of material having their own unique set of advantages and disadvantages, there is increasing interest in exploring and using hybrid or composite scaffolds, which are biomaterials comprised of multiple phases and materials [34]. In general, composites have higher biological capacity because it is comprised of two or more materials, where one material's weakness is supported by the strength of another material. Depending on the intended use or desired properties, researchers may combine bioactive ceramics with synthetic or natural biomaterials, or more commonly, they may combine synthetic biomaterials with natural biomaterials. Researchers may even combine materials of the same category, such as a protein phase with a carbohydrate phase from the natural biomaterial category. This allows researchers to create ideal bioinks or scaffolds that would otherwise be not viable when used alone. For example, synthetic materials may often create local acidity through its byproducts as it degrades, in addition to being bioinert. Conversely, natural materials have excellent bioactivity, though it lacks mechanical properties. By combining a synthetic material such as PEG with a natural material such as collagen, a cell-inductive scaffold with improved mechanical properties can be created [35].

10.3.3 Regulators

The third and final component to the tissue engineering triad is the regulators, which consists of signaling molecules, notably growth factors (GFs). These biological molecules signal cells to undergo proliferation, morphogenesis, differentiation, migration, and survival [57]. While GFs exist naturally in the human body, for the purpose of tissue engineering, an exogenous source is also required. They are typically incorporated into scaffolds and are released as the material degrades. The release of GFs should be controlled spatiotemporally to adequately guide proper cellular growth, differentiation, morphology, and function. The release of GFs should also be steady as to prevent unwanted diffusion and therefore unwanted outcomes [57, 58]. A gradient of GF diffusion in conjunction with physical contact of ECM and other inductive cues establishes the microenvironment necessary to induce these effects on the embedded cells.

There are several types of GFs that are being used in craniofacial regeneration research. These include transforming growth factor beta (TGF-β), fibroblast growth factor (FGF), vascular endothelial growth factor (VEGF), platelet-derived growth factor (PDGF), insulin-like growth factor (IGF), and bone morphogenic proteins (BMP-2,4,6,7) [35, 59]. While many GFs show promising results in vivo and in vitro, there are currently only two GFs that are FDA-approved for clinical use: recombinant human PDGF-BB (rhPDGF-BB) and recombinant human BMP-2 (rhBMP-2).

PDGF has several isoforms (PDGF-A, PDGF-B, PDGF-C, and PDGF-D) and becomes active when it dimerizes. These dimeric isoforms include PDGF-AA, PDGF-BB, PDGF-AB, PDGF-CC, and PDGF-DD. PDGF-BB has the highest activity as it is capable of binding all dermic forms of PDGF receptors and thus is the isoform that has translated into clinical use. In 2005, Nevins et al. reported in a pivotal randomized control trial study involving 180 subjects the clinical application of rhPDGF-BB in periodontal tissue regeneration, more specifically, its effectiveness in inducing radiographic bone fill, and clinical attachment level gain, and reduction in probing depth when used in conjunction with β-TCP [59–61].

BMPs are the second FDA-approved GF for clinical use. Currently, the only BMP isoform approved by the FDA for clinical use is rhBMP-2 (InFUSE Bone Graft®, Medtronic and Wyeth). rhBMP-2 is infused in an absorbable collagen scaffold and is capable of guiding bone regeneration via inducing MSCs to differentiate into osteoblasts. rhFGF-2 is also noteworthy as it currently used for periodontal regeneration in Japan, and has shown to have beneficial outcomes in patients with lower limb ischemia [55, 59].

10.4 3D Bioprinter Technology

10.4.1 Inkjet 3D Bioprinting

Inkjet bioprinting uses thermal or piezoelectric processes in the nozzle head to dispense droplets of bioink (Fig. 10.3a). The thermal induced inkjet nozzles pass a current through a resistor to create a bubble by vaporizing the nearby fluid, and therefore building up pressure in the nozzle head resulting in droplet ejection. The piezoelectric inkjet nozzles apply voltage to the piezo element to create a pulse, which produces volumetric changes in the nozzle head resulting

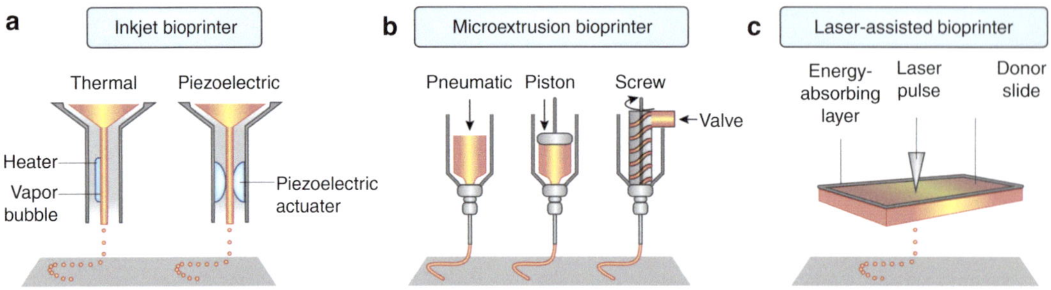

Fig. 10.3 Types of 3D bioprinting technologies: (**a**) Inkjet 3D bioprinting. Droplets dispensed by thermal or piezoelectric processes in the nozzle head. (**b**) Extrusion 3D bioprinting. The bioink is extruded through a nozzle due to pneumatic or mechanical (piston/screw driven) pressure. (**c**) Laser-assisted 3D bioprinting. A laser is focused on an absorbing substrate to generate pressure that propel cell-containing bioink onto a collector substrate. (Reproduced with permission from Murphy and Atala, 2014 [6])

in droplet ejection. The droplet ejection process from the piezoelectric nozzle head allows for more control of the droplet shape and size and has a greater tolerability for heat-sensitive materials (such as cells) when compared to the thermal induced nozzle head, but the vibrational frequencies from the piezoelectric process can cause cell membrane damage [31, 62]. Thermal inkjet printers have advantages in availability, higher print speed, and lower cost of parts fabrication [62]. When compared to other bioprinting technologies, inkjet bioprinting has advantages in high print speeds, low cost, and wide availability [11, 63].

Several considerations must be made with regard to cell viability when choosing to use inkjet-based bioprinters. While a wide range of bioinks can be used with inkjet bioprinting (including various combinations of cells, ceramics, polymers, and proteins), a limitation however in the type of bioink used is the viscosity requirement: to prevent the continuous flow of the material once a droplet is ejected and to prevent high ejection pressures (which can damage the cells), a low viscosity fluid between 3.5 and 12 mPa s is required [64]. This limitation is achieved by using low concentration solutions, which can increase the possibilities of cells drying and dying, and a low viscosity fluid will have greater difficulty in forming larger 3D structures [11, 31, 65]. The mechanical impact of the cells leaving the nozzle head and hitting the collector surface also affects cell viability. Another limitation is that cell aggregation within the bioink can affect droplet formation and trajectory, resulting in the poor precision of bioink droplet placement and potentially affecting the distribution of cells in the final construct [65]. Despite these considerations, observations of inkjet-based bioprinting have reported good cell viabilities (over 90%), and a resolution of greater than 50 μm [11, 31, 65].

10.4.2 Light-Assisted 3D Bioprinting

Light-assisted 3D bioprinting includes stereolithography (SLA) and laser-induced forward transfer (LIFT). While selective laser sintering (SLS) is another light-based 3D printing technology, it is not compatible for bioprinting due to its methodology of melting the polymer (and subjecting cells to high temperatures) to create a 3D construct [2]. Light-assisted 3D printing is noncontact and nozzle free and therefore has the advantage that materials with higher viscosities can be used (1–300 mPa s) without the issue of nozzle clogging [64].

SLA, more specifically, consists of directing a light source (UV or visible light) over a photopolymerizable fluid (Fig. 10.3b). Once a layer of the polymer is completed, the printing platform is lowered to allow for photopolymerization of a new layer. An advantage of SLA printing is its high fabrication accuracy and low printing time [66]. The resolution and cell viability are >50 μm and >85%, respectively [65]. The materials that are used for SLA consist of photocrosslinkable hydrogels such as those containing an acryloyl or alkenyl functional group, or a photoinitiator such as lithium phenyl-2,4,6-trimethylbenzoylphosphinate (LAP) or the benzophenone/tertiary amine system [67]. A major cell viability consideration for SLA bioprinting is the exposure to laser energy, which can be harmful to cells [65, 66]. Regardless, the specific material requirements of being photocrosslinkable are a disadvantage of the SLA methodology due to a lack of compatible materials, and the potential cytotoxicity to cells due to the photoinitiators that are added to the hydrogels [63].

LIFT consists of three main components: a light source, a ribbon (transparent glass, metal, and bioink), and a collection plate. The light source vaporizes the metal layer and creates a high-pressure bubble resulting in the production of bioink droplets that are deposited onto the collection plate (Fig. 10.3c). As with inkjet printing, LIFT is droplet-based and has similar considerations for cell viability such as the mechanical impact of the bioink hitting the collector surface, and the less accurate positioning of cells [31]. An advantage of LIFT is high precision and resolution (>20 μm) with high cell viability (>95%); however, LIFT is often costly and time-consuming due to the use of high viscosity materials which are required to obtain a highly precise shape [64, 66, 68]. The materials that have been used with LIFT include polymers, ceramics, proteins, and cells of varying viscosities (in contrast to the low viscosity requirement of inkjet printing) [69].

10.4.3 Extrusion 3D Bioprinting

Extrusion 3D printing is the most commonly used 3D bioprinting technique and is a pressure-driven system. The bioink is continuously extruded (in contrast to the droplet-based system in inkjet and LIFT) through a nozzle due to pneumatic or mechanical (piston/screw driven) pressure. A more complex construct can be created by using multiple nozzles, each carrying different a bioink [70]. Fused deposition modeling is a type of extrusion printing that heats and melts the material as it is extruded through the nozzle. It can be used to create scaffolds for tissue engineering. However, this technique is unable to bioprint cells due to high temperatures reached during the printing process.

The considerations for cell viability when choosing to use extrusion-based bioprinters include dispensing pressure and shear stress. While the pressure-assisted system in extrusion printing allows for the printing of very high cell densities, higher viscosity fluids, and more homogenous cell distributions, shear stress is a factor that can affect cell viability and is increased as the viscosity of the fluid is increased [31]. Furthermore, as the dispensing pressure increases, there is greater cellular distortion, all of which can result in low cell viability (>40%) [66]. In addition, the absence of droplet control (compared to inkjet and LIFT) results in a lower resolution (>100 μm) [65]. The higher viscosity bioinks used in extrusion printing can include natural polymers such as collagen, gelatin, alginate, hyaluronic acid, as well as synthetic polymers such as polyvinyl alcohol (PVA) and polyethylene glycol (PEG) [71].

The types of 3D bioprinting technologies are summarized in Table 10.2.

10.5 3D Bioprinting Clinical Applications

Although various 3D printing methods are widely applied to the manufacturing of biocompatible scaffolds and constructs to support complex functional living tissue in clinical trials, the use of 3D bioprinting to generate functional craniofacial tissues remains at an experimental stage. This section reviews key areas for clinical application of 3D bioprinting at the tooth level, periodontal support tissue level, craniofacial, and maxillofacial tissue level.

Table 10.2 Types of 3D bioprinting technologies

	Inkjet [72]	Light-assisted [65, 68] Stereolithography (SLA)	Light-assisted [65, 68] Laser-induced forward transfer (LIFT)	Extrusion [31, 73]
Description	Uses thermal or piezoelectric processes in the nozzle head to dispense droplets of bioink	A light source is directed layer by layer over a photopolymerizable fluid	A light source is directed over a ribbon to create a high-pressure bubble resulting in bioink droplets that are received onto the collection plate	The bioink is extruded through a nozzle due to pneumatic or mechanical (piston/screw driven) pressure
Materials	• Low viscosity hydrogels, ceramics, proteins, and cells	• Photocrosslinkable hydrogels	• Varying viscosities of hydrogels, ceramics, proteins, and cells	• Higher viscosity hydrogels, polymers, ceramics, proteins, and cells
Considerations for cell viability	• Mechanical impact of bioink hitting surface • Heat energy (thermal) • Vibrating frequencies (piezoelectric) • Less accurate positioning of cells • Higher possibility of cell aggregate formation	• Potential cytotoxicity of photoinitiators • Laser energy exposure	• Mechanical impact of bioink hitting surface • Less accurate positioning of cells	• Shear stress • Dispensing pressure
Advantages	• High print speeds • Low cost • Wide availability	• Nozzle-free • Highest fabrication accuracy • Low print time	• Nozzle-free • High precision • High resolution • High cell viabilities	• Homogeneous distribution of cells • Can print high cell density • Can use high viscosity fluid
Disadvantages	• Poor precision in droplet placement • Low viscosity bioink requirements	• Lack of compatible materials	• Costly • Ribbon preparation is time-consuming	• Low resolution • Low cell viability
Resolution/cell viability	>50 μm >90%	>50 μm >85%	>20 μm >95%	>100 μm >40%

10.5.1 Dental Pulp and Whole-Tooth Regeneration

The dental pulp is a highly vascularized and innervated tissue enclosed within the root canal that plays a crucial role in providing sensation, nutrition, and innervation to the tooth [74]. After trauma, dental caries, and iatrogenic exposure of the pulp, there is an unmet clinical need to regenerate the pulp and reestablish innervation and vascularization. The ultimate goal of dental pulp regeneration is the formation of reparative dentin, vascular supply, and pulp neurotization [75].

Current strategies in pulp regeneration have been largely unsuccessful, although researchers are exploring the use of hydrogels to support dental pulp stem cells (DPSCs), mimic the native pulp chamber microenvironment, and recapitulate cell proliferation and differentiation into functional tissue [76]. However, the main limitation of this strategy, consisting of simple scaffolds loaded with cells and growth factors, is the inability to control multicellular spatial orientation, and subsequent cellular interactions and function [77].

With 3D bioprinting, researchers can achieve enhanced spatial control by printing cells to specific locations in the tissue-engineered construct to achieve desired cellular interactions. In addition, the use of bioink with tunable mechanical properties, optimized rheological properties to enhance printability, and inclusion of growth factors may further potentiate cell function. For the generation of vascularized constructs mimicking the human dental pulp, extrusion-based bioprinting is the preferred method. In these methods, sacrificial template material composed of dissolvable or removal material can be extruded and subsequently replaced with a cell-laden hydrogel or aggregate of cells to create vascular channels [74].

Currently, there is a lack of evidence supporting the use of 3D bioprinting for regenerative endodontics in patients [75]. Several in vitro studies have made progress toward developing biomaterials and bioinks that allow for control of stem cells and endothelial cells to promote pulp regeneration. For instance, Khayat et al. (2017) developed a photocrosslinkable GelMA hydrogel to encapsulate hDPSCs/HUVECs to promote revascularization and regenerate human dental pulp tissue [78]. Similarly, Yu et al. (2019) demonstrated that alginate/gelatin scaffold hydrogel is suitable for growth of hDPSCs [79]. Researchers have also combined extracellular matrix-derived scaffolds with natural polymers to develop a novel bioink with cytocompatibility and natural odontogenic capacity. The hydrogel consisting of alginate and dentin matrix was shown to have the ability to enhance odontogenic differentiation of stem cells from the apical papilla. In addition, 3D bioprinting was used to induce odontoblast at specific positions by localizing growth factors between the pulp tissue and wall of the pulp cavity [80]. To further enhance cell differentiation, growth factors can be conjugated to the biomaterial scaffold. Park et al. (2020) demonstrate that a bone morphogenetic protein (BMP) peptide-tethered GelMA-based bioink formulation can accelerate the differentiation of hDPCs in a 3D bioprinted dental construct [81]. Together, development of 3D bioprinting technology and its main components, including the bioprinters and bioinks, will enable predictable dental pulp regeneration and accelerate its clinical translation to ultimately help treating patients [74].

When it comes to whole-tooth regeneration, two strategies have been proposed: (1) reconstruction of tooth germ and autologous transplantation and (2) 3D printing of tooth mimicking tissue-engineered constructs [75]. 3D printing has been applied to fabricating anatomically mimicking human molar and rat incisal scaffolds with PCL and HA with interconnecting microchannels. Upon stimulation with stromal-derived factor 1 (SDF-1) and bone morphogenetic protein 7 (BMP-7), PDL and new bone regeneration were demonstrated in the rat model [82]. In the future, 3D bioprinting technology will boost the precise and controlled manufacturing of bioengineered teeth to one day benefit patients in the clinical arena as a biomimetic dental implant.

10.5.2 Periodontal Regeneration

Periodontal regeneration is the regeneration of tooth supporting structures including periodontal ligament (PDL), cementum, and alveolar bone, lost due to periodontal disease. Notably, key advances in the field of periodontal tissue engineering in developing biomaterial scaffolds, enhancing growth factor delivery systems, and optimizing cell delivery systems paved the road for these elements to be integrated with 3D bioprinting [75, 82–84].

Previously, 3D printing has been demonstrated to be an effective approach in periodontal tissue engineering due to its ability to manufacture scaffolds with precision. Polyphasic biomaterial scaffolds composed of three distinct compartments were developed using 3D printing to guide various periodontal ligament fiber orientations to mimic native periodontal attachment apparatus [85–89]. In addition, growth factors release from polymeric scaffolds can be tuned spatially and temporally to promote the optimal growth of cementum, PDL, and bone [90].

When it comes to clinical application, Rasperini et al. (2015) pioneered the first in human use of a 3D-printed bioresorbable polycaprolactone (PCL) scaffold adapted to the patient's periodontal defect in combination with human platelet growth-derived growth factor (rh-PDGF-BB) to stimulate periodontal regeneration. Although the long-term follow-up showed graft failure, this study contributed significantly toward the clinical translation of 3D printing for periodontal tissue engineering. The authors proposed areas of improvement including the use of fast resorbing material with highly porous structure, which may contribute to improved tissue ingrowth and vascularization [91].

The main drawback of 3D printing is that it only allows control over the external properties of the scaffolds, and macroarchitecture of the printer construct, but does not allow precise distribution of individual cells. More specifically, stem and progenitor cells may be seeded onto the scaffold but cannot penetrate the scaffold uniformly [75].

Although 3D bioprinting technology is not currently used clinically for periodontal regeneration, it offers several advantages worth investigating. 3D bioprinting allows deposition of single cells or multicellular constructs to precise locations and enables the use of a wide range of biomaterial and bioinks that can be functionalized with growth factors.

Recent progress has been made to utilize 3D bioprinting for periodontal tissue engineering. Notably, several bioinks were optimized for 3D bioprinting of constructs with PDLSCs including gelatin-methacryloyl (GelMA), GelMA/PEG, and sodium alginate (SA)/gelatin (Gel)/nano-hydroxyapatite (na-HA) to ensure cell viability, proliferation, and differentiation [92–94]. In addition, the influence of bioprinting parameters including photoinitiator concentration, UV exposure, pressure, and dispensing needle diameter were fine-tuned [93]. The next step in periodontal tissue engineering research would be to explore the use of 3D bioprinting to fabricate biomimetic polyphasic scaffolds with various cells deposited precisely into each compartment and stimulated with specific growth factors [95, 96].

10.5.3 Craniofacial and Maxillofacial Regeneration

10.5.3.1 Craniomaxillofacial Bone

Craniomaxillofacial bone defects are common and result from trauma, tumor resection, infection, or congenital malformation. In addition, alveolar bone resorption after tooth loss may result in atrophic maxillary and mandibular ridge and maxillary sinus pneumatization that require reconstructive surgery [97]. Regeneration of craniofacial bone defect is challenging due to the complexity of the anatomical structures, bone biomechanics, and microenvironment. 3D bioprinting has been used to generate heterogeneous tissue-engineered bone constructs with customized architecture, cellular composition, and growth factor incorporation [98, 99].

Currently, the implementation of personalized scaffolding technologies for craniofacial bone regeneration shows promise for clinical translation. With advances in 3D bioprinting to allow for fabrication of personalized biomaterial matrices functionalized with biologics or genes with precise and spatially controlled delivery of cells, patients with debilitating bone defects will benefit from this transformative technology. Additional preclinical animal studies and human clinical trials with long-term results are needed to ensure safety and efficacy of this technology for routine use in clinical practice [100].

10.5.3.2 Cartilage

Cartilaginous tissues in the craniofacial area primarily consist of the temporomandibular joint (TMJ) disc, the auricular cartilage, and the nasal cartilage [101]. 3D printing has been used to mimic the 3D architecture of these cartilages. Previous studies have used extrusion 3D printing to fabricate cell-laden hydrogels using various natural and synthetic polymers to encapsulate chondrocytes and MSCs capable of synthesizing native cartilaginous ECM [102].

Several biomaterials have been studied as bioink to regenerate cartilaginous tissue including GelMA, alginate, collagen, and PCL [103–106]. For instance, GelMA in combination with hyaluronic acid and co-deposition with thermoplastics such as PCL may allow engineered constructs to match native human cartilage mechanical and geometrical properties [103].

Although 3D printing technology is not currently used clinically, it has been applied to regenerate TMJ discs in animal models. Using a micro-precise spatiotemporal delivery system with heterogeneous fibrocartilaginous matrix and region-dependent viscoelastic properties, Taradfer et al. (2016) have demonstrated significant healing of perforated TMJ discs in a rabbit model [107, 108]. In addition, 3D printing and sacrificial layer technology were applied to regenerate both the auricular cartilage and adipose tissue using PCL and cell-laden hydrogel. This study showcases that the aforementioned technique can be used to regenerate tissues and organs with complex morphology and multiple types of cells in addition to enhancing cartilage growth with chondrocyte adipose-derived stem cell co-culture [70, 109].

Several key challenges remain in the field of cartilage regeneration, which may be addressed using 3D bioprinting. Future research aimed at mimicking structural and biomechanical properties of cartilage combined with precise deposition of bioink, and cells will enhance integration of native cartilage to the tissue engineered cartilage.

10.5.3.3 Salivary Gland

Salivary gland hypofunction with subjective xerostomia is a clinical condition caused by radiotherapy for head and neck cancers and other systemic conditions such as Sjogren's syndrome. Consequently, saliva output is greatly reduced putting patients at risk of rampant dental caries, impaired speech, mastication, and swallowing. Despite various therapeutic strategies to repair and regeneration salivary glands and regain salivary flow, this remains an unmet clinical need. Recently, 3D bioprinting has been used to fabricate an innervated salivary gland (SG) like organoid from hDPSC and implanted into an ex vivo model. After implantation, the SG-like organoid significantly stimulated epithelial and neuronal growth in the damaged SG. This is an important step toward the regeneration of salivary gland to treat patients with radiotherapy-induced and Sjogren syndrome-induced xerostomia [25].

10.5.3.4 Nerve

Peripheral facial nerve injuries lead to dysfunction of facial muscles, impaired sensation, and painful neuropathies. Reconstruction of these nerve defects has been commonly performed using autologous nerve graft, which may be hindered by donor site morbidity and limited availability of donor nerves [110]. Recently, a novel scaffold-free 3D bioprinting approach was successfully used to fabricate nerve constructs by using GMSC spheroids, which were implanted and promoted the repair and regeneration of rat facial nerve defects [26]. This is a promising step toward using an easily accessible, minimally invasive source of stem cells that can be used in conjunction with 3D bioprinting to address the increasing clinical demand for nerve repair and regeneration.

10.6 Limitations and Areas of Research

Despite considerable advances in the recent years, the field of 3D bioprinting remains in the early stages of development. Most studies have been performed in vitro followed by a limited amount of in vivo animal studies. Significant work remains before 3D bioprinting technology can be predictably applied to address unmet clinical needs in craniofacial regenerative medicine and enter the clinical arena.

Several key areas of improvement and future research are critical at the level of the bioprinters, bioinks, and cell sources to ensure the scalability and clinical application of 3D bioprinting (Table 10.3). First, faster printing speed must be achieved in order to manufacture tissues and organs of clinically relevant size in a time efficient manner. Second, printing resolution must be enhanced to better biomimic the native tissue microarchitecture, which promotes the functionality of the printed tissue. Third, the ability to predictably print microvasculature must be devel-

Table 10.3 Areas of future research for 3D bioprinting

Areas of research [6]	Focus and priorities [6]
Bioprinter technology	• Increase compatibility with physiologically and clinically relevant biomaterials and cells • Enhance printing resolution and speed • Scale up for commercial and industrial manufacturing
Biomaterials and bioink	• Enhance mechanical properties of materials to support tissue constructs of clinically relevant sizes • Development of smart and programmable materials to allow for spatiotemporal control
Cell sources	• Improve understanding of required cell types to mimic native heterogeneous tissue • Minimally invasive, reproducible, and viable cell sources • Enhance control over cell proliferation and differentiation with biologics and small molecules
Tissue vascularization	• Enhance resolution to print microvasculature that withstands physiological hydrostatic and osmotic pressures • Develop new methods to print vascular networks with structural integrity to allow surgical anastomosis
Tissue innervation	• Ability to print heterogeneous tissues with integrated innervation • Generate inducible innervation after transplantation with biologics signaling
Tissue maturation	• Create bioreactors that allow for rapid tissue maturation • Develop quality control assessment protocol preimplantation

oped in order to maintain high cell viability of printed tissues over a long period of time allowing the construct to be integrated in vivo. Finally, new generations of bioinks with tunable mechanical, rheological, and biological properties must be formulated in order to achieve a fine balance between tissue printability, structure, and function to support larger 3D printer organs for clinical use [6, 111].

However, significant progress must be made in preclinical animal studies and human clinical trials before widespread adoption to address unmet medical needs in the clinical arena. For 3D bioprinting to be approved by regulatory authorities (i.e., FDA), large animal preclinical studies demonstrating safety and efficacy combined with human clinical studies with long-term follow-up are required.

10.7 Future Perspectives and Summary

With rapid advances in 3D bioprinting, interdisciplinary collaboration between biologists, engineers, and clinicians is crucial to spearhead this powerful technology to overcome clinical challenges and resolve unmet clinical needs in craniofacial regeneration.

In summary, 3D bioprinting has the potential to limit the use of animals in drug discovery and testing; reduce the need to harvest autologous tissues to repair and regenerate craniofacial, oral, and dental defects; decrease the risk of rejection; and enhance the generation of artificial craniofacial tissues and organs such as salivary glands. With the emergence of novel techniques including 4D bioprinting using smart and programmable materials to guide tissue regeneration, the future of craniofacial regenerative medicine is promising for both patients and clinicians.

References

1. Nyberg EL, Farris AL, Hung BP, Dias M, Garcia JR, Dorafshar AH, et al. 3D-printing technologies for craniofacial rehabilitation, reconstruction, and regeneration. Ann Biomed Eng. 2017;45(1):45–57.
2. Obregon F, Vaquette C, Ivanovski S, Hutmacher DW, Bertassoni LE. Three-dimensional bioprinting for regenerative dentistry and craniofacial tissue engineering. J Dent Res. 2015;94(9_Suppl):143S–52S.
3. Dwivedi R, Mehrotra D. 3D bioprinting and craniofacial regeneration. J Oral Biol Craniofac Res. 2020;10(4):650–9.
4. Mustoe TA, Corral CJ. Soft tissue reconstructive choices for craniofacial reconstruction. Clin Plast Surg. 1995;22(3):543–54.
5. Urken ML, Weinberg H, Buchbinder D, Moscoso JF, Lawson W, Catalano PJ, et al. Microvascular free flaps in head and neck reconstruction: report of 200 cases and review of complications. Arch Otolaryngol Head Neck Surg. 1994;120(6):633–40.

6. Murphy SV, Atala A. 3D bioprinting of tissues and organs. Nat Biotechnol. 2014;32(8):773–85.
7. Shafiee A, Atala A. Printing technologies for medical applications. Trends Mol Med. 2016;22(3):254–65.
8. Skardal A, Atala A. Biomaterials for integration with 3-D bioprinting. Ann Biomed Eng. 2015;43(3):730–46.
9. Phillippi JA, Miller E, Weiss L, Huard J, Waggoner A, Campbell P. Microenvironments engineered by inkjet bioprinting spatially direct adult stem cells toward muscle- and bone-like subpopulations. Stem Cells. 2008;26(1):127–34.
10. Langer R, Vacanti JP. Tissue engineering. Science. 1993;260(5110):920–6.
11. Kačarević ŽP, Rider PM, Alkildani S, Retnasingh S, Smeets R, Jung O, et al. An introduction to 3D bioprinting: possibilities, challenges and future aspects. Materials. 2018;11(11):2199.
12. Pirraco RP, Marques AP, Reis RL. Cell interactions in bone tissue engineering. J Cell Mol Med. 2010;14(1–2):93–102.
13. Fishero B, Kohli N, Das A, Christophel J, Cui Q. Current concepts of bone tissue engineering for craniofacial bone defect repair. Craniomaxillofac Trauma Reconstr. 2015;8(1):23–30.
14. Mao JJ, Giannobile WV, Helms JA, Hollister SJ, Krebsbach PH, Longaker MT, et al. Craniofacial tissue engineering by stem cells. J Dent Res. 2006;85(11):966–79.
15. Abou Neel EA, Chrzanowski W, Salih VM, Kim H-W, Knowles JC. Tissue engineering in dentistry. J Dent. 2014;42(8):915–28.
16. Warren SM, Fong KD, Chen CM, Loboa EG, Cowan CM, Lorenz HP, et al. Tools and techniques for craniofacial tissue engineering. Tissue Eng. 2003;9(2):187–200.
17. Huang GT-J, Gronthos S, Shi S. Mesenchymal stem cells derived from dental tissues vs. those from other sources: their biology and role in regenerative medicine. J Dent Res. 2009;88(9):792–806.
18. Gronthos S, Mankani M, Brahim J, Robey PG, Shi S. Postnatal human dental pulp stem cells (DPSCs) in vitro and in vivo. Proc Natl Acad Sci. 2000;97(25):13625–30.
19. Gronthos S, Brahim J, Li W, Fisher LW, Cherman N, Boyde A, et al. Stem cell properties of human dental pulp stem cells. J Dent Res. 2002;81(8):531–5.
20. Shi S, Gronthos S. Perivascular niche of postnatal mesenchymal stem cells in human bone marrow and dental pulp. J Bone Miner Res. 2003;18(4):696–704.
21. Miura M, Gronthos S, Zhao M, Lu B, Fisher LW, Robey PG, et al. SHED: stem cells from human exfoliated deciduous teeth. Proc Natl Acad Sci. 2003;100(10):5807–12.
22. Seo B-M, Miura M, Gronthos S, Mark Bartold P, Batouli S, Brahim J, et al. Investigation of multipotent postnatal stem cells from human periodontal ligament. Lancet. 2004;364(9429):149–55.
23. Sonoyama W, Liu Y, Yamaza T, Tuan RS, Wang S, Shi S, et al. Characterization of the apical papilla and its residing stem cells from human immature permanent teeth: a pilot study. J Endod. 2008;34(2):166–71.
24. Emmerson E, Knox SM. Salivary gland stem cells: a review of development, regeneration and cancer. Genesis. 2018;56(5):e23211.
25. Adine C, Ng KK, Rungarunlert S, Souza GR, Ferreira JN. Engineering innervated secretory epithelial organoids by magnetic three-dimensional bioprinting for stimulating epithelial growth in salivary glands. Biomaterials. 2018;180:52–66.
26. Zhang Q, Nguyen PD, Shi S, Burrell JC, Cullen DK, Le AD. 3D bio-printed scaffold-free nerve constructs with human gingiva-derived mesenchymal stem cells promote rat facial nerve regeneration. Sci Rep. 2018;8(1):6634.
27. Heo DN, Hospodiuk M, Ozbolat IT. Synergistic interplay between human MSCs and HUVECs in 3D spheroids laden in collagen/fibrin hydrogels for bone tissue engineering. Acta Biomater. 2019;95:348–56.
28. Zhuang P, Sun AX, An J, Chua CK, Chew SY. 3D neural tissue models: from spheroids to bioprinting. Biomaterials. 2018;154:113–33.
29. Zanoni M, Cortesi M, Zamagni A, Arienti C, Pignatta S, Tesei A. Modeling neoplastic disease with spheroids and organoids. J Hematol Oncol. 2020;13(1):97.
30. Gungor-Ozkerim PS, Inci I, Zhang YS, Khademhosseini A, Dokmeci MR. Bioinks for 3D bioprinting: an overview. Biomater Sci. 2018;6(5):915–46.
31. Adhikari J, Roy A, Das A, Ghosh M, Thomas S, Sinha A, et al. Effects of processing parameters of 3D bioprinting on the cellular activity of bioinks. Macromol Biosci. 2021;21(1):2000179.
32. Tevlin R, McArdle A, Atashroo D, Walmsley GG, Senarath-Yapa K, Zielins ER, et al. Biomaterials for craniofacial bone engineering. J Dent Res. 2014;93(12):1187–95.
33. Datta P, Ozbolat V, Ayan B, Dhawan A, Ozbolat IT. Bone tissue bioprinting for craniofacial reconstruction. Biotechnol Bioeng. 2017;114(11):2424–31.
34. O'Brien FJ. Biomaterials & scaffolds for tissue engineering. Mater Today. 2011;14(3):88–95.
35. Thrivikraman G, Athirasala A, Twohig C, Boda SK, Bertassoni LE. Biomaterials for craniofacial bone regeneration. Dent Clin N Am. 2017;61(4):835–56.
36. Maroulakos M, Kamperos G, Tayebi L, Halazonetis D, Ren Y. Applications of 3D printing on craniofacial bone repair: a systematic review. J Dent. 2019;80:1–14.
37. Christopher B, Patrick L. Biopolymers. In: Standard handbook of biomedical engineering & design. New York: McGraw-Hill Education; 2003.
38. Qu H, Fu H, Han Z, Sun Y. Biomaterials for bone tissue engineering scaffolds: a review. RSC Adv. 2019;9:11.
39. Visscher DO, Farré-Guasch E, Helder MN, Gibbs S, Forouzanfar T, van Zuijlen PP, et al. Advances in bioprinting technologies for craniofacial reconstruction. Trends Biotechnol. 2016;34(9):700–10.

40. Juhasz JA, Best SM. Bioactive ceramics: processing, structures and properties. J Mater Sci. 2012;47(2):610–24.
41. Fahmy MD, Jazayeri HE, Razavi M, Masri R, Tayebi L. Three-dimensional bioprinting materials with potential application in preprosthetic surgery. J Prosthodont. 2016;25(4):310–8.
42. Wen Y, Xun S, Haoye M, Baichuan S, Peng C, Xuejian L, et al. 3D printed porous ceramic scaffolds for bone tissue engineering: a review. Biomater Sci. 2017;5(9):1690–8.
43. Poologasundarampillai G, Nommeots-Nomm A. 3 - Materials for 3D printing in medicine: metals, polymers, ceramics, hydrogels. In: Kalaskar DM, editor. 3D printing in medicine. Sawston, UK: Woodhead Publishing; 2017. p. 43–71.
44. Bose S, Roy M, Bandyopadhyay A. Recent advances in bone tissue engineering scaffolds. Trends Biotechnol. 2012;30(10):546–54.
45. Yazdimamaghani M, Razavi M, Vashaee D, Moharamzadeh K, Boccaccini AR, Tayebi L. Porous magnesium-based scaffolds for tissue engineering. Mater Sci Eng C. 2017;71:1253–66.
46. Osidak EO, Kozhukhov VI, Osidak MS, Domogatsky SP. Collagen as bioink for bioprinting: a comprehensive review. Int J Bioprint. 2020;6(3):270.
47. Chen G, Lv Y. Decellularized bone matrix scaffold for bone regeneration. In: Turksen K, editor. Decellularized scaffolds and organogenesis: methods and protocols. New York, Springer; 2018. p. 239–54.
48. Heath DE. A review of decellularized extracellular matrix biomaterials for regenerative engineering applications. Regen Eng Transl Med. 2019;5(2):155–66.
49. Akter F. Chapter 2 - Principles of tissue engineering. In: Akter F, editor. Tissue engineering made easy. San Diego: Academic Press; 2016. p. 3–16.
50. Deepthi S, Venkatesan J, Kim S-K, Bumgardner JD, Jayakumar R. An overview of chitin or chitosan/nano ceramic composite scaffolds for bone tissue engineering. Int J Biol Macromol. 2016;93:1338–53.
51. Shams S, Silva EA. Chapter 4 - Bioengineering strategies for gene delivery. In: Fernandes TG, Diogo MM, Cabral JMS, editors. Engineering strategies for regenerative medicine. San Diego: Academic Press; 2020. p. 107–48.
52. Kim B-S, Baez CE, Atala A. Biomaterials for tissue engineering. World J Urol. 2000;18(1):2–9.
53. Yang S, Leong K-F, Du Z, Chua C-K. The design of scaffolds for use in tissue engineering. Part I. Traditional factors. Tissue Eng. 2001;7(6):679–89.
54. Neumann A, Kevenhoerster K. Biomaterials for craniofacial reconstruction. GMS Curr Top Otorhinolaryngol Head Neck Surg. 2009;8:Doc08.
55. Tollemar V, Collier ZJ, Mohammed MK, Lee MJ, Ameer GA, Reid RR. Stem cells, growth factors and scaffolds in craniofacial regenerative medicine. Genes Dis. 2016;3(1):56–71.
56. Mobaraki M, Ghaffari M, Yazdanpanah A, Luo Y, Mills DK. Bioinks and bioprinting: a focused review. Bioprinting. 2020;18:e00080.
57. Chen F-M, Zhang M, Wu Z-F. Toward delivery of multiple growth factors in tissue engineering. Biomaterials. 2010;31(24):6279–308.
58. Bittner SM, Guo JL, Mikos AG. Spatiotemporal control of growth factors in three-dimensional printed scaffolds. Bioprinting. 2018;12:e00032.
59. Kuroda Y, Kawai T, Goto K, Matsuda S. Clinical application of injectable growth factor for bone regeneration: a systematic review. Inflamm Regen. 2019;39(1):20.
60. Nevins M, Giannobile WV, McGuire MK, Kao RT, Mellonig JT, Hinrichs JE, et al. Platelet-derived growth factor stimulates bone fill and rate of attachment level gain: results of a large multicenter randomized controlled trial. J Periodontol. 2005;76(12):2205–15.
61. Colciago A, Celotti F, Casati L, Giancola R, Castano SM, Antonini G, et al. In vitro effects of PDGF isoforms (AA, BB, AB and CC) on migration and proliferation of SaOS-2 osteoblasts and on migration of human osteoblasts. Int J Biomed Sci. 2009;5(4):380–9.
62. Shirazi SFS, Gharehkhani S, Mehrali M, Yarmand H, Metselaar HSC, Adib Kadri N, et al. A review on powder-based additive manufacturing for tissue engineering: selective laser sintering and inkjet 3D printing. Sci Technol Adv Mater. 2015;16(3):033502.
63. Rider P, Kačarević ŽP, Alkildani S, Retnasingh S, Barbeck M. Bioprinting of tissue engineering scaffolds. J Tissue Eng. 2018;9:204173141880209.
64. Jeong H-J, Nam H, Jang J, Lee S-J. 3D bioprinting strategies for the regeneration of functional tubular tissues and organs. Bioengineering. 2020;7(2):32.
65. Zheng Z, Eglin D, Alini M, Richards GR, Qin L, Lai Y. Visible light-induced 3D bioprinting technologies and corresponding bioink materials for tissue engineering: a review. Engineering. 2020;7:966–78.
66. Bishop ES, Mostafa S, Pakvasa M, Luu HH, Lee MJ, Wolf JM, et al. 3-D bioprinting technologies in tissue engineering and regenerative medicine: current and future trends. Genes Dis. 2017;4(4):185–95.
67. Kumar H, Kim K. Stereolithography 3D bioprinting. Methods Mol Biol. 2020;2140:93–108.
68. Guillotin B, Souquet A, Catros S, Duocastella M, Pippenger B, Bellance S, et al. Laser assisted bioprinting of engineered tissue with high cell density and microscale organization. Biomaterials. 2010;31(28):7250–6.
69. Li J, Chen M, Fan X, Zhou H. Recent advances in bioprinting techniques: approaches, applications and future prospects. J Transl Med. 2016;14(1):271.
70. Lee J-S, Hong JM, Jung JW, Shim J-H, Oh J-H, Cho D-W. 3D printing of composite tissue with complex shape applied to ear regeneration. Biofabrication. 2014;6(2):024103.

71. Tappa K, Jammalamadaka U. Novel biomaterials used in medical 3D printing techniques. J Funct Biomater. 2018;9(1):17.
72. Crook JM, editor. 3D bioprinting: principles and protocols, Methods in molecular biology, vol. 2140. New York: Springer; 2020.
73. Jiang T, Munguia-Lopez JG, Flores-Torres S, Kort-Mascort J, Kinsella JM. Extrusion bioprinting of soft materials: an emerging technique for biological model fabrication. Appl Phys Rev. 2019;6(1):011310.
74. Bertassoni LE. Progress and challenges in microengineering the dental pulp vascular microenvironment. J Endod. 2020;46(9S):S90–100.
75. Ma Y, Xie L, Yang B, Tian W. Three-dimensional printing biotechnology for the regeneration of the tooth and tooth-supporting tissues. Biotechnol Bioeng. 2019;116(2):452–68.
76. Yang J, Yuan G, Chen Z. Pulp regeneration: current approaches and future challenges. Front Physiol. 2016;7:58.
77. Tao O, Wu DT, Pham HM, Pandey N, Tran SD. Nanomaterials in craniofacial tissue regeneration: a review. Appl Sci. 2019;9(2):317.
78. Khayat A, Monteiro N, Smith EE, Pagni S, Zhang W, Khademhosseini A, et al. GelMA-encapsulated hDPSCs and HUVECs for dental pulp regeneration. J Dent Res. 2017;96(2):192–9.
79. Yu H, Zhang X, Song W, Pan T, Wang H, Ning T, et al. Effects of 3-dimensional bioprinting alginate/gelatin hydrogel scaffold extract on proliferation and differentiation of human dental pulp stem cells. J Endod. 2019;45(6):706–15.
80. Athirasala A, Tahayeri A, Thrivikraman G, França CM, Monteiro N, Tran V, et al. A dentin-derived hydrogel bioink for 3D bioprinting of cell laden scaffolds for regenerative dentistry. Biofabrication. 2018;10(2):024101.
81. Park JH, Gillispie GJ, Copus JS, Zhang W, Atala A, Yoo JJ, et al. The effect of BMP-mimetic peptide tethering bioinks on the differentiation of dental pulp stem cells (DPSCs) in 3D bioprinted dental constructs. Biofabrication. 2020;12(3):035029.
82. Kim JH, Park CH, Perez RA, Lee HY, Jang JH, Lee HH, et al. Advanced biomatrix designs for regenerative therapy of periodontal tissues. J Dent Res. 2014;93(12):1203–11.
83. Ivanovski S, Vaquette C, Gronthos S, Hutmacher DW, Bartold PM. Multiphasic scaffolds for periodontal tissue engineering. J Dent Res. 2014;93(12):1212–21.
84. Liu J, Ruan J, Weir MD, Ren K, Schneider A, Wang P, et al. Periodontal bone-ligament-cementum regeneration via scaffolds and stem cells. Cell. 2019;8(6):537.
85. Park CH, Rios HF, Jin Q, Bland ME, Flanagan CL, Hollister SJ, et al. Biomimetic hybrid scaffolds for engineering human tooth-ligament interfaces. Biomaterials. 2010;31(23):5945–52.
86. Park CH, Rios HF, Jin Q, Sugai JV, Padial-Molina M, Taut AD, et al. Tissue engineering bone-ligament complexes using fiber-guiding scaffolds. Biomaterials. 2012;33(1):137–45.
87. Vaquette C, Fan W, Xiao Y, Hamlet S, Hutmacher DW, Ivanovski S. A biphasic scaffold design combined with cell sheet technology for simultaneous regeneration of alveolar bone/periodontal ligament complex. Biomaterials. 2012;33(22):5560–73.
88. Park CH, Rios HF, Taut AD, Padial-Molina M, Flanagan CL, Pilipchuk SP, et al. Image-based, fiber guiding scaffolds: a platform for regenerating tissue interfaces. Tissue Eng Part C Methods. 2014;20(7):533–42.
89. Park CH, Kim K-H, Lee Y-M, Giannobile WV, Seol Y-J. 3D printed, microgroove pattern-driven generation of oriented ligamentous architectures. Int J Mol Sci. 2017;18(9):1927.
90. Lee CH, Hajibandeh J, Suzuki T, Fan A, Shang P, Mao JJ. Three-dimensional printed multiphase scaffolds for regeneration of periodontium complex. Tissue Eng Part A. 2013;20(7–8):1342–51.
91. Rasperini G, Pilipchuk SP, Flanagan CL, Park CH, Pagni G, Hollister SJ, et al. 3D-printed bioresorbable scaffold for periodontal repair. J Dent Res. 2015;94(9 Suppl):153S–7S.
92. Ma Y, Ji Y, Huang G, Ling K, Zhang X, Xu F. Bioprinting 3D cell-laden hydrogel microarray for screening human periodontal ligament stem cell response to extracellular matrix. Biofabrication. 2015;7(4):044105.
93. Thattaruparambil Raveendran N, Vaquette C, Meinert C, Samuel Ipe D, Ivanovski S. Optimization of 3D bioprinting of periodontal ligament cells. Dent Mater. 2019;35(12):1683–94.
94. Tian Y, Liu M, Liu Y, Shi C, Wang Y, Liu T, et al. The performance of 3D bioscaffolding based on a human periodontal ligament stem cell printing technique. J Biomed Mater Res A. 2020; https://doi.org/10.1002/jbm.a.37114.
95. Vaquette C, Saifzadeh S, Farag A, Hutmacher DW, Ivanovski S. Periodontal tissue engineering with a multiphasic construct and cell sheets. J Dent Res. 2019;98(6):673–81.
96. Staples RJ, Ivanovski S, Vaquette C. Fibre guiding scaffolds for periodontal tissue engineering. J Periodontal Res. 2020;55(3):331–41.
97. Nguyen TT, Wu DT, Ramamoorthi M, Syrbu J, Tran SD. 17 - Scaffolds for maxillary sinus augmentation. In: Mozafari M, Sefat F, Atala A, editors. Handbook of tissue engineering scaffolds, vol. 1. Sawston, UK: Woodhead Publishing; 2019. p. 369–86.
98. Zhang L, Yang G, Johnson BN, Jia X. Three-dimensional (3D) printed scaffold and material selection for bone repair. Acta Biomater. 2019;84:16–33.
99. Genova T, Roato I, Carossa M, Motta C, Cavagnetto D, Mussano F. Advances on bone substitutes through 3D bioprinting. Int J Mol Sci. 2020;21(19):7012.

100. Yu N, Nguyen T, Cho YD, Kavanagh NM, Ghassib I, Giannobile WV. Personalized scaffolding technologies for alveolar bone regenerative medicine. Orthod Craniofac Res. 2019;22(Suppl 1):69–75.
101. Lin Y, Lin H, Ramamoorthi M, Wu DT, Zhang Z, Tran SD. 21 - Scaffolds for temporomandibular joint disc engineering. In: Mozafari M, Sefat F, Atala A, editors. Handbook of tissue engineering scaffolds, vol. 1. Sawston, UK: Woodhead Publishing; 2019. p. 437–55.
102. Tao O, Kort-Mascort J, Lin Y, Pham HM, Charbonneau AM, ElKashty OA, et al. The applications of 3D printing for craniofacial tissue engineering. Micromachines. 2019;10(7):480.
103. Schuurman W, Levett PA, Pot MW, van Weeren PR, Dhert WJA, Hutmacher DW, et al. Gelatin-methacrylamide hydrogels as potential biomaterials for fabrication of tissue-engineered cartilage constructs. Macromol Biosci. 2013;13(5):551–61.
104. Rhee S, Puetzer JL, Mason BN, Reinhart-King CA, Bonassar LJ. 3D bioprinting of spatially heterogeneous collagen constructs for cartilage tissue engineering. ACS Biomater Sci Eng. 2016;2(10):1800–5.
105. Park SH, Yun BG, Won JY, Yun WS, Shim JH, Lim MH, et al. New application of three-dimensional printing biomaterial in nasal reconstruction. Laryngoscope. 2017;127(5):1036–43.
106. Messaoudi O, Henrionnet C, Bourge K, Loeuille D, Gillet P, Pinzano A. Stem cells and extrusion 3D printing for hyaline cartilage engineering. Cell. 2020;10(1):2.
107. Tarafder S, Koch A, Jun Y, Chou C, Awadallah MR, Lee CH. Micro-precise spatiotemporal delivery system embedded in 3D printing for complex tissue regeneration. Biofabrication. 2016;8(2):025003.
108. Legemate K, Tarafder S, Jun Y, Lee CH. Engineering human TMJ discs with protein-releasing 3D-printed scaffolds. J Dent Res. 2016;95(7):800–7.
109. Morrison RJ, Nasser HB, Kashlan KN, Zopf DA, Milner DJ, Flanangan CL, et al. Co-culture of adipose-derived stem cells and chondrocytes on three-dimensionally printed bioscaffolds for craniofacial cartilage engineering. Laryngoscope. 2018;128(7):E251–7.
110. Volk GF, Pantel M, Guntinas-Lichius O. Modern concepts in facial nerve reconstruction. Head Face Med. 2010;6(1):25.
111. Yu J, Park SA, Kim WD, Ha T, Xin Y-Z, Lee J, et al. Current advances in 3D bioprinting technology and its applications for tissue engineering. Polymers (Basel). 2020;12(12):2958.

3D-Printed Metal Implants for Maxillofacial Restorations

11

Bhanupratap Gaur, Samrat Sagar, Suraj Naik, Nakul Parasharami, Adil Gandevivala, Srivalli Natrajan, Rupesh Ghyar, and Ravi Bhallamudi

Contents

11.1 Maxillofacial and Orbital Restoration

More than 200,000 maxillofacial and orbital reconstruction surgeries were performed globally in the year 2019 [1]. The increasing need for such surgeries is driven by complex traumatic injuries in high-velocity accidents that result in the mutilation of the facial and orbital skeleton. In addition, these surgeries are aesthetically demanding. Any implants used to meet these requirements must ensure good fitment and restore the original contour of the affected region. Moreover, they need to have superior quality and reliability. At the same time, they should be economical and can be made available within reasonable time. The current techniques for maxillofacial and orbital restorations are briefly discussed here.

11.1.1 Introduction

The use of bone graft continues to be the most common method for handling complex maxillofacial defects. Based on the source, graft materials are classified as autografts, allografts, and xenografts. These serve as scaffolds for hard tissue regeneration and ensure good quality, size, and bone shape after healing.

B. Gaur · S. Sagar · S. Naik · R. Ghyar
R. Bhallamudi (✉)
Mechanical Engineering Department, Indian Institute of Technology Bombay, Mumbai, India
e-mail: bhanupratap.gaur@iitb.ac.in; rupesh.ghyar@betic.org; b.ravi@iitb.ac.in

N. Parasharami
Lokmanya Hospital for Special Surgery, Pune, India

A. Gandevivala · S. Natrajan
Mahatma Gandhi Mission Dental College and Hospital, Navi Mumbai, India

© The Author(s), under exclusive license to Springer Nature Switzerland AG 2022
P. K. Chaudhari et al. (eds.), *3D Printing in Oral Health Science*,
https://doi.org/10.1007/978-3-031-07369-4_11

Autografts are considered the "gold standard" due to the absence of adverse immunological response [2]. The bone of the desired size can be harvested from a donor site and grafted at the recipient site after manually after reshaping the same. It provides support, rigidity, and stability at the grafted site, promoting new bone formation and uneventful wound healing. The primary graft harvesting sites of maxillofacial region are chin, lateral ramus, maxillary tuberosity, and anterior palate [3]; other common sites include fibula, iliac crest, calvarial, and rib [4].

The main drawbacks of bone grafting are additional iatrogenic trauma to the patient and associated donor site morbidity. Mismatch of bone porosity and strength with that at the grafting site also influences the outcome during and after the reconstruction surgery [5]. Bony substitutes from other individuals of the same species (allograft) and other species (xenograft) have biocompatibility issues, risk of cross-infection, and ultimately rejection. To overcome these limitations, synthetic substitutes with alloplastic materials have increasingly found a place in bony reconstructions.

The alternative approach to bone grafting is using standard metal plates of different shapes, sizes, and thicknesses. These usually need to be cut, bent, and twisted to conform to the bone's topology at the implantation site; however, perfect conformity is rarely achieved [6]. Many a time these implants are weakened due to residual stresses, and its fatigue life is usually compromised due to bending and unbending a few times during or before surgery (Fig. 11.1a, b). Further, the hospitals need to maintain an inventory of such implant system to match the diverse requirements of patients.

These limitations can be overcome by leveraging the latest advancements in medical imaging, computer-aided design, biocompatible materials, and additive manufacturing technologies to create patient-customized implants. Since these implant systems are customized as per the need of the patient, there are fewer chances of post-surgery loosening, breaking, exposure, or infection [7, 8].

11.1.2 Maxillofacial Restoration

Maxillofacial bone restorations using bone grafts or implants help to regain the original structure, function, and aesthetics of the defected facial skeleton. The defects can be caused by infections, tumors, wounds, congenital deformities, or acquired defects [9].

The degree of comminution, displacement, and amount of bone loss determines the type of surgery. Maxillofacial fractures are classified as Le Fort, naso-orbito-ethmoid, zygomaticomaxillary complex, mandibular, frontal sinus, alveolar process, and nasal bone fractures [10]. Bone tumors in this region can be malignant or benign maxillofacial bone tumors, odontogenic tumors, odontogenic cysts, fibro-osseous and osteochondromatous lesions, giant bone lesions, simple bone cysts, and hematolymphoid tumors [11]. Congenital deformities of the maxillofacial region include craniofacial stenosis, facial clefts, as well as Treacher Collins, oto-mandibular, and craniofacial stenosis. Growth-related deformations include prognathism, hemimandibular hyperplasias, and maxillary hypoplasia [12]. These conditions might arise due to maxillofacial trauma, craniofacial pathologies, and congenital malformation of the head and neck region.

During the surgery, the bone grafts or implants are placed and positioned at the defect and held in place by reconstruction plates anchored by screws onto the residual native bone on either side of the defect (Fig. 11.1c, d). A bone plate used for restoration must have adequate length, width, and thickness. Ideally it should be symmetrically loaded from both sides with the fracture/defect at the center. Their contour should match the topology of bone and must neutralize bending, shearing, twisting, and compressive loads on the bone during static and dynamic stages of function [13].

11.1.3 Orbital Reconstruction and Restoration

Orbital fractures result from physical impacts, primarily due to road accidents and sports. The primary objective of orbital bone restoration is to

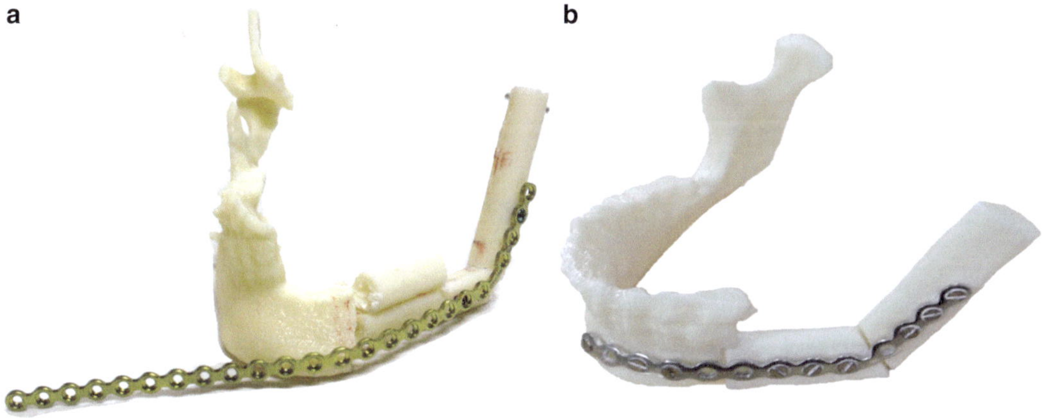

Fig. 11.1 (**a**) Mandibular plate being bent on a double barrel fibula graft model, (**b**) mandibular plate made conformal to single barrel fibula graft model by manually bending operations

keep the eye globe at the right place and restore the orbital volume necessary for the proper functioning of the eye [14].

Orbital bone defects are usually treated using meshes and plates. The plates are primarily used to fix broken bone pieces and meshes to restore the orbital floor and volume. Titanium orbital meshes are available as stock implants and preformed meshes with one, two, and three wall extensions. The bone plates are available in various thicknesses, shapes, and sizes. The selection of appropriate bone plate of mesh is based on the type and location of the defect, then cut, bend, and twist them to match the topology during the preoperative planning and/or during actual surgery [15].

11.1.4 Current Challenges

There are several challenges in maxillofacial and orbital reconstruction surgeries depending on the implants and the surgical methods along with the longevity of restorations. Implant-related issues rely on the material, design, manufacturing, and usability of the implants under consideration.

The anatomical complexity of the maxillofacial and orbital region makes it difficult to restore the original aesthetics and functionality. Any displacement and resorption of the implants can cause infection, which may require revision surgeries.

As mentioned earlier, 3D-printed patient-customized implants can overcome the above challenge. However, they require a high level of expertise and coordination between medical and engineering teams [16]. Compared to standard implants, 3D-printed implants are associated with longer lead time and higher costs [17]. It is not easy to capture the fine details required for designing these implants from computed tomography (CT), scans, which can cause dimensional inaccuracy in the manufactured implant [18]. This can potentially lead to improper restoration and volumetric mismatch, resulting in impaired vision [19]. Most of these implants are metallic, causing artifacts in CT imaging required for follow-up [20]. Some of these challenges can be addressed by evolving a scientific approach and the best practices for implant design and manufacturing, which are described next sections.

11.2 Design of Patient-Customized Maxillofacial Implants

Customized implants are designed based on the medical image data of a particular patient. First, the image data is segmented to obtain the region of interest (ROI) and is converted into a 3D CAD model using suitable medical modeling software. The model is used for designing the implant, keeping in mind manufacturability and other considerations. This is followed by design verification and validation before manufacturing. The key steps involved are described here.

11.2.1 Maxillofacial Imaging and Modeling

Medical images are commonly acquired using CT, since it is more suitable for bone than magnetic resonance imaging (MRI) and saved in DICOM (Digital Imaging and Communications in Medicine) format. The quality of CT images depends on the detector configuration, tube current, tube potential, reconstruction algorithm, patient positioning, scan range, slice thickness, and pitch [21].

For imaging maxillofacial and orbital regions, special expertise is required considering thin walls and underlying air spaces. In addition, it involves a trade-off between image quality and radiation dose to patients [22]. The latter can cause radiation-induced tissue damage such as skin erythema, baldness, and even tissue death; and stochastic effects such as mutations lead to cancer.

The DICOM files are 2D images (pixels), which are converted into 3D images (voxels) using suitable medical modeling software. The region of interest (ROI) is obtained by segmenting out unwanted regions and using region-growing algorithms. The thresholding uses the Hounsfield value of the type of tissue under consideration [23]. Each boundary pixel is examined during region growing; if no edges are detected, the pixel is added to the central region [24]. All unwanted anatomical features that are not in contact with the ROI are removed by creating a mask. The masks can be edited to join unattached regions to ROI and separate regions that are not required.

The 3D voxel model is converted into stereolithography (STL) file and transferred to medical CAD software. The STL file is first repaired (such as closing open faces); then the number of triangles is optimized (to reduce the file size), and the surface is smoothened (Fig. 11.2).

11.2.2 Implant Design Considerations

The implant is expected to provide the desired function, comfort, durability, and aesthetics. The relevant design considerations, which are based on mechanical, biological, and biomaterial factors and manufacturability, are mentioned below.

- **Mechanical Factors**: Maxillofacial implants must have the adequate load-bearing capability and yet prevent stress shielding of the adjacent bone; this requires the Young's modulus of the implant to be close to that of the bone [25]. The type, magnitude, direction, and duration of loading should be considered while designing an implant [26]. Stretch-dominated failure is considered better than bend-dominated failures of implants [27]. Their design should minimize osteolysis and implant loosening [28]. Articulating load-bearing implants such as the temporomandibular joint should have enough fatigue strength to overcome cyclic forces due to masticating actions [29], and resistance to wear of rubbing surfaces between mandible condyle and mandibular fossa in the temporal bone [30]. The joint motion range should be similar to that of the natural joint and replicate all the degrees of freedom exhibiting by it [31].
- **Biological Factors**: An implant is expected to perform its intended function for a long time, ideally the patient's remaining life. It should be designed for osteointegration, which helps in preventing loosening, necrosis, and infection. The suitable coating enables this and surface treatments such as grit blasting, acid etching, and anodizing [32], and inducing porosity in design. Porous implants should have permeability [32], tortuosity [33], topology [34], and interconnectivity [35] to replicate the physical, biological, and structural properties of bone. These can be achieved by varying the relative density, pore size, strut size, and lattice unit structure of the implant [36].
- **Biomaterial Factors**: The implant material must be biocompatible—it should not be toxic

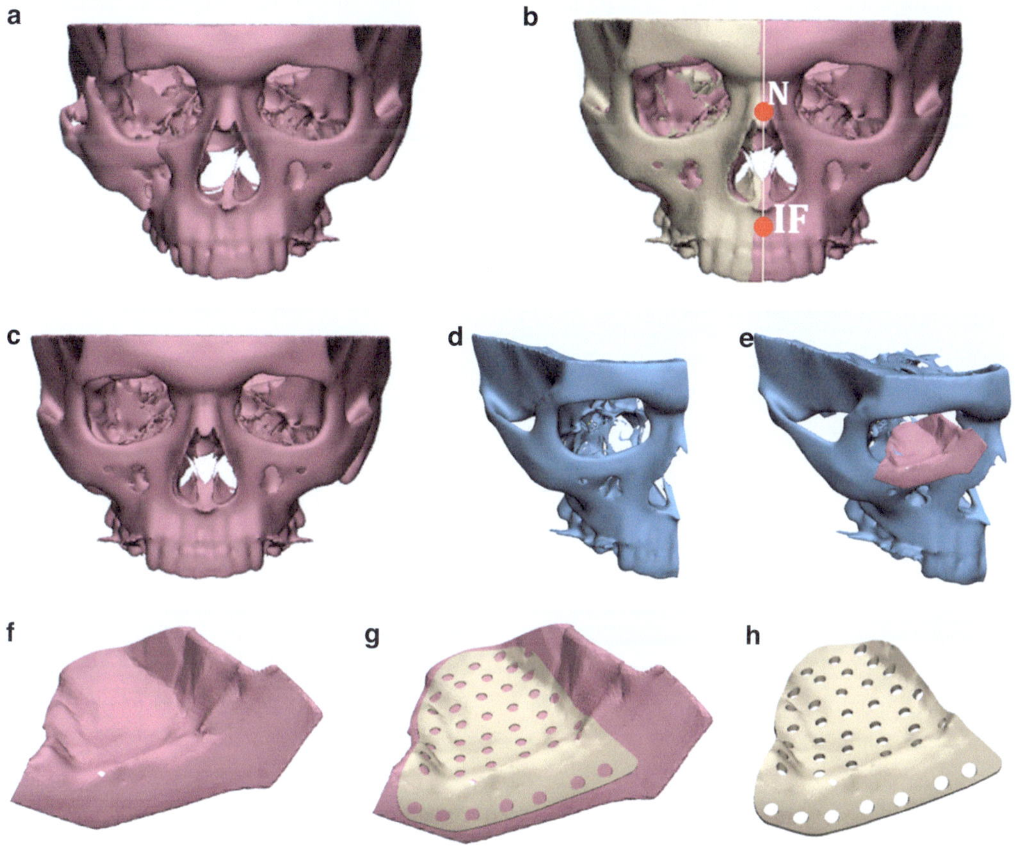

Fig. 11.2 (**a**) CAD model of defective anatomy with right orbital fracture, (**b**) left-side healthy orbit superimposed over the right orbit by mirroring about the midsagittal plane using cephalometric method, (**c**) left and right halves merged to obtain the original facial structure, (**d**) right orbit (region of interest), (**e**, **f**) extracted orbital floor on which implant is to be developed, (**g**) design of implant with geometry matching that of the extracted orbital floor, (**h**) final model of customized orbital implant

or cause any systemic effect while simultaneously getting readily accepted by the body and promoting osteointegration [37]. It has to be resistant to corrosion during sterilization and while inside the human body. Various types of biomimetic and biocompatible coatings such as hydroxyapatite [38], alumina, calcium phosphate [39], and titanium oxide [40], as well as surface texture alterations [41], enhance the above properties of implants. Implants can also be made of biodegradable materials that have gradual solubility [42]. While bone does not generally bond with metal, its attachment can be enhanced by inducing porosity into implants [43]. This can be easily achieved using additive manufacturing techniques.

- **Manufacturability Factors**: A well-designed implant is useless unless it is produced with the desired quality at an affordable cost. While additive manufacturing provides great freedom in terms of the external shape and internal topology, it also imposes certain constraints such as the need for support structures (for overhanging features) that have to be removed later. These can be avoided by designing overhanging portions with an inclination angle greater than 45° [44]. Porous implants need to have provision to remove unmelted powder particles from the pores. The design of orbital implants that have thin and delicate features must be compatible with the dimensional accuracy that the selected manufacturing process can achieve and enable post-processing operations such as heat treatment, bead blasting, grinding, and grinding polishing and machining [45]. Other design constraints (considering additive manufacturing) include overall size limits, surface quality, fabrication time, and total cost.

11.2.3 Design Verification and Validation

Design review coupled with verification and validation enables early identification and mitigation of potential flaws that could affect implant quality and reliability (Fig. 11.3). This is applicable and useful for standard implants, as per FDA. In 3D-printed customized implants, it is difficult to standardize the design process since the implant geometry varies from patient to patient. Hence, design verification and validation become even more important [46].

Design verification is a process of checking each stage (during the design process) whether the design output is compatible with input requirements within an acceptable range [46]. The design verification for 3D-printed implants can include the following steps.

- *Virtual Assembly*: The 3D CAD model of the implant is superimposed over the anatomical model of the patient to check the geometric fit (shape and size). Software tools such as 3-Matic (Materialise, Belgium) show the positive and negative deviation between the two models. The implant design is then corrected to minimize discrepancy [46]. Figure 11.4 illustrates this for a customized orbital implant.
- *Finite Element Analysis (FEA)*: It is a proven tool to reduce design cycle time and is performed using commercially available software such as Abaqus, Ansys, and Comsol. It involves preprocessing of implant model (mesh generation, imposing loads, boundary conditions, and material properties), computation of strains and stresses, and post-processing to view the results. The results can be improved by the coarse meshing of noncritical regions and finer meshing of critical areas [47]. The FEA has been successfully employed

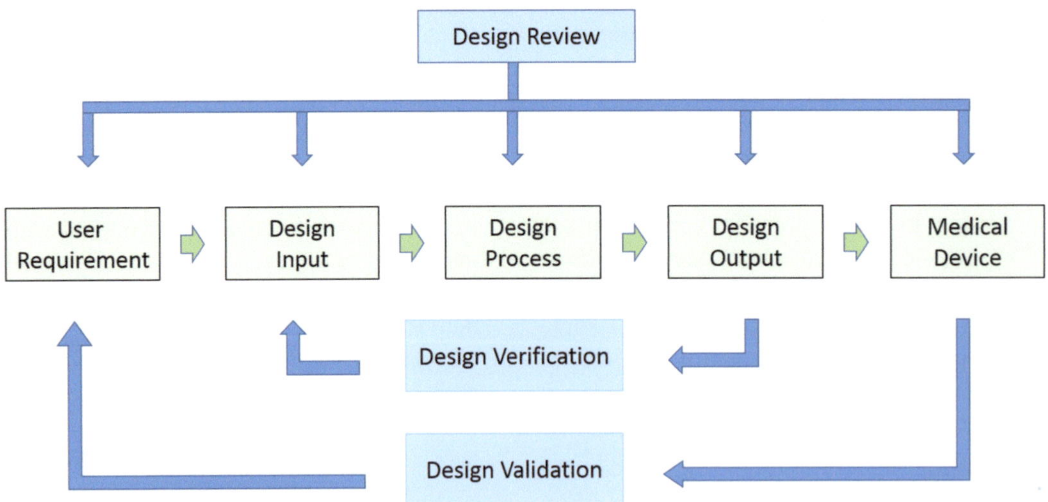

Fig. 11.3 Design control model for developing a medical device

to design customized mandibular implants under chewing conditions [48].

- *Rapid Prototyping (RP)*: This involves fabricating plastic prototypes of the implant and the anatomical model of the patient, using 3D printing techniques such as fused deposition modeling [47]. The models are assembled to verify the function and fit. They also aid clinicians in visualizing the defects and rehearsing the surgery. The implant design is revised, if needed, to achieve the desired results [48]. Figure 11.5a, b show the plastic orbital implant and assembly on the orbital model.

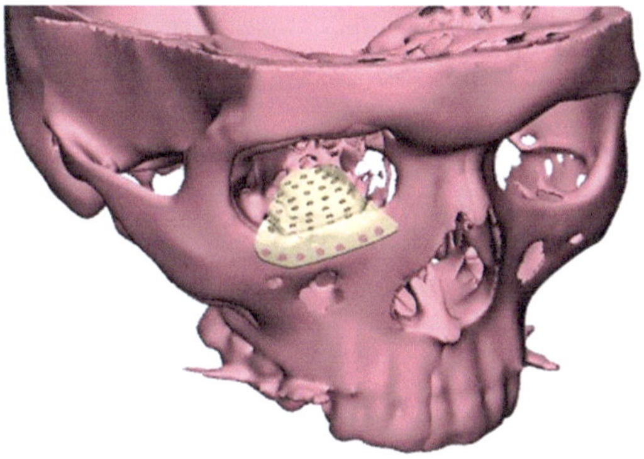

Fig. 11.4 Virtual assembly of an orbital implant on the anatomical model

For design validation, the manufactured implant is 3D scanned, and the scanned model is superimposed on the original CAD model to check that any deviations are within the acceptable range [49]. Further validation can include mechanical testing of the implant prototype and checking the postoperative CT images of the patient [47]. Figure 11.6 shows the postoperative CT images of orbital floor reconstruction surgery, which indicate good conformance of the implant with the orbital floor.

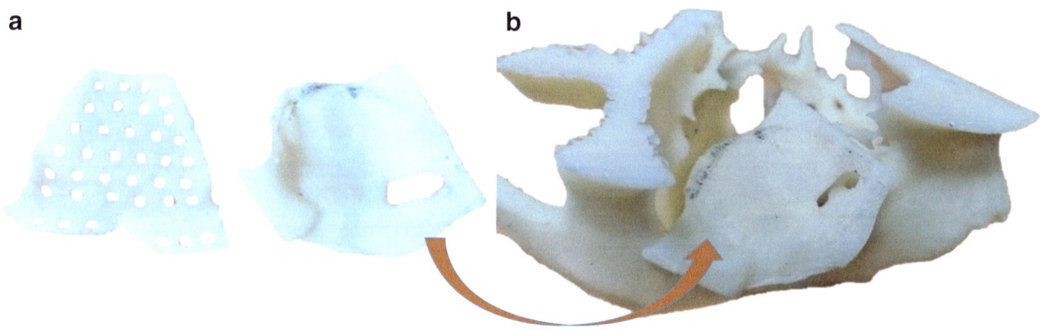

Fig. 11.5 (**a**) 3D-printed polymer orbital implants, (**b**) implant assembled on 3D-printed anatomical model

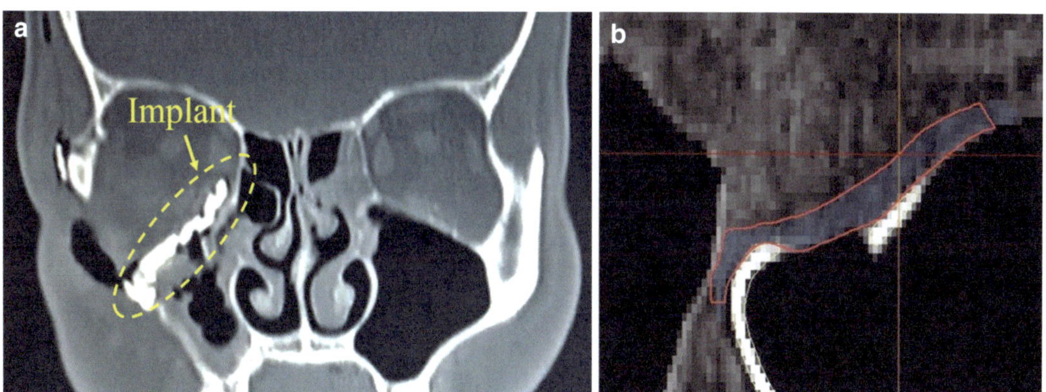

Fig. 11.6 Postoperative CT images showing the position of orbital implant in (**a**) coronal plane and (**b**) sagittal plane

11.3 Fabrication of Customized Implants

Customized maxillofacial and orbital implants having complex geometries can be fabricated in metallic materials using 3D printing techniques, followed by post-processing (such as machining and heat treatment) and quality checks (inspection and testing). The relevant materials, processes, and equipment are briefly described in this section.

11.3.1 Biocompatible Materials

Some of the biocompatible metallic materials are suitable for fabricating orthopedic implants. Stainless steel (S.S. 316L), cobalt-chromium (Co-Cr), commercially pure titanium (CP-Ti), and titanium-6 aluminum-4 vanadium extra low interstitial (Ti-6Al-4V ELI) are among the most widely used metallic materials for maxillofacial and orbital implants [50]. Table 11.1 compares these biocompatible metallic materials for use in maxillofacial and orbital implants [50–53].

Table 11.1 Comparison of S.S. 316L, Co-Cr, and Ti-6Al-4V ELI as biocompatible materials

Material	Advantages	Challenges
S.S. 316L	Oldest and most used Easily available High machinability Low cost	Medium biocompatibility High Young's modulus Stress shielding Medium corrosion and wear resistance
Co-Cr	High biocompatibility Load-bearing ability Corrosion and wear resistance	High Young's modulus Stress shielding Poor osteointegration
Ti-6Al-4V ELI	High biocompatibility Corrosion resistance Low Young's modulus (in metals) High osteointegration	Presence of aluminum and vanadium Poor machinability

11.3.2 Metal 3D Printing of Implants

Additive manufacturing, more popularly known as 3D printing, allows rapid fabrication of complex and organic shaped components in small order quantities, making it an attractive route for customized devices. For metal implants, powder bed 3D printing technologies such as selective laser sintering (SLS), selective laser melting (SLM), direct metal laser sintering (DMLS), and electron beam melting (EBM) can be used [54]. The energy source (laser or electron beam) melts a layer of metal powder up to a small depth (typically 20–100 μm). Once the layer solidifies, subsequent layers are spread and melted till the entire part is built up (Fig. 11.7). These technologies are briefly described here.

- **Laser-Based 3D Printing**: In these processes, the energy density influences the melt pool quality, which affects the mechanical and physical properties of the fabricated part [55]. The energy density is directly proportional to laser power; it is inversely proportional to laser velocity, laser diameter, hatch distance, and layer thickness. Pressurized inert gas flows over the melt pool and low substrate temperature (below 200 °C); the resulting martensite (α') makes the implants hard and brittle. Fast cooling rates, up to 10^6 K s^{-1} [56], also result in thermal stresses and warpage [57]. This makes it difficult to fabricate thin implants such as orbital floors, requiring optimal orientation and process parameters.
- **EBM-Based 3D Printing**: In these processes, energy density is directly proportional to the beam current and cathode voltage; it is inversely proportional to scanning velocity, spacing, and layer thickness [58]. The building chamber's preheating up to 700 °C causes slow cooling of the printed implants, resulting in alpha (α) and beta (β) microstructures, low residual stresses, and high elastic properties. The metal powder used for EBM is coarser compared to that used for laser-based technologies, leading to rougher surfaces [59].

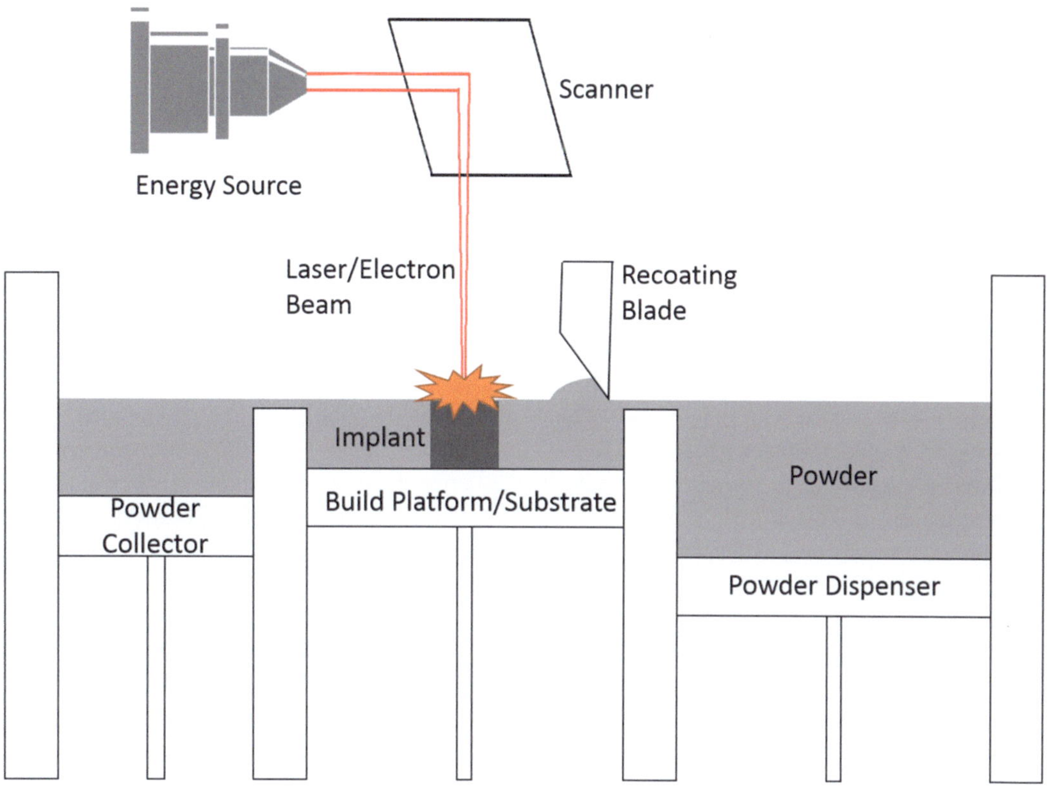

Fig. 11.7 Working principle of powder bed fusion technology

In general, the selection of a 3D printing technique can be made based on raw material to be used, physical form of the raw material, surface quality, accuracy of printing, cost, and lead time. The post-processing techniques and equipment also may need to be considered [60], described next.

11.3.3 Post-processing of Implants

The 3D-printed implants usually need to be processed further to achieve the desired surface roughness, density, hardness, strength, microstructure, etc., to meet the specific functional requirements of maxillofacial and orbital cases. These processes include the following.

- **Heat Treatment**: It is primarily meant for stress-relieving thermally stressed implants and annealing for microstructure improvement [61]. Implants made using laser-based technologies always have α' microstructure, which has to be converted to $\alpha + \beta$ form before using in any patient [62]. A vacuum furnace or a muffle furnace is employed for the heat treatment operations, in which the thermal cycles for stress relieving and annealing are determined as per the standards [63].
- **Support Removal**: Implants are removed from the substrate (on which they are built) using the wire EDM process. The remaining support structures on the implant are removed using bench vice, pliers, nibblers, hammer, chisel, files, etc. This must be done carefully, without scratching or damaging the implant.
- **Surface Smoothening**: The 3D-printed implants possess varying surface roughnesses, with side surfaces rougher than top ones [64]. To make all surfaces uniformly smooth, abrasive blasting, tumbling, and grinding operations are performed [45]. The abrasive

particles used operations must be biocompatible and non-reacting with the implant material.

- **Polishing/Buffing**: The implants that need to be very smooth (such as articulating surfaces in TMJ and hip joint) to prevent wear, metal debris, and ions that cause harmful effects to tissues [65] are polished and buffed to obtain the required surface finish.
- **Pickling and Anodizing**: The above operations result in surface contamination with shop floor grease, dust, inorganic and organic fluids, abrasive particles, and oxidation and corrosion. Pickling thoroughly cleans the surfaces by dissolving the impurities. This is immediately followed by anodizing to improve the aesthetics, make passive surfaces, redefine the oxide layer, or impart nanotubes on the surface [66]. It is also useful for color-coding of implants to indicate different sizes and thicknesses.
- **Cleaning and Packaging**: A multistage ultrasonic cleaner with distilled water, isopropyl alcohol, and other cleaning agents at elevated or room temperatures removes dirt from the outer surfaces of the implants and even from deep holes and cracks using cavitation action [67]. This is immediately followed by sterilizing and packaging in a 10,000-class cleanroom or equivalent as per standards and relevant guidelines.

11.3.4 Inspection and Testing

Inspection 3D-printed implants, being class III devices, pose a high risk to human safety. Hence their quality and reliability need to be assured through various inspection and testing protocols. Furthermore, since these implants are made one-off (customized for a single patient), nondestructive techniques (NDT) are employed for the purpose.

Inspection of 3D-printed implants includes visual, internal, and external testing and surface roughness tests. Some limitations are imposed due to part complexity, thin layer build, and high surface roughness. The roughness makes the fluorescent liquid penetrant test, electromagnetic test, and ultrasonic test very difficult. In addition, the textured surfaces are often mistakenly identified as cracks. This makes visual inspection subjective.

Due to the above limitations, implant manufacturers rely on digital radiography (DR) and computed tomography (CT) to inspect 3D-printed implants. These techniques are useful for gross defect detection and geometry validation. However, layer-level defects are difficult to detect due to the low resolution of both methods.

Another method to check the integrity of 3D-printed parts is Process Compensated Resonance Testing (PCRT). Its mass, stiffness, and geometry determine the resonant frequency of a part. In the presence of a structural defect, the paralysis of a part changes, changing its resonant frequencies. In some studies, PCRT has been used to detect unacceptable levels of cracks, porosity, lack of fusion, and suboptimal heat treatment in additively manufactured parts [68, 69]. In addition, it is a relatively fast process, requires no chemicals, and produces no waste.

Testing As suggested by US-FDA, performance tests conducted on 3D-printed medical devices must be the same as the devices made using a conventional manufacturing process. The testing may involve checking the material properties such as yield strength, ultimate strength, creep, fatigue, and abrasive wear. For orthopedic implants, fatigue life (the number of loading cycles until failure) and how the implant influences the adjacent tissue are also important [70].

In general, additive manufacturing produces anisotropy in part depending on the build direction and location within the build space. This should be considered in mechanical testing of the device since the properties will change based on the orientation and build location. For example, in powder bed fusion systems, the distance between the energy source and different locations within the build space (i.e., center vs. corner) can cause variation in the mechanical properties of devices built in those locations. A baseline study should be carried out to determine the effect of process parameters on mechanical testing. If found significant, the worst-case samples must be identified for mechanical testing.

Ideally, the implants should undergo performance testing. It is however difficult to replicate body forces on the implants considering their complex geometry. Since destructive testing of customized implants is not economically feasible, FDA suggests using test coupons or representative test samples of the device. An adequate justification should be provided as to how a coupon represents the final device in terms of critical design elements and post-printing processes. The test coupons or samples are also useful for process validation and identifying the worst-case conditions, including orientation and build location [71].

11.4 Case Studies

One maxillofacial and one orbital restoration case study are presented here to illustrate the design, fabrication, and testing procedures described earlier. The maxillofacial case study covers presurgery steps, and the orbital case study covers the implantation and post-surgery steps.

11.4.1 Maxillofacial Restoration

- **Background**: A female patient, aged 34 years, had deformed mandible bone and was suffering from a benign tumor for which she had undergone multiple surgeries. The CT scan at the time of the study showed that mandibular reconstruction had been carried out using an iliac crest graft. The CT scanning had a pixel size of 0.32 mm; the slice thickness as well as slice increment were 0.67 mm each. A customized implant was developed for this patient, as described here.
- **Segmenting**: The region of interest was obtained by segmenting the DICOM file of CT scanning, creating a 3D model of the region, and exporting as an STL file. For this purpose, a medical modeling software (Mimics 15.0, Materialise, Belgium) was employed, and the steps mentioned in Sect. 11.2.2 were followed (Fig. 11.8). As a result, the fractured segment of the reconstructed bone (iliac crest) on the left side and the healthy native mandible on the right side can be seen in the figure.
- **Surgery Planning and Implant Design**: The STL file was imported into a design software (3-matic 7.0, Materialise, Belgium) for virtual surgery planning (Fig. 11.9). The surgeon marked the boundary lines for mandibular resection. The healthy right side

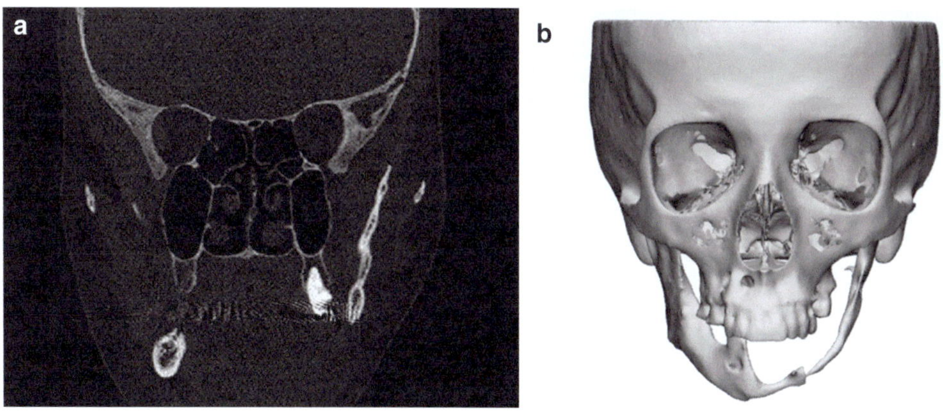

Fig. 11.8 (**a**) Coronal view of the CT scan, (**b**) 3D model of the segmented region of interest showing the defect in the left region of the mandible

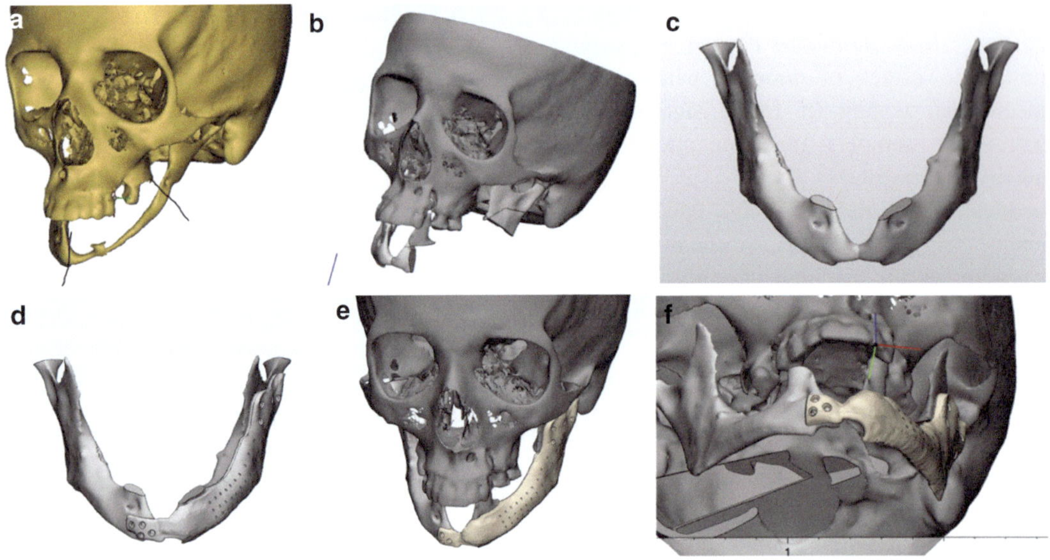

Fig. 11.9 (**a**) Markings by the surgeon to decide the resection area of the mandible, (**b**) region of interest (ROI) after the resection, (**c**) mirroring of the right side of the mandible on the left side, (**d**) conversion of the mirrored portion into the implant body, (**e**) ROI with virtual implant, (**f**) provision of screws for proper positioning and fixing the implant on the mandible

mandible was mirrored and superimposed on the diseased left side to achieve anatomical symmetry and aesthetics. The model thus generated acted as a framework to design a customized implant for mandible restoration. Screw holes were incorporated in the implant for fixing it to the mandible. The implant is expected to reduce operation time and associated complications.

- **Stress Analysis**: The CAD files of the mandible framework and the implant were imported into the FEA software Ansys Workbench 2020 (Ansys Inc, Pennsylvania, USA) for stress analysis (Fig. 11.10a). The models were assigned the cortical bone and Ti6Al4V ELI alloy properties, respectively (Table 11.2). For approximating the screw connection between bone and implant, the contacts between them were defined as bonded, and other boundary conditions were applied [48]. Table 11.3 shows the magnitude and direction of muscular forces (involving masseter, medial pterygoid, and temporalis) used in the analysis to simulate the chewing operation. Both condyles were fixed to simulate the biting conditions, and the molar movement was constrained in *Z*-direction, allowing displacement in *X* and *Y* directions (Fig. 11.10b). The FEA results (Fig. 11.10c) showed that the maximum von Mises stress (383 MPa) was in the implant region, but it was significantly less than the yield strength (790 MPa) of the implant material. Furthermore, the maximum strain value (0.0042) in the model was found to be away from the screw holes (Fig. 11.10d), indicating safe transfer of load through the screws, indicating a low probability of screw loosening and implant instability. A mesh convergence study was carried out on five element sizes (Fig. 11.11a), based on which a mesh size of 1.6 mm was selected to optimize the computational time and error (Fig. 11.11b).
- **3D Printing and Post-processing**: The implant was 3D-printed in Ti64-ELI alloy using DMLS technology (model M280, EOS GmbH, Germany), using the default printing parameters as described in Sect. 11.3.2. The as-built implant after removal from the build chamber is shown in Fig. 11.12a. Figure 11.12b

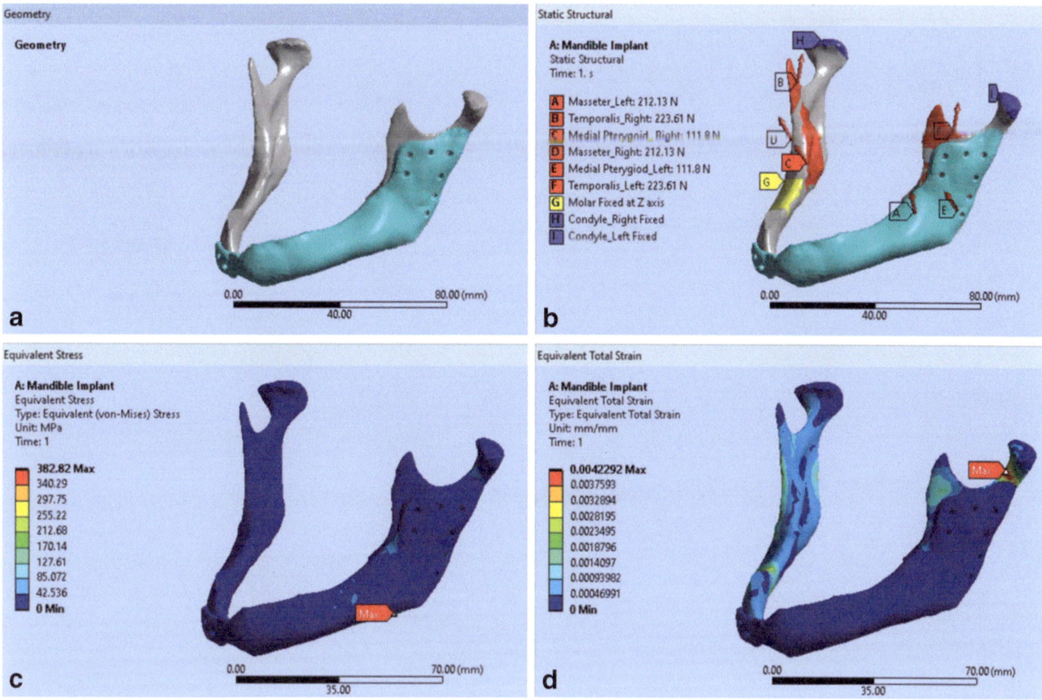

Fig. 11.10 (**a**) CAD files of mandible and implant imported in Ansys Workbence software, (**b**) boundary conditions and forces applied on the assembly, (**c**) stress distribution, (**d**) strain distribution

Table 11.2 Material properties used in FEA [48]

Material	Young's modulus (MPa)	Poisson's ratio	Yield strength (MPa)
Ti-6Al-4V ELI (Grade 23)	113,800	0.342	790
Cortical bone	13,700	0.3	122

Table 11.3 Magnitude and direction of muscle forces in Newton [48]

Muscle forces	*X*-direction	*Y*-direction	*Z*-direction
Masseter	50	−50	200
Temporalis	0	100	200
Medial pterygoid	0	−50	100

shows the implant after removing it from the build platform using wire EDM and manual support removal as described in Sect. 11.3.3. The implant was checked for fitment on a 3D-printed mandibular anatomical model (Fig. 11.12c) to validate the design. A similar model was also supplied to the surgeon for preoperative planning. Finishing operations included grinding, centrifugal tumbling, buffing and polishing, followed by ultrasonic cleaning, pickling, and anodizing the implant, as shown in Fig. 11.12d.

11.4.2 Orbital Restoration

- **Background**: This case involved a 34-year-old male patient with a complex orbital floor fracture on the right side. The CT scanning was carried out with a pixel size of 0.5 mm, and both slice thickness and slice increment had values of 0.625 mm. The medical model segmentation, virtual surgery planning, implant design, stress analysis, 3D printing, and post-processing steps were similar to the maxillofacial case study described earlier and are omitted here for brevity. However, the implantation and post-surgery follow-up are described here.
- **Surgery Planning and Implantation**: The surgeon used the 3D-printed anatomical model and orbital implant for preoperative

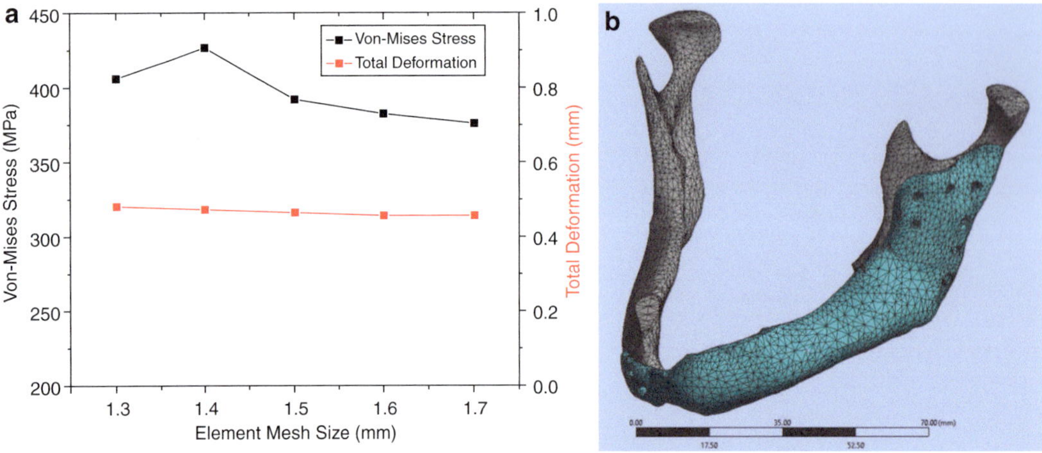

Fig. 11.11 (**a**) Mesh convergence study, (**b**) model meshing (11,857 nodes, 39,718 elements)

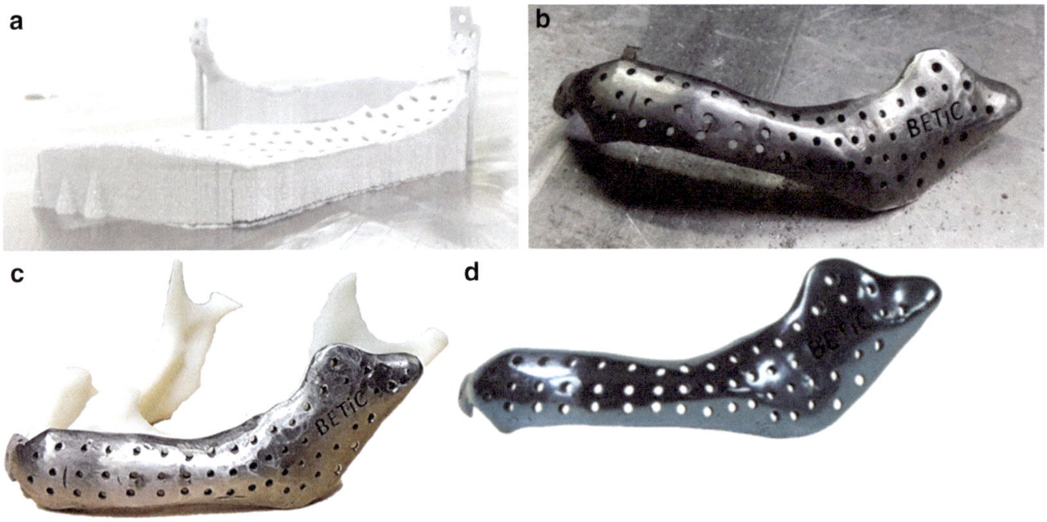

Fig. 11.12 (**a**) 3D-printed implant on the build plate along with support structure, (**b**) semifinished implant, (**c**) design validation of the implant, (**d**) final device ready for implantation

surgery planning as discussed in Sect. 11.2.3 and Fig. 11.5. The orbital implant was surgically placed into the patient's eye to support the eye globe and restore the orbital volume. The actual surgery was performed using a trans-conjunctival approach to reach the orbital floor. After exposure to the floor, the globe was gently retracted up to expose the fracture of the rim and defect in the floor. The orbital implant was then placed in the correct position as planned earlier and fastened to the rim using titanium screws (Fig. 11.13). It was also ensured that the implant edges were seated on a sound bone to avoid the prolapse of the implant into the underlying airspace and cantilevering. Finally, the operated area

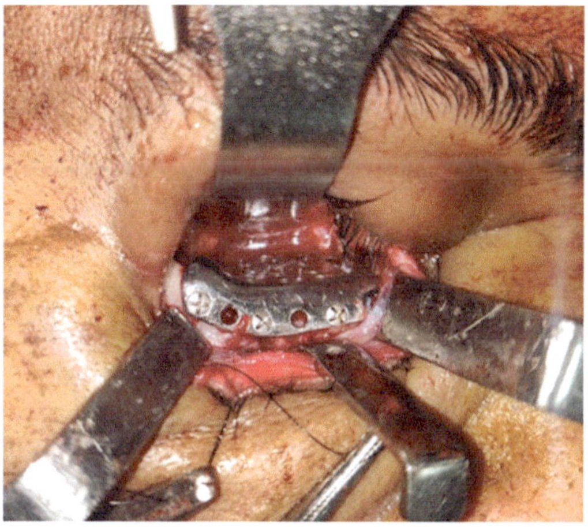

Fig. 11.13 Surgical implantation of the patient-customized orbital implant

was covered back with the soft tissues and stitched.

- **Post-surgery Follow-Up**: The restoration surgery was followed up to ascertain the outcome and track the condition of the patient. Figure 11.14 shows the pre- and post-surgery photographs of the patient. It is clear that the sunken eye globe was restored in terms of position and projection. A post-operative CT scan was performed to check the position of the implant and track the condition of the construct (Fig. 11.15). The images confirm the correct placement of the customized orbital implant at the intended location, aligning the affected eye with the normal one.

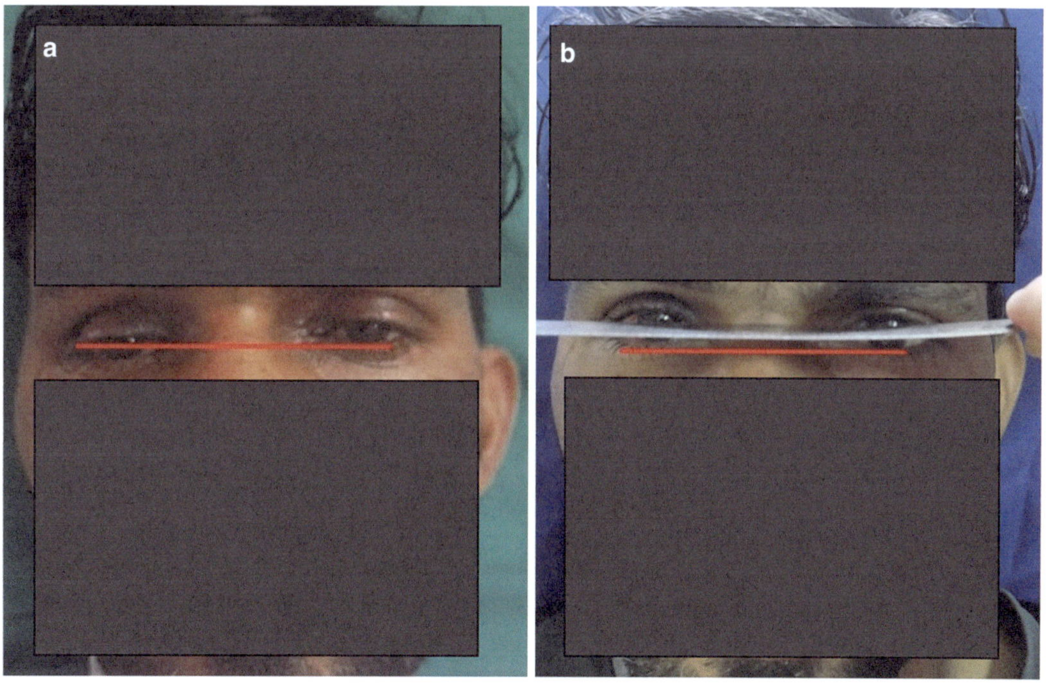

Fig. 11.14 Photographs of the patient, (**a**) presurgery with sunken eye globe and, (**b**) post-surgery with the eye globe raised back to the original level

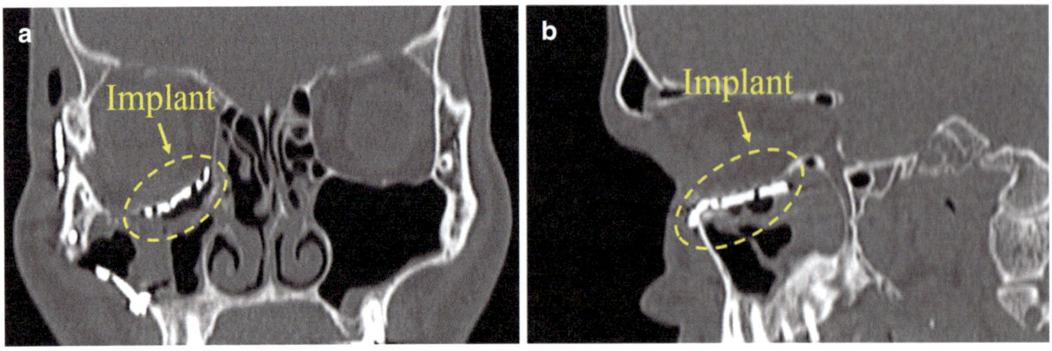

Fig. 11.15 Postoperative CT images showing the orbital implant position in (**a**) coronal plane and (**b**) sagittal plane for validation. The implant is conformal to the bone and is supporting the eye globe as required

11.5 Regulations and Future

All medical devices are regulated, certified, and licensed by the regulatory bodies of the respective countries. They allow developers to manufacture, import, export, store, and sell the devices in the local market. While most countries have their own regulatory agencies and procedures, some also recognize or follow the relevant guidelines laid out by US-FDA for medical devices [72]. The regulatory landscape in the context of customized 3D-printed maxillofacial and orbital implants is briefly described here, followed by new trends in this field.

11.5.1 Regulatory Landscape

As per US-FDA, medical devices are categorized as Class I (low risk), Class II (moderate risk), and Class III (high risk), based on the intended use and potential level of risk to the life of patients. Most implantable devices, including maxillofacial and orbital implants, come under Class III, requiring premarket approvals and stringent control by regulatory bodies [73].

The above requirements apply to mass-produced standard implants. While patient-specific customized implants are exempted from regulations, they are expected to comply with standard practices to ensure the safety of the patients. The standard practices include good manufacturing practices (GMP) along with quality management system (QMS), good clinical practices (GCP), and good laboratory practices (GLP).

The manufacture of customized 3D-printed orthopedic implants is required to be carried out in accordance with relevant international standards [62, 74–80]. Therefore, the device developers need to maintain a "medical device file" or dossier. It includes signed records of product description, intended use, market need assessment, ideas and proofs-of-concept, project feasibility assessment, prototypes and final product designs, safety and efficacy analysis, and risk analysis [81]. In addition, design verification and validation, design reviews, quality system assessment, technology transfer, etc. also need to be maintained so that they can be presented to the regulatory body as and when required [82].

The dossier should also contain the procedures, equipment, and results of mechanical, electrical (not required for passive implants), and biological tests. Mechanical testing ascertains the suitability of a medical device to bear all kinds of mechanical loads as per international standards. In the case of 3D-printed implants, the testing of its porosity, density, hardness, and microstructure ensures that it has properties close to the wrought material used in standard implants [83, 84]. Maxillofacial and orbital implants are Class III medical devices required to undergo biological evaluation through in vivo and in vitro tests. This ensures that the raw material, manufacturing process, and post-processing operations do not adversely affect the living cells.

To prove that an implant is free of side effects and is safe for the patients, human clinical trials are performed on a controlled number of volunteer patients after obtaining approvals from the institution ethics committee and regulatory bodies. These results are also recorded in the medical device dossier to support the reliability of the implant. In addition, a medical device quality management system as per ISO 13485 adds significant weight to the application submitted by the device developer to the region's regulatory body.

11.5.2 Future Trends

Clinicians are always looking for innovative medical devices with improved accuracy, efficacy, quality, reliability, usability, and affordability. For this purpose, they are increasingly collaborating with scientists and engineers, and such teams are leveraging the technologies and advancements in medical imaging, computer-aided design, additive manufacturing, virtual reality (VR), augmented reality (AR), tissue engineering, and regenerative medicine. Some of these developments and practical applications are already underway.

In particular, VR and AR are becoming popular for diagnosis, preoperative planning, surgery, and training through real-time navigation. They enable creating a virtual, augmented, haptic, and supportive environment for performing surgery in less time with improved outcome quality. Hospitals and clinical setups are adopting these technologies for dental implantology, orthognathic surgeries, orbital floor restorations (in case of blowout fractures), mandibular and maxillary restoration after tumor, and other bone deformities resections [85, 86]. The use of virtual surgical planning and 3D printing for patient-specific implants is becoming the gold standard in orbital reconstruction to restore pupillary level, orbital volume, aesthetics, and function. Intraoperative CT scans and real-time navigation enable precise surgical guides and implants, leading to better surgical outcomes. However, many of these technologies are still evolving and yet to reach the majority of hospitals, especially those in small towns [87]. Their affordability for patients in economically weaker sections of society is also a challenge that needs to be addressed.

Among the biomaterials available today for orthopedic implants, it is difficult to find one that fulfills all the biomechanical requirements of natural bone. Research trends in new biomaterials include 3D-printed metal-polymer-based porous maxillofacial implants having the same structure and mechanical properties as the original bone, capable of withstanding deglutition and masticatory stresses [87]. More biomaterials are also being continuously developed, although proving their biocompatibility is a long and arduous task.

In this context, tissue engineering and regenerative medicine hold much promise. These are driven by the development of new biomaterials coupled with new processing equipment to grow, heal, and replace damaged tissues. The new tissues are grown at the defective sites using scaffolds, cell culture (such as mesenchymal stem cells and bone marrow-derived mesenchymal stem cells), and growth factors, leading to bone regeneration similar to the parent bone at that site [88]. This is expected to significantly change the treatment protocols in the domain of maxillofacial and orbital restorations [89]. In the future, in situ 3D bioprinting technology may enable direct deposition of biomaterials at the implantation site to create the desired constructs [90]. This opens a wide door for research collaborations, innovative solutions, and clinical applications.

Acknowledgments The authors acknowledge the funding support from R.G. Science & Technology Commission, Mumbai, and the Department of Science & Technology, New Delhi, to establish the medical device development and medical metal printing facilities at Biomedical Engineering and Technology Innovation Centre (BETIC), IIT Bombay. The authors also wish to thank Mr. Mahesh Dhoka and his team at Incredible AM Pvt Ltd, Pune for manufacturing the orbital implant.

References

1. ASPS Public Relations. American Society of Plastic Surgeons. Plastic Surgery Statistics Report. 2020 [cited 2020 Jun 27]. p. 1–25. Available from: https://www.plasticsurgery.org/documents/News/Statistics/2019/plastic-surgery-statistics-full-report-2019.pdf.
2. Carlisle ER, Fischgrund JS. Bone graft and fusion enhancement. In: Surgical management of spinal deformities. Amsterdam: Elsevier Inc.; 2009. p. 433–48.
3. Khojasteh A, Esmaeelinejad M, Aghdashi F. Regenerative techniques in oral and maxillofacial bone grafting. In: Motamedi MHK, editor. A textbook of advanced oral and maxillofacial surgery, vol. 2. London: IntechOpen; 2015.
4. Mahsut DY. Bone graft types. In: Bone grafting - recent advances with special references to cranio-maxillofacial surgery. London: IntechOpen; 2018. p. 13.
5. Elsalanty ME, Genecov DG. Bone grafts in craniofacial surgery. Craniomaxillofac Trauma Reconstr. 2009;2(3–4):125–34.
6. Martola M, Lindqvist C, Hänninen H, Al-Sukhun J. Fracture of titanium plates used for mandibular reconstruction following ablative tumor surgery. J Biomed Mater Res Part B Appl Biomater. 2007;80B(2):345–52.
7. Hernandez Rosa J, Villanueva NL, Sanati-Mehrizy P, Factor SH, Taub PJ. Review of maxillofacial hardware complications and indications for salvage. Craniomaxillofac Trauma Reconstr. 2016;9(2):134–40.
8. Ahangar P, Cooke ME, Weber MH, Rosenzweig DH. Current biomedical applications of 3D printing and additive manufacturing. Appl Sci. 2019;9(8):1713.
9. Gupta DA, Verma DA, Islam DJ, Agarwal DS. Maxillofacial defects and their classification: a review. Int J Adv Res. 2016;4(6):109–14.
10. Gómez Roselló E, Quiles Granado AM, Artajona Garcia M, Juanpere Martí S, Laguillo Sala G, Beltrán Mármol B, et al. Facial fractures: classification and highlights for a useful report. Insights Imaging. 2020;11(1):49.
11. Sivapathasundharam B, Biswas P, Preethi S. The World Health Organization classification of odontogenic and maxillofacial bone tumors: an appraisal. J Oral Maxillofac Pathol. 2019;23(2):178.
12. Pelotti P, Marchetti C, Bonetti M, Monti C. Modern imaging of maxillofacial deformities. Radiol Med. 1994;88(6):725.
13. Perren SM. Basic aspects of internal fixation. In: Manual of internal fixation in small animals. Berlin: Springer; 1998. p. 3–56.
14. Bell RB, Al-Bustani SS. Orbital fractures. In: Current therapy in oral and maxillofacial surgery. 1st ed. Amsterdam: Elsevier; 2012. p. 304–23.
15. Chen CH, Chen CT, Wang PF, Wang YT, Hsu PH, Lin CL. A novel anatomical thin titanium mesh plate with patient-matched bending technique for orbital floor reconstruction. J Cranio Maxillofac Surg. 2018;46(9):1526–32.
16. Schlittler F, Vig N, Burkhard JP, Lieger O, Michel C, Holmes S. What are the limitations of the non-patient-specific implant in titanium reconstruction of the orbit? Br J Oral Maxillofac Surg. 2020;58(9):e80–5.
17. Alasseri N, Alasraj A. Patient-specific implants for maxillofacial defects: challenges and solutions. Maxillofac Plast Reconstr Surg. 2020;42(1):15.
18. Boyette JR, Pemberton JD, Bonilla-Velez J. Management of orbital fractures: challenges and solutions, Clinical ophthalmology, vol. 9. London: Dove Medical Press Ltd; 2015. p. 2127–37.
19. Kozakiewicz M, Elgalal M, Walkowiak B, Stefanczyk L. Technical concept of patient-specific, ultrahigh molecular weight polyethylene orbital wall implant. J Cranio Maxillofac Surg. 2013;41(4):282–90.
20. Kalender WA, Hebel R, Ebersberger J. Reduction of CT artifacts caused by metallic implants. Radiology. 1987;164(2):576–7.
21. Raman SP, Mahesh M, Blasko RV, Fishman EK. CT scan parameters and radiation dose: practical advice for radiologists. J Am Coll Radiol. 2013;10:840–6.
22. Trattner S, Pearson GDN, Chin C, Cody DD, Gupta R, Hess CP, et al. Standardization and optimization of computed tomography protocols to achieve low-dose. J Am Coll Radiol. 2014;11(3):271–8.
23. Sun S, Zhang R. Region of interest extraction of medical image based on improved region growing algorithm. In: Proceedings of the 2017 International Conference on Material Science, Energy and Environmental Engineering (MSEEE 2017). Paris, France: Atlantis Press; 2017. p. 360–4.
24. Kamdi S, Krishna RK. Image segmentation and region growing algorithm. Int J Comput Technol Electron Eng. 2012;2(1):103–7.
25. Pilliar RM, Cameron HU, Binnington AG, Szivek J, Macnab I. Bone ingrowth and stress shielding with a porous surface coated fracture fixation plate. J Biomed Mater Res. 1979;13(5):799–810.
26. Turkyilmaz I, Hoders AB. Immediate loading in implant dentistry. In: Current concepts in dental implantology. London: InTech; 2015.
27. Deshpande VS, Ashby MF, Fleck NA. Foam topology: bending versus stretching dominated architectures. Acta Mater. 2001;49(6):1035–40.
28. Patil N, Goodman SB. Wear particles and osteolysis. In: Orthopaedic bone cements. Amsterdam: Elsevier Ltd; 2008. p. 140–63.
29. Rodriguez E, Ramirez-Martinez A. Fatigue and wear analysis for temporomandibular joint prosthesis by finite element method, Lecture notes in computational vision and biomechanics. New York: Springer; 2020. p. 317–34.
30. Van Loon JP, Verkerke GJ, De Vries MP, De Bont LGM. Design and wear testing of a temporomandibular joint prosthesis articulation. J Dent Res. 2000;79(2):715–21.
31. Helland MM. Anatomy and function of the temporomandibular joint. J Orthop Sports Phys Ther. 1980;1(3):145–52.

32. Blecha LD, Rakotomanana L, Razafimahery F, Terrier A, Pioletti DP. Targeted mechanical properties for optimal fluid motion inside artificial bone substitutes. J Orthop Res. 2009;27(8):1082–7.
33. Condie R, Bose S, Bandyopadhyay A. Bone cell-materials interaction on Si microchannels with bioinert coatings. Acta Biomater. 2007;3(4):523–30.
34. Yoo DJ. Advanced porous scaffold design using multi-void triply periodic minimal surface models with high surface area to volume ratios. Int J Precis Eng Manuf. 2014;15(8):1657–66.
35. Babaie E, Bhaduri SB. Fabrication aspects of porous biomaterials in orthopedic applications: a review. ACS Biomater Sci Eng. 2018;4(1):1–39.
36. Taniguchi N, Fujibayashi S, Takemoto M, Sasaki K, Otsuki B, Nakamura T, et al. Effect of pore size on bone ingrowth into porous titanium implants fabricated by additive manufacturing: an in vivo experiment. Mater Sci Eng C. 2016;59:690–701.
37. Hench LL, Wilson J. An introduction to bioceramics. In: Hench LL, editor. An introduction to bioceramics. 2nd ed. Florida: World Scientific; 1993.
38. Oonishi H, Yamamoto M, Ishimaru H, Tsuji E, Kushitani S, Aono M, et al. The effect of hydroxyapatite coating on bone growth into porous titanium alloy implants. J Bone Joint Surg (Br). 1989;71(2):213–6.
39. Jemat A, Ghazali MJ, Razali M, Otsuka Y. Surface modifications and their effects on titanium dental implants, vol. 2015. New York: BioMed Research International, Hindawi Publishing Corporation; 2015.
40. Drnovšek N, Rade K, Milačič R, Štrancar J, Novak S. The properties of bioactive TiO_2 coatings on Ti-based implants. Surf Coat Technol. 2012;209:177–83.
41. Hayashi K, Inadome T, Tsumura H, Nakashima Y, Sugioka Y. Effect of surface roughness of hydroxyapatite-coated titanium on the bone-implant interface shear strength. Biomaterials. 1994;15(14):1187–91.
42. Razavi M, Fathi M, Savabi O, Mohammad Razavi S, Hashemi Beni B, Vashaee D, et al. Controlling the degradation rate of bioactive magnesium implants by electrophoretic deposition of akermanite coating. Ceram Int. 2014;40(3):3865–72.
43. Abbasi N, Hamlet S, Love RM, Nguyen NT. Porous scaffolds for bone regeneration. J Sci Adv Mater Devices. 2020;5:1–9.
44. Hussein A, Hao L, Yan C, Everson R, Young P. Advanced lattice support structures for metal additive manufacturing. J Mater Process Technol. 2013;213(7):1019–26.
45. Sing SL. Concepts of selective laser melting for orthopaedic implants. In: Selective laser melting of novel titanium-tantalum alloy as orthopaedic biomaterial. Singapore: Springer; 2019. p. 9–36.
46. Morrison RJ, Kashlan KN, Flanangan CL, Wright JK, Green GE, Hollister SJ, et al. Regulatory considerations in the design and manufacturing of implantable 3D-printed medical devices. Clin Transl Sci. 2015;8(5):594–600.
47. Aitchison G, Hukins DW, Parry J, Shepherd DE, Trotman S. A review of the design process for implantable orthopedic medical devices. Open Biomed Eng J. 2009;3(1):21–7.
48. Moiduddin K. Implementation of computer-assisted design, analysis, and additive manufactured customized mandibular implants. J Med Biol Eng. 2018;38(5):744–56.
49. Mandolini M, Caragiuli M, Brunzini A, Mazzoli A, Pagnoni M. A procedure for designing custom-made implants for forehead augmentation in people suffering from Apert syndrome. J Med Syst. 2020;44(9):146.
50. Nag S, Banerjee R. Fundamentals of medical implant materials. In: Narayan R, editor. Materials for medical devices. Novelty, OH: ASM International; 2012. p. 6–17.
51. Saini M, Singh Y, Arora P, Arora V, Jain K, Singh SM, et al. Implant biomaterials: a comprehensive review. A Compr Rev World J Clin Cases. 2015;3(1):52–7.
52. Albrektsson T, Johansson C. Osteoinduction, osteoconduction and osseointegration. Eur Spine J. 2001;10:96–101.
53. Gomes CC, Moreira LM, Santos VJSV, Ramos AS, Lyon JP, Soares CP, et al. Assessment of the genetic risks of a metallic alloy used in medical implants. Genet Mol Biol. 2011;34(1):116–21.
54. Sing SL, An J, Yeong WY, Wiria FE. Laser and electron-beam powder-bed additive manufacturing of metallic implants: a review on processes, materials and designs. J Orthop Res. 2016;34(3):369–85.
55. Olakanmi EO, Cochrane RF, Dalgarno KW. A review on selective laser sintering/melting (SLS/SLM) of aluminium alloy powders: processing, microstructure, and properties, Progress in materials science, vol. 74. Amsterdam: Elsevier Ltd; 2015. p. 401–77.
56. Pauly S, Löber L, Petters R, Stoica M, Scudino S, Kühn U, et al. Processing metallic glasses by selective laser melting. Mater Today. 2013;16(1–2):37–41.
57. Kruth J, Vandenbroucke B, Vaerenbergh J, Mercelis P. Benchmarking of different SLS/SLM processes as Rapid Manufacturing techniques. In: International Conference Polymers & Moulds Innovations. Gent Belgium; 2005. p. 525.
58. Kurzynowski T, Madeja M, Dziedzic R, Kobiela K. The effect of EBM process parameters on porosity and microstructure of Ti-5Al-5Mo-5V-1Cr-1Fe alloy. Scanning. 2019;2019:1–12.
59. Vayssette B, Saintier N, Brugger C, Elmay M, Pessard E. Surface roughness of Ti-6Al-4V parts obtained by SLM and EBM: effect on the high cycle fatigue life. In: Procedia engineering. Amsterdam: Elsevier Ltd; 2018. p. 89–97.
60. Dalgarno K. The impact and potential for 3D printing and bioprinting in the medical devices industry. London: Newcastle University; 2020.
61. Xiao Z, Chen C, Zhu H, Hu Z, Nagarajan B, Guo L, et al. Study of residual stress in selective laser melting of Ti6Al4V. Mater Des. 2020;193:108846.
62. ASTM F136-13. Standard specification for wrought Titanium-6Aluminum-4Vanadium ELI (Extra Low

Interstitial) alloy for surgical implant applications (UNS R56401). West Conshohocken, PA: ASTM International; 2013. [cited 2020 Nov 6]
63. AMSH81200. Heat treatment of titanium and titanium alloys. SAE International. 2001 [cited 2020 Nov 6]. Available from: https://www.sae.org/standards/content/amsh81200/.
64. Majeed A, Ahmed A, Salam A, Sheikh MZ. Surface quality improvement by parameters analysis, optimization and heat treatment of AlSi10Mg parts manufactured by SLM additive manufacturing. Int J Light Mater Manuf. 2019;2(4):288–95.
65. Jones LC, Tsao AK, Topoleski LDT. Factors contributing to orthopaedic implant wear. In: Wear of orthopaedic implants and artificial joints. Sawston, UK: Woodhead Publishing Limited; 2013. p. 310–50.
66. Minagar S, Berndt CC, Wang J, Ivanova E, Wen C. A review of the application of anodization for the fabrication of nanotubes on metal implant surfaces. Acta Biomater. 2012;8:2875–88.
67. Kanegsberg B, Kanegsberg E. Parameters in ultrasonic cleaning for implants and other critical devices. J ASTM Int. 2006;3(4) https://doi.org/10.1520/JAI13387.
68. Weaver GJ. Additive manufacturing and inspection difficulties. 2018.
69. Edger S. Non-destructive testing of additive manufactured parts. 2020.
70. Friis E. Mechanical testing of orthopaedic implants. 2017. p. 1–260.
71. Food and Drug Administration. Technical considerations for additive manufactured devices. 2016. 28 p.
72. Cheng M. Medical device regulation global overview and guiding principles. Geneva: WHO; 2003.
73. U.S. Food and Drug Administration. Classify your medical device. FDA. 2020 [cited 2021 Jan 25]. Available from: https://www.fda.gov/medical-devices/overview-device-regulation/classify-your-medical-device.
74. ASTM F2924-14. Standard specification for additive manufacturing Titanium-6 Aluminum-4 Vanadium with powder bed fusion. West Conshohocken, PA: ASTM International; 2014 [cited 2020 Nov 6]. Available from: https://www.astm.org/Standards/F2924.htm
75. ASTM F1472-14. Standard specification for wrought Titanium-6Aluminum-4Vanadium alloy for surgical implant applications (UNS R56400). West Conshohocken, PA: ASTM International; 2014 [cited 2020 Nov 6]. Available from: https://www.astm.org/Standards/F1472.htm.
76. ASTM F3001-14. Standard specification for additive manufacturing Titanium-6 Aluminum-4 Vanadium ELI (Extra Low Interstitial) with powder bed fusion. West Conshohocken, PA: ASTM International; 2014 [cited 2020 Nov 6]. Available from: https://www.astm.org/Standards/F3001.htm.
77. ISO/ASTM52910-18. Additive manufacturing—design—requirements, guidelines and recommendations. West Conshohocken, PA: ASTM International; 2018. Available from: www.astm.org
78. ISO/ASTM 52904. Additive manufacturing—process characteristics and performance—practice for metal powder bed fusion process to meet critical applications. ASTM International. 2019 [cited 2021 Jan 25]. Available from: https://www.iso.org/standard/74637.html.
79. ASTM F3184-16. Standard specification for additive manufacturing stainless steel alloy (UNS S31603) with powder bed fusion. West Conshohocken, PA: ASTM International; 2016. Available from: www.astm.org
80. ASTM F3301-18a. Standard for additive manufacturing – post processing methods – standard specification for thermal post-processing metal parts made via powder bed fusion. West Conshohocken, PA: ASTM International; 2018. Available from: www.astm.org
81. ISO 14971:2019(en). Medical devices—application of risk management to medical devices. International Organization of Standards. 2019 [cited 2021 Jan 25]. p. 36. Available from: https://www.iso.org/obp/ui/#iso:std:iso:14971:ed-3:v1:en.
82. French-Mowat E, Burnett J. How are medical devices regulated in the European Union? J R Soc Med. 2012;105:22–8.
83. ASTM F3122-14. Standard guide for evaluating mechanical properties of metal materials made via additive manufacturing processes. West Conshohocken, PA: ASTM International; 2014. Available from: www.astm.org
84. ISO/ASTM DTR 52905. Additive manufacturing of metals—non-destructive testing and evaluation—defect detection in parts. 2018 [cited 2021 Jan 25]. Available from: https://www.iso.org/standard/71988.html.
85. Badiali G, Ferrari V, Cutolo F, Freschi C, Caramella D, Bianchi A, et al. Augmented reality as an aid in maxillofacial surgery: validation of a wearable system allowing maxillary repositioning. J Cranio Maxillofac Surg. 2014;42(8):1970–6.
86. Ayoub A, Pulijala Y. The application of virtual reality and augmented reality in Oral & Maxillofacial Surgery. BMC Oral Health. 2019;19(1):238.
87. Parthasarathy J. 3D modeling, custom implants and its future perspectives in craniofacial surgery. Ann Maxillofac Surg. 2014;4(1):9.
88. Hosseinkhani M, Mehrabani D, Karimfar MH, Bakhtiyari S, Manafi A, Shirazi R. Tissue engineered scaffolds in regenerative medicine. World J Plast Surg. 2014;3(1):3–7.
89. Salah M, Tayebi L, Moharamzadeh K, Naini FB. Three-dimensional bio-printing and bone tissue engineering: technical innovations and potential applications in maxillofacial reconstructive surgery. Maxillofac Plast Reconstr Surg. 2020;42(1):18.
90. Ozbolat IT, Peng W, Ozbolat V. Application areas of 3D bioprinting, Drug discovery today, vol. 21. Amsterdam: Elsevier Ltd; 2016. p. 1257–71.

3D Printing in Endodontics

12

Sidhartha Sharma, Vijay Kumar, Amrita Chawla, and Ajay Logani

Contents

12.1 Introduction

Visualising and comprehending the internal root structure and tooth shape are critical for successful endodontic treatment [1]. As a result, the emergence of 3D imaging, 3D virtual planning, and 3D printing has generated considerable attention. The surface scanner and cone-beam computed tomography (CBCT) are used to acquire the digital data. Following that, the data is processed and designed using a software tool. Finally, the object is fabricated utilising additive manufacturing processes, which entail selective curing or binding of material in consecutive vertical layers that fuse together. It is possible to manufacture identical objects with geometrically complex shapes and varying cross-sectional form, density, colour, or mechanical properties [2]. Endodontics has benefited from the acceptance and application of 3D printing in treatment, education, and research [3]. A total of 73 articles were extracted following a full-text reading utilising the following phrases and Boolean operators in a PubMed literature search until August 31, 2021: (((3d printing) *or* (rapid prototyping)) *or* (directed)) *or* (targeted)) *and* (endodontics). Five of them were reviews focusing on endodontics (Table 12.1). The remaining articles contain case reports, clinical research and in vitro experiments (Tables 12.2, 12.3, 12.4 and 12.5) about the following applications mentioned in Fig. 12.1.

S. Sharma (✉) · V. Kumar · A. Chawla · A. Logani
Division of Conservative Dentistry and Endodontics, Centre for Dental Education and Research, All India Institute of Medical Sciences, New Delhi, India
e-mail: sid.aiims@aiims.edu

© The Author(s), under exclusive license to Springer Nature Switzerland AG 2022
P. K. Chaudhari et al. (eds.), *3D Printing in Oral Health Science*,
https://doi.org/10.1007/978-3-031-07369-4_12

Table 12.1 PubMed-based review articles on 3D printing in endodontics

Serial number	Authors	Year	Journal	Title
1.	van der Meer W.J. et al. [51]	2016	*Journal of Dentistry*	3D Computer-aided treatment planning in endodontics
2.	Anderson J. et al. [1]	2018	*International Endodontic Journal*	Endodontic applications of 3D printing
3.	Shah P. and Chong B.S. [2]	2018	*Clinical Oral Investigation*	3D imaging, 3D printing and 3D virtual planning in endodontics
4.	Cheng Q. and Xia W.W. [52]	2019	*Zhonghua Kou Qiang Yi Xue Za Zhi*	Research and application of three-dimensional printed template in endodontics
5.	C. Moreno-Rabié [53]	2020	*International Endodontic Journal*	Clinical applications, accuracy and limitations of guided endodontics: a systematic review
6.	Lio F. et al. [54]	2021	*J Biol Regul Homeost Agents*	Guided endodontic microsurgery in apicoectomy: a review

12.2 Guided Endodontic Access

Luxation injuries account for 15–61% of traumatic dental injuries and most frequently involve the anterior maxillary teeth [4]. It has an effect on the periodontium as well as the pulp. The fate of the pulp tissue is determined by the severity of the trauma and the developmental status of the injured tooth's root [5]. Pulp necrosis occurs as a result of severe injuries and a closed apex [6]. Mild injuries or immature teeth result in the survival of the pulp and subsequent revascularisation or healing of the pulp. As a result, dentin deposition is accelerated, and pulp canal calcification (PCC) or obliteration occurs (PCO) [7]. PCC occurs at an incidence rate of 10% following a dental trauma and is most frequently seen in lateral luxations (71%), followed by extrusions (61%) [7]. It can also occur due to chronic caries, restorations, orthodontic treatment, ageing, and procedures involving vital pulp therapy. PCC (pulp space narrowing or obliteration) is most severe coronally and decreases towards the root apex [8]. It is a sign of pulp healing and, with the exception of discoloration, is asymptomatic. Endodontic treatment is necessary only when clinical or radiographic evidence of pulp necrosis exists which is a late complication of PCC that occurs several years after the initial diagnosis [9]. Locating and negotiating such canals require a specialist with anatomical knowledge, mental visualisation in three dimensions, and a steady hand to drill in the anticipated canal opening direction. According to the AAE case difficulty assessment criteria, this becomes quite difficult and is classified as a high difficulty category [10]. The access opening may cause significant tooth destruction, perforations, canal deviation, and instrument separation [11]. It is recommended to use a surgical operating microscope and cone-beam computed tomography (CBCT). CBCT assists the operator in visualising the affected tooth in three dimensions and is a reliable method of detecting root canal anatomy. However, the operator must interpret and create a mental 3D map to perform the practical task free hand. It all comes back to the operator's ability to locate the canal without causing iatrogenic errors. This prolongs the duration of the treatment and still results in overpreparation. As a result, a novel channeled technique was developed for apically extended access cavity preparations based on the design and principles of guided implant surgeries [12]. It uses 3D printing technology to create customised guides and is referred to as guided or targeted endodontics (Fig. 12.2).

The procedure is simple and begins with acquiring a limited field CBCT image set to the smallest voxel size specified by the planning programme. The parameters chosen should result in the least amount of radiation exposure possible to the patient while still obtaining the best imaging result possible to facilitate optimal planning. The CBCT images are stored as Digital Imaging and Communication (DICOM) files. The surface scans are performed using a 3D intraoral (direct)

Table 12.2 PubMed articles on guided endodontic access

Serial number	Authors	Year	Journal	Type of study	Tooth involved	Drill used
1.	Álvaro Zubizarreta Macho et al. [55]	2015	*The Journal of the American Dental Association*	Case report	Maxillary lateral incisor	Diamond burr surface (882 314 012, Komet Medical)
2.	Krastl G. et al. [13]	2016	*Dental Traumatology*	Case report	Maxillary central incisor	Straumann drill for tempimplants
3.	M.S. Zehnder et al. [18]	2016	*International Endodontic Journal*	In vitro study	Single-rooted teeth	Straumann drill for tempimplants
4.	J. Buchgreitz et al. [56]	2016	*International Endodontic Journal*	Ex vivo study	Multiple teeth	Spiral Bur
5.	Wicher J. van der Meer et al. [51]	2016	*Journal of Dentistry*	Case report	Anterior teeth	Munce bur number 2
6.	Thomas Connert et al. [20]	2017	*Journal of Endodontics*	Ex vivo study	Mandibular anterior teeth	Specially designed bur had a total length of 28 mm with a working length of 20 mm, and the diameter was 0.85 mm
7.	Jesús Mena-Álvarez et al. [57]	2017	*Journal of Esthetic and Restorative Dentistry*	Case report	Maxillary central incisor	A dental diamond bur featuring a 1.2 mm diameter and length of 14 mm
8.	J. Buchgreitz et al. [58]	2018	*International Endodontic Journal*	Observational study	Multiple teeth	Modified Spiral Bur
9.	Warley Luciano Fonseca Tavares et al. [59]	2018	*Journal of Endodontics*	Case report	Anterior teeth	Neodent drill for tempimplants
10.	Sônia T. de O. Lara-Mendes et al. [60]	2018	*Journal of Endodontics*	Case report	Maxillary molars	Prototyped guide
11.	Sônia T. O. Lara-Mendes et al. [61]	2018	*Journal of Endodontics*	Case report	Maxillary central incisor	Neodent drill for tempimplants
12.	Ankit Nayak et al. [62]	2018	*Journal of Engineering in Medicine*	In vitro study	Single-rooted teeth	Spiral bur
13.	A. Torres et al. [63]	2018	*International Endodontic Journal*	Case report	Maxillary lateral incisor	Size 1 Munce Discovery bur
14.	J. Buchgreitz et al. [64]	2019	*Journal of Endodontics*	Case report	Maxillary first molar	Spiral Bur
15.	Bruna de Athayde Casadei et al. [65]	2019	*Australian Endodontic Journal*	Case report	Maxillary right second premolar	Neodent drill for tempimplants
16.	Thomas Connert et al. [19]	2019	*Journal of Endodontics*	In vitro study	Incisors	Muller bur

(continued)

Table 12.2 (continued)

Serial number	Authors	Year	Journal	Type of study	Tooth involved	Drill used
17.	Jana Kostunov et al. [66]	2019	*Journal of Endodontics*	In vitro pilot study	Acrylic typodont teeth (incisors, premolars, and molars)	Special guided endodontic burs
18.	Lucas Moreira Maia et al. [67]	2019	*Journal of Endodontics*	Case report	Molar and premolars	Neodent drill for tempimplants
19.	Cyril Perez et al. [68]	2019	*Australian Endodontic Journal*	Case report	Maxillary first molar	Drill 0.75 mm in diameter and 22 mm in length (FFDM-Pneumat®, Bourges, France)
20.	Jörg Philipp Tchorz et al. [69]	2019	*International Journal of Computerized Dentistry*	Case report	Mandibular central incisor	Carbide bur with a diameter of 1.2 mm
21.	Marco Antônio Z. Loureiro et al. [70]	2020	*Journal of Endodontics*	Ex vivo study	Mandibular incisors and maxillary molars	Milling cutter 1.3 mm in diameter and 20-mm long
22.	Ralf Krug et al. [71]	2020	*International Journal of Computerized Dentistry*	In vitro study	Incisors	3D-printed template and bur
23.	Amanda Stephanie Silva et al. [72]	2020	*Journal of Esthetic and Restorative Dentistry*	Case report	Left upper lateral Incisor	Drill with a diameter of 1.3 mm and a length of 20 mm
24.	Ralf Krug et al. [15]	2020	*Head and Face Medicine*	Case report	Multiple teeth	Customized drill (diameter = 1.0 mm, steco-system-technik GmbH)
25.	Warley Luciano et al. [73]	2020	*Photodiagnosis and Photodynamic Therapy*	Case report	Maxillary anterior	Not mentioned
26.	Wadim Leontiev et al. [74]	2021	*Journal of Endodontics*	In vitro study	Anteriors and premolars	Carbide bur (Endoseal drill)
27.	Marc Llaquet Pujol et al. [75]	2021	*Journal of Endodontics*	Case report	Maxillary anteriors	Customized (length = 21 mm and diameter = 1 mm) cylindrical diamond bur
28.	Jacob C. Simon et al. [76]	2021	*Journal of Endodontics*	Pilot study	Posterior teeth	Laser at 100 Hz
29.	Randolph Todd et al. [77]	2021	*The Journal of the American Dental Association*	Case report	Maxillary lateral incisor	A 24 mm bur
30.	Andres Torres et al. [17]	2021	*Journal of Endodontics*	Case report	Maxillary second premolar	Size 2 Munce discovery bur

Table 12.3 PubMed articles on 3D-printed surgical guides

Serial number	Author	Year	Journal	Article type	Tooth involved
1.	Harold M. Pinsky et al. [78]	2007	*Journal of Endodontics*	In vitro study	Multiple teeth
2.	James K. Bahcall [79]	2014	*Compendium of Continuing Education in Dentistry*	Not known	Not known
3.	S. Patel et al. [26]	2017	*International Endodontic Journal*	Case report	Soft tissue retraction for maxillary central incisor
4.	Georg D. Strbac et al. [22]	2017	*Journal of Endodontics*	Case reports	Maxillary first molar and second premolar
5.	C. Michelle Giacomino et al. [23]	2018	*Journal of Endodontics*	Case reports	Multiple teeth
6.	Shangzu Ye et al. [80]	2018	*BMC Oral Health*	Case reports	Maxillary lateral incisor and canine
7.	So-Yeon Ahn et al. [81]	2018	*Journal of Endodontics*	Case report	Mandibular first molar
8.	L. Peng et al. [82]	2018	*Journal of Peking University*	In vitro study	Upper anterior tooth
9.	Kim J.E. [25]	2019	*Restorative Dentistry and Endodontics*	Case report	Maxillary central incisor
10.	Shira Ackerman et al. [83]	2019	*Journal of Endodontics*	In vitro study	Multiple teeth 48 roots
11.	Eveline Sutter et al. [84]	2019	*International Journal of Computerized Dentistry*	Case report	Mandibular first molar
12.	Yuehong Fan et al. [85]	2019	*Journal of Endodontics*	In vitro study	Multiple teeth 42 roots
13.	P. Avantaggiato et al. [86]	2020	*Journal of Biological Regulators and Homeostatic Agents*	Case report	Maxillary first molar
14.	Roy George et al. [87]	2020	*Australian Endodontic Journal*	Case report	Mandibular second premolar
15.	Warley Luciano et al. [88]	2020	*Journal of Oral and Maxillofacial Surgery*	Case report	Maxillary second premolar
16.	Hawkins T.K. [89]	2020	*International Endodontic Journal*	Original article	Surgical simulation model
17.	Gary Benjamin et al. [90]	2021	*Journal of Endodontics*	Case reports	Multiple teeth
18.	Arianne Galino et al. [24]	2021	*Journal of Endodontics*	Clinical research	Multiple teeth

or laboratory (indirect) surface scanner, and the data is saved as surface tessellation language (STL) files. Since the planning process requires a three-dimensional model of the teeth's roots (the CBCT dataset) and a three-dimensional model of the teeth's crowns (the digital imprint), these two models are combined and superimposed. The planning software enables the creation of a virtual drill path leading to the target position in the root where the canal is first visible. Additionally, a virtual sleeve is created for guidance with a slightly larger internal diameter, and a length determined by the patient's mouth opening (about 5–7 mm) [12]. After superimposing the virtual bur on the tooth, direct access to the root canal's apical third is gained. Finally, a virtual template is created and printed using 3D software. The guide utilises the dentition to provide stable anatomical fixation (Fig. 12.3).

This technology has been mentioned in numerous case reports where it has been used to treat pulp canal obliteration and anatomical variations in permanent anterior teeth [13–15]. Anterior teeth are easily accessible and provide

Table 12.4 PubMed articles of 3D-printed tooth model

Serial number	Author	Year	Journal	Article type	Tooth involved
1.	Ronald Ordinola-Zapata et al. [91]	2014	*Journal of Applied Oral Science*	In vitro study	Two mandibular molars
2.	Chanhee Byun et al. [36]	2015	*Journal of Endodontics*	Case report	Maxillary central incisor
3.	Saif Alarab Mohmmed et al. [92]	2016	*Dental Materials*	In vitro study	40 root canal models
4.	Rahmi Eken et al. [33]	2016	*Journal of Endodontics*	A finite element analysis study	Mandibular second premolar
5.	L. Robberecht et al. [93]	2016	*European Journal of Dental Education*	In vitro study	Multiple teeth
6.	Tuba Gok et al. [39]	2017	*Journal of Endodontics*	Case report	Mandibular molar with a C1 root canal configuration
7.	David Christofzik et al. [32]	2018	*PLOS ONE*	In vitro study	30 specimens of the 9 different simulated root canal anatomies
8.	Xiao Liang et al. [94]	2018	*Journal of Biomedical Nanotechnology*	In vitro study	Premolars with a single root canal
9.	M. Reymus et al. [27]	2018	*International Endodontic Journal*	Original scientific article	Multiple teeth
10.	Christian Höhne et al. [95]	2019	*Journal of Dental Education*	In vitro study	First permanent molar
11.	Anıl Özgün Karatekin et al. [96]	2019	*PLOS ONE*	Research article	Mandibular molars which had fused roots and a longitudinal groove on lingual or buccal surface of the root
12.	Przemysław Kustra et al. [97]	2020	*European Journal of Dental Education*	In vitro study	Mandibular central incisor, maxillary second premolar and maxillary first molar

space for guide templates to be placed. Similar procedures, with some modifications, have been documented in posterior teeth. In 2018, Shi et al. reported on a case of pulp canal obliteration in a mandibular molar [16]. A three-dimensional printed template was created for this patient using the CBCT dataset. The coronal and apical portion of the root canal was penetrated with ultrasonic tip ET20 and ET25, respectively. The treatment was successful, and at a 6-month follow-up, the patient was asymptomatic. Torres et al. (2021) used a sleeveless guide to manage calcified maxillary first premolar [17]. The open system used guiding rails consisting of two cylinders placed against each other on the sides of the tooth to guide the handpiece head. Additionally, a leg system or adapter was firmly attached to the handpiece head, which fitted on the rails and guided the handpiece during treatment. The absence of sleeves allowed working with shorter burs, water cooling during drilling and improved visibility.

Zehnder et al. [18] demonstrated that the technique is accurate, with deviations of planned and prepared access cavities ranging from 0.16 to 0.21 mm for various aspects at the bur's base and 0.17–0.47 mm at the bur's tip. Angle deviations, on average, were 1.81°. As a result, they concluded that 'Guided Endodontics' enabled precise access cavity preparation up to the apical third of the root. Connert et al. [19] conducted an in vitro investigation in which they compared a guided endodontic technique to conventional access preparation employing three operators with varying levels of experience. They concluded that guided therapy enabled operators to locate 92% of canals regardless of their skill

Table 12.5 Articles on autotransplantation using 3D printing in PubMed-indexed endodontic journals

Serial number	Authors	Year	Journal	Title
1.	Park Y.S. et al. [98]	2012	*Journal of Endodontics*	Autotransplantation with simultaneous sinus floor elevation
2.	Lee S.J. et al. [46]	2012	*Restorative Dentistry and Endodontics*	Minimizing the extra-oral time in autogenous tooth transplantation: use of computer-aided rapid prototyping (CARP) as a duplicate model tooth
3.	Jang J.H. et al. [99]	2013	*Journal of Endodontics*	Autotransplantation of immature third molars using a computer-aided rapid prototyping model: a report of four cases
4.	Lee Y.S. et al. [43]	2014	*International Endodontic Journal*	Autotransplantation of mesiodens for missing maxillary lateral incisor with cone-beam CT-fabricated model and orthodontics
5.	Strbac G.D. et al. [100]	2016	*Journal of Endodontics*	Guided autotransplantation of teeth: a novel method using virtually planned three-dimensional templates
6.	Abella F. et al. [101]	2018	*Journal of Endodontics*	Outcome of autotransplantation of mature third molars using three-dimensional-printed guiding templates and donor tooth replicas
7.	Oh S. et al. [102]	2018	*Journal of Endodontics*	Virtual simulation of autotransplantation using three-dimensional printing prototyping model and computer-assisted design program
8.	Kim K. et al. [103]	2019	*Australian Endodontic Journal*	Clinical application of 3D technology for tooth autotransplantation: a case report
9.	Strbac G.D. et al. [104]	2020	*Journal of Endodontics*	Guided osteotomy and guided autotransplantation for treatment of severely impacted teeth: a proof-of-concept report
10.	Lucas-Taule E. et al. [105]	2020	*Journal of Endodontics*	Fully guided tooth autotransplantation using a multidrilling axis surgical stent: proof of concept
11.	Sato M. et al. [106]	2021	*Journal of Endodontics*	Use of three-dimensional-printed guide in hemisection and autotransplantation of a fusion tooth: a case report

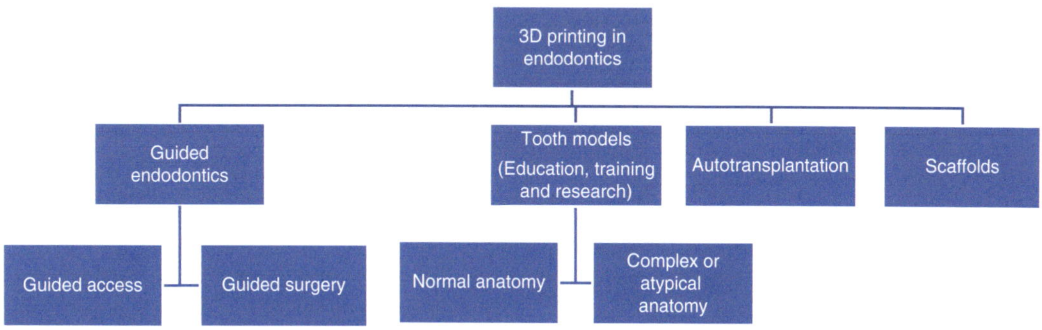

Fig. 12.1 Applications of 3D printing in endodontics

level, a statistically significant increase over the conventional technique (42%). Thus, guided endodontics is a promising approach that provides a highly predictable outcome while reducing the danger of iatrogenic harm and chairside time. However, it does have some drawbacks. The approach is only effective on straight canals and patients with adequate mouth opening [20]. The lack of dedicated long shank endodontic burs, the development of heat and the possibility of developing cracks in the tooth are all challenges that must be resolved.

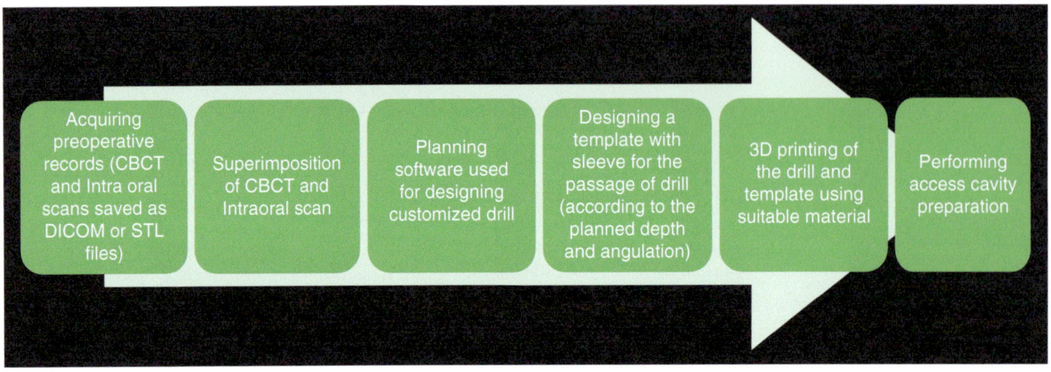

Fig. 12.2 Steps of guided endodontic access preparation

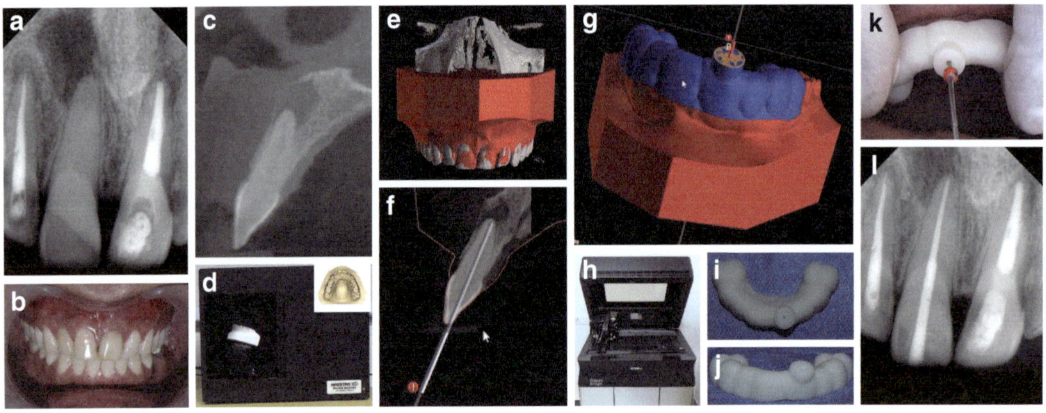

Fig. 12.3 Representative case. (**a**, **b**) Pulp canal calcification; (**c**) CBCT scan; (**d**) surface scan; (**e**) merging; (**f**) virtual drill planning; (**g**) guide designing; (**h**–**j**) 3D printing; (**k**) drilling; (**l**) post-obturation radiographs

12.3 Guided Endodontic Microsurgery

Endodontic microsurgery (EMS) is a procedure to treat persistent and recurrent apical periodontitis when the orthograde approach to the apical portion of the root canal is not possible. Incorporating the preoperative 3D scan, dental operating microscope, miniature instruments, ultrasonic root-end preparation tips and calcium silicate materials has made EMS's success rate comparable to nonsurgical endodontic retreatment [21]. However, the direct intraoral localisation of root apex based on the CBCT scans is challenging, and a significant amount of bone removal is inevitable with freehand surgical procedures. Moreover, the gingival flap elevation and cortical bone removal increase the surgery duration, which further increases the trauma. Further surgical procedures in the mandibular posterior region are challenging due to the proximity of the neurovascular structure. Taking a cue from guided implant surgeries, a 3D-printed surgical guide was introduced, allowing the operator to create a small targeted wound in otherwise complex areas of the oral cavity to execute the entire surgical procedure. This method is known as guided or targeted EMS [22, 23]. It ensures precise and conservative bone removal and root resection and reduces surgical time and postoperative healing issues.

The procedure begins with obtaining the intraoral scans of the region of interest. The study cast is made by taking alginate impression. The

CBCT scan is acquired, and the data is overlapped with digital data obtained from the patient's intraoral scan, which is exported to the planning software. The surgical template is designed such that it covers two teeth on either side. An appropriate guide drill (e.g. trephine bur) is selected targeting the root apex. The template is designed such that it has sleeves to accommodate the drill. The position and direction of the sleeve are determined based on the angulation of the drill. It should allow accessibility for a minimum of 3 mm of preparation depth [24, 25]. The drill should be designed so that it accurately targets the target root while avoiding damage to nearby roots or essential structures. The created guide template is exported as an STL file and then fabricated using a 3D printer.

3D printing has also been utilised for flap retraction by Patel et al. in 2017 [26]. The authors performed endodontic microsurgery for maxillary left central incisor using a custom-made retractor based on diagnostic CBCT scans. The 3D printed retractor enhanced visualisation and soft tissue handling during the surgery.

12.4 Tooth Models for Education, Training and Research

Pre-clinical training constitutes an integral part of the endodontic postgraduate curriculum. It helps the residents to acquire knowledge and develop fine motor hand skills, which will subsequently translate into their clinical practice and reduce iatrogenic errors. The extracted natural tooth remains the mainstay of pre-clinical exercise in restorative dentistry and endodontics. The advantages include low cost and actual simulation of natural conditions like the hardness of enamel/dentin and the intricacies of the root canal system. However, the extracted tooth has its limitations. It includes restricted availability, the possibility of cross-infection and variability in the root canal morphology that hampers standardised teaching and learning. Resin blocks have also been utilised for pre-clinical training. Being transparent, they allow the individual treatment steps to be visualised. However, they simulate a single root canal and do not accurately represent the complex root canal morphology. Hence, it provides oversimplified training [27].

In recent years, rapid prototyping has been used to quickly fabricate a scale model of a physical part or assembly using three-dimensional computer-aided design (CAD) data. Construction of the part or assembly is usually done using 3D printing or 'additive layer manufacturing' technology [28]. Similarly, the digitisation of human teeth based on a cone-beam computed tomography (CBCT) or micro-computed tomography (μCT) scan and the subsequent reproduction by a 3D printer facilitate the manufacturing of the artificial tooth model (Fig. 12.4). The 3D models can be printed to illustrate only the external features of the tooth, or by using the same technique, a tooth model with a hollow interior can be printed. Space can be filled with red ink to visualise the shape of the pulp. Such a 3D-printed model is gaining acceptance for pre-clinical education globally. The advantage includes ease of availability, visualisation of internal anatomy, ability to standardise the morphology and realistic simulation of the actual root canal configuration [29].

Several studies in the literature have compared extracted teeth to artificial teeth in endodontic skills training. A systematic review by Decurcio et al. concluded that the use of artificial 3D printed teeth for pre-clinical endodontic training achieved similar educational outcomes compared to extracted teeth [30]. Numerous manufacturers have developed tooth models using 3D printing. Examples include DEPT (Instituto Tecnologico de Producto Infantily Ocio, Alicante, Spain), DRSK (DRSK Group, Hassleholm, Sweden), Nissin (Nissin Dental Products Inc., Kyoto, Japan), TrueTooth (Dental Engineering Laboratories, Santa Barbara, CA, USA), DENTALIKE (Dentsply Sirona, Ballaigues, Switzerland), Real-T Endo (Acadental Inc., Kansas, USA), etc. [29]. The digitisation of teeth by a micro-CT is regarded as the reference standard in dental 3D imaging due to its ability to produce high-resolution images. However, the limited access to micro CTs and the requirement of an extracted tooth for scanning restrict its usage [27]. Al-Rawi et al. compared the accuracy of CBCT- and micro CT-based 3D-printed models of human

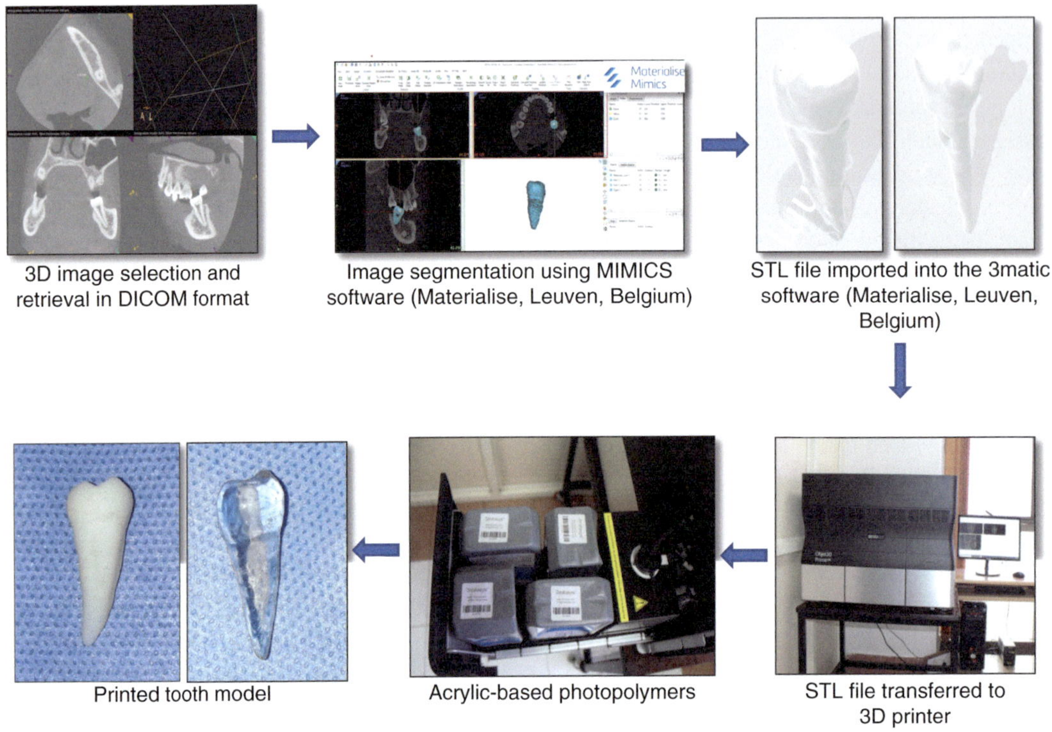

Fig. 12.4 Process of making 3D-printed tooth models

teeth. They reported that CBCT did produce realistic three-dimensional reconstructions of the teeth at a resolution of 133 m voxels, but only when a limited field of view was used [31]. Routine CBCT scans suffice for STL data in 3D printing of models negating the use of extracted tooth, which is a prerequisite for micro CT-based printing.

The applications of 3D printed models include:

Normal Tooth Anatomy

- To scale the teeth for teaching purposes
- In pre-clinical research to study the shaping ability [32] and stress values [33] of different rotary file systems
- To print an in vitro model to evaluate irrigation techniques [34, 35]
- To mass manufacture teeth for destructive analysis
- 3D-printed models can be developed to simulate the clinical experience for students:
  – Pulpotomy model
  – Revitalisation model
  – Complex anatomy tooth model for a root canal treatment
  – Instrument separation model

Atypical Tooth Anatomy

- To compile a database of 3D tooth models exhibiting anomalies
- To reconstruct challenging clinical cases such as:
  – Dilacerated root [36]
  – External invasive resorption
  – Molar with three distal roots [37]
  – Type 3 dens invaginatus [38]
- In pre-clinical research to study different obturation techniques for C-shaped canals [39]

Dental Trauma

- Pre-clinical study model simulating dental trauma [40]

12.5 Autotransplantation

Autotransplantation of teeth is a viable option for replacing a single tooth in young patients [41]. The transplanted teeth have been documented to have a favourable long-term prognosis. Third molar autotransplantation is an excellent treatment option for severely damaged teeth caused by trauma or caries, particularly in young patients. First molars are the most vulnerable teeth since they are the first permanent teeth to erupt, whereas third molars are the last teeth to develop and are typically nonfunctional. As a result, they are an ideal donor tooth until patients reach the age of about 18 years. The procedure is timed so that the donor tooth's root development is approximately 50–75% complete [42]. Because routine extraction of malpositioned wisdom teeth is common, third molars are an excellent source of suitable donor teeth. It is also possible to autotransplant teeth with complete root formation if endodontic treatment is performed before or within 2 weeks of the procedure. Autotransplantation is contraindicated in medically compromised patients, those with poor dental hygiene and periodontitis, without a suitable donor tooth, and with insufficient bone at the receptor site. When there is an insufficient bone at the receptor location, augmentation of the alveolar process may assist autotransplantation. The effectiveness of the treatment is dependent upon the preservation of periodontal ligament (PDL) cells and an acceptable adaptation of the donor tooth to the recipient site. Extra-oral time and trauma to the PDL during the procedure have a significant impact on the outcome. Traditionally, the transplanted tooth is used to prepare the recipient site, requiring multiple attempts at 'fitting'. Adjustments to the alveolar bone socket lengthen the extra-oral time and increase the risk of PDL cell damage [1].

The addition of three-dimensional planning has increased the predictability of existing autotransplantation techniques. It incorporates CBCT and rapid prototyping, allowing for preoperative planning and fabrication of a replica of the donor tooth. The replica is then used to prepare the alveolar socket at the recipient site before extracting the donor tooth, reducing extra-alveolar time and improving donor tooth fit. The procedure begins with a preoperative evaluation of the treatment's prognosis. If autotransplantation is determined to be a viable treatment option, a CBCT scan is performed with the occlusal plane parallel to the floor, as recommended by the manufacturer. The DICOM-formatted scan volumes are exported and imported into an image analysis software to segment the donor tooth. The donor tooth's root and crown are separated from the surrounding bone using a local greyscale threshold value and an image gradient. A manual selection of the root's most apical portion is required to complete the segmentation process based on the sagittal slides. Additional processing may be necessary to remove artefacts from the images. The donor tooth's 3D surface mesh is created, stored, and exported as an STL file. The replica of the donor tooth is created using 3D printing and a biocompatible resin material. The recipient site is prepared by inserting the printed replica of the donor tooth into the alveolar socket until it fits perfectly. Additionally, the 3D-printed tooth is used to check the occlusion and make necessary adjustments. After obtaining complete satisfaction, the donor tooth is extracted gently to avoid damaging the cementum, periodontal ligament, or apex of the tooth. To avoid postoperative forces, the donor tooth is placed in a slight infra-occlusion. With minimal extra-oral time, an immediate good fit of the donor tooth is achieved. Sutures across the occlusal surface are required to secure the donor tooth in its new location.

This technique has been utilised exhaustively in both anterior and posterior teeth. The case report by Lee in 2014 used a mesiodens as a donor tooth to replace the lateral incisor [43]. The ectopic premolar was used by Cahuana-Bartra in 2020 [44], while Shahbazian et al. placed a premolar in the alveolar socket of the maxillary central incisor [45]. There are multiple reports of replacing first and second molars with a wisdom tooth. Materials used to print donor teeth varied from biocompatible resins to cobalt chrome prototypes. This technique has brought down the extra-oral time of donor tooth to less than a minute, causing exponential rise in the success of the treatment. More prospective studies similar to Lee and Kim in 2012 with 182 patients [46] were to be conducted to establish rapid prototyping as an integral part of autotransplantation.

12.6 Regenerative Endodontics

In endodontics, the 3D printing principle can be used to deliver stem cells, pulp scaffolds, injectable calcium phosphates, growth factors, and gene therapy [47]. Scaffolds used in regenerative dental pulp applications must replicate the milieu found in the root canal and provide mechanical support; therefore, they must be optimised both temporally and spatially for optimal cell attachment and proliferation. The use of conventional scaffolds presents a number of difficult-to-solve issues, including constraints imposed by the scaffold shape and low reproducibility. Three-dimensional (3D) bioprinting, a rapid prototyping technique that uses computer-aided design blueprints to construct a complex tissue model rapidly, can be utilised. This technology enables precise control over scaffolds' surface morphology and internal microstructure, facilitates cell distribution and extension regulation and promotes cell growth in the 3D space of simulated tissue in vivo. Yu et al. compared the growth of human dental pulp stem cells on Alg-Gel versus 3D-printed Alg-Gel scaffolds and discovered that the 3D-printed Alg-Gel scaffolds were more conducive to cell adhesion and proliferation [48]. The technique of 3D cell printing can be used to replace pulp tissue. The pulp tissue structure can be recreated by dispensing layers of suspended cells in the hydrogel. This enables the cells to be precisely positioned to mimic natural pulp tissue, with odontoblast-like cells on the periphery and fibroblasts in the centre, surrounded by a network of vascular and neural cells. Athirasala et al. used a 3D printing-inspired method to engineer pre-vascularised, cell-laden hydrogel pulp-like tissue constructs in full-length root canals for dental pulp regeneration [49]. Using a bioprintable bioink suitable for hand-held in situ bioprinting, Duarte Campos et al. demonstrated strong vasculogenesis [50]. Future research should focus on dental pulp regeneration in vivo.

12.7 Summary

3D printing technology has a vast application in endodontic teaching, patient care and research. It has made endodontic treatment faster, predictable and safer. It allows for virtual planning that improves the preparedness of the operator before attempting complex procedures. The chances of human error are minimal with this technology, as proper planning guarantees accurate execution. The endodontist should be well versed with this technological advancement to harness its full potential and improve the standard of patient care.

References

1. Anderson J, Wealleans J, Ray J. Endodontic applications of 3D printing. Int Endod J. 2018;51(9):1005–18.
2. Shah P, Chong BS. 3D imaging, 3D printing and 3D virtual planning in endodontics. Clin Oral Invest. 2018;22:641–54.
3. Drăgan OC, Fărcăşanu AŞ, Câmpian RS, Turcu RV. Human tooth and root canal morphology reconstruction using magnetic resonance imaging. Clujul Med. 2016;89(1):137–42.
4. Andreasen JO. Etiology and pathogenesis of traumatic dental injuries. A clinical study of 1,298 cases. Scand J Dent Res. 1970;78:329–42.
5. Andreasen FM, Zhijie Y, Thomsen BL. Relationship between pulp dimensions and development of pulp necrosis after luxation injuries in the permanent dentition. Endod Dent Traumatol. 1986;2:90–8.
6. Andreasen JO. Luxation of permanent teeth due to trauma. A clinical and radiographic follow-up study of 189 injured teeth. Scand J Dent Res. 1970;78:273–86.
7. Andreasen FM, Yu Z, Thomsen BL, Andersen PK. Occurrence of pulp canal obliteration after luxation injuries in the permanent dentition. Endod Dent Traumatol. 1987;3:103–15.
8. Smith JW. Calcific metamorphosis: a treatment dilemma. Oral Surg Oral Med Oral Pathol Oral Radiol. 1982;54:441–4.
9. Jacobsen I, Kerekes K. Long-term prognosis of traumatized permanent anterior teeth showing calcify-

ing processes in the pulp cavity. Scand J Dent Res. 1977;85:588–98.
10. https://www.aae.org/specialty/wp-content/uploads/sites/2/2019/02/19AAE_CaseDifficultyAssessmentForm.pdf. Accessed 22 Oct 2021.
11. Cvek M, Granath L, Lundberg M. Failures and healing in endodontically treated non-vital anterior teeth with posttraumatically reduced pulpal lumen. Acta Odontol Scand. 1982;40:223–8.
12. Kinariwala N, Samaranayake L. Guided endodontics. Cham, Switzerland AG: Springer Nature; 2021.
13. Krastl G, Zehnder MS, Connert T, Weiger R, Kuhl S. Guided Endodontics: a novel treatment approach for teeth with pulp canal calcification and apical pathology. Dent Traumatol. 2016;32:240–6.
14. Connert T, Zehnder MS, Amato M, Weiger R, Kühl S, Krastl G. Microguided Endodontics: a method to achieve minimally invasive access cavity preparation and root canal location in mandibular incisors using a novel computer-guided technique. Int Endod J. 2018;51(2):247–55.
15. Krug R, Volland J, Reich S, Soliman S, Connert T, Krastl G. Guided endodontic treatment of multiple teeth with dentin dysplasia: a case report. Head Face Med. 2020;16(1):27.
16. Shi X, Zhao S, Wang W, Jiang Q, Yang X. Novel navigation technique for the endodontic treatment of a molar with pulp canal calcification and apical pathology. Aust Endod J. 2018;44(1):66–70.
17. Torres A, Lerut K, Lambrechts P, Jacobs R. Guided endodontics: use of a sleeveless guide system on an upper premolar with pulp canal obliteration and apical periodontitis. J Endod. 2021;47(1):133–9.
18. Zehnder MS, Connert T, Weiger R, Krastl G, Kühl S. Guided endodontics: accuracy of a novel method for guided access cavity preparation and root canal location. Int Endod J. 2016;49(10):966–72.
19. Connert T, Krug R, Eggmann F, Emsermann I, ElAyouti A, Weiger R, Kühl S, Krastl G. Guided endodontics versus conventional access cavity preparation: a comparative study on substance loss using 3-dimensional-printed teeth. J Endod. 2019;45(3):327–31.
20. Connert T, Zehnder MS, Weiger R, Kühl S, Krastl G. Microguided endodontics: accuracy of a miniaturized technique for apically extended access cavity preparation in anterior teeth. J Endod. 2017;43(5):787–90.
21. Kang M, In Jung H, Song M, Kim SY, Kim HC, Kim E. Outcome of nonsurgical retreatment and endodontic microsurgery: a meta-analysis. Clin Oral Investig. 2015;19:569–82.
22. Strbac GD, Schnappauf A, Giannis K, Moritz A, Ulm C. Guided modern endodontic surgery: a novel approach for guided osteotomy and root resection. J Endod. 2017;43(3):496–501.
23. Giacomino CM, Ray JJ, Wealleans JA. Targeted endodontic microsurgery: a novel approach to anatomically challenging scenarios using 3-dimensional-printed guides and trephine burs-a report of 3 cases. J Endod. 2018;44(4):671–7.
24. Buniag AG, Pratt AM, Ray JJ. Targeted endodontic microsurgery: a retrospective outcomes assessment of 24 cases. J Endod. 2021;47(5):762–9.
25. Kim JE, Shim JS, Shin Y. A new minimally invasive guided endodontic microsurgery by cone beam computed tomography and 3-dimensional printing technology. Restor Dent Endod. 2019;44(3):e29.
26. Patel S, Aldowaisan A, Dawood A. A novel method for soft tissue retraction during periapical surgery using 3D technology: a case report. Int Endod J. 2017;50(8):813–22.
27. Reymus M, Fotiadou C, Kessler A, et al. 3D printed replicas for endodontic education. Int Endod J. 2019;52:123–30.
28. Mahindru DV, Mahendru P. Review of rapid prototyping-technology for the future. Glob J Comput Sci Technol Graphics Vision. 2013;13:27–38.
29. Gancedo-Caravia L, Bascones J, Garcıa-Barbero E, Arias A. Suitability of different tooth replicas for endodontic training: perceptions and detection of common errors in the performance of postgraduate students. Int Endod J. 2020;53:562–72.
30. Decurcio DA, Lim E, Chaves GS, et al. Pre-clinical endodontic education outcomes between artificial versus extracted natural teeth: a systematic review. Int Endod J. 2019;52:1153–61.
31. Al-Rawi B, Hassan B, Vandenberge B, Jacobs R. Accuracy assessment of three-dimensional surface reconstructions of teeth from cone beam computed tomography scans. J Oral Rehabil. 2010;37:352–8.
32. Christofzik D, Bartols A, Faheem MK, Schroeter D, Groessner-Schreiber B, Doerfer CE. Shaping ability of four root canal instrumentation systems in simulated 3D-printed root canal models. PLoS One. 2018;13(8):e0201129.
33. Eken R, Sen OG, Eskitascioglu G, Belli S. Evaluation of the effect of rotary systems on stresses in a new testing model using a 3-dimensional printed simulated resin root with an oval-shaped canal: a finite element analysis study. J Endod. 2016;42(8):1273–8.
34. Mohmmed SA, Vianna ME, Penny MR, Hilton ST, Knowles JC. The effect of sodium hypochlorite concentration and irrigation needle extension on biofilm removal from a simulated root canal model. Aust Endod J. 2017;43(3):102–9.
35. Mohmmed SA, Vianna ME, Penny MR, Hilton ST, Mordan NJ, Knowles JC. Investigations into in situ Enterococcus faecalis biofilm removal by passive and active sodium hypochlorite irrigation delivered into the lateral canal of a simulated root canal model. Int Endod J. 2018;51(6):649–62.
36. Byun C, Kim C, Cho S, Baek SH, Kim G, Kim SG, Kim SY. Endodontic treatment of an anomalous anterior tooth with the aid of a 3-dimensional printed physical tooth model. J Endod. 2015;41(6):961–5.
37. Lee SJ, Jang KH, Spangberg LS, Kim E, Jung IY, Lee CY, Kum KY. Three-dimensional visualization of a mandibular first molar with three distal

roots using computer-aided rapid prototyping. Oral Surg Oral Med Oral Pathol Oral Radiol Endod. 2006;101(5):668–74.
38. Kfir A, Telishevsky-Strauss Y, Leitner A, Metzger Z. The diagnosis and conservative treatment of a complex type 3 dens invaginatus using cone beam computed tomography (CBCT) and 3D plastic models. Int Endod J. 2013;46(3):275–88.
39. Gok T, Capar ID, Akcay I, Keles A. Evaluation of different techniques for filling simulated C-shaped canals of 3-dimensional printed resin teeth. J Endod. 2017;43(9):1559–64.
40. Zafar S, Renner MP, Zachar JJ. Dental trauma simulation training using a novel 3D printed tooth model. Dent Traumatol. 2020;36(6):641–7.
41. Sharma S, Logani A. Third molar autotransplantation: a predictable treatment option in the era of dental implants. J Pierre Fauchard Acad (India). 2016;30:21–3.
42. Verweij JP, Anssari Moin D, Wismeijer D, van Merkesteyn JPR. Replacing heavily damaged teeth by third molar autotransplantation with the use of cone-beam computed tomography and rapid prototyping. J Oral Maxillofac Surg. 2017;75(9):1809–16.
43. Lee Y, Chang SW, Perinpanayagam H, Yoo YJ, Lim SM, Oh SR, Gu Y, Ahn SJ, Kum KY. Autotransplantation of mesiodens for missing maxillary lateral incisor with cone-beam CT-fabricated model and orthodontics. Int Endod J. 2014;47(9):896–904.
44. Cahuana-Bartra P, Cahuana-Cárdenas A, Brunet-Llobet L, Ayats-Soler M, Miranda-Rius J, Rivera-Baró A. The use of 3D additive manufacturing technology in autogenous dental transplantation. 3D Print Med. 2020;6(1):16.
45. Shahbazian M, Jacobs R, Wyatt J, Denys D, Lambrichts I, Vinckier F, Willems G. Validation of the cone beam computed tomography-based stereolithographic surgical guide aiding autotransplantation of teeth: clinical case-control study. Oral Surg Oral Med Oral Pathol Oral Radiol. 2013;115(5):667–75.
46. Lee SJ, Kim E. Minimizing the extra-oral time in autogeneous tooth transplantation: use of computer-aided rapid prototyping (CARP) as a duplicate model tooth. Restor Dent Endod. 2012;37(3):136–41.
47. Murray PE, Garcia-Godoy F, Hargreaves KM. Regenerative endodontics: a review of current status and a call for action. J Endod. 2007;33:377–90.
48. Yu H, Zhang X, Song W, Pan T, Wang H, Ning T, Wei Q, Xu HHK, Wu B, Ma D. Effects of 3-dimensional bioprinting alginate/gelatin hydrogel scaffold extract on proliferation and differentiation of human dental pulp stem cells. J Endod. 2019;45(6):706–15.
49. Athirasala A, Lins F, Tahayeri A, Hinds M, Smith AJ, Sedgley C, Ferracane J, Bertassoni LE. A novel strategy to engineer pre-vascularized full-length dental pulp-like tissue constructs. Sci Rep. 2017;7(1):3323.
50. Duarte Campos DF, Zhang S, Kreimendahl F, Köpf M, Fischer H, Vogt M, Blaeser A, Apel C, Esteves-Oliveira M. Hand-held bioprinting for *de novo* vascular formation applicable to dental pulp regeneration. Connect Tissue Res. 2020;61(2):205–15.
51. van der Meer WJ, Vissink A, Ng YL, Gulabivala K. 3D computer aided treatment planning in endodontics. J Dent. 2016;45:67–72.
52. Cheng Q, Xia WW. Research and application of three-dimensional printed template in endodontics. Zhonghua Kou Qiang Yi Xue Za Zhi. 2019;54(1):67.
53. Moreno-Rabié C, Torres A, Lambrechts P, Jacobs R. Clinical applications, accuracy and limitations of guided endodontics: a systematic review. Int Endod J. 2020;53(2):214–31.
54. Lio F, Mampieri G, Mazzetti V, Leggeri A, Arcuri L. Guided endodontic microsurgery in apicoectomy: a review. J Biol Regul Homeost Agents. 2021;35(3 Suppl. 1):47–55.
55. Macho ÁZ, Ferreiroa A, Rico-Romano C, Alonso-Ezpeleta LÓ, Mena-Álvarez J. Diagnosis and endodontic treatment of type II dens invaginatus by using cone-beam computed tomography and splint guides for cavity access: a case report. J Am Dent Assoc. 2015;146(4):266–70.
56. Buchgreitz J, Buchgreitz M, Mortensen D, Bjørndal L. Guided access cavity preparation using cone-beam computed tomography and optical surface scans - an ex vivo study. Int Endod J. 2016;49(8):790–5.
57. Mena-Álvarez J, Rico-Romano C, Lobo-Galindo AB, Zubizarreta-Macho Á. Endodontic treatment of dens evaginatus by performing a splint guided access cavity. J Esthet Restor Dent. 2017;29(6):396–402.
58. Buchgreitz J, Buchgreitz M, Bjørndal L. Guided root canal preparation using cone beam computed tomography and optical surface scans - an observational study of pulp space obliteration and drill path depth in 50 patients. Int Endod J. 2019;52(5):559–68.
59. Tavares WLF, Viana ACD, de Carvalho Machado V, Henriques LCF, Sobrinho APR. Guided endodontic access of calcified anterior teeth. J Endod. 2018;44(7):1195–9.
60. Lara-Mendes ST d O, Barbosa CFM, Santa-Rosa CC, Machado VC. Guided endodontic access in maxillary molars using cone-beam computed tomography and computer-aided design/computer-aided manufacturing system: a case report. J Endod. 2018;44(5):875–9.
61. Lara-Mendes STO, Barbosa CFM, Machado VC, Santa-Rosa CC. A new approach for minimally invasive access to severely calcified anterior teeth using the guided endodontics technique. J Endod. 2018;44(10):1578–82.
62. Nayak A, Jain PK, Kankar PK, Jain N. Computer-aided design-based guided endodontic: a novel approach for root canal access cavity preparation. Proc Inst Mech Eng H. 2018;232(8):787–95.
63. Torres A, Shaheen E, Lambrechts P, Politis C, Jacobs R. Microguided Endodontics: a case report of a max-

illary lateral incisor with pulp canal obliteration and apical periodontitis. Int Endod J. 2019;52(4):540–9.
64. Buchgreitz J, Buchgreitz M, Bjørndal L. Guided endodontics modified for treating molars by using an intracoronal guide technique. J Endod. 2019;45(6):818–23.
65. de Athayde Casadei B, Lara-Mendes ST d O, Barbosa CFM, Araújo CV, de Freitas CA, Machado VC, Santa-Rosa CC. Access to original canal trajectory after deviation and perforation with guided endodontic assistance. Aust Endod J. 2020;46(1):101–6.
66. Kostunov J, Rammelsberg P, Klotz A-L, Zenthöfer A, Schwindling FS. Minimization of tooth substance removal in normally calcified teeth using guided endodontics: an in vitro pilot study. J Endod. 2021;47(2):286–90.
67. Maia LM, de Carvalho Machado V, da Silva NRFA, Brito Júnior M, da Silveira RR, Moreira Júnior G, Sobrinho APR. Case reports in maxillary posterior teeth by guided endodontic access. J Endod. 2019;45(2):214–8.
68. Perez C, Finelle G, Couvrechel C. Optimisation of a guided endodontics protocol for removal of fibre-reinforced posts. Aust Endod J. 2020;46(1):107–14.
69. Tchorz JP, Wrbas K-T, Hellwig E. Guided endodontic access of a calcified mandibular central incisor using a software-based three-dimensional treatment plan. Int J Comput Dent. 2019;22(3):273–81.
70. Loureiro MAZ, Elias MRA, Capeletti LR, Silva JA, Siqueira PC, Chaves GS, Decurcio DA. Guided endodontics: volume of dental tissue removed by guided access cavity preparation-an ex vivo study. J Endod. 2020;46(12):1907–12.
71. Krug R, Reich S, Connert T, Kess S, Soliman S, Reymus M, Krastl G. Guided endodontics: a comparative in vitro study on the accuracy and effort of two different planning workflows. Int J Comput Dent. 2020;23(2):119–28.
72. Silva AS, Santos ACC, Caneschi C d S, Machado VC, Moreira AN, Morgan LFDSA, Tavares WLF. Adaptable fiberglass post after 3D guided endodontic treatment: novel approaches in restorative dentistry. J Esthet Restor Dent. 2020;32(4):364–70.
73. Tavares WLF, Ferreira MVL, Machado V d C, Braga T, Amaral RR, Cohen S. Antimicrobial photodynamic therapy and guided endodontics: a case report. Photodiagn Photodyn Ther. 2020;31:101935.
74. Leontiev W, Bieri O, Madörin P, Dagassan-Berndt D, Kühl S, Krastl G, Krug R, Weiger R, Connert T. Suitability of magnetic resonance imaging for guided endodontics: proof of principle. J Endod. 2021;47(6):954–60.
75. Pujol ML, Vidal C, Mercadé M, Muñoz M, Ortolani-Seltenerich S. Guided endodontics for managing severely calcified canals. J Endod. 2021;47(2):315–21.
76. Simon JC, Kwok JW, Vinculado F, Fried D. Computer-controlled CO(2) laser ablation system for cone-beam computed tomography and digital image guided endodontic access: a pilot study. J Endod. 2021;47(9):1445–52.
77. Todd R, Resnick S, Zicarelli T, Linenberg C, Donelson J, Boyd C. Template-guided endodontic access. J Am Dent Assoc. 2021;152(1):65–70.
78. Pinsky HM, Champleboux G, Sarment DP. Periapical surgery using CAD/CAM guidance: preclinical results. J Endod. 2007;33(2):148–51.
79. Bahcall JK. Using 3-dimensional printing to create presurgical models for endodontic surgery. Compend Contin Educ Dent. 2014;35(8):e29–30.
80. Ye S, Zhao S, Wang W, Jiang Q, Yang X. A novel method for periapical microsurgery with the aid of 3D technology: a case report. BMC Oral Health. 2018;18(1):85.
81. Ahn SY, Kim NH, Kim S, Karabucak B, Kim E. Computer-aided design/computer-aided manufacturing-guided endodontic surgery: guided osteotomy and apex localization in a mandibular molar with a thick buccal bone plate. J Endod. 2018;44(4):665–70.
82. Peng L, Wang ZH, Sun YC, Qu W, Han Y, Liang YH. Computer aided design and three-dimensional printing for apicoectomy guide template. Beijing Da Xue Xue Bao. 2018;50(5):905–10.
83. Ackerman S, Aguilera FC, Buie JM, Glickman GN, Umorin M, Wang Q, Jalali P. Accuracy of 3-dimensional-printed endodontic surgical guide: a human cadaver study. J Endod. 2019;45(5):615–8.
84. Sutter E, Lotz M, Rechenberg D-K, Stadlinger B, Rücker M, Valdec S. Guided apicoectomy using a CAD/CAM drilling template. Int J Comput Dent. 2019;22(4):363–9.
85. Fan Y, Glickman GN, Umorin M, Nair MK, Jalali P. A novel prefabricated grid for guided endodontic microsurgery. J Endod. 2019;45(5):606–10.
86. Avantaggiato P, Piva A, Salamini A, Avantaggiato A, Terroni R, Carinci F. Technical note: Surgical guide for computer-aided endodontic surgery. J Biol Regul Homeost Agents. 2020;34(1 Suppl. 1):161–4.
87. George R, Cameron A, Meer M. Streamlining and simplification of surgical stent fabrication for micro-endodontic surgery. Aust Endod J. 2020;46(3):445–51.
88. Tavares WLF, Fonseca FO, Maia LM, de Carvalho Machado V, Silva NRFA, Moreira Junior G, Sobrinho APR. 3D apicoectomy guidance: optimizing access for apicoectomies. J Oral Maxillofac Surg. 2020;78(3):357.e1–8.
89. Hawkins TK, Wealleans JA, Pratt AM, Ray JJ. Targeted endodontic microsurgery and endodontic microsurgery: a surgical simulation comparison. Int Endod J. 2020;53(5):715–22.
90. Benjamin G, Ather A, Bueno MR, Estrela C, Diogenes A. Preserving the neurovascular bundle in targeted endodontic microsurgery: a case series. J Endod. 2021;47(3):509–19.
91. Ordinola-Zapata R, Bramante CM, Duarte MAH, Cavenago BC, Jaramillo D, Versiani MA. Shaping

ability of reciproc and TF adaptive systems in severely curved canals of rapid microCT-based prototyping molar replicas. J Appl Oral Sci. 2014;22(6):509–15.
92. Mohmmed SA, Vianna ME, Penny MR, Hilton ST, Mordan N, Knowles JC. A novel experimental approach to investigate the effect of different agitation methods using sodium hypochlorite as an irrigant on the rate of bacterial biofilm removal from the wall of a simulated root canal model. Dent Mater. 2016;32(10):1289–300.
93. Robberecht L, Chai F, Dehurtevent M, Marchandise P, Bécavin T, Hornez J-C, Deveaux E. A novel anatomical ceramic root canal simulator for endodontic training. Eur J Dent Educ. 2017;21(4):e1–6.
94. Liang X, Liao W, Cai H, Jiang S, Chen S. 3D-printed artificial teeth: accuracy and application in root canal therapy. J Biomed Nanotechnol. 2018;14(8):1477–85.
95. Höhne C, Schmitter M. 3D printed teeth for the preclinical education of dental students. J Dent Educ. 2019;83(9):1100–6.
96. Karatekin AÖ, Keleş A, Gençoğlu N. Comparison of continuous wave and cold lateral condensation filling techniques in 3D printed simulated C-shape canals instrumented with Reciproc Blue or Hyflex EDM. PLoS One. 2019;14(11):e0224793.
97. Kustra P, Dobroś K, Zarzecka J. Making use of three-dimensional models of teeth, manufactured by stereolithographic technology, in practical teaching of endodontics. Eur J Dent Educ. 2021;25(2):299–304.
98. Park Y-S, Baek S-H, Lee W-C, Kum K-Y, Shon W-J. Autotransplantation with simultaneous sinus floor elevation. J Endod. 2012;38(1):121–4.
99. Jang J-H, Lee S-J, Kim E. Autotransplantation of immature third molars using a computer-aided rapid prototyping model: a report of 4 cases. J Endod. 2013;39(11):1461–6.
100. Strbac GD, Schnappauf A, Giannis K, Bertl MH, Moritz A, Ulm C. Guided autotransplantation of teeth: a novel method using virtually planned 3-dimensional templates. J Endod. 2016;42(12):1844–50.
101. Abella F, Ribas F, Roig M, Sánchez JAG, Durán-Sindreu F. Outcome of autotransplantation of mature third molars using 3-dimensional-printed guiding templates and donor tooth replicas. J Endod. 2018;44(10):1567–74.
102. Soram O, Kim S, Lo HS, Choi J-Y, Kim H-J, Ryu G-J, Kim S-Y, Choi K-K, Kim D-S, Jang J-H. Virtual simulation of autotransplantation using 3-dimensional printing prototyping model and computer-assisted design program. J Endod. 2018;44(12):1883–8.
103. Kim K, Choi H-S, Pang N-S. Clinical application of 3D technology for tooth autotransplantation: a case report. Aust Endod J. 2019;45(1):122–8.
104. Strbac GD, Schnappauf A, Bertl MH, Vasak C, Ulm C, Giannis K. Guided osteotomy and guided autotransplantation for treatment of severely impacted teeth: a proof-of-concept report. J Endod. 2020;46(11):1791–8.
105. Lucas-Taulé E, Llaquet M, Muñoz-Peñalver J, Somoza J, Satorres-Nieto M, Hernández-Alfaro F. Fully guided tooth autotransplantation using a multidrilling axis surgical stent: proof of concept. J Endod. 2020;46(10):1515–21.
106. Sato M, Garcia-Sanchez A, Sanchez S, Chen I-P. Use of 3-dimensional-printed guide in hemisection and autotransplantation of a fusion tooth: a case report. J Endod. 2021;47(3):526–31.

3D Printing: Limitations, Safety, and Regulatory Considerations for Oral Health Science

13

Rong-Fu Kuo, Yu-Sheng Lin, Tai-Hua Yang, and Anh-Thu Nguyen

Contents

13.1 Introduction

3D printing, also known as additive manufacturing, was reported by "The Economist" as the technology of the fourth industrial revolution. It is one of the most well-known technologies in digital dentistry since the twentieth century. Comparing to the conventional fabrication methods, additive manufacturing takes the advantage of making complex and irregular morphology to be able produced precisely. In addition to the automation subtractive manufacturing, 3D printing can overcome some limitations of machining for inwardly chamfering with varieties of material. Moreover, additive manufacturing has a spectacular potential to economically and

R.-F. Kuo (✉)
Department of Biomedical Engineering, National Cheng Kung University, Tainan, Taiwan

Medical Device Innovation Center, National Cheng Kung University, Tainan, Taiwan

Y.-S. Lin
Department of Mechanical Engineering, Southern Taiwan University of Science and Technology, Tainan, Taiwan
e-mail: yushenglin@stust.edu.tw

T.-H. Yang
Department of Biomedical Engineering, National Cheng Kung University, Tainan, Taiwan
e-mail: yangtaihua@mail.ncku.edu.tw

A.-T. Nguyen
Department of Dental Technology-Faculty of Odonto-Stomatology, University of Medicine and Pharmacy, Ho Chi Minh, Vietnam
e-mail: anhthu.nguyen@ump.edu.vn

© The Author(s), under exclusive license to Springer Nature Switzerland AG 2022

P. K. Chaudhari et al. (eds.), *3D Printing in Oral Health Science*,
https://doi.org/10.1007/978-3-031-07369-4_13

efficiently reduce the whole design process to a final product. This technology allows dentists to fabricate in situ without sending a design to a dental workshop and shortens the lead time for patient [1, 2].

History reveals that the first industrial revolution began in Britain in the late eighteenth century, with the mechanization of the textile industry. With the help of nature power generated by cotton mill and water wheel, the factory can therefore complete some tasks with less manual work. The second industrial revolution came in the early twentieth century, when Henry Ford mastered the moving assembly line and ushered in the age of mass production. The third revolution is a combination of mechanical and analogue electronic technology to become digital electronics, which leads to a computed processor and furthermore the automation. The first three industrial revolutions made people less manual work but large production. Comparatively, the fourth revolution creates personalization and variety rather than do large-scale production of the same products [3]. That is exactly how 3D printing has been applied in the state of oral health science.

13.1.1 Role of 3D Printing in Digital Dentistry Workflow

Currently, oral health science moves toward automation, precision, patient-oriented, and cloud digitalization. The importance of digital dentistry provides the opportunity for the public to share the equally good-quality treatment and upon efficiency. When the patient walks into the outpatient, a quick diagnosis can be done via equipment like intraoral, desktop, and face scanners [4], cone beam computed tomography (CBCT). The diagnosis creates a data flow of the personalized information, followed by software for computer-aided design/computer-aided manufacturing (CAD/CAM) and fabrication procedures like milling and 3D printing. A cloud-based dental workflow also shortens the distance of shipping, delivery, and design houses in the world, everywhere the dental workshops can deal with the cases and dentists can discuss the treatment without the barrier of information. The digital revolution is changing the world: computers and digital devices are making what were previously manual tasks easier, faster, cheaper, and more predictable [5, 6].

From Fig. 13.1, it is obvious that 3D printing saves more time than traditional procedure. It only takes four steps to finish the framework in comparison with six steps of traditional technique.

There are many factors that affect the quality of 3D-printed products. Type of a printer, material property, automation, and cost all will affect the quality of finished products. Table 13.1 lists dental applications for three different typical materials manufactured by a 3D printing process.

13.1.2 Emergence of 3D Printing Technologies and Its Procedure

3D printing plays an important role in digital dentistry. The method can be referred to rapid prototyping technologies long time ago, but the variety of material and machines have been dramatically improved upon over the recent years due to the open-source society [7]. The concept is to accumulate layer by layer with a specific material, performing 2D patterns to become a 3D model [8, 9]. Many 3D printers can be purchased at a reasonable price for a clinic. The rapid popularity impacts various medical fields, including oral health science. In oral health science, everyone has different teeth morphology; thus, in specialties such as prosthodontics and orthodontics, 3D printing can easily accommodate for these varying dental features. Table 13.2 lists the most popular 3D printing methods, such as Fused Deposition Modeling (FDM), Stereolithography, Digital Light Processing (DLP), Poly Jet, Selective Laser Sintering (SLS), Selective Laser Melting (SLM), Electron Beam Melting (EBM), Bio-printing, etc. 3D printing methods are often categorized by the types of material. For example, FDM is typically using thermoplastic polymer and can be seen as the most popularly used.

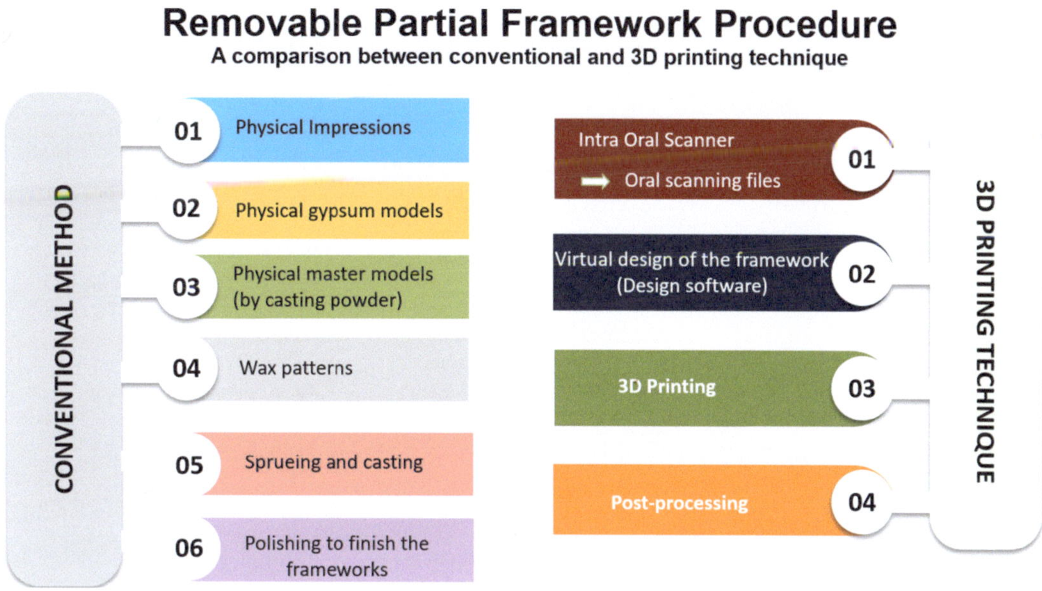

Fig. 13.1 Comparison between traditional procedure and 3D printing procedure to fabricate partial framework

Table 13.1 Dental applications for different 3D printing process

Material	Applied technology	Type of printer	Printed products	Equipment configuration considerations	Scene
Photopolymer	Stereolithography	SLA/DLP	Dental Mold	Ventilation, post-processing, cost, automation, fast, benefit	Dental Clinic
Metallic powder	SLM/EBM	3D Printer—Metal	Denture	Clean room, powder handling, mass customization	GMP Factory
Ceramic powder	SLM/EBM	3D Printer—Ceramics	Crowns Veneers	Clean room, powder handling, mass customization	GMP Factory

SLA stereolithography, *DLP* digital light processing, *SLM* selective laser melting, *EBM* electronic beam melting

Stereolithography, DLP and Poly Jet are all using photon resin sensitive to the specific spectrum light. Metal material can consider SLS, SLM, and EBM. There are many materials still under development with their specialties and uses.

13.1.3 Clinical Applications of 3D Printing in Oral Health Science

13.1.3.1 3D-Printed Devices in Prosthodontics

One of the main uses of 3D printing in oral health science is in prosthodontics, applications such as interim crowns, bridges, partial denture, and complete denture [11]. Especially in all over the world which is entering an aging society, the age-related issue of oral health would gradually lead the problem of food intake and nutrition absorption. With the skills of 3D printing and reconstruction technology, prosthesis can be more easily fitted intraorally and produced directly in the clinic. If there is a need of temporary crown, it can be provided in less than 10 min. This technology can replace the heavy labor workflow as well as satisfy the requirement of precision to avoid patient's illness. Currently, the interim crowns are often printed using stereolithography or SLM that depends on the requirement of resolution or strength. Loads of printers based on the same technologies are present in the market but differ in the machine strength, automation, and software, and therefore the prices varied. The popular brands are Formlabs [11], Micraft,

Table 13.2 A comparison of popular 3D printing methods [10]

3D printing method	Material	Advantages	Disadvantages	Resolution
Fused Deposition Modeling (FDM)	Continuous filaments of thermoplastic polymers Continuous fiber-reinforced polymers	Cheap, easy manipulation	Rough surface	50–200 μm
Stereolithography	A resin with photoactive monomers [11–14]	Smooth surface, high resolution	Restrictions of material	10–200 μm
Digital Light Processing (DLP)		Faster	Restrictions of material	
Poly Jet		Smooth surface, high resolution, large production	Expansive	
Selective Laser Sintering (SLS) Selective Laser Melting (SLS)	Compacted fine powders Metals, alloys, and limited polymers (SLS or SLM) ceramic and polymers (3DP)	High strength and duration	Slow printing Expensive High porosity in the binder method (3DP)	80–250 μm
Electron Beam Melting (EBM)		Even higher strength		250 μm
Bioprinting	Bio-ink	Cell culture	Difficult in cell living	N/A

NextDent, 3D systems [12], Envision TEC, and 3D Stratasys.

To verify the quality of printed crown and denture as compared to the conventional methods, many studies demonstrated that the accuracy is as good as the current subtractive manufacturing like the computer-aided milling [13, 15, 16]. Some additive manufacturing procedures such as stereolithography, laser sintering, or printing of materials like wax, resins, or metals have shown to be even more precise than the subtractive manufacturing [17]. In fact, the results can be divided into system errors and random errors. Influences generated by the digital workflow such as 3D intraoral optical scanners, stone casts, design software, and machine parameters can all result in the system errors [18–20]. Apart from the workflow, material, post-process, and printed geometry may also affect the dimensional accuracy and cause random errors [21–23]. Currently, this mainly relies on the veteran experience of technicians and cases [24].

13.1.3.2 Orthodontics and Invisible Aligner

With the prevalence of correct oral health concepts and raised medical aesthetics, orthodontics has become very popular, especially in younger generations. In fabrication using directly 3D printing or thermoforming upon 3D-printed models, clear aligner is definitely the spokesperson of recent development in orthodontic care [25, 26]. 3D printing-assisted fabrication of clear aligners is commonly known under the brand name "Invisalign." These thermoformed clear aligners on sequential 3D-printed dental models (molds) are normally transparent, thin, comfortable, and precise to fit in individual patients, as shown in Fig. 13.2. In some places, patient can place the order of an orthodontic treatment online, and the dental workshop will deliver a scanner or molding toolkit for patients to assess by themselves and send the results back [27]. The orthodontist can then digitally align your teeth in a sequential order and 3D print the molds. A series of aligners is then made, with each aligner moving patient's teeth ever so slightly per aligner. The entire process does not require necessity of frequent recall or follow-up visits at the clinic or hospital as compared to the conventional fixed orthodontic treatment.

13.1.3.3 Periodontal Tissue Constructs: Scaffolds

Bioprinting is one of the fascinating technologies that would be considered as an important treatment for the next generation in many medical fields. Bioprinting involves the usage of additive manufacturing technology to create structures made up of organic materials to replace or self-

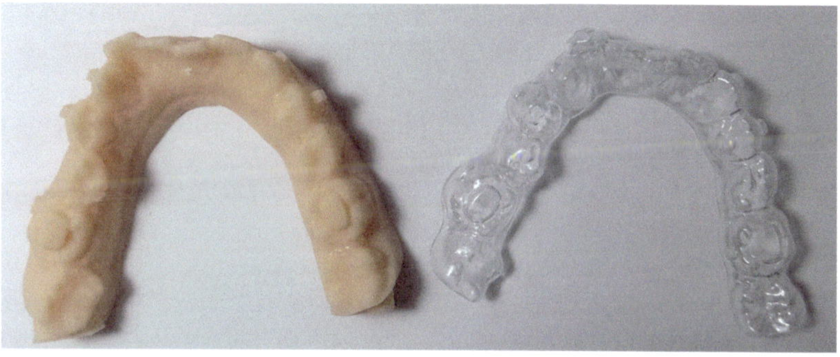

Fig. 13.2 Clear aligner with a thermoformed process after the printed dental model

repair wounds, also known as tissue engineering. This already has the innate need for biocompatibility as bioprinting almost always has the purpose of fabricating materials that can be implanted inside the human body. Bioprinting in dentistry usually involves the scaffolds or other structures that may be inserted into the gums of the teeth [28, 29].

3D printing enables one to accommodate defects of complex geometries and structures. This technology is ongoing the clinical studies of 3D-printed scaffolds [30]. After scanning the patient defect by cone-beam CT, the clinicians can use the data to create biodegradable polycaprolactone (PCL) scaffolds with high adaptability for cells to live in. This gives a promising future for the utilization of 3D printing for fabricating complicated and unpredictable geometries.

13.2 Limitations of 3D Printing in Dentistry

From the printer's point of view, influencing factors cover a wide range from the nature of the printing material to the optimized process parameters. Printing materials have a concern of their storage conditions and valid period. As for printers, users generally do not calibrate on their own but outsource maintenance to a third-party service company. In addition, the stability and successful rate of each printer are very important to purchase considerations.

Any printing service center needs to be GMP certified to provide separate manufacturing activities for dental clinics. The dentist must scan patient's teeth and send out a 3D data file. According to the dentist's treatment requirements, a dental technician then start to design and next select a printer that meets the required product characteristics. The supporting structure of the printing product must also be designed correctly to improve the printing success rate. The higher the precision degree, the longer the printing time.

Design software related to a clinical indication could be considered as a medical device. The maintenance fee of the software becomes higher. For printing dental crowns, the design time for an experienced technician is basically within a few minutes. Only after accumulating a large amount of design data, machine learning algorithms are used to model and learn the experience of humans. Thus, automatic design then can be carried out in a 3D printing workflow.

From the above description, it can be seen that digital printing is only a one-step manufacturing process to obtain its final product. Simple products such as dental models can be directly obtained through a digital 3D model obtained from an intraoral scanner. To obtain dental products requires professional or technical personnel to design, operate, and maintain the overall production process.

13.3 Pros and Cons of 3D Printing in Dentistry

13.3.1 Pros of 3D Printing in Dentistry

Benefits of the 3D printing technology include being efficient in both time and expenses. Hospitalization and operation times can be reduced through the use of 3D-printed templates and models in surgery planning, and the periodontal tissue constructs make surgical operations more efficient and accurate. Drill and cutting guides fabricated using 3D printing are another beneficial laboratory technology for improving surgical operations.

Dentists receive a variety of patients in their work period. Commonly, not one patient has the same jaw, tooth, or implant, so each procedure done would be unique to each patient. Precision would then be required to execute reproduction and reconstruction of convoluted geometries for reconstruction or restoration. The rapid prototyping nature of 3D printing gives a definite advantage in this sense, as compared to the slow and wasteful milling processes [31]. The possibility of high-quality restoration without the compromise of having a very difficult process places 3D printing operations on top of conventional CAD/CAM processes [32].

For the longest time, dentists require the assistance of technicians to create the models used for their operation and education. This complicates the process of surgery planning and only adds additional steps which take up more time in an operation schedule. Additive manufacturing offers a way for dentists to quickly customize and fabricate the needed models. This saves up the expenses for either acquiring the tools and materials needed for handmade models or simplifies the logistics in giving an order to a dental technician. This technology does not endanger the employment of the said technicians as they themselves need to be the ones that are educated and updated to provide better service to their industry.

13.3.1.1 Comfortable Experience for Patients

3D printing is a component of the digital dental process, which has grown in popularity in recent years. The technique normally starts with a digital impression taken via oral scanning rather than the traditional approach of physical impression with impression material. This prevents the patient from uncomfortable, even painful experiences with conventional physical impressions such as vomiting when the material overflows too deeply into the palate or the compressive force damages the mucosa. Thorsten Grunheid et al. [33] conducted a study to evaluate oral scanning technique took (using direct chairside oral scanners (Lava COS; 3M ESPE, St Paul, Minn)). They compared "digital impressions" taken by oral scanners to alginate impressions in terms of accuracy, time-savings, and patient acceptance and found that although both methods were accurate, oral scanning required less chairside time as well as reduced uncomfortable feeling for patients.

Furthermore, while physical impressions may require retaking in some circumstances, digital impressions can be saved and used at any time, saving time, money, and effort.

13.3.1.2 Time-Saving

The principle of saving time can be applied to dentists, dental technicians, and, most importantly, patients.

There are several phases in the traditional method, and none of them can be bypassed. To make a porcelain crown, for example, the dentist grinds the teeth crown in the patient's mouth before taking an impression. This impression is sent to the dental lab for further processing. Before the porcelain teeth can be finalized and returned to the dentist, they must go through four to five steps in the dental laboratory. In some situations, an intermediate stage of crown try-in in the patient's mouth is required, requiring the patient to return to the clinic one more time. Even the shipment time should be considered.

The digital process is much simpler when starting with taking digital impressions by oral scanning; the impression file will be rapidly delivered to the dental laboratory with just one mouse click for prosthetic design (CAD). The CAD file is then transferred to the 3D printer to complete the restoration. The printing, in instance, is totally automated, allowing the dental technician to accomplish other tasks while waiting, resulting in increased productivity.

Today, an increasing number of dentists are investing in 3D printers to create chairside models, and this trend will continue in the future. As a result of this tendency, restorations are being produced at a faster rate than ever before. Patients don't have to travel or wait for restorations because they can acquire prosthesis in one visit.

13.3.1.3 Cost-Effective

As previously said, 3D printing and digital dentistry reduce time, which in turn saves money.

Furthermore, the digital method with 3D printing eliminates many intermediate processes that require a variety of expensive machines and materials, particularly the metal casting or plastic injection of dentures molding process. The laboratories can save money on this expense by using 3D printing.

This also reduces the amount of discarded materials, helping to safeguard the environment and providing a clean working environment for dental professionals, as there is no gypsum dust or grinding dust to worry about.

13.3.1.4 Accuracy and Precision

It is an indisputable fact that traditional dental prostheses, like many other handicrafts, rely heavily on the talents and knowledge of dentists and dental technicians in terms of aesthetics and functionality, especially in the tiniest of details.

When all of these subjective aspects are digitized by particular parameters set on dental software and machines, the digital process with 3D printing will help to minimize them.

Furthermore, by taking a digital impression, errors caused by materials such as alginate distortion due to weather will be reduced.

There is a number of factors that contribute to the survival of a fixed restoration and in which marginal and internal fit of fixed prosthetic constructions will play a vital role [34]. Marginal fit is a term to mention about the measured gap between the margin of the restoration and the finish line of the prepared tooth on patient's mouth [35]. Because any gap between them creates an excellent environment for biofilm deposition, which leads to the development of caries and periodontal disease, the measured number of the most perfect marginal fit will be zero. According to the American Dental Association, the optimal fit for a margin will be between 25 and 40 μm. However, with present production technology, achieving these numbers is very challenging [36]. The marginal fit resulted from various factors including the materials used, the type of finish line used and the manufacturing process used. The procedure for fabricating the prosthesis will be discussed in this chapter. Saurabh Chaturvedy et al. [37] studied the marginal fit of fixed construction created using the most prevalent technologies at the time: the traditional method, 3D printing, and milling. The findings of this study revealed that 3D current printed crowns had a better marginal and internal fit than milled and molded crowns. Other research conducted by Nawal Alharbi et al. [38] with the marginal fit of 3D-printed contemporary crown ranged from 28 to 40 μm compared to 38 to 56 μm of those were manufactured by milling.

13.3.1.5 Superior Aesthetics

The excellent aesthetic value of 3D printing is an indisputable benefit.

First and foremost, as previously stated, 3D-printed restorations are extremely rapid, which means that the patient will not experience the no-teeth phenomena or be forced to wear a temporary prosthesis for an extended period time while awaiting permanent restorations. This is especially important for patients who are concerned about their looks or who engage in communicative activities.

> The goal of an aesthetic makeover is to develop a peaceful and stable masticatory system, where the teeth, tissues, muscles, skeletal structures and joints all function in harmony (Peter Dawson) [39].

In order to assess a prosthesis in terms of functioning and aesthetics, it must be considered in conjunction with other elements such as lips, cheeks, and motions. The grin is one of the most important actions in prosthesis aesthetics. To test this in the past, we had to make a restoration and trial it in the patient's mouth, which took a long time and cost a lot of money. Today's dental restorations may accomplish this harmonious aesthetic in the most effective and cost-efficient way, thanks to the advent of Smile Design at the treatment planning stage. Smile design is only the beginning; to communicate the dentist's, technician's, and patient's expectations from the design files to the final restoration, 3D printing is the best option.

13.3.2 Cons of 3D Printing in Dentistry

3D printing technology is expensive because it has the ability to produce extremely precise outcomes. Hardware and software, like scanners and dental design software, are required throughout the procedure. Furthermore, processes like stereolithography and selective laser melting necessitate the use of expensive materials and energy. These costs will be added to the original estimate. These costs will increase the original price, which may be out of reach for some patients.

The types of materials that can be used by various 3D printing processes are likewise limited. Light curable liquid polymers containing resins, for example, are commonly employed in stereolithography and can cause skin irritation and inflammation when inhaled or touched [40]. Tissue engineering necessitates the use of specialized machines, which adds to the printing costs [41].

Post-processing is a time-consuming operation, which reduces the likelihood of 3D printing being used. Fine metal powders, and much finer nanoparticle trash, pose major health and safety risks. As a result, post-production equipment is required, which takes up space and adds to the costs.

13.3.2.1 Environmental Impacts and Occupational Health

This aspect of 3D printing was covered in a study by the National Science Foundation in October 2014 [42]. This study examined the effects of 3D printing on the environment based on three factors: energy consumptions, waste management, and air pollution.

To compare the total energy consumption of the three most common manufacturing methods: bulk-forming, subtractive, and additive processes, Hae-Sung Yoon et al. did a very complete review as well as a reasonably detailed case study [43]. These three approaches are examined and measured with various techniques for each to ensure fair and accurate findings (e.g., SLA, SLS for 3D printing; injection molding and metal casting for conventional method and milling, grinding, turning of subtractive process). Furthermore, the energy consumed for each stage of each procedure is thoroughly measured before an aggregate value is given. The energy consumption of the three techniques stated above is summarized in Table 13.3, and it is clear that 3D spends the most energy.

We can easily recognize that 3D technology requires a significant amount of energy to operate, much more than that used for milling and drilling machines. Despite the fact that different fabrication processes exist, a study (Atkins Project) [44] conducted at Loughborough University in the United Kingdom found that some 3D printing processes require 50–100 times more electrical energy than injection molding machines to produce the same object of the same weight.

Another stumbling block when it comes to environmental issues is that 3D printing is overly reliant on plastic, which is not ecologically friendly. In some circumstances, an intermediary function template is necessary, and these function templates are not recyclable and are dumped into the environment after usage.

The printer will release ultrafine particles (UFPs) and volatile organic compounds (VOCs) during operation, which will contaminate the environment and endanger human health [45]. According to Morawska et al. [46], ultrafine par-

Table 13.3 Energy consumption of conventional techniques as well as additive and subtractive processes in manufacturing (adapted from [43])

Process	Specific energy consumption (kWh kg^{-1})
Additive processes: SLA, SLS, FDM	14.5–346.4
Subtractive processes: milling, turning, grinding	6.8–188
Conventional technique: metal casting, injection molding	0.19–7.78

ticles (UFPs) are those with an aerodynamic diameter of less than 0.1 m and are produced by the 3D printing process, with typical emission rates ranging from 2×10^8 to 2×10^{12} min^{-1} [47–52]. It will be deposited in the alveoli of the lungs and then travel into the intracellular space, causing inflammatory reactions throughout the body. Exposure to particulate matter causes chronic inflammation, which leads to lung fibrosis, epithelial cell proliferation, tumor formation, and finally cancer, according to Orberdorster mouse model (1996) [53].

In addition to the respiratory system, UFPs can affect the cardiovascular system, causing symptoms such as elevated heart rate, sympathetic imbalance, altered hemostasis, and so on [46].

Aside from UFPs, volatile organic chemicals emitted by 3D printers include styrene from ABS and HIPS (10–110 g min^{-1}), lactide from polylactide PLA (4–5 g min^{-1}), and polylactide from ABS and HIPS (10–110 g min^{-1}). VOCs (volatile organic compounds) are organic chemical molecules that quickly exasperate at ambient temperature [54].

VOCs can have a variety of detrimental health impacts on those who are exposed to them, ranging from simple skin sensitivities to severe inflammatory reactions and even cancer [55–57].

For a long time (the early 1990s), a robust link between sensory complaints and VOC exposure was discovered in Berglind's study [58]. In addition, Boeglin et al. [59] undertook a large-scale epidemiological analysis (92 counties in Indiana, USA) into the occurrence of various malignancies, including brain, skin, and endocrine system tumors, among others. According to study findings, VOC exposure can be considered one of the direct causes of cancer, even in tumors that are not directly exposed to organs.

13.3.2.2 Limit Options of 3D Printing Materials

Polymers and metals are the two most often utilized material categories in dental 3D printing at the moment. Table 13.4 [60] discusses the most prevalent 3D printing materials used in dentistry for the creation of restorations, jaw models, guide troughs, and other dental applications. Dental porcelain, on the other hand, is the ultimate hope of 3D printing. Due to its similar features to actual teeth, such as compressive strength, thermal conductivity, color durability, and high aesthetic value, dental porcelain is currently a top priority material in dentistry, particularly in aesthetic dentistry [61]. However, because it is too hard, it makes it difficult to manipulate [62], and that is also a challenge for 3D printing with porcelain [63] when compared to the same sort of restoration created by traditional or milling technologies; 3D-printed porcelain restorations often have low mechanical characteristics, and the surface is not as fine [64]. The main reason for this is that dental porcelain has a high processing temperature (ranging from 800 to 1400 °C) [65] which requires complicated processing and cracks formation during product cooling after conduction, increasing the intraluminal porosity of the final restoration and thereby lowering its quality [33]. However, a new study by Hongyu Xing et al. [66] reveals that SLA technology can print porcelain, resulting in crowns with the same aesthetic and toughness as milled crowns (800–1000 MPa) [64].

3D printing technology will likely overcome present constraints in printing porcelain materials in the future, allowing for the production of rapid, beautiful, and high-quality porcelain restorations.

Furthermore, another aspect of materials to consider is the investment in machinery for various material lines, as we cannot print with a single machine system for all material lines.

13.3.2.3 Changing the Role of Dental Technician in Dentistry Workflow

Many authors are concerned that the advancement of digital dentistry in general, and 3D printing in particular, as well as the tendency of producing chairside models at clinics, would eventually supplant the work of dental technicians, resulting in the "disappearance" of dental laboratories.

A study [67] on the impact of digital technologies such as CAD/CAM and 3D printing has

Table 13.4 Materials used for 3D printing in dentistry (adapted from [60])

Category	Material	Pros	Cons	Application	3D printing technique
Polymer	Vinyl polymer: Poly(methyl methacrylate) (PMMA)	Tunable properties Biocompatibility Ease of processing Low cost Lightweight Stability in the oral environment Aesthetic properties	• Undesirable degradation for long-term using • Poor surface properties • Weak mechanical properties	Denture base	SLS, SLA
	Styrene polymers: ABS	Less tolerance High impact strength Rigidity	High melting point; unpleasant fumes Made out of oil, so more damaging to the environment Deforms when not being print on a heated surface Hot plastic fumes when printing		SLA, FDM, SLS
	Polyesters: • PC • PCL • PLA	• Mechanically robust, amorphous, and transparent • Biodegradable and biocompatible polyester • High in vivo stability • Low melting point (~63 °C) • Highly biocompatible and possesses tunable physicochemical properties	Bisphenol A release making it a potentially harmful substance Prints degrade over time; rough texture	Orthodontic brackets, denture bases and prefabricated provisional crowns Bone tissue regeneration such as alveolar bone augmentation Drill guides for surgical insertion of dental implants, and provisional restorations for protecting teeth after crown preparations	• Electro hydrodynamic jetting SLA, FDM
Metal-based materials	Titanium and its alloys	• Corrosion resistance, biocompatibility, strength, and lightweight • Manufacturing orthopedic implants	• Potential concerns with toxicity of the vanadium or aluminum present in the alloy • More embrittlement during fabrication/heat treatment	• Removable partial dentures and overdentures • Fixed dental prostheses	• SLA • SLM • EBM
	Cobalt-based alloys	Superior corrosion and wear resistance, excellent mechanical properties, biocompatibility and low rigidity	Allergic reaction due to their nickel content	Removable partial dentures and over dentures Fixed dental prostheses	• SLA • SLM • EBM

Table 13.4 (continued)

Category	Material	Pros	Cons	Application	3D printing technique
	Stainless steel	Bioinert, corrosion resistance		Removable partial dentures and over dentures Fixed dental prostheses	• SLA • SLM • FDM
Ceramics	Zirconia	• Ability to accommodate peri-implant soft tissues well • Decreases inflammatory response, plaque accumulation, reduce bacterial population and modify fibroblast adhesion and proliferation		Dental bridges	• SLA • DLP • Inkjet 3D printing • Thermoplastic printing
	Alumina	Wear-resistant and biocompatible		Ceramic abutments, endodontic posts, orthodontic brackets, dental implants, and crowns	• SLA • DLP • Inkjet 3D printing Thermoplastic printing.

FDM fused deposition modeling, *SLA* stereolithography, *SLS* selective laser sintering, *SLM* selective laser melting

demonstrated that the advent of these techniques not only does not replace labor but also demand it. They must, however, possess high qualifications in a variety of areas, including specific expertise in a given subject, materials science, mechanics, computer technology, and foreign languages, in order to meet the job's requirements. The function of dental technicians in the closed working process of dentistry is changing as a result of this trend. As a result, technicians and dental laboratories must evolve as a matter, of course, to keep up with the trends.

13.4 Safety of 3D Printing in Dentistry

Although all printed materials must pass the ISO 10993 [68] biocompatibility test before they are approved for marketing, the storage conditions and material properties change over time, and there will be safety risks, even the sterilization process sometimes changes the product properties [69]. In vitro tests on light-cured 3D printing products using zebrafish embryonic cells showed that there is still some slight toxicity, but there is no strong clinical evidence of carcinogenesis for a long time [70]. Medical device regulations are still possible when the controllable risk is less than the clinical benefit obtained, the principle of use. The safety items related to 3D printing are listed below.

13.4.1 Risk of Fire and Burn

As we all know, 3D printers work with extremely high temperatures, from the printed surface to the hot end. The material is melted into granules and then printed layer by layer to create the desired object shape. The print head is where the material is hot and its temperature will range from 190 to 300 °C depending on the type of material being printed [71].

As a result, incorrect contact with the printer during operation may cause burns to the user, or the printer may overheat during operation, resulting in a fire.

- **Solutions**

  Wear high-quality protective gloves when operating.

  The printer should have a protective cabinet.

  Avoid using the printer continuously for too long; follow the manufacturer's instructions in this regard.

  It is recommended to use a printer equipped with a heat-exhaust system and reduce the temperature when the printer is too hot or there is a risk of fire.

  Paste the active printer warning.

  Avoid leaving flammable materials around the printer; it is best to keep the printer in a well-ventilated location.

  Equip with smoke alarm and firefighting system in the printer room.

  Limit the use of self-assembled printers.

13.4.2 Pinch Points Injuries

Pinch point hazard is a phrase for a frequent mechanical danger caused by one or more objects (machine parts) moving toward each other and crushing or shearing anything in their path [72].

Printers work on an open-loop system, which means they have a lot of mechanical movement along the *X*, *Y*, and *Z* axes, resulting in pinch spots that might trap humans or hurt them. Lesions in the limbs are the most prevalent, although they can also occur in other areas of the body.

Printers work on an open-loop system, which means they have a lot of mechanical movement along the *X*, *Y*, and *Z* axes, which can cause users to become stuck or hurt if they come into touch with them. Lesions in the limbs are the most prevalent, although they can also occur in other areas of the body.

- **Solutions**

  When operating the machine, users need to pay attention to labor protection principles: neat clothes and hair (for women) should be wrapped neatly in a helmet, and avoid wearing jewelry to minimize the risk of injury due to entangled or dropped on the machine.

  Observe to determine the range of movement of machine components before use to avoid this range.

  Avoid going into non-business areas; focus mentally while working.

13.4.3 Scraper Blades

To remove the print product from the printer, scraper blades are usually utilized. Scraper blades are frequently highly sharp and can cause injury to users, especially when working in areas where the printed object has firmly adhered to the printing surface. Wounds can be anything from a minor cut that doesn't require a bandage to a serious wound that necessitates numerous stitches in the hospital.

- **Solutions**

  Printers with a detachable print surface can be purchased. As a result, there is no need to employ scraper blades to avoid harm.

  When handling the blade, keep it away from the body to avoid "slipping," which causes the blade to follow the movement and cut into the flesh.

  When handling, special protective gloves might be worn.

13.4.4 Electric Shocks from Printer

Printers, like other equipment, are powered by electricity, and they consume a significant amount of it (this issue is mentioned on Sect. 13.3.2, energy consumption of 3D printing). As a result, operating the printer carries the risk of electric shock.

- **Solutions**

  When operating the machine, users need to pay attention to labor protection principles: neat clothes and hair (for women) should be wrapped neatly in a helmet, and avoid wearing jewelry to minimize the risk of

injury due to entangled or dropped on the machine.
Observe to determine the range of movement of machine components before use to avoid this range.
Avoid going into non-business areas; focus mentally while working.

13.4.5 Removing Support Structures from Finished Prints

Printed objects need a supporting structure underneath. These support structures are typically printed with low ink density as well as thinner tube wall thicknesses so that they can be easily detached from the finished object.

There are some designs that will create cutouts that separate these two components (supporting objects and structures) first or allow the support to be printed with a different material.

However, no matter which method is applied, removing this structure at the post-processing step has certain risk factors such as:

- This part of the structure is often sharp and can cut the hand causing injury to the operator.
- In addition, removal with an abrasive will produce fine dust particles (usually plastic dust) that can get into the eyes or cause skin irritation by contact in some individuals.
- To print an object, a lot of supporting structures are required (Fig. 13.3), which are often very thin with sharp edges that can cause injuries to the technician at the post-processing stage from scratches to more serious injuries.

- **Solutions**
  There is software available today that minimizes the size or use of support structures while still providing good quality of the finished product. Users should learn and consider using the software.
  There should be a dedicated room and desk when manipulating post-processing steps with ventilation systems and physical barriers to prevent dust.

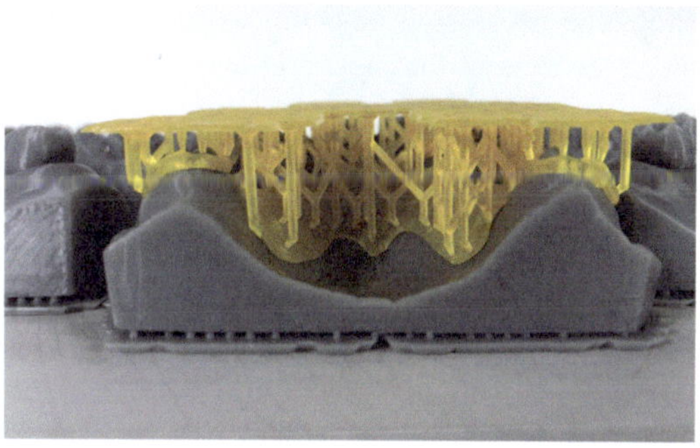

Fig. 13.3 Resin removable partial framework with support structures

In addition, once again, the operator needs to have good protective equipment such as gloves and goggles to minimize damage to the body when working.

13.4.6 Dangers from Materials: Health Impacts from Ultrafine Particles (UFPs) and Volatile Organic Compounds (VOC)

Printers produce a lot of dangerous gases and ultrafine particles when they're running. Many studies have shown a link between these toxic emissions and health issues ranging from simple skin allergies to more serious cancers, depending on a variety of factors such as the materials and equipment utilized [73, 74]. This problem is carefully discussed in Sect. 13.3.2.

Several solutions to this problem, which can influence both the computer and the user, are presented in this section. These measures can be classified into four categories in general: plants, increased ventilation, personal protective equipment (PPEs), and engineering controls are all things that can be done [41].

To begin with, most vegetation absorbs carbon dioxide and produces oxides during photosynthesis, helping to air purification and providing a healthy habitat for other living organisms in the same area. Some plants have also been discovered to be capable of absorbing UFPs and VOCs produced by 3D printers, according to recent research. There are two types of plants that should be considered: to begin with, since formaldehyde (one of the most common VOCs) may be efficiently absorbed from the atmosphere. The areca palm (or butterfly palm) is the second plant that can remove toluene and xylene from the atmosphere [75].

Second, improve the room's ventilation to produce a well-ventilated environment. The simplest method is to place the printer in a room with numerous windows; nevertheless, it is vital to guarantee that the air entering the room is similarly clean. Air conditioners are commonly installed in offices nowadays, which we may combine with air purifiers to lower the amount of VOCs and ultrafine particles produced by the 3D printer when it is in use [75].

Increased personal protection equipment (PPE) while working is the third method: wear a mask to protect your respiratory system and to prevent hazardous gas and fine particle inhalation. Wear gloves to avoid allergies caused by direct contact with materials. Furthermore, goggles are an essential piece of safety equipment for protecting eyes from mechanical and chemical damage while working [76]. However, it is critical to utilize protective equipment that is tailored to each individual to achieve a tight fit and eliminate leaks. Furthermore, users must be instructed on how to correctly use PPE to get the most out of it.

Finally, to isolate the printer from its surroundings, a self-contained enclosure with integrated exhaust and fume hood might be employed. Fans or air filtration systems with HEPA [77] and carbon filters can assist in filtering out UFPs and VOCs emitted from the printer.

13.5 Regulatory Issues in 3D Printing in Dentistry: International and National Perspective

13.5.1 The United States

3D printing is still not a standard manufacturing process that includes greater freedom of design, simple customization, and novel material properties. According to the workflow of 3D printing, there are many variables that may affect the quality and safety of finished products. These are their form and shape of printing materials (powder/particles/monomers), type of printers, printing process parameters, post-processing, cleaning, and sterilization. Before printing, in the digital design phase, patient's 3D model obtained by intraoral scanning is used for design. The software in design is also a regulated medical device. Because of the diversity of 3D printers, FDA tends to regulate finished device and software. Manufacturers know the best condition of printing process and must provide verification and validation data of their registered device. The definition of a "manufacturer" could be printing as dental labs, dental department in a hospital, dental technicians, and dentists/dental clinics. Thus, this complicates the regulatory framework.

Due to the high advanced development of digital dentistry, the process of 3D printing shifts from dental lab to clinic gradually. This trend drives the harmonization of 3D printing workflow to a daily practice of treatment process in a dental clinic or hospital. How a clinic implements the requirement of a quality system needs to be regulated according to the level of risk. FDA treats the materials used for 3D printing as finished devices and regulates them at the point of care. FDA provides a guidance, "Technical Considerations for Additively Manufactured Medical Devices," for manufacturers to follow [78]. Most dental devices fall into Class I or Class II category. The examples of 3D-printed Class I and Class II devices have been presented in Table 13.5.

The Class I [79] medical devices are defined by US FDA as "not intended for use in supporting or sustaining life or of substantial importance in preventing impairment to human health, and

Table 13.5 Classification of 3D printing dental medical devices (adapted from [80–85])

Class I	Surgical guides, custom trays
Class II	Denture bases, temporary crowns and bridges, mouth guards, and aligners

they may not present a potential unreasonable risk of illness or injury." The Class II [79] medical devices are defined as "devices for which general controls are insufficient to provide reasonable assurance of the safety and effectiveness of the device." A medical device must assure its safety and effectiveness based on the control level of safety and effectiveness. If a medical device is classified as Class I or II, and if it is not exempt, then a 510(k) will be required to market the device in the USA.

For registration, Class II dental equipment must follow the 510(k) process. Applicants must show that their device is at least as safe and effective as a legally marketed, or predicate, device under the 510(k) method. 3D-printed denture teeth, on the other hand, may be exempt from the 510(k) process.

"A 510(k) is a premarket submission made to FDA to demonstrate that the device to be marketed is as safe and effective, that is, substantially equivalent, to a legally marketed device" [78]. Any individual, organization, or company seeking FDA 510(k) approval must first conduct a comparative analysis of their products against one or more similar items already on the market. Products must be compared specifically in terms of quality and safety. Unless the device was marketed before May 28, 1976, anyone intending to sell medical equipment in the United States must submit a 510(k) at least 90 days before it is put on the market. Furthermore, with licensed devices on the market, if a change or modification is made to a lawfully marketed device, such change could have a major impact on the device's safety or effectiveness [78].

13.5.2 EU

There are provisions for customized products in the "Guidance on legislation: Borderlines with Medical Devices," and the scope of application includes dental appliances such as dental alloys and dental ceramics. The EU's customized products must meet the description of Article 11(6) of the 93/42/EEC Medical Device Directive and the requirements of Annex VIII. However, those major manufacturers of 3D printing machines do not need to obtain device-related product certification, but they still need to pass CE product certification for the market. Therefore, 3D-printed medical devices are currently managed by the requirements of customized products in the European Union [86].

In MDR (Medical Device Regulation) 2017/745 [87], a custom-made product is a medical device that has "specific design characteristics" that make it suitable "for the sole use of a particular patient exclusively to meet their conditions and needs." The new regulation goes on to exclude two categories of medical device product from the custom-made definition:

1. Mass-produced medical device products which are adapted to the specific requirements of a patient
2. Mass-produced products manufactured as per a written prescription

Therefore, "custom-made" can only be applied to products made from scratch.

The result of such change is a new regulatory uncertainty for the manufacturers of 3D-printed medical device. The manufacturer could be a hospital or an outsourcing vendor. However, manufacturers are advised to follow existing medical device manufacturing standards. It is also very likely future CE mark requirements will be parallel to its US counterpart.

Since the European Union Medical Device Directive (MDD, 93/42/EEC) (2016) includes more than 16 annexes, which deal with medical device-related issues, such as the basic requirements for patient health and safety, it depends on the type of medical devices, involving materials, packaging (sterile or non-sterile), labeling, and waste disposal procedures. The customized device should have a unique identification number, detailed information about the manufacturer, the name of a professionally qualified doctor, the name of the patient, and the intended indication of the device. If the product is implantable, this information should be available for 10–15 years.

13.5.3 China

At present, the technology of 3D printing teeth in mainland China is mature, and it is rapidly popularized in major dental hospitals, while the technology of 3D printing cells, soft tissues, organs, etc. still needs further research and development. The use of 3D printing equipment for manufacturing in hospitals has become widespread. However, because medical institutions themselves do not have a production quality system for medical equipment products, medical institutions engaged in 3D printing technology in practice use GMP (Good Manufacturing Practice) [88]-compliant medical equipment manufacturers to commission on-demand customers. It complies with the "Regulations on the Supervision and Administration of Medical Devices" reviewed and approved by the State Food and Drug Administration. It is used to support life-sustaining medical devices that are potentially dangerous to the human body and must be strictly controlled for their safety and effectiveness. It can be seen that although the products manufactured by 3D printing technology are different from the products produced in factories in the traditional sense, they still belong to the definition of "product" stipulated in the "Product Quality Law", "Products used for sale."

13.5.4 Japan

The "Layered Manufacturing Medical Device Development Guide 2015 (General)" was issued for the utilization of layered manufacturing technology, which is useful for the development of clinically necessary models, surgical instruments, defective restorations, and high-adaptability grafts. The so-called multilayer manufacturing technology refers to the additive manufacturing technology that is automatically manufactured with three-dimensional data according to ASTM (American Society of Testing and Materials) F2792 [89] standards. In addition, it can be expected to be used in application cases of medical devices.

13.5.5 India

Although the importance of intelligent manufacturing has been addressed from research, India has not yet provided their official 3D printing guidance. India's Drugs and Cosmetics Act, 1940, expands the definition of "drug" to the diagnosis, prevention, or treatment of a disease. This is same as the definition of medical devices. The finished products from 3D printing then can be regulated by the India's Drugs and Cosmetics Act, 1940 [90].

13.6 3D Printing in Dental Training and Education

The emergence of CAD/CAM technology made it become a staple in dental laboratories. What was once only possible through rigorous and fine artisan processes can now be produced digitally. This in turn brings a requirement for dentists and dental technicians to become acquainted and updated with the technology of 3D printing, and rapid prototyping in general. Training and education are needed for them to overcome the learning curve associated with the technology [91].

Dentistry's educational sectors have long acknowledged the importance of mastering this new technology. It provides excellent prospects for high-precision replication of orofacial anatomy and complex geometry, which can be used to instruct students and practitioners in various maxillofacial operations.

13.6.1 Benefits

The introduction of 3D printing has improved dentistry education, particularly preclinical training in universities and educational institutions. As we all know, proper preclinical preparation for dentistry students reduces errors in patient manipulation. To do this, educators must establish a practice environment that is as close to that of the students as possible. Prior to 3D printing, however, this was difficult, if not impossible, to achieve in some specializations due to a lack of specimens to serve as learning materials, non-standard status, or a lack of diversity.

All of this is solved through 3D printing, which allows for the creation of an infinite number of instructional function templates. Educators can use real data from patients to generate straightforward, colorful, and easily accessible function models for a range of illnesses that can be used as the foundation of training materials. Because the function template can be reprinted many times based on the recorded data, students will be exposed to a range of scenarios and will be able to manipulate them until they are adept [91, 92].

Furthermore, because the patient's (extracted tooth) specimens are employed as learning materials, the 3D model eliminates the risk of cross-contamination. It is obvious that genuine teeth, no matter how skillfully handled, pose some risks, and students are frequently advised to use extreme caution when managing them to avoid unfavorable outcomes. Education and training professionals can standardize situations and establish a uniform, fair training environment for each student by creating a variety of preclinical settings on a 3D model. This is critical in competency-based tests, particularly for individuals who have completed preclinical training and are ready to begin direct patient care.

The process of making a model in 3D usually includes five basic steps as shown in Fig. 13.4, starting with the selection and data collection of the anatomical structure to be printed (1); then using 3D compositing software to create or produce a 3D model with properties from the collected data (2); the next step is to convert this design to a format suitable for 3D printers (3); this step is closely related to the selection of select the printer and materials (4); and finally the post-processing step of the product (5) [93].

13.6.2 Applications of 3D Printing in Education Today

3D printing is now used in many areas of dentistry education, particularly those that need high practical skills and precision, such as maxillofacial surgery, due to the benefits outlined above.

Kröger et al. constructed a variety of 3D models, including a prosthodontic model for veneer

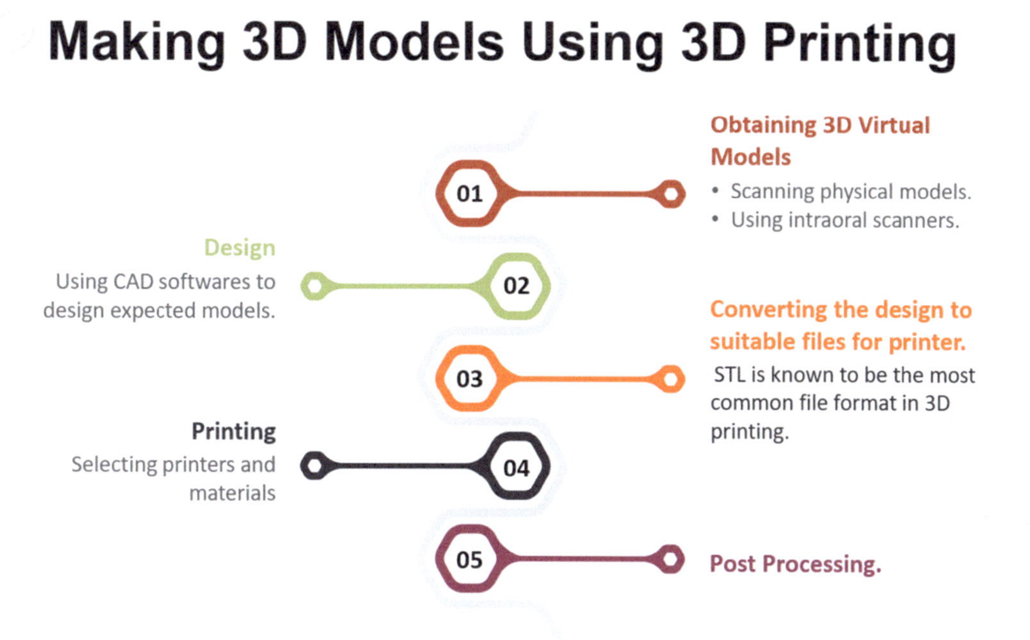

Fig. 13.4 Making models using 3D printing. (Adapted from [93])

preparation, a model for dental bonding practice, and a model with carious teeth and an insufficient crown. All of the models were created using data from real patients, resulting in highly realistic features [94]. In 2019, Christian Hohne et al. created a model of teeth that resembles the structure of natural teeth (various layers of dentin and enamel) and was utilized in crown preparation with the goal of providing students with realistic knowledge and feel when practicing [95].

The following part will describe 3D function prototyping in certain specific areas to further illustrate the application of 3D printing in dentistry training.

- **Printed Model Teeth for Endodontic Bench Exercises** [96, 97]

  For example, in endodontic training, students formerly practiced on dry teeth extracted from people, cadavers, or rosin models. The drawback is that the source of dry teeth is becoming increasingly scarce as individuals become more aware of the importance of oral health care, resulting in a decrease in the number of teeth extracted. Furthermore, dry teeth do not always satisfy the condition and qualities that educators desire, and sterilization and preservation of dry teeth will alter the tooth's characteristics, making it difficult or even judgmental. When operating on real patients, students must be able to spot errors. Models made from plastic can be expensive, while students need to manipulate many times to master. The introduction of 3D printing and micro-CT scanning has aided in the solution of this challenge by allowing for the creation of tooth models that accurately imitate desired real-tooth properties at a low cost and in an endless quantity. Micro-CT technology can photograph discrete tissue samples with slices as tiny as 0.5 μm (0.0005 mm) from people with even better resolution scans. These image files are in the (.STL) format and are used to produce models for endodontic practical training classes in universities or dental training units using 3D printers (Fig. 13.5a–f).
- **Printing in Dental Implant Education** [98, 99]

  Implant is a complex array in dentistry because it requires practitioners to master the

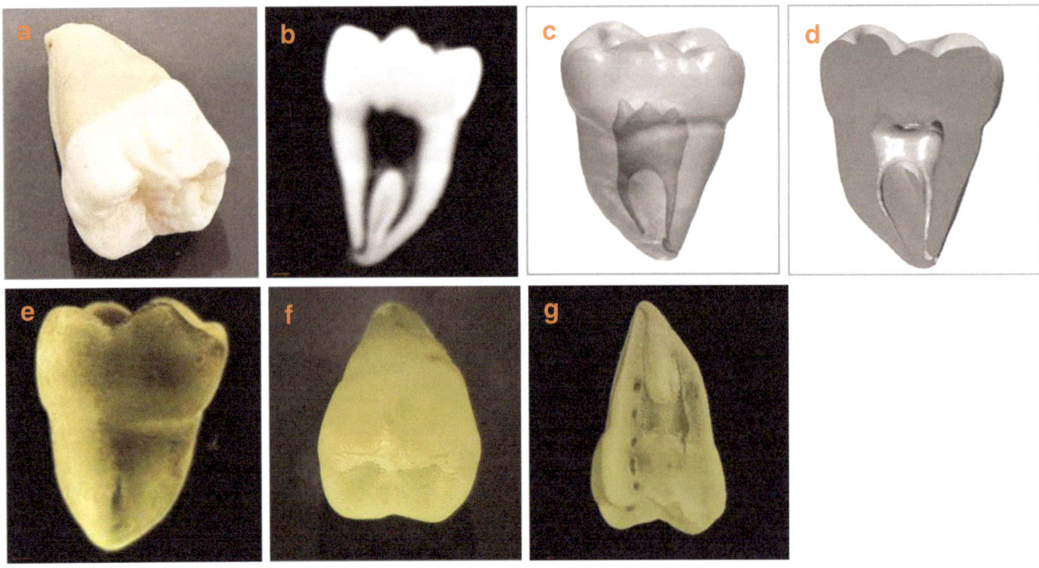

Fig. 13.5 Canned extracted tooth using Kavo OP Pro (CT). The scanned files then were converted to the 3D printable files for 3D printing to achieve 3D tooth model with root canals. (**a**) Extracted tooth from patients; (**b**) scanning file; (**c**) scanning file; (**d**) scanning file; (**e**) 3D printing tooth; (**f**) 3D printing tooth; (**g**) longitudinal section of the 3D printing tooth

knowledge of anatomical structures to be able to operate correctly. It is not allowed to make any mistake; even the smallest will cause serious problems then resulting in unpredictable and life-threatening consequences.

Normally, the observation of anatomical structures is quite difficult, especially with internal structures. Printing 3D models can benefit in visualizing vividly the entire anatomical structure of the upper and lower jaw. In particular, those are directly involved in implant procedures such as mandibular foramen, the inferior alveolar nerve, the mental foramina, the mental nerve, the lingual nerve, and the incisive canal and its associated neurovascular bundle.

Besides this, 3D-printed models can also be used to explain the patients about their dental condition and upcoming implant treatment plans, so that they can have a clearer and more precise understanding.

13.6.3 Dental Education in the Future

Ayman M. Khalifah used the term "digital student" [100] to refer to students who were born in a period when technology was being applied to education in a significant degree, and it is apparent that they profit immensely from the trend. Dentistry education and dental students are no exception to this tendency, particularly under such difficult circumstances as the COVID-19 outbreak brought to the world in 2020, when all institutions were forced to close and students were forced to learn remotely.

Extended reality (XR) is a generic term that refers to all situations in which there is a combination of actual and virtual aspects, as well as user-machine interaction through computer technologies and wearable devices [40]. 3D printing is a part of XR technology that differs only in the finished product's appearance and how it aids students in their learning. Educators will have two alternatives after creating a virtual 3D model with software: print it out as a physical model or integrate it into the XR gadget, depending on the purpose and circumstances. In terms of training, actual skills training, and dental practice, XR easily outperforms the competition.

13.7 Future Trend of 3D Printing in Dentistry

Dental restoration based on software design has become popular. The cost of 3D printing and dental restoration and treatment is an important factor that dental clinics and laboratories need to consider. Many dental clinics and laboratories have introduced digital dental technology to improve efficiency and cost reduction. Digital dental technology combined with 3D printing brings high precision, low cost, high efficiency, and dental material that conform to the standardized production chain for the dental industry. The more important significance of digital dental technology including 3D printing is to reduce the time for doctors to manually make dental products such as models and dentures and to return their energy to the diagnosis of oral diseases and the implementation of oral surgery itself. For dental technicians, although they are far away from the doctor's office, as long as the patient's oral information is obtained, accurate dental products can be customized according to the doctor's requirements.

In the application of 3D printing technology, the machine is waiting on the same production line, and the finished product is from scratch, which is very different from the existing traditional manufacturing process. How to conduct quality control, design verification, validation, and testing requires more in-depth consideration. The biggest breakthrough of 3D printing technology to the industry lies in the innovative products that were originally limited by the manufacturing process. Therefore, the generation of technology has a new opportunity. In the face of the massive development of innovative products in the future and the application of composite materials, it will be difficult in the management system.

3D printing is used more and more widely, and its medical uses are becoming more and more diversified. Future innovation trends include:

1. 3D printing production of anatomical structure objects for medical education and training purposes
2. Remake preclinical verification tests for human pathological phenomena
3. Customization of implantable medical objects and surgical equipment
4. 3D printing production of individual case models for surgical planning
5. Design files with management information
6. Color printing
7. Porous material, composites, biodegradable devices, functional devices
8. More integrated design software
9. Remote control for automation in design and manufacturing
10. Mass production
11. Intelligent printing process adopted for chairside smart workflow

13.8 Summary

The continuous innovation of 3D printing technology in various aspects including printing speed, new material development, material biocompatibility, clinical process adaption, printing accuracy, mass production, and subsequent material and surface treatment after printing makes the design and printing of complex structures possible. Thus, it can replace some traditional subtractive manufacturing processes. At present, regulatory guidelines from various countries have been released gradually, and regulatory management strategy has become increasingly clear. The development regulatory pathway of the digital dentistry industry then can be followed whether it is applied at the manufacturer side or clinical site. Based on the effective management and control of risks, 3D printers and printing materials are innovated constantly. As the price has reduced to an acceptable range for the industrial players, it also benefits from the innovative functions and simplification of collaborative design software. For example, titanium and ceramic customized 3D printing products for clinical applications will become more and more popular. Innovative cloud and chairside services will further promote the serviceability of digital dentistry. With AI, the biggest beneficiaries of smart dentistry will be patients.

References

1. Gali S, Meleppura K, Nayak VM, Shaw M. A technique of designing a 3-dimensionally printed elastomeric impression mixing tip for reuse: a step toward eco-friendly dentistry. J Prosthet Dent. 2020; https://doi.org/10.1016/j.prosdent.2020.05.035.
2. Kholgh Eshkalak S, Rezvani Ghomi E, Dai Y, Choudhury D, Ramakrishna S. The role of three-dimensional printing in healthcare and medicine. Mater Des. 2020;194:108940.
3. Jayavelu J. 1 - Advancements in three-dimensional printing for the medical device industry. In: Timiri Shanmugam PS, Chokkalingam L, Bakthavachalam P, editors. Trends in development of medical devices. San Diego: Academic Press; 2020. p. 3–16.
4. Nikoyan L, Patel R. Intraoral scanner, three-dimensional imaging, and three-dimensional printing in the dental office. Dent Clin N Am. 2020;64(2):365–78.
5. Kuo R, Fang K, Su F. Interface oral health science 2016, vol. 2016. Singapore: Springer; 2017. p. 165–71.
6. Turkyilmaz I, Wilkins GN. 3D printing in dentistry – exploring the new horizons. J Dent Sci. 2021;16:1037–8.
7. Kim GD, Oh YT. A benchmark study on rapid prototyping processes and machines: quantitative comparisons of mechanical properties, accuracy, roughness, speed, and material cost. Proc Inst Mech Eng Part B J Eng Manuf. 2008;222:201–15.
8. Jani G, Johnson A, Marques J, Franco A. Three-dimensional (3D) printing in forensic science—an emerging technology in India. Ann 3D Print Med. 2021;1:100006.
9. Katkar RA, Taft RM, Grant GT. 3D volume rendering and 3D printing (additive manufacturing). Dent Clin N Am. 2018;62:393–402.
10. Revilla-León M, Sadeghpour M, Özcan M. A review of the applications of additive manufacturing technologies used to fabricate metals in implant dentistry. J Prosthodont. 2020;29:579–93.
11. Tahayeri A, Morgan MC, Fugolin AP, Bompolaki D, Athirasala A, Pfeifer CS, Ferracane JL, Bertassoni LE. 3D printed versus conventionally cured provisional crown and bridge dental materials. Dent Mater. 2018;34:192–200.

12. Formlabs. (04 May). [Online]. Available from: https://dental.formlabs.com/indications/crown-and-bridge-models/.
13. DS. (04 May). [Online]. Available from: https://www.3dsystems.com/3d-printers/dmp dental 100.
14. Scotti CK, Velo MM d AC, Rizzante FAP, Nascimento TR d L, Mondelli RFL, Bombonatti JFS. Physical and surface properties of a 3D-printed composite resin for a digital workflow. J Prosthet Dent. 2020;124:614.e1–5.
15. Kenning KB, Risinger DC, English JD, Cozad BE, Harris LM, Ontiveros JC, Kasper FK. Evaluation of the dimensional accuracy of thermoformed appliances taken from 3D printed models with varied shell thicknesses: an in vitro study. Int Orthod. 2021;19:137–46.
16. Kim SY, Shin YS, Jung HD, Hwang CJ, Baik HS, Cha JY. Precision and trueness of dental models manufactured with different 3-dimensional printing techniques. Am J Orthod Dentofac Orthop. 2018;153:144–53.
17. Park JM, Jeon J, Koak JY, Kim SK, Heo SJ. Dimensional accuracy and surface characteristics of 3D-printed dental casts. J Prosthet Dent. 2020; https://doi.org/10.1016/j.prosdent.2020.07.008.
18. Mai HN, Lee KB, Lee DH. Fit of interim crowns fabricated using photopolymer-jetting 3D printing. J Prosthet Dent. 2017;118:208–15.
19. Homsy FR, Özcan M, Khoury M, Majzoub ZAK. Marginal and internal fit of pressed lithium disilicate inlays fabricated with milling, 3D printing, and conventional technologies. J Prosthet Dent. 2018;119:783–90.
20. Lin CH, Lin YM, Lai YL, Lee SY. Mechanical properties, accuracy, and cytotoxicity of UV-polymerized 3D printing resins composed of Bis-EMA, UDMA, and TEGDMA. J Prosthet Dent. 2020;123:349–54.
21. Atieh MA, Ritter AV, Ko CC, Duqum I. Accuracy evaluation of intraoral optical impressions: a clinical study using a reference appliance. J Prosthet Dent. 2017;118:400–5.
22. Ellakany P, Alharbi F, El Tantawi M, Mohsen C. Evaluation of the accuracy of digital and 3D-printed casts compared with conventional stone casts. J Prosthet Dent. 2020; https://doi.org/10.1016/j.prosdent.2020.08.039.
23. Ishida Y, Miyasaka T. Dimensional accuracy of dental casting patterns created by 3D printers. Dent Mater J. 2016;35:250–6.
24. Raghunath N, Pandey PM. Improving accuracy through shrinkage modelling by using Taguchi method in selective laser sintering. Int J Mach Tools Manuf. 2007;47:985–95.
25. Jindal P, Juneja M, Siena FL, Bajaj D, Breedon P. Mechanical and geometric properties of thermoformed and 3D printed clear dental aligners. Am J Orthod Dentofac Orthop. 2019;156:694–701.
26. Jindal P, Worcester F, Siena FL, Forbes C, Juneja M, Breedon P. Mechanical behaviour of 3D printed vs thermoformed clear dental aligner materials under non-linear compressive loading using FEM. J Mech Behav Biomed Mater. 2020;112:104045.
27. US OP. Agile aligner delivery Orthodontic Practice US [Internet]. Orthodontic Practice US. [cited 2021 May 10] Available from: https://orthopracticeus.com/agile-aligner-delivery/.
28. Gu Y, Van Dessel J, Politis C, Sun Y. 3D printing and 3D printed scaffolds. In: Comput Oral Maxillofac Surg. Amsterdam: Elsevier; 2021. p. 183–200.
29. Sordi MB, Cruz A, Fredel MC, Magini R, Sharpe PT. Three-dimensional bioactive hydrogel-based scaffolds for bone regeneration in implant dentistry. Mater Sci Eng C. 2021;124:112055.
30. Rasperini G, Pilipchuk SP, Flanagan CL, Park CH, Pagni G, Hollister SJ, Giannobile WV. 3D-printed bioresorbable scaffold for periodontal repair. J Dent Res. 2015;94:153S–7S.
31. Dawood A, Marti Marti B, Sauret-Jackson V, Darwood A. 3D printing in dentistry. Br Dent J. 2015;219(11):521–9. https://doi.org/10.1038/sj.bdj.2015.914. Erratum in: Br Dent J. 2016 Jan 22;220(2):86
32. Keßler A, Dosch M, Reymus M, Folwaczny M. Influence of 3D-printing method, resin material, and sterilization on the accuracy of virtually designed surgical implant guides. J Prosthet Dent. 2021; https://doi.org/10.1016/j.prosdent.2020.08.038.
33. Grünheid T, McCarthy SD, Larson BE. Clinical use of a direct chairside oral scanner: an assessment of accuracy, time, and patient acceptance. Am J Orthod Dentofac Orthop. 2014;146:673–82.
34. Heboyan AG. Marginal and internal fit of fixed prosthodontic constructions: a literature review. Int J Dent Res Rev. 2019; https://doi.org/10.28933/ijdrr-2019-06-1105.
35. Jang Y, Sim JY, Park JK, Kim WC, Kim HY, Kim JH. Evaluation of the marginal and internal fit of a single crown fabricated based on a three dimensional printed model. J Adv Prosthodont. 2018;10:367–73.
36. Campos RE, Maristela C. Marginal gap evaluation in non-cemented crown restorations. J Dent Oral Disord. 2020;6(4):1136.
37. Chaturvedi S, Alqahtani NM, Addas MK, Alfarsi MA. Marginal and internal fit of provisional crowns fabricated using 3D printing technology. Technol Health Care. 2020;28:635–42.
38. Alharbi N, Alharbi S, Cuijpers VMJI, Osman RB, Wismeijer D. Three-dimensional evaluation of marginal and internal fit of 3D-printed interim restorations fabricated on different finish line designs. J Prosthodont Res. 2018;62:218–26.
39. https://www.dentistrytoday.com/news/industrynews/item/5143-occlusion-pioneer-dr-peter-dawson-1930%202019#:~:text=2019%20Dentistry%20Today,Dr.,diagnosing%20and%20treating%20temporomandibular%20disorders.
40. Singh S, Prakash C, Singh R. 3D printing in biomedical engineering. New York: Springer; 2020. https://doi.org/10.1007/978-981-15-5424-7.

41. Mak SL, Tang WF, Li CH, Wu MY, Lai CW. A study on the ways to solve hazardous chemical emission from 3D printing process. In: ISPCE-CN 2020 - IEEE Int Symp Prod Compliance Eng, vol. 2020; 2020. p. 1–5.
42. Rejeski D, Huang Y. An NSF workshop report; 2015.
43. Yoon HS, Lee JY, Kim HS, Kim MS, Kim ES, Shin YJ, Chu WS, Ahn SH. A comparison of energy consumption in bulk forming, subtractive, and additive processes: review and case study. Int J Precis Eng Manuf Green Technol. 2014;1:261–79.
44. https://www.lboro.ac.uk/service/publicity/news-releases/2009/03_ATKINS.html.
45. Zhang Q, Pardo M, Rudich Y, Kaplan-Ashiri I, Wong JPS, Davis AY, Black MS, Weber RJ. Chemical composition and toxicity of particles emitted from a consumer-level 3D printer using various materials. Environ Sci Technol. 2019;53:12054–61.
46. Morawska L, Ristovski Z, Jayaratne ER, Keogh DU, Ling X. Ambient nano and ultrafine particles from motor vehicle emissions: characteristics, ambient processing and implications on human exposure. Atmos Environ. 2008;42:8113–38.
47. Azimi P, Zhao D, Pouzet C, Crain NE, Stephens B. Emissions of ultrafine particles and volatile organic compounds from commercially available desktop three-dimensional printers with multiple filaments. Environ Sci Technol. 2016;50:1260–8.
48. Yi J, LeBouf RF, Duling MG, Nurkiewicz T, Chen BT, Schwegler-Berry D, Virji MA, Stefaniak AB. Emission of particulate matter from a desktop three-dimensional (3D) printer. J Toxicol Environ Heal Part A Curr Issues. 2016;79:453–65.
49. Gu J, Wensing M, Uhde E, Salthammer T. Characterization of particulate and gaseous pollutants emitted during operation of a desktop 3D printer. Environ Int. 2019;123:476–85.
50. Steinle P. Characterization of emissions from a desktop 3D printer and indoor air measurements in office settings. J Occup Environ Hyg. 2016;13:121–32.
51. Zontek TL, Ogle BR, Jankovic JT, Hollenbeck SM. An exposure assessment of desktop 3D printing. J Chem Health Saf. 2017;24:15–25.
52. Stabile L, Scungio M, Buonanno G, Arpino F, Ficco G. Airborne particle emission of a commercial 3D printer: the effect of filament material and printing temperature. Indoor Air. 2017;27:398–408.
53. Oberdörster G, Oberdörster E, Oberdörster J. Nanotoxicology: an emerging discipline evolving from studies of ultrafine particles. Environ Health Perspect. 2005;113:823–39.
54. Anand SS, Philip BK, Mehendale HM. Volatile organic compounds. In: Encycl. Toxicol. 3rd ed. Amsterdam: Elsevier; 2014. p. 967–70.
55. Rumchev K, Brown H, Spickett J. Volatile organic compounds: do they present a risk to our health? Rev Environ Health. 2007;22:39–55.
56. Maroni M, Seifert B, Lindvall T. Indoor air quality, a comprehensive reference book. Amsterdam: Elsevier; 1995.
57. Mølhave L. Organic compounds as indicators of air pollution. Indoor Air. 2003;13:12–9.
58. Mølhave L, Bach B, Pedersen OF. Human reactions to low concentrations of volatile organic compounds. Environ Int. 1986;12:167–75.
59. Boeglin ML, Wessels D, Henshel D. An investigation of the relationship between air emissions of volatile organic compounds and the incidence of cancer in Indiana counties. Environ Res. 2006;100:242–54.
60. Khorsandi D, Fahimipour A, Abasian P, et al. 3D and 4D printing in dentistry and maxillofacial surgery: printing techniques, materials, and applications. Acta Biomater. 2021;122:26–49.
61. Ligon SC, Liska R, Stampfl J, Gurr M, Mülhaupt R. Polymers for 3D printing and customized additive manufacturing. Chem Rev. 2017;117:10212–90.
62. Shenoy A, Shenoy N. Dental ceramics: an update. J Conserv Dent. 2010;13:195.
63. Prasad S, Kader NA, Sujatha G, Raj T, Patil S. 3D printing in dentistry. J 3D Print Med. 2018;2:89–91.
64. Galante R, Figueiredo-Pina CG, Serro AP. Additive manufacturing of ceramics for dental applications: a review. Dent Mater. 2019;35:825–46.
65. Babu PJ, Alla RK, Alluri VR, Datla SR, Konakanchi A. Dental ceramics: Part I – An overview of composition, structure and properties. Am J Mater Eng Technol. 2015;3:13–8.
66. Xing H, Zou B, Li S, Fu X. Study on surface quality, precision and mechanical properties of 3D printed ZrO_2 ceramic components by laser scanning stereolithography. Ceram Int. 2017;43:16340–7.
67. Milan Šušić University of Business Studies Banja Luka: the importance of new technologies and their impact on industrial development, service activities and unemployment.
68. https://en.wikipedia.org/wiki/ISO_10993.
69. Ferraris S, Pan G, Cassinelli C, Mazzucco L, Verne E, Spriano S. Effects of sterilization and storage on the properties of ALP-grafted biomaterials for prosthetic and bone tissue engineering applications. Biomed Mater. 7(5):054102.
70. MacDonald NP, Zhu F, Hall CJ, Reboud J, Crosier PS, Patton EE, Wlodkowic D, Cooper JM. Assessment of biocompatibility of 3D printed photopolymers using zebrafish embryo toxicity assays. Lab Chip. 2016;16:291–7.
71. Sirjani E, Cragg PJ, Dymond MK. Glass transition temperatures, melting temperatures, water contact angles and dimensional precision of simple fused deposition model 3D prints and 3D printed channels constructed from a range of commercially available filaments. Chem Data Collect. 2019;22:100244.
72. https://en.wikipedia.org/wiki/Pinch_point_hazard.

73. Zhang Q, Wong JPS, Davis AY, Black MS, Weber RJ. Characterization of particle emissions from consumer fused deposition modeling 3D printers. Aerosol Sci Technol. 2017;51:1275–86.
74. Väisänen AJK, Hyttinen M, Ylönen S, Alonen L. Occupational exposure to gaseous and particulate contaminants originating from additive manufacturing of liquid, powdered, and filament plastic materials and related post-processes. J Occup Environ Hyg. 2019;16(3):258–71.
75. Wood RA, Burchett MD, Alquezar R, Orwell RL, Tarran J, Torpy F. The potted-plant microcosm substantially reduces indoor air VOC pollution: I. Office field-study. Water Air Soil Pollut. 2006;175:163–80.
76. Wojtyła S, Klama P, Baran T. Is 3D printing safe? Analysis of the thermal treatment of thermoplastics: ABS, PLA, PET, and nylon. J Occup Environ Hyg. 2017;14(6):D80–5. https://doi.org/10.1080/15459624.2017.1285489.
77. https://en.wikipedia.org/wiki/HEPA.
78. https://www.fda.gov/medical-devices/overview-device-regulation/classify-your-medical-device.
79. https://www.fda.gov/medical-devices/regulatory-controls/general-controls-medical-devices.
80. https://www.accessdata.fda.gov/scripts/cdrh/cfdocs/cfpcd/classification.cfm?ID=1304.
81. https://www.accessdata.fda.gov/scripts/cdrh/cfdocs/cfpcd/classification.cfm?ID=1096.
82. https://www.accessdata.fda.gov/scripts/cdrh/cfdocs/cfpcd/classification.cfm?ID=1097.
83. https://www.accessdata.fda.gov/scripts/cdrh/cfdocs/cfPCD/classification.cfm?ID=1095.
84. https://www.accessdata.fda.gov/scripts/cdrh/cfdocs/cfPCD/classification.cfm?ID=OBR.
85. https://www.accessdata.fda.gov/scripts/cdrh/cfdocs/cfPCD/classification.cfm?ID=NXC.
86. Medicine and health care products, regulatory and agency: Guidance on legislation Borderlines with medical devices and other products in Great Britain. In relation to the UK Medical Device Regulations 2002 (SI 2002 No 618, as amended). September 2021. https://assets.publishing.service.gov.uk/government/uploads/system/uploads/attachment_data/file/1023171/Rev_Borderlines_with_medical_devices_and_other_products-_V1_4_80_.pdf.
87. The European Union Medical Device Regulation of 2017. https://eumdr.com/.
88. Federal Register of July 21, 1978 (43 FR 31 508). https://www.federalregister.gov/documents/1996/10/07/96-25720/medical-devices-current-good-manufacturing-practice-cgmp-final-rule-quality-system-regulation.
89. Standard Terminology for Additive Manufacturing Technologies. https://web.mit.edu/2.810/www/files/readings/AdditiveManufacturingTerminology.pdf.
90. https://legislative.gov.in/sites/default/files/A1940-23.pdf.
91. Ford S, Minshall T. Invited review article: where and how 3D printing is used in teaching and education. Addit Manuf. 2019;25:131–50.
92. Oberoi G, Nitsch S, Edelmayer M, Janjić K, Müller AS, Agis H. 3D printing-encompassing the facets of dentistry. Front Bioeng Biotechnol. 2018;6:172. https://doi.org/10.3389/fbioe.2018.00172.
93. Garcia J, Yang ZL, Mongrain R, Leask RL, Lachapelle K. 3D printing materials and their use in medical education: a review of current technology and trends for the future. BMJ Simul Technol Enhanc Learn. 2018;4:27–40.
94. Kröger E, Dekiff M, Dirksen D. 3D printed simulation models based on real patient situations for hands-on practice. Eur J Dent Educ. 2017;21(4):e119–25. https://doi.org/10.1111/eje.12229. Epub 2016 Jul 29
95. Höhne C, Schwarzbauer R, Schmitter M. 3D printed teeth with enamel and dentin layer for educating dental students in crown preparation. J Dent Educ. 2019;83(12):1457–63. https://doi.org/10.21815/jde.019.146.
96. https://www.aegisdentalnetwork.com/id/2016/01/3d-printing-for-education-and-training-in-endodontics.
97. Reymus M, Fotiadou C, Kessler A, Heck K, Hickel R, Diegritz C. 3D printed replicas for endodontic education. Int Endod J. 2019;52:123–30.
98. Georgantza A, Loomer P, Suzuki T, Froum S, Cho S.-C, Yu Y. The use of 3D printing in dental implant education. Dent Learn. 2016.
99. Suzuki T. The use of 3D printing in dental implant education. Dent Learn. 2016. pp. 1–12.
100. Pennefather P, Krebs C. Exploring the role of xR in visualisations for use in medical education. Adv Exp Med Biol. 2019;1171:15–23. https://doi.org/10.1007/978-3-030-24281-7_2.

Future of 3D Printing in Oral Health Sciences

14

Dinesh Bhatia, Harnoor Dhillon, and Laishram Bijaya Devi

Contents

14.1 Introduction

Customization is integral to dentistry. In the dental laboratory, any appliance or restoration made for the patient is unique. The process of manual fabrication is time-consuming and laborious. It often takes multiple appointments to deliver a single appliance/restoration, and despite the effort put into making it, chances are, the result is not as expected. There is a need to overcome this kind of unpredictability, and 3D printing can play an important role.

3D printing has attracted significant attention in dentistry with increased availability and affordability. It provides personalized dental care without the hassle of frequent visits to a clinic, reduces wastage, and allows room for adjustment according to patient needs. It helps to develop the appliance with active patient feedback and participation to relieve any concerns before fabrication. The freedom to customize (in a short time), the potential to involve and educate the patient

D. Bhatia (✉)
Biomedical Engineering Department, North Eastern Hill University, Shillong, Meghalaya, India

H. Dhillon
Division of Orthodontics and Dentofacial Deformities, Center for Dental Education and Research, All India Institute of Medical Sciences, New Delhi, India

L. B. Devi
Department of Orthodontics and Dentofacial Orthopedics, Dental College, Regional Institute of Medical Sciences (RIMS), Imphal, India

© The Author(s), under exclusive license to Springer Nature Switzerland AG 2022
P. K. Chaudhari et al. (eds.), *3D Printing in Oral Health Science*,
https://doi.org/10.1007/978-3-031-07369-4_14

and address potential setbacks, gives longevity and greater acceptability to the treatment provided.

3D printing technologies are augmented by using scanners for data acquisition and creating a digital file that can be manipulated to produce the desired result. Recent surveys show widespread use of three-dimensional imaging and a belief in increased usage of 3D technologies in the future [1–5]. The increased application of 3D imaging such as intraoral scanning and cone beam computed tomography (CBCT) will promote the adaptation of 3D printing in clinical, educational, or research settings by easing and accelerating workflow for clinicians and oral health researchers.

A PubMed search of the keywords "3D printing" and "dentistry" provided 1263 results in the year range of 2000–2020 with an uptick in the research during the past 5 years (Fig. 14.1). The publications are in all major fields of dentistry, but only ten among them were clinical trials. Reviews—systematic or otherwise—totaled to around 139. This reflects the need for more rigorous experimental and original research in this area. Few studies compare 3D-printed materials and conventional ones to assess their long-term outcomes. Chaas et al. showed that after 14.7 years, 3D-printed crowns made by metal laser sintering had a survival rate of 81% with no technical complications affecting the function of the restoration [6]. Such clinical studies are the need of the hour. They will promote confidence and guide the adoption of 3D printing into clinical practice.

Digital technologies were of immense help to dentists during the pandemic. The use of intraoral scans reduced the need for repeated impression making. The interaction of dental personnel with virtual images reduced contact with potentially infected fluids from patients. Scarce interactions for procedures and fewer visits limited the spread of infection while providing necessary care, during the pandemic. 3D printers were also used to manufacture face shields and mouth masks that are routinely used in dental clinics [7, 8].

The global market for dental materials is expected to be valued at over a billion dollars in the USA alone and is expected to grow at a high rate [9]. The global dental 3D printing market is believed to be valued at nearly US$3.4 billion by the end of 2025 [10]. Continued development in technology can be expected due to constant endeavors to ease the workflow of dentists and increase patient comfort [9]. Keeping the clinical, research, and economic outlook in mind, there is no doubt that 3D printing is here to stay in dentistry for a long time.

3D printing has been used for multiple applications in the oral sciences. It has been tested for fabricating restorations and complete dentures, developing surgical guides to place accurate incisions and fabricating dental models for creating orthodontic aligners [11]. It provides an easier workflow with reduced labor costs and processing time. 3D bioprinting allows easier production

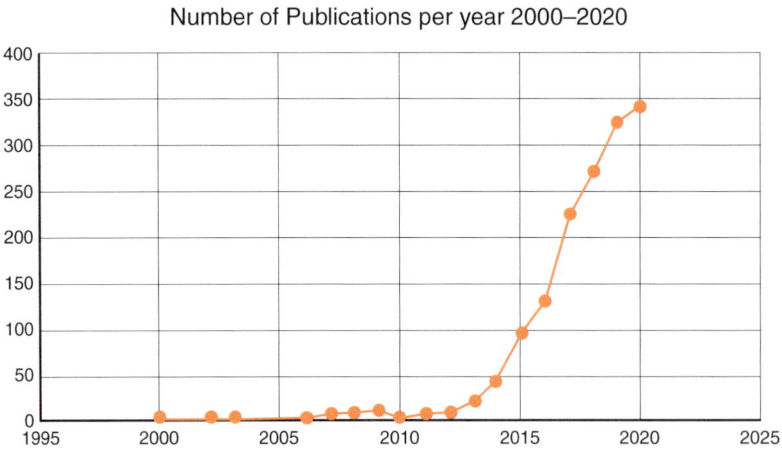

Fig. 14.1 The number of studies published over the last 20 years in PubMed when searched with keywords "3D printing" and "dentistry"

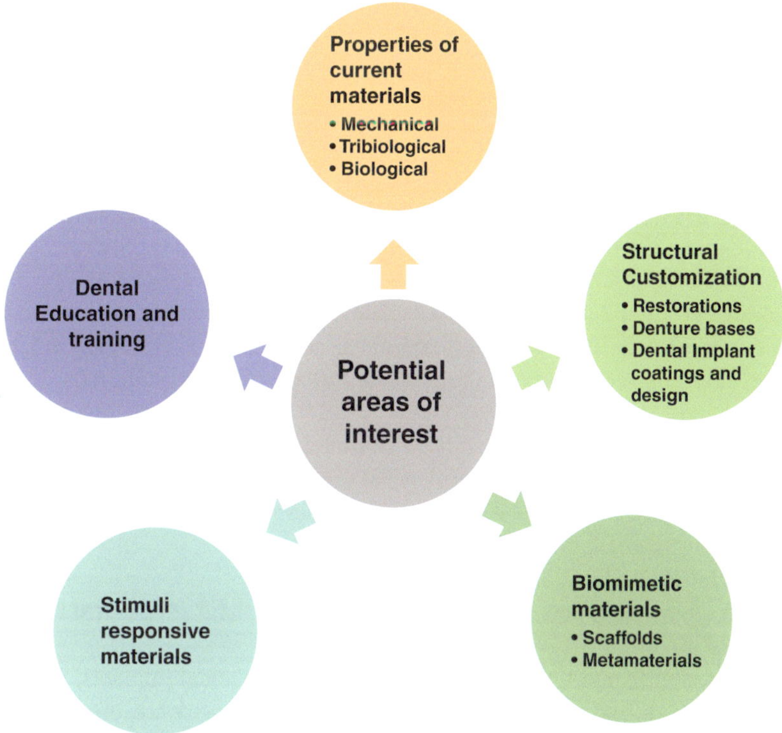

Fig. 14.2 Areas of interest in 3D printing in the oral sciences

of scaffolds for regenerative techniques in the oral sciences [12]. Though it is becoming more prevalent, there are certain areas in 3D printing that have the potential to be explored further in the oral sciences. A brief overview of these is presented in Fig. 14.2. This chapter aims to explain their application in different branches of the oral sciences.

14.2 Future of 3D Printing in Prosthetic and Restorative Dentistry

3D printing offers multiple advantages over the process of subtractive milling which is used to fabricate restorations and prostheses. These include:

- Reduced material lost to milling
- No obvious microcrack propagation
- Allows production of complex geometries with ease
- No limit on resolution according to the size of a milling bur [13–15]

3D printing is used in restorative dentistry for the fabrication of tooth models and dies, fabrication of wax patterns, single-tooth restorations, fixed 3-unit bridges, and implants. RPD frameworks, resin patterns, and denture bases have been successfully prototyped with different materials [16–20].

14.2.1 Fixed Restorations

Fixed intraoral restorations provide rehabilitation or replacement of damaged teeth. These include single tooth complete or partial coverage crowns, fixed dental prosthesis, implant-supported restorations, etc. Fixed restorations made with conventional methods such as the lost wax technique have withstood the test of time [21–23]. 3D-printed restorations must be evaluated and perform as well or better than conventional materials in vivo if they are to be adopted.

Recent research shows that 3D-printed restorations perform similarly to conventional ones. Single-tooth restorations, made by both techniques, were compared for marginal adaptation.

The 3D-printed metal crowns were within the clinically acceptable limit and comparable with conventional crowns [24–26]. A drawback in 3D-printed restorations was a higher gap in the occluso-axial and occlusal regions [25, 27–29] that could be due to optical rounding of line angles when using a digital light scanner [30]. Developing better scanning technology will help to reduce this discrepancy and provide a better fit compared to the techniques used today. The mechanical properties of 3D-printed metal copings and single-tooth restorations need further evaluation.

Ceramics, a preferred restorative material due to their superior esthetics and mechanical properties, have been printed by extrusion techniques. Posterior crowns fabricated with zirconia using inkjet printing have high accuracy, with strength and fracture toughness comparable to conventionally produced zirconia [15]. Ceramic veneers fabricated with rapid prototyped zirconia have a superior load-bearing capacity than either milled or heat-pressed ceramics [31]. Despite the superior esthetics and mechanical advantages of ceramics, metals are economical. Another disadvantage is the wear of natural teeth seen with the use of these restorations.

In a clinical setup, the technique of robocasting, which allows fabrication of a model using viscous solutions of porcelain followed by post-fabrication processing, has potential for application. This process involves the fabrication of a solid free-form "green" model by extrusion of a viscous slurry onto a platform at extreme low extrusion pressure. The model undergoes post-fabrication processing to remove the binder in the ink by burning off the organic additives, followed by sintering to densify the structure and give rise to the final form to be fabricated—such as a restoration [32, 33]. It was used to manufacture low-cost accurate dental prosthesis with zirconia and alumina, in a minimal amount of time [34]. The restorations made had an accurate intaglio (internal, tooth facing) surface, but the fidelity of the external surface to the STL file was inadequate. Further improvements in the technology that focus on accuracy and adaptability and increased variety and availability of inks used in the fabrication process will aid its use in the oral sciences. Using hybrid inks with ceramics and resins can help lower costs and also improve the tribological properties of these restorations.

Studies that evaluate the mechanical properties of these restorations, their adaptation, dimensional accuracy, biocompatibility, survival in the oral environment, and eventually economic feasibility are required. The development of ceramic or metal restorations by robocasting can make it an economical and accurate alternative in dental clinics. It can aid in fabricating single or multiple tooth restorations, implants, and connectors for partial dentures. Newer biocompatible polymers can be used to fabricate restorations with longer viability, antibacterial, and biomimetic properties in the oral cavity.

14.2.2 Dental Implants

Dental implants, commonly made from titanium or zirconia, offer a conservative and long-lasting option for the replacement of missing teeth. They are currently produced commercially in standard designs and compositions. However, fabrication of implants by commercial entities at a central facility, materials for molds, and manufacturing processes such as milling, storage, and transportation increase the cost of production.

3D printing can provide in-office dental implants made from either zirconia or titanium. Dental implants have been rapid-prototyped using laser techniques with metal [35]. They show promising results in clinical studies—a survival rate of 94.5% for 3D-printed implants placed in the oral cavity and a 94.3% implant-crown success rate in one study [36]. Zirconia implants have also been made with DLP using yttria-stabilized zirconia and photosensitive monomer which was cured layer by layer as the implant was built [37].

Surface coatings and nanostructural modifications of dental implants improve their function and biocompatibility, specifically, rapid healing and osseointegration [38–40]. Studies have used drugs, nanoparticles, and other bioactive substances coated on implants to obtain better results [41–43]. These coatings are added to a previously manufactured implant substrate. They must adhere to the surface and withstand forces applied to thc implants. 3D printing can allow the addition of a particular substance onto the implant as part of the surface. Bioactive substances can be embedded in the ink used to print the final layer of the fabricated implant. The potential for customization here is immense. Patients who need antibiotics, osseointegration, or other similar nanoparticle coatings can get custom implants according to their physiological needs and probably at a lower cost.

In addition to these materials, rapid prototyping makes it possible to use polymers for the manufacturing of dental implants. Polyether ether ketone (PEEK) is an extensively studied polymer for biomedical applications. Its low modulus and biocompatibility make it one of the preferred polymers for dental implants. It has been used to design them in vitro with a favorable stress distribution and a wider design range, made from fiber-reinforced PEEK [44]. A recent finite element study indicated that PEEK may have an advantage over conventional metal implants in thermal adaptation to changing temperatures in bone tissue [45]. PEEK is inherently bio-inert and has been modified to improve its properties [46]. It is a promising alternative to zirconia and metal for in-office 3D-printed implants.

Rapid prototyping can produce customized implant designs according to the patient's bone and incorporate modifications as a part of the implant and not a physical coating. It can fabricate materials for the patient, specific to their needs by combining different inks for printing. The application of rapid prototyping in dental offices will reduce expenses by saving time and resources needed for commercial manufacture and transportation. The implants thus made will be patient-specific, more economical, and have better longevity in vivo.

14.2.3 Dentures

Conventional fabrication of a denture base necessitates multiple visits to a dentist: repeated impression making, try-in procedures, numerous adjustments at every step during, and post-fabrication. Although milling has a rapid processing time, it does not necessarily reduce visits to the dental office. Unkovskiy et al. prototyped dentures with a complete digital workflow that offers an alternative with only two visits [47]. However, they were unable to provide a predictable esthetic outcome, and the base had poor retention. Their proof-of-concept study incorporated an additional step of digital relining which included a chairside try-in procedure. This helped increase retention of the bases and allowed examination of esthetics and function before final fabrication. Soft tissues in the oral cavity are mobile, and achieving a functional impression with an intraoral scanner is difficult. Improvements in scanning techniques or the development of aides that allow digital impressions of soft tissues in their functional position will help to advance this method of fabrication.

Rapid-prototyped complete dentures are partially polymerized as they are fabricated. As a result, they may contain lower amounts of residual monomer compared to milled dentures that release as much monomer as conventional dentures [48]. However, this makes them susceptible to distortion when removed from their supports on the platform. They also have a post-fabrication processing cycle which involves the removal of an outer layer of unprocessed resin or a final cure cycle for unpolymerized resin. Complete denture bases made using rapid prototyping were found to have better accuracy after processing and hydration for 24 h than those fabricated by injection molding or milling [49], but an in vitro study that compared milled and 3D-printed complete dentures found that the trueness of artificially aged rapid-prototyped dentures was less than milled dentures in the examined areas [50]. It must be noted that the dentures printed in the second study were made with a layer thickness of 100 μm. Previous studies have shown that layer thickness can affect the accuracy of fabricated models and fixed prostheses [51, 52]. There is no reason to believe oral appliances will not be affected by this parameter. The questions that arise now are monomer release, distortion, the effect of layer thickness, and build angle, which need to be answered with further experimental and comparative studies.

Choi and colleagues studied the bond strength of teeth to different denture bases (heat polymerized, milled, and 3D printed) [53]. It was lower in the milled and 3D-printed bases compared to conventional heat-cured bases. In the previously mentioned study [50], milled denture bases had prefabricated teeth attached to them, while rapid-prototyped dentures were made as one object (the teeth and the base were a single unit). Rapid prototyping can lead to a stronger bond between the teeth and the base if both are made simultaneously and not as separate components. This needs to be determined with experimental models.

The oral cavity is not immediately suitable for impression making after extraction. A delay in fabrication of complete dentures—due to delayed healing, or schedule in a busy dental practice—can affect the patient adversely. The resulting edentulous period can deteriorate the quality of alveolar ridges and masticatory function leading to an unhealthy diet, social disability, and an adverse effect on the quality of life [54]. An intraoral scan is a noninvasive method of mapping the oral cavity to print immediate dentures that can be modified to suit the patient. Neumeier and Neumeier used a digital workflow for the fabrication of immediate digital dentures and a surgical guide before alveoloplasty [55]. This can be adopted into a clinical setup to provide digital records with jaw relations before extraction and can be used to fabricate a replacement or a new denture that is economical and efficient.

As dentures are subject to cyclic masticatory forces, high flexural strength is a desirable property in materials used for the fabrication of denture bases. A recent study that compared the flexural strength of denture bases fabricated by four methods (conventional heat polymerization, injection molding, milling, and rapid prototyping) found that 3D-printed materials had the least flexural strength, but within clinically acceptable limits [56]. Build orientation can affect the mechanical properties of 3D-printed dental materials [57]. This factor needs to be evaluated further to objectively determine its effects on appliance fabrication.

3D-printed denture bases can be customized to individual needs. A patient with high bite forces, for example, can be given a denture base and teeth fabricated with additives to improve their mechanical properties. Those who are immunocompromised or susceptible to infections can have an antibiotic material, such as silver nanoparticles, incorporated into the base. These additions can be done in two ways—as a layer between the inks used for printing of denture bases or by manufacturing specialized inks for specific purposes. 3D-printed polymers—styrenes, polyamides, polyether ketones, and polycarbonates—have been modified to improve their biocompatibility, strength, and antibacterial action [58, 59].

Fabrication of teeth with esthetics similar to natural is easy with rapid prototyping. A gradual change in the shade of the resin in each layer can help imitate the hues of dentin and provide a more translucent layer of enamel on the outside. This can be achieved by multi-head deposition. Soft tissue esthetics such as gingival pigmentation can be similarly customized. This control is achieved by techniques that use jetting of materials onto a platform such as polyjet printing. Using this technique, the dentist will be able to decide the composition of each build layer at a microscopic level and modify the appliance according to the patient's needs. To cater to this requirement, companies are setting foot into this area, such as Stratasys with its recently launched J5 DentaJet™ (Stratasys Ltd©, USA) that allows printing in multiple materials and colors.

Further developments in the fabrication process can aim to improve the accuracy of the intaglio (tissue-facing) surface by incorporating functional assessment of oral tissues and integrate esthetic evaluation into the procedure. This will allow the fabrication of complete dentures with reduced clinical visits, which will be especially useful where resources such as transport, shipping facilities, or dental laboratories are scarce.

14.3 Future of 3D Printing in Orthodontics

The advent of digital scanning and printing techniques has created a paradigm shift in orthodontics. The principles of tooth movement remain the same, but their execution has changed. 3D printing affects every aspect of orthodontics—from aligners and wires to customized compensations (tip and torque) in brackets. These are made possible in the orthodontic clinic itself without outsourcing to a third party. These changes, however, are not without their requirements and investments. The role of the specialist will change from placing the appropriate wire available to fabricating the ideal wire, which will need a deeper understanding of material behavior, or deciding the morphology of the attachment to move teeth effectively.

As intraoral scanners replace dental impression materials and allow integration of dental models with 3D imaging, diagnosis and treatment planning in orthodontics will become simpler and accurate. Adding 3D printing offers many advantages: reduced material and labor costs, in-house fabrication of appliances, and models that can be used repeatedly. However, it ushers in a new era of digital orthodontics where the setup will need different equipment and skills from both the laboratory technical staff and the orthodontist. Since the technology is not taught in most dental schools across the world, especially in developing nations, it will require training of personnel. These added costs could be offset by saving on perishable materials, faster fabrication leading to a greater patient interest in treatment, and reduced need for manpower.

14.3.1 In-House Aligners

A shift in orthodontic treatment planning occurred from a focus on hard tissues to the soft tissues and esthetic at the beginning of the twenty-first century [60, 61]. Today, there is a similar shift in orthodontic appliances. Patients desire an esthetic and functional occlusion with removable invisible appliances. A large segment of patients requiring orthodontic treatment, especially adults, find braces unacceptable [62]. Currently, most orthodontists use proprietary software affiliated with commercial entities (such as Align Technology, USA) that print successive models and fabricate aligners remotely. These are then delivered to the orthodontist with little to no room for adjustment. In addition to material and software, logistics contribute to the price. 3D printing allows fabricating in-house aligners where the orthodontist has complete control over the sequence of tooth movement.

Fabrication of aligners can be carried out in two ways: thermoforming on printed models or direct printing. The former is more common as thermoformed retainers are already used in orthodontics. The direct printing of aligners can be cost and time-efficient. They can be created virtually and printed—in contrast to the former, where digital models are manipulated, printed, and followed by aligner adaptation. Compared to 3D-printed models, which are complete solids, printed aligners are akin to shells. Thus, they use less material and time. However, research in this area is scarce compared to the thermoforming technique.

Aligners must have high accuracy for adequate adaptation, tooth movement, and patient comfort. For thermoformed aligners, this means printing an accurate cast from a digital model. A study found 3D-printed dental models comparable to conventional stone casts in dimensional accuracy [63]. It explored layer thickness, build orientation, position on the build platforms, and whether the model was hollow or solid as potential factors. As mentioned previously, layer thickness (which also determines resolution) affects accuracy [51, 52]. Orthodontic models do not need as accurate a reproduction as intra- or extra-coronal prosthesis—a variation of 0.20–0.50 mm is acceptable [64, 65]. A recent study by Sherman et al. determined that the casts printed by varying their build orientation, position, and fill (solid or hollow) are comparable in accuracy [64]. The highest layer thickness used was 100 μm. A thicker layer will reduce printing time but compromise accuracy. Studies that used polyjet printers with a layer thickness of 16 μm found them to be more accurate than other printing techniques, though the overall accuracy was not significantly different from stone casts [65–67]. Further research on optimizing printing, i.e., minimal time with maximum accuracy, will help establish guidelines for orthodontists who wish to use 3D printing in their practice.

Direct printing of aligners will also help reduce errors. Any possible discrepancy can come from intraoral scanning and printing only, in contrast to the thermoforming technique where intraoral scanning, 3D printing of the model, and adaptation to the model—all contribute. Jindal et al. used Dental Long-Term (LT) resin (Form Labs, USA) which is an approved biocompatible material for direct printing of aligners [68]. They found that the mechanical and geometric properties of these aligners were superior to thermoformed ones. Another material used directly for 3D printing is the E-guard (EnvisionTEC, Germany) recommended for fabricating mouth guards and retainers that can be used to make aligner trays [69]. Cytotoxicity of trays that are either thermoformed or 3D printed was found within acceptable limits [69]. However, studies on their clinical performance and patient acceptability are lacking. In addition to these materials, novel materials are also being studied which may provide aligners with better control on tooth movement as a result of their architecture or as a response to stimuli in the oral cavity [70].

14.3.2 Presurgical Orthopedics

Nasoalveolar molding (NAM) is a procedure that reduces the size of the congenital defect in infants born with cleft lip and/or palate. It uses a series of successive plates to approximate segments that failed to fuse during development. These plates are changed in size and shape to accommodate the morphology of the defect and growth of the upper jaw. Conventionally, they are fabricated by making impressions every week to create a plaster model of the patient. The model is altered manually to fabricate a plate that molds the segments further each time. The procedure needs frequent visits to a doctor for a long time. This greatly increases the burden of care on the patients and their guardians.

Advances in 3D technology, intraoral scanning combined with 3D printing, can help the orthodontist to deliver a set of plates in a single visit. There is published literature that shows this is an effective approach [71, 72]. Digitization allows the orthodontist to manipulate virtual models so that the entire or a part of the sequence of plates is generated in one appointment. The patient can change plates weekly with clinical visits spaced out over several weeks or scheduled only when a difficulty is encountered. In addition to 3D printing, studies have also focused on developing algorithms that can predict the growth of the maxilla and incorporate it into the sequence instead of a doctor performing the task manually [73]. The combination of an automated approach combined with 3D printing will lead to higher adoption of this treatment in areas where access to specialized health care is limited. It is also possible to have a remotely located specialist decide the sequence of plates based on an intraoral scan and send it to a healthcare center closer to the patient's location for 3D printing.

14.3.3 3D Printing in Conventional Orthodontics

Despite the increased availability and popularity of aligners, they are expensive and the mechanics of tooth movement associated with them needs further research. Conventional orthodontic appliances are still forerunners in orthodontic treatment. Orthodontists use commercially manufactured metal or ceramic appliances and wires for the treatment of patients. Stock-made brackets with varying inbuilt prescriptions of tip and torque and wires with predetermined widths are available. The orthodontist bonds the brackets on the tooth crown to apply force and move teeth.

Bonding of orthodontic brackets is done in two ways—directly, by placing them onto the tooth surface and curing the adhesive or indirectly using trays for bonding. The procedure for indirect bonding is lengthier but allows better visualization and accurate placement of brackets on the tooth. Brackets are placed on a plaster cast using an adhesive followed by the fabrication of a customized tray for each tooth or segment of the dental arch. Bracket position is evaluated and corrected by the practitioner in the lab and not the chairside. 3D printing can reduce the time needed to fabricate indirect bonding trays. A clinician can place virtual brackets on a digital model or use a scanned plaster model with previously positioned brackets. These can be used to design and print a tray for the indirect bonding of a patient. A study that used virtual brackets positioned on digital models to print the trays and compared them to conventional laboratory procedures found reduced active production time in the digital model group [74].

3D printing can customize brackets and wires for patients. There are commercial systems (e.g., Insignia™, ORMCO, USA) through which the orthodontist can acquire patient-specific archwires and brackets. Again, as in commercial aligners, the clinician invests time and money, but partial control rests with a third party. Rapid prototyping the components, metal or ceramic, can aid the orthodontist in fabricating customized in-house appliances (both brackets and archwires). This will enable better treatment outcomes and also reduce the cost and duration of the treatment. The limitation here is the high cost associated with printing metal and ceramic. Resins with mechanical properties to withstand masticatory forces and forces applied during orthodontic treatment can be developed to overcome this issue.

14.4 3D Printing in Cranio-Maxillofacial Surgery

14.4.1 Orthognathic Surgery

Correction of skeletal jaw discrepancies, deformations, or injuries requires orthognathic surgery. Planning for surgery traditionally includes an assessment of plaster models of the jaws and 2D cephalograms of the face along with clinical examination. However, 2D imaging has limitations: projection errors and inaccurate identification, especially, in patients with facial asymmetry [75, 76]. The advent of 3D imaging and cone beam computed tomography (CBCT) made thorough surgical planning with predictable and accurate results possible [77–80]. 3D printing, combined with 3D imaging, can lead to a complete overhaul of procedures that facilitate the surgeon.

Occlusal splints guide surgeons in repositioning the jaws during orthognathic surgery by establishing optimal occlusion. Traditional fabrication of the splints involves using a facebow, plaster models, and an articulator. Several manual steps and different materials are involved in this process, all subject to error. 3D printing can be used to manufacture accurate splints when used in combination with 3D virtual surgical planning. Studies have already assessed and validated their use in patients [81, 82]. A recent randomized controlled trial compared the use of conventionally fabricated splints, polyjet printed splints, and digitally planned and 3D printed cutting guides along with splints. The researchers achieved better results by using repositioning guides along with the splints than the other two groups [83].

3D printing can eliminate the dependence on occlusion while making orthognathic surgery more predictable. 3D repositioning guides used by surgeons for the placement of cuts and repositioning bony segments can replace splints. A prospective clinical study found that this splintless system was more accurate in surgical correction, with smaller errors in all three axes [84]. Besides, 3D-printed osteotomy guides allow a precise transfer of the virtual surgical plan to the operating table. These guides have been prototyped for genioplasties, maxillary repositioning, and designed in vitro for bilateral sagittal split osteotomies [85–88]. They were found suitable and stable for use by both in vitro and in vivo studies. These guides help surgeons demarcate areas safe for osteotomy based on the patient's image data. Randomized controlled trials with a higher number of patients comparing 3D-printed guides to conventional splints are needed to establish their efficacy.

14.4.2 Patient-Specific Implants and Plates

Unlike the guides for surgery, which are aids and do not have demands other than accuracy, patient-specific implants have different requirements. They are placed permanently in vivo and are expected to be biocompatible and bear functional stresses. Although autologous implants are the gold standard, they can be scarce and involve multiple surgical procedures. Similar implants can be customized for a patient from titanium or alloplastic materials to replace a segment of a bone or fix bony segments together. A complete virtual workflow for surgical procedures is described by Volpe and colleagues [89]. This allows a single-step rehabilitation of the patient without undergoing separate surgeries for removal and implant placement. 3D printing can be easily incorporated into such digitized workflows. It allows precise anatomic replication, offers biocompatible materials, and can easily mimic the hollow structures and complex anatomy present in bones, especially, the nasomaxillary complex.

Titanium implants have been 3D printed using laser sintering techniques for patients of orthognathic surgery and maxillofacial rehabilitation. A recent study that examined patient-specific titanium plates made to reposition and fix the maxilla found them more accurate than 3D-printed occlusal splints in transferring the virtual surgical plan in vivo [90]. 3D-printed titanium implants have also been used by Melville et al. in a case for restoring the maxilla with a free fibula graft with satisfactory results in the follow-up [91]. A similar use in the mandible has been studied in detail by Goodson and colleagues [92]. They highlight the numerous advantages of using in-house 3D-printed implants for surgeons: biomechanical superiority, better customization, freedom of design, reduced time (for surgery and implant production), and reduction in complications/recovery time. These advantages help to remove limitations imposed upon surgeons by conventional materials and methods leading to the development of novel techniques for patients who do not fit the descriptions of regular cases. An example is a technique used by Qassemar et al. who devised a procedure to benefit patients ineligible for bone-free flaps [93].

Patient-specific implants for the temporomandibular joint fabricated by third parties (e.g., TMJ Concepts, USA) have shown favorable results over 20 years [94], but the process of acquiring these implants is tedious. The company first approves the patient for their device and prints a stereolithographic model which is reviewed by the surgeon and sent back. They then create the implant design, which is also reviewed by the surgeon and sent back. Finally, the implants are shipped to the surgeon [95]. This repeated back and forth can be eliminated by in-house 3D printing. Not only does it avoid dependence on a third party, but a surgeon can modify designs on a case-to-case basis leading to increased availability of the implants for patients—even those deemed unsuitable by commercial entities. It can also allow rapid prototyping using materials other than metals. 3D-printed titanium condyles have been successfully used in a patient with no postoperative complications [96]. This procedure can be developed as the norm in clinics where similar cases are treated regularly.

The cranium and the orbit are complex and challenging skeletal structures to rehabilitate. They may present with irregular defects that are not easily repaired with stock materials when the autologous graft is unavailable. These grafts need to be accurate to avoid postoperative complications such as enophthalmos or diplopia. Cranioplasty with 3D-printed titanium plates has shown promising results [97, 98]. Apart from titanium, polymethylmethacrylate (PMMA), polyetheretherketone (PEEK), and hydroxyapatite (HA) are also used for the procedure [97]. They can be easily 3D printed leading to a low-cost, customized option for patients where metal printing is not available. Reconstruction of the orbital floor using in-house 3D-printed polylactic acid implants showed no postoperative complications in the follow-up period. Their accuracy was comparable to the commercial systems available [99]. There is a need to test the clinical feasibility and success of these implants in a larger group of patients. They can also be compared to other 3D-printed materials such as titanium, which had a 17% failure rate in one study [100].

Most of the applications published in the literature where 3D printing was used outsourced the job to official medical device manufacturers [101]. Despite low-cost manufacturing possible, these have not been widely adopted in clinical settings. Since titanium is the material of choice for most applications, the ability to 3D print metals in clinics will greatly aid the adoption of 3D printing. Alternatively, 3D-printed biomimetic materials can also be developed for clinical use.

14.4.3 Scaffolds for Bone Regeneration

The gold standard for replacing bone defects in the maxillofacial region is autologous grafts. However, these require multiple invasive surgeries as the bone is removed from one region of the body to provide for another. Synthetic substitutes such as PMMA and titanium are passive fillers. They do not allow active regeneration of the bone and do not mimic the architecture and biological qualities of bone. Rapid prototyping can offer dual advantages: a customized external shape and controlled internal architecture. The implants can be made bioactive and ideally should be resorbable and eventually replaced by indigenous bone after some time.

Scaffolds that have been used in bone tissue engineering include ceramic biomaterials (e.g., hydroxyapatite, tricalcium phosphate, osteocalcium phosphate, and bioglass), natural polymers (e.g., collagen, chitosan, and gelatin), and synthetic polymers (e.g., polylactic acid, polyglycolic acid, polycaprolactone) and decellularized extracellular matrix [102]. They are used along with mesenchymal stem cells and growth factors for osteoinduction and osteoconduction. The combination of a scaffold along with stem cells and biological factors is used to devise a bio-ink which can then be used for bio-printing of tissues.

Scaffolds need to have a porous structure to allow migration of cells and flow of growth factors to all parts of the defect. Pore size and architecture affect the biological and mechanical properties of a scaffold [103]. Pore size can influence osteogenic signal expression and differentiation. The large pore size is not conducive to the bone formation, while a very small pore size can inhibit vascularization. Roohani-Esfahani and colleagues 3D printed a glass-ceramic scaffold using inkjet printing [104]. Their experiments demonstrated that scaffolds fabricated with a hexagonal structure were comparable to the human cortical bone in strength. This was higher than the other architectures tested by them. Using the correct microstructural attributes for an implant will allow it to mimic the human bone's characteristics [105]. 3D printing has the advantage of manufacturing with varying pore size and architecture to provide optimal mechanical properties in different parts of the graft similar to the natural tissue structure of human bone.

14.5 Dental Education and Training

Dental education and training determine the quality of dental service available to society at large. The best teaching practices must be incorporated into the curriculum to create good dental professionals. Dental education requires understanding normal versus abnormal anatomy and knowledge of the oral hard and soft tissues. These concepts are not easy to understand by theoretical explanation. Therefore, dental professionals are taught using histology specimens, sectioned extracted teeth, or patient records. These biologic samples come from patients who consent to their extracted teeth, case records, or their tissue specimens being used for these purposes. As a result, the exposure received by students largely depends on the chance of a particular condition presenting at their institution.

Similarly, preclinical training in tooth preparation for restorative purposes uses a model that has the external anatomy of a normal tooth with no carious lesion. The ideal shape and forms of tooth preparation taught using them can educate students about basic concepts such as resistance and strength, but a clinical setting rarely calls for it. These preclinical specimens also fail to replicate the internal architecture of teeth. In clinics, where a 0.5 mm bridge of dentin determines the difference between a restoration and endodontic treatment, knowledge of the tooth's internal anatomy is invaluable.

3D printing can allow the fabrication of anatomically accurate teeth with the entire structure replicating a human tooth. Höhne and Schmitter created accurate replicas of teeth with the pulp chamber depicted in red and a carious lesion represented on the tooth visually and radiographically [106]. They provided these specimens to dental students for tooth preparation and found that the specimens were effective in helping students understand the concepts of treatment. As materials and techniques in 3D printing advance, it might be possible to print teeth that, in addition to the specimens made by them, will have mechanical properties resembling natural human teeth. Techniques such as multi-head deposition will allow using materials with different properties for each layer. The teeth thus fabricated will be able to replicate dentin and enamel more accurately providing a better tactile sense to students for both tissues.

3D-printed specimens can also aid in providing practical skills for those training to work in dental laboratories. A research group 3D-printed ideal tooth preparations for teaching first-year dental students, and as standards of comparison for the students' tooth preparation [107]. These preparations can also be used to train laboratory personnel for the fabrication of single tooth or 3-unit fixed prosthesis especially ceramic restorations which are very technique sensitive.

With progress in 3D bioprinting, it will be possible to create sample histologic sections along with clinical specimens depicting normal and abnormal characteristics seen in different diseases. Prototyped teeth can be used to teach anatomic malformations and methods of dealing with the same. The teaching curriculum can be modified to depict problem scenarios, including multiple details and specimens, that are most frequently encountered by a clinician, resulting in a better quality of care.

14.6 Low-Cost Digital Workflow

Intraoral scanning and additive manufacturing offer a noninvasive and accurate means to visualize oral tissues. Traditional methods of record making such as dental impressions can have adverse effects such as inciting a gag reflex in the patient. At times, the presence of a wound or an intraoral lesion does not allow impression making. 3D imaging overcomes these barriers. Currently, the most common form of 3D imaging used for diagnosis is CBCT. This modality exposes the patient to ionizing radiation and is expensive. It is not recommended for daily use. In comparison, 3D imaging offers a less harmful and noninvasive option for the visualization of oral tissues. Studies found it viable for clinical use [108, 109], but the initial cost of the technology can be a barrier for widespread adoption by clinics.

Additive manufacturing involves a relatively high initial investment. A study by Resnick and colleagues on the economic aspect of using a 3D printer in an oral and maxillofacial practice found that using the scanner reduces the cost per patient to less than half incurred by using a conventional method for impressions [110]. They estimate that the time needed to offset the initial investment would be little more than a year if a practice treats at least two patients per day. However, this cost will likely be higher and the time needed to recover the investment longer in developing countries where intraoral scanners are not yet commonplace and the cost of dental treatment is minimal.

A 3D printer needs STL files of the appliance or restoration that needs to be printed. This will require either an intraoral or a desktop scanner. The cost of replacing the material used and the shelf life for storage also play an important role in the decision to buy a printer for the practice. The adoption of a digital workflow—from the intraoral scanner to rapid prototyping—will be costly but could significantly offset the material and time costs. The number of materials that are handled in a dental clinic, especially expendables, will be reduced with the elimination of multiple steps of impression making, pouring casts, and making bases. The patients will also be able to see their virtual models immediately to understand their treatment needs or plans.

Increased application of digital workflows in dentistry will also help to provide a safer and more hygienic work environment as well. Handling materials such as alginate and dental plaster produce aerosols that are harmful to personnel in the dental operatory. Adopting a digital workflow will eliminate these steps by virtual processing of the patient records and data. Another method to reduce costs is the generalization of the software and devices available. An increase in the number of manufacturers and software providers leading to more options will lead to greater interest from the dental community. Creating awareness about the technology will also allow practitioners and researchers to innovate methods of reducing costs leading to a wider acceptance of rapid prototyping. Kamio et al. were able to create fairly accurate models of the dentition using a general desktop 3D printer instead of a commercial one [111]. There is a need for studies that explore these aspects of digital dentistry to help aid its adoption.

14.7 Metamaterials and 4D Printing

To be used in the human body, an ideal prosthesis (hard or soft tissue) must have an important quality—adaptation. The human body is a dynamic system, and thus materials that are used as a substitute for natural tissues need to mimic their behavior closely. A 3D-printed biomimetic structure that does not respond to stimuli in vivo, e.g., a bone graft that does not have the elasticity of natural bone, is bound to fail eventually and is of little value to the patient. To help resolve this obstacle, the next generation of rapid prototyping technology has modified the nature of materials used, giving rise to 4D printing. While 3D printing fabricates static models, 4D printing uses materials that respond to stimuli such as temperature, pH, pressure, or the presence of water to alter their shape over some time. The materials used for 4D printing are called smart materials due to this property. Although it has not been explicitly researched for dental applications, the technique has many potential uses in the field.

Shape memory metal alloys are already used in dentistry. Nickel-titanium (NiTi) alloys have been used as endodontic files, orthodontic archwires, and arch expanders. Studies on these alloys indicate better biocompatibility, light and porous structure, and control over mechanical properties when manufactured by additive manufacturing [112–114]. 3D printing such materials can control their property of superelasticity and limit it to the desired regions of activation. NiTi alloys have also been 3D printed with hydroxyapatite using laser sintering [115]. This alloy was tested for the fabrication of implants and other biomedical applications.

Liu et al. demonstrated that strategic placement of a thermosensitive and a nonsensitive polyacrylamide gel can lead to shape changes in a 3D-printed biomaterial [116]. Heat variation in areas of inflammation or areas of increased vascularization can be used as a stimulus. These materials can be used for drug delivery in the periapical region, sockets with delayed healing, or areas of periodontal inflammation. Guo et al. fabricated a thermosensitive hydrogel made of acrylamide and agarose that responded to an external force by changing its shape [117]. Such hydrogels can be used to create materials that activate and adapt to the contours of oral structures when subject to oral temperatures. They can be used in the form of endodontic fillers, self-adapting denture bases or implants, and bone grafts.

Rapid prototyped polymers with controlled degradation in the body have been used as smart drug delivery systems [118–121]. Similar smart systems can be used to deliver drugs in the oral cavity, for example, fluoride release in response to changes in pH (to prevent caries) or antibiotics in the periodontal ligament when a change in the local microbiota or inflammatory markers is detected.

The pressure-responsive printed materials have multiple applications in the oral sciences. The oral cavity has tissues that constantly adapt to changing masticatory loads. The position of teeth in the arch is such that the muscular forces from the tongue and the cheek are balanced. 3D-printed pressure-sensitive materials can be used in orthodontics to fabricate devices for arch expansion or fabricating aligners that apply controlled force according to desired movements. It can be used to fabricate restorations that can absorb trauma from occlusion till the cause is corrected. Printing of materials responsive to proteins in biologic tissues will lead to the fabrication of materials that adapt to the shape of the prepared canals, to the periodontal defect, or lead to osteoinduction starting at the borders of a large bony defect. Surgical prosthetic implants can be fabricated that change their shape according to the force applied on the graft during function. As an example, a TMJ condylar graft made using such material will be able to provide the optimal function with minimal effects on the surrounding tissues.

Metamaterials are man-made materials designed with periodic architectures to either mimic other materials or provide properties that are not found in nature. These materials have a complex design with small repeating architectures. They are cast using high-resolution 3D printing techniques such as two-photon polymerization and microstereolithography [122]. High-resolution 3D printing has been used to study the relationship between material microstructure, architecture, and its properties in both metals and ceramics [123, 124] both of which are commonly used for dental restorations. It can help fabricate strong, ultralight, and resilient products made of inherently brittle materials such as ceramics.

High-resolution electrohydrodynamic printing (EHDP) has also been used to print tissue engineering scaffolds. It was used by different groups of researchers for successfully printing scaffolds with hydrogels, polymers, or a combination of both [122]. Bas et al. were able to manipulate the properties of PCL scaffolds infiltrated with hydrogels by varying porosities and structure [125], similar to the observations by Estafani and colleagues [104]. 3D-printed titanium alloy metamaterials have also been studied and characterized as bone replacement implants for load-bearing applications [126].

Metamaterials can also be made using stimuli-responsive materials used for 4D printing. Materials such as hydrogels and shape memory polymers can be exploited for dental use. Zhang et al. fabricated a bilayer metamaterial that could undergo programmable, reversible deformations by tuning the lattice geometry and hydrogel distribution [127]. Gladman and colleagues used a flexible biocompatible ink that could morph into different shapes when immersed in water based on a variation of printing parameters such as filament size, orientation, and spacing [128]. Such materials may ultimately find use as programmable aligners, distraction devices, or arch expanders in the future. Metamaterials with shape-changing properties that are thermally or magnetically activated are also being studied. The current challenges seen with this field are those of scaling the prints to higher dimensions without compromising on resolution, improving speed, time, and economic characteristics of the prints [70, 122].

14.8 Summary

Personalization, i.e., the customization of every aspect of medical care for each patient, is the future of medicine. The oral sciences are no exception. From dental education and rehabilitation to skeletal reconstruction, additive manufacturing can offer choices that are not currently available to clinicians. Research in this area appears to be in its initial stages but shows an upward trend. Conducting clinical studies will help us realize the true extent of its benefits. With advances in 3D and 4D printing, we can manufacture synthetic and natural substitutes for lost tissues with complete control over their properties and activity in vivo. However, these advances can only occur in tandem with the development and adoption of imaging techniques. At present, the main barrier for widespread adoption of these techniques appears to be cost and knowledge. With advances in technology, research, and increased exposure—as was seen with computers—this economic barrier is sure to dissolve. The result will be an availability of affordable personalized dental medicine in all corners of the world, and additive manufacturing will play a huge role in this development.

References

1. Rice DD, Abramovitch K, Olson GW, Christiansen EL. Data management practices of cone-beam computed tomography volumes: an exploratory user survey. Oral Surg Oral Med Oral Pathol Oral Radiol. 2019;128:e100–7.
2. Shah PH, Venkatesh R. Dental students' knowledge and attitude towards cone-beam computed tomography: an Indian scenario. Indian J Dent Res. 2016;27:581–5.
3. Shim Y-S, Kim A-H, Choi J-E, An S-Y. Use of three-dimensional computed tomography images in the dental care of children and adolescents in Korea. Technol Health Care. 2014;22:333–7.
4. Setzer FC, Hinckley N, Kohli MR, Karabucak B. A survey of cone-beam computed tomographic use among endodontic practitioners in the United States. J Endod. 2017;43:699–704.
5. Tran D, Nesbit M, Petridis H. Survey of UK dentists regarding the use of CAD/CAM technology. Br Dent J. 2016;221:639–44.
6. Chaar MS, Passia N, Kern M. Long-term clinical outcome of posterior metal-ceramic crowns fabricated with direct metal laser-sintering technology. J Prosthodont Res. 2020;64:354–7.
7. Belhouideg S. Impact of 3D printed medical equipment on the management of the Covid19 pandemic. Int J Health Plann Manag. 2020;35:1014–22.
8. Pedraja J, Maestre JM, Rabanal JM, Morales C, Aparicio J, Del Moral I. Role of 3D printing for the protection of surgical and critical care professionals in the COVID-19 pandemic. Rev Esp Anestesiol Reanim. 2020;67:417–24.
9. Montazerian M, Zanotto ED. Bioactive and inert dental glass-ceramics. J Biomed Mater Res A. 2017;105:619–39.
10. Lin L, Fang Y, Liao Y, Chen G, Gao C, Zhu P. 3D printing and digital processing techniques in dentistry: a review of literature. Adv Eng Mater. 2019;21:1801013.
11. Dawood A, Marti Marti B, Sauret-Jackson V, Darwood A. 3D printing in dentistry. Br Dent J. 2015;219:521–9.
12. Athirasala A, Tahayeri A, Thrivikraman G, França CM, Monteiro N, Tran V, et al. A dentin-derived hydrogel bioink for 3D bioprinting of cell-laden scaffolds for regenerative dentistry. Biofabrication. 2018;10:024101.
13. Azari A, Nikzad S. The evolution of rapid prototyping in dentistry: a review. Rapid Prototyp J. 2009;15:216–25.
14. Strub JR, Rekow ED, Witkowski S. Computer-aided design and fabrication of dental restorations: current systems and future possibilities. J Am Dent Assoc. 2006;137:1289–96.
15. Ebert J, Ozkol E, Zeichner A, Uibel K, Weiss O, Koops U, et al. Direct inkjet printing of dental prostheses made of zirconia. J Dent Res. 2009;88:673–6.
16. Gao B, Wu J, Zhao X, Tan H. Fabricating titanium denture base plate by laser rapid forming [Internet]. undefined. 2009 [cited 2021 Jan 14]. Available from: https://www.proquest.com/docview/214011570.
17. Inokoshi M, Kanazawa M, Minakuchi S. Evaluation of a complete denture trial method applying rapid prototyping. Dent Mater J. 2012;31:40–6.
18. Sun Y, Lü P, Wang Y. Study on CAD&RP for a removable complete denture. Comput Methods Prog Biomed. 2009;93:266–72.
19. Eggbeer D, Bibb R, Williams R. The computer-aided design and rapid prototyping fabrication of removable partial denture frameworks. Proc Inst Mech Eng H IMECHE. 2005;219:195–202.
20. Wu J, Wang X, Zhao X, Zhang C, Gao B. A study on the fabrication method of removable partial denture framework by computer-aided design and rapid prototyping. Rapid Prototyp J. 2012;18:318–23.
21. Wassermann A, Kaiser M, Strub JR. Clinical long-term results of VITA In-Ceram Classic crowns and fixed partial dentures: a systematic literature review. Int J Prosthodont. 2006;19:355–63.
22. Ozer F, Mante FK, Chiche G, Saleh N, Takeichi T, Blatz MB. A retrospective survey on long-term survival of posterior zirconia and porcelain-fused-to-metal crowns in private practice [Internet]. Quintessence Int; 2014 [cited 2021 Jan 13]. Available from: https://pubmed.ncbi.nlm.nih.gov/24392493/.
23. Reitemeier B, Hänsel K, Range U, Walter MH. Prospective study on metal-ceramic crowns in private practice settings: 20-year results. Clin Oral Investig. 2019;23:1823–8.
24. Quante K, Ludwig K, Kern M. Marginal and internal fit of metal-ceramic crowns fabricated with a new laser melting technology. Dent Mater. 2008;24:1311–5.
25. Tamac E, Toksavul S, Toman M. Clinical marginal and internal adaptation of CAD/CAM milling, laser sintering, and cast metal-ceramic crowns. J Prosthet Dent. 2014;112:909–13.
26. Huang Z, Zhang L, Zhu J, Zhang X. Clinical marginal and internal fit of metal-ceramic crowns fabricated with a selective laser melting technology. J Prosthet Dent. 2015;113:623–7.
27. Freifrau von Maltzahn N, Bernhard F, Kohorst P. Fitting accuracy of ceramic veneered Co-Cr crowns produced by different manufacturing processes. J Adv Prosthodont. 2020;12:100–6.
28. Gholamrezaei K, Vafaee F, Afkari BF, Firouz F, Seif M. Fit of cobalt-chromium copings fabricated by the selective laser melting technology and casting method: a comparative evaluation using a profilometer. Dent Res J (Isfahan). 2020;17:200–7.
29. Ullattuthodi S, Cherian KP, Anandkumar R, Nambiar MS. Marginal and internal fit of cobalt-chromium copings fabricated using the conventional and the direct metal laser sintering techniques: a comparative in vitro study. J Indian Prosthodont Soc. 2017;17:373–80.

30. Alharbi N, Wismeijer D, Osman RB. Additive manufacturing techniques in prosthodontics: where do we currently stand? A critical review. Int J Prosthodont. 2017;30:474–84.
31. Ioannidis A, Bomze D, Hämmerle CHF, Hüsler J, Birrer O, Mühlemann S. Load-bearing capacity of CAD/CAM 3D-printed zirconia, CAD/CAM milled zirconia, and heat-pressed lithium disilicate ultra-thin occlusal veneers on molars. Dent Mater. 2020;36:e109–16.
32. Wang J, Shaw LL, Cameron TB. Solid freeform fabrication of permanent dental restorations via slurry micro-extrusion. J Am Ceram Soc. 2006;89:346–9.
33. Peng E, Zhang D, Ding J. Ceramic robocasting: recent achievements, potential, and future developments. Adv Mater. 2018;30:1802404.
34. Silva NRFA, Witek L, Coelho PG, Thompson VP, Rekow ED, Smay J. Additive CAD/CAM process for dental prostheses. J Prosthodont. 2011;20:93–6.
35. Oliveira TT, Reis AC. Fabrication of dental implants by the additive manufacturing method: a systematic review. J Prosthet Dent. 2019;122:270–4.
36. Tunchel S, Blay A, Kolerman R, Mijiritsky E, Shibli JA. 3D printing/additive manufacturing single titanium dental implants: a prospective multicenter study with 3 years of follow-up [Internet]. Int J Dent. Hindawi; 2016 [cited 2021 Jan 29]. p. e8590971. Available from: https://www.hindawi.com/journals/ijd/2016/8590971/.
37. Osman RB, van der Veen AJ, Huiberts D, Wismeijer D, Alharbi N. 3D-printing zirconia implants; a dream or a reality? An in-vitro study evaluating the dimensional accuracy, surface topography, and mechanical properties of printed zirconia implant and discs. J Mech Behav Biomed Mater. 2017;75:521–8.
38. Rasouli R, Barhoum A, Uludag H. A review of nanostructured surfaces and materials for dental implants: surface coating, patterning, and functionalization for improved performance. Biomater Sci. 2018;6:1312–38.
39. Salou L, Hoornaert A, Louarn G, Layrolle P. Enhanced osseointegration of titanium implants with nanostructured surfaces: an experimental study in rabbits. Acta Biomater. 2015;11:494–502.
40. Schünemann FH, Galárraga-Vinueza ME, Magini R, Fredel M, Silva F, Souza JCM, et al. Zirconia surface modifications for implant dentistry. Mater Sci Eng C Mater Biol Appl. 2019;98:1294–305.
41. Civantos A, Martínez-Campos E, Ramos V, Elvira C, Gallardo A, Abarrategi A. Titanium coatings and surface modifications: toward clinically useful bioactive implants. ACS Biomater Sci Eng. 2017;3:1245–61.
42. Kunrath MF, Lombardo EM. Drug delivery surfaces for oral implant applications: current view and perspectives. J Oral Maxillofac Surg. 2019;77:2370–1.
43. Gupta S, Noumbissi S, Kunrath MF. Nano modified zirconia dental implants: advances and the frontiers for rapid osseointegration. Med Devices Sens. 2020;3:e10076.
44. Lee W-T, Koak J-Y, Lim Y-J, Kim S-K, Kwon H-B, Kim M-J. Stress shielding and fatigue limits of polyether-ether-ketone dental implants. J Biomed Mater Res B Appl Biomater. 2012;100B:1044–52.
45. Arshad SR, Hashemi A, Oskui IZ. Does peek dental implant has a thermal advantage over zirconia or titanium implants? J Mech Med Biol. 2020;20:2050005.
46. Rahmitasari F, Ishida Y, Kurahashi K, Matsuda T, Watanabe M, Ichikawa T. PEEK with reinforced materials and modifications for dental implant applications. Dent J. 2017;5:35.
47. Unkovskiy A, Wahl E, Zander AT, Huettig F, Spintzyk S. Intraoral scanning to fabricate complete dentures with functional borders: a proof-of-concept case report. BMC Oral Health. 2019;19:46.
48. Steinmassl P-A, Wiedemair V, Huck C, Klaunzer F, Steinmassl O, Grunert I, et al. Do CAD/CAM dentures really release less monomer than conventional dentures? Clin Oral Investig. 2017;21:1697–705.
49. Lee S, Hong S-J, Paek J, Pae A, Kwon K-R, Noh K. Comparing the accuracy of denture bases fabricated by injection molding, CAD/CAM milling, and rapid prototyping method. J Adv Prosthodont. 2019;11:55–64.
50. Kalberer N, Mehl A, Schimmel M, Müller F, Srinivasan M. CAD-CAM milled versus rapidly prototyped (3D-printed) complete dentures: an in vitro evaluation of trueness. J Prosthet Dent. 2019;121:637–43.
51. Scherer MD. Digital dental model production with high accuracy 3D printing. 17. https://www.additive-x.com/media/sector_resources/Digital-Dental-Model-Production_white-paper.pdf.
52. Kaleli N, Ural Ç, Us YÖ. Evaluation of marginal discrepancy in metal frameworks fabricated by sintering-based computer-aided manufacturing methods. J Adv Prosthodont. 2020;12:124–30.
53. Choi JJE, Uy CE, Plaksina P, Ramani RS, Ganjigatti R, Waddell JN. Bond strength of denture teeth to heat-cured, CAD/CAM, and 3D printed denture acrylics. J Prosthodont. 2020;29:415–21.
54. Emami E, de Souza RF, Kabawat M, Feine JS. The impact of edentulism on oral and general health. Int J Dent [Internet]. 2013 [cited 2021 Jan 14]; 2013. Available from: https://www.ncbi.nlm.nih.gov/pmc/articles/PMC3664508/.
55. Neumeier TT, Neumeier H. Digital immediate dentures treatment: a clinical report of two patients. J Prosthet Dent. 2016;116:314–9.
56. Prpić V, Schauperl Z, Ćatić A, Dulčić N, Čimić S. Comparison of mechanical properties of 3D-printed, CAD/CAM, and conventional denture base materials. J Prosthodont. 2020;29:524–8.
57. Alharbi N, Osman R, Wismeijer D. Effects of build direction on the mechanical properties of 3D-printed complete coverage interim dental restorations. J Prosthet Dent. 2016;115:760–7.
58. Jockusch J, Özcan M. Additive manufacturing of dental polymers: an overview on pro-

cesses, materials, and applications. Dent Mater J. 2020;2020:2019–123.
59. Chen S-G, Yang J, Jia Y-G, Lu B, Ren L. TiO_2 and PEEK reinforced 3D printing PMMA composite resin for dental denture base applications. Nanomaterials. 2019;9:1049.
60. Ackerman JL, Proffit WR, Sarver DM. The emerging soft tissue paradigm in orthodontic diagnosis and treatment planning. Clin Orthod Res. 1999;2:49–52.
61. Sarver DM, Ackerman JL. About face—the re-emerging soft tissue paradigm. Am J Orthod Dentofac Orthop. 2000;117:575–6.
62. Rosvall MD, Fields HW, Ziuchkovski J, Rosenstiel SF, Johnston WM. Attractiveness, acceptability, and value of orthodontic appliances. Am J Orthod Dentofac Orthop. 2009;135:276–7.
63. Ellakany P, Al-Harbi F, El Tantawi M, Mohsen C. Evaluation of the accuracy of digital and 3D-printed casts compared with conventional stone casts. J Prosthet Dent. 2022;127(3):438–44.
64. Sherman SL, Kadioglu O, Currier GF, Kierl JP, Li J. Accuracy of digital light processing printing of 3-dimensional dental models. Am J Orthod Dentofac Orthop. 2020;157:422–8.
65. Brown GB, Currier GF, Kadioglu O, Kierl JP. Accuracy of 3-dimensional printed dental models reconstructed from digital intraoral impressions. Am J Orthod Dentofac Orthop. 2018;154:733–9.
66. Hazeveld A, Huddleston Slater JJR, Ren Y. Accuracy and reproducibility of dental replica models reconstructed by different rapid prototyping techniques. Am J Orthod Dentofac Orthop. 2014;145:108–15.
67. Murugesan K, Anandapandian PA, Sharma SK, Vasantha Kumar M. Comparative evaluation of dimension and surface detail accuracy of models produced by three different rapid prototype techniques. J Indian Prosthodont Soc. 2012;12:16–20.
68. Jindal P, Juneja M, Siena FL, Bajaj D, Breedon P. Mechanical and geometric properties of thermoformed and 3D printed clear dental aligners. Am J Orthod Dentofac Orthop. 2019;156:694–701.
69. Ahamed SF, Kumar SM, Kumar, ASA K, Dharshini KI. Cytotoxic evaluation of directly 3D printed aligners and Invisalign. Eur J Mol Clin Med. 2020;7:1129–40.
70. Montgomery SM, Kuang X, Armstrong CD, Qi HJ. Recent advances in additive manufacturing of active mechanical metamaterials. Curr Opin Solid State Mater Sci. 2020;24:100869.
71. Batra P, Gribel BF, Abhinav BA, Arora A, Raghavan S. OrthoAligner "NAM": a case series of presurgical infant orthopedics (PSIO) using clear aligners. Cleft Palate Craniofac J. 2020;57:646–55.
72. Bous RM, Kochenour N, Valiathan M. A novel method for fabricating nasoalveolar molding appliances for infants with cleft lip and palate using 3-dimensional workflow and clear aligners. Am J Orthod Dentofac Orthop. 2020;158:452–8.
73. Schiebl J, Bauer FX, Grill F, Loeffelbein DJ. RapidNAM: algorithm for the semi-automated generation of nasoalveolar molding device designs for the presurgical treatment of bilateral cleft lip and palate. IEEE Trans Biomed Eng. 2020;67:1263–71.
74. Comparative study between the overall production time of digitally versus conventionally produced indirect orthodontic bonding trays [Internet]. [cited 2021 Feb 10]. Available from: http://www.turkjorthod.org/en/comparative-study-between-the-overall-production-time-of-digitally-versus-conventionally-produced-indirect-orthodontic-bonding-trays-131064.
75. Edwards SP. Computer-assisted craniomaxillofacial surgery. Oral Maxillofac Surg Clin North Am. 2010;22:117–34.
76. Lonic D, Pai BC-J, Yamaguchi K, Chortrakarnkij P, Lin H-H, Lo L-J. Computer-assisted orthognathic surgery for patients with cleft lip/palate: from traditional planning to three-dimensional surgical simulation. PLoS One. 2016;11:e0152014.
77. De Riu G, Meloni SM, Baj A, Corda A, Soma D, Tullio A. Computer-assisted orthognathic surgery for correction of facial asymmetry: results of a randomized controlled clinical trial. Br J Oral Maxillofac Surg. 2014;52:251–7.
78. Van Hemelen G, Van Genechten M, Renier L, Desmedt M, Verbruggen E, Nadjmi N. Three-dimensional virtual planning in orthognathic surgery enhances the accuracy of soft tissue prediction. J Cranio Maxillofac Surg. 2015;43:918–25.
79. Shirota T, Shiogama S, Watanabe H, Kurihara Y, Yamaguchi T, Maki K, et al. Three-dimensional virtual planning and intraoperative navigation for two-jaw orthognathic surgery. J Oral Maxillofac Surg Med Pathol. 2016;28:530–4.
80. Pietruski P, Majak M, Swiatek-Najwer E, Popek M, Szram D, Zuk M, et al. Accuracy of experimental mandibular osteotomy using the image-guided sagittal saw. Int J Oral Maxillofac Surg. 2016;45:793–800.
81. Shaheen E, Sun Y, Jacobs R, Politis C. Three-dimensional printed final occlusal splint for orthognathic surgery: design and validation. Int J Oral Maxillofac Surg. 2017;46:67–71.
82. Ying B, Ye N, Jiang Y, Liu Y, Hu J, Zhu S. Correction of facial asymmetry associated with vertical maxillary excess and mandibular prognathism by combined orthognathic surgery and guiding templates and splints fabricated by rapid prototyping technique. Int J Oral Maxillofac Surg. 2015;44:1330–6.
83. Comparison of three different types of splints and templates for maxilla repositioning in bimaxillary orthognathic surgery: a randomized controlled trial. Int J Oral Maxillofac Surg [Internet]. 2020 [cited 2021 Feb 17]; Available from: https://www.sciencedirect.com/science/article/pii/S0901502720303751.
84. Karanxha L, Rossi D, Hamanaka R, Giannì AB, Baj A, Moon W, et al. Accuracy of splint vs splintless technique for virtually planned orthognathic surgery: a voxel-based three-dimensional analysis. J Cranio Maxillofac Surg. 2021;49:1–8.

85. Kang S-H, Lee J-W, Lim S-H, Kim Y-H, Kim M-K. Validation of mandibular genioplasty using a stereolithographic surgical guide: in vitro comparison with a manual measurement method based on preoperative surgical simulation. J Oral Maxillofac Surg. 2014;72:2032–42.
86. Arcas A, Vendrell G, Cuesta F, Bermejo L. Advantages of performing mentoplasties with customized guides and plates generated with 3D planning and printing. Results from a series of 23 cases. J Cranio Maxillofac Surg. 2018;46:2088–95.
87. Kim J-W, Kim J-C, Jeong C-G, Cheon K-J, Cho S-W, Park I-Y, et al. The accuracy and stability of the maxillary position after orthognathic surgery using a novel computer-aided surgical simulation system. BMC Oral Health. 2019;19:1–13.
88. Philippe B. Accuracy of the position of cutting and drilling guide for sagittal split guided surgery: a proof of concept study. Br J Oral Maxillofac Surg. 2020;58:940–6.
89. Volpe Y, Furferi R, Governi L, Uccheddu F, Carfagni M, Mussa F, et al. Surgery of complex craniofacial defects: a single-step AM-based methodology. Comput Methods Prog Biomed. 2018;165:225–33.
90. Rückschloß T, Ristow O, Müller M, Kühle R, Zingler S, Engel M, et al. Accuracy of patient-specific implants and additive-manufactured surgical splints in orthognathic surgery—a three-dimensional retrospective study. J Cranio Maxillofac Surg. 2019;47:847–53.
91. Melville JC, Manis CS, Shum JW, Alsuwied D. Single-unit 3D-printed titanium reconstruction plate for maxillary reconstruction: the evolution of surgical reconstruction for maxillary defects-a case report and review of current techniques. J Oral Maxillofac Surg. 2019;77:874.e1–874.e13.
92. Goodson AM, Kittur MA, Evans PL, Williams EM. Patient-specific, printed titanium implants for reconstruction of mandibular continuity defects: a systematic review of the evidence. J Craniomaxillofac Surg. 2019;47:968–76.
93. Qassemyar Q, Assouly N, Temam S, Kolb F. Use of a three-dimensional custom-made porous titanium prosthesis for mandibular body reconstruction. Int J Oral Maxillofac Surg. 2017;46:1248–51.
94. Twenty-year follow-up study on a patient-fitted temporomandibular joint prosthesis: the Techmedica/TMJ concepts device. J Oral Maxillofac Surg. 2015;73:952–60.
95. Our Process – TMJ Concepts [Internet]. [cited 2021 Feb 19]. Available from: https://tmjconcepts.cóm/surgeons-office-staff/our-process/.
96. Ackland DC, Robinson D, Redhead M, Lee PVS, Moskaljuk A, Dimitroulis G. A personalized 3D-printed prosthetic joint replacement for the human temporomandibular joint: from implant design to implantation. J Mech Behav Biomed Mater. 2017;69:404–11.
97. Williams LR, Fan KF, Bentley RP. Custom-made titanium cranioplasty: early and late complications of 151 cranioplasties and review of the literature. Int J Oral Maxillofac Surg. 2015;44:599–608.
98. Park E-K, Lim J-Y, Yun I-S, Kim J-S, Woo S-H, Kim D-S, et al. Cranioplasty enhanced by three-dimensional printing: custom-made three-dimensional-printed titanium implants for skull defects. J Craniofac Surg. 2016;27:943–9.
99. Abbate V, Iaconetta G, Califano L, Pansini A, Bonavolontà P, Romano A, et al. Self-made rapid prototyping technique for orbital floor reconstruction: showcases for technical description. J Craniofac Surg. 2019;30:2106–10.
100. Bachelet J-T, Cordier G, Porcheray M, Bourlet J, Gleizal A, Foletti J-M. Orbital reconstruction by patient-specific implant printed in porous titanium: a retrospective case series of 12 patients. J Oral Maxillofac Surg. 2018;76:2161–7.
101. Louvrier A, Marty P, Barrabé A, Euvrard E, Chatelain B, Weber E, et al. How useful is 3D printing in maxillofacial surgery? J Stomatol Oral Maxillofac Surg. 2017;118:206–12.
102. Zhang Y, Wu D, Zhao X, Pakvasa M, Tucker AB, Luo H, et al. Stem cell-friendly scaffold biomaterials: applications for bone tissue engineering and regenerative medicine. Front Bioeng Biotechnol [Internet]. 2020 [cited 2021 Feb 20];8. Available from: https://www.ncbi.nlm.nih.gov/pmc/articles/PMC7767872/.
103. Kim K, Yeatts A, Dean D, Fisher JP. Stereolithographic bone scaffold design parameters: osteogenic differentiation and signal expression. Tissue Eng Part B Rev. 2010;16:523–39.
104. Roohani-Esfahani S-I, Newman P, Zreiqat H. Design and fabrication of 3D printed scaffolds with a mechanical strength comparable to cortical bone to repair large bone defects. Sci Rep. 2016;6:19468.
105. Bobbert FSL, Lietaert K, Eftekhari AA, Pouran B, Ahmadi SM, Weinans H, et al. Additively manufactured metallic porous biomaterials based on minimal surfaces: a unique combination of topological, mechanical, and mass transport properties. Acta Biomater. 2017;53:572–84.
106. Höhne C, Schmitter M. 3D printed teeth for the preclinical education of dental students. J Dent Educ. 2019;83:1100–6.
107. Boonsiriphant P, Al-Salihi Z, Holloway JA, Schneider GB. The use of 3D printed tooth preparation to assist in teaching and learning in preclinical fixed prosthodontics courses. J Prosthodont. 2019;28:e545–7.
108. Aragón ML, Pontes LF, Bichara LM, Flores-Mir C, Normando D. Validity and reliability of intraoral scanners compared to conventional gypsum models measurements: a systematic review. Eur J Orthod. 2016;38:429–34.
109. García-Gil I, Cortés-Bretón-Brinkmann J, Jiménez-García J, Peláez-Rico J, Suárez-García M-J. Precision and practical usefulness of intraoral scanners in implant dentistry: a systematic literature review. J Clin Exp Dent. 2020;12:e784.

110. Is it cost-effective to add an intraoral scanner to an oral and maxillofacial surgery practice? J Oral Maxillofac Surg. 2019;77:1687–94.
111. Kamio T, Hayashi K, Onda T, Takaki T, Shibahara T, Yakushiji T, et al. Utilizing a low cost desktop 3D printer to develop a "one-stop 3D printing lab" for oral and maxillofacial surgery and dentistry fields. 3D Print Med. 2018;4:1–7.
112. Dadbakhsh S, Speirs M, Humbeeck JV, Kruth J-P. Laser additive manufacturing of bulk and porous shape-memory NiTi alloys: from processes to potential biomedical applications. MRS Bull. 2016;41:765–74.
113. Taheri Andani M, Saedi S, Turabi AS, Karamooz MR, Haberland C, Karaca HE, et al. Mechanical and shape memory properties of porous Ni50.1Ti49.9 alloys manufactured by selective laser melting. J Mech Behav Biomed Mater. 2017;68:224–31.
114. Lu H, Yang C, Luo X, Ma H, Song B, Li Y, et al. Ultrahigh-performance TiNi shape memory alloy by 4D printing. Mater Sci Eng A. 2019;763:138166.
115. Shishkovskii IV, Yadroitsev IA, Smurov IY. Selective laser sintering/melting of nitinol–hydroxyapatite composite for medical applications. Powder Metall Met Ceram. 2011;50:275.
116. Liu J, Erol O, Pantula A, Liu W, Jiang Z, Kobayashi K, et al. Dual-gel 4D printing of bioinspired tubes. ACS Appl Mater Interfaces. 2019;11:8492–8.
117. Guo J, Zhang R, Zhang L, Cao X. 4D printing of robust hydrogels consisted of agarose nanofibers and polyacrylamide. ACS Macro Lett. 2018;7:442–6.
118. Salmoria GV, Klauss P, Zepon KM, Kanis LA. The effects of laser energy density and particle size in the selective laser sintering of polycaprolactone/progesterone specimens: morphology and drug release. Int J Adv Manuf Technol. 2013;66:1113–8.
119. Melocchi A. Expandable drug delivery system for gastric retention based on shape memory polymers development via 4D printing and extrusion. 2019.
120. Melocchi A, Inverardi N, Uboldi M, Baldi F, Maroni A, Pandini S, et al. Retentive device for intravesical drug delivery based on water-induced shape memory response of poly(vinyl alcohol): design concept and 4D printing feasibility. Int J Pharm. 2019;559:299–311.
121. Gioumouxouzis CI, Tzimtzimis E, Katsamenis OL, Dourou A, Markopoulou C, Bouropoulos N, et al. Fabrication of an osmotic 3D printed solid dosage form for controlled release of active pharmaceutical ingredients. Eur J Pharm Sci. 2020;143:105176.
122. Mao M, He J, Li X, Zhang B, Lei Q, Liu Y, et al. The emerging frontiers and applications of high-resolution 3D printing. Micromachines (Basel). 2017 [cited 2021 Apr 15];8. Available from: https://www.ncbi.nlm.nih.gov/pmc/articles/PMC6190223/.
123. Jiang Y, Wang Q. Highly-stretchable 3D-architected mechanical metamaterials. Sci Rep. 2016;6:34147.
124. Jang D, Meza LR, Greer F, Greer JR. Fabrication and deformation of three-dimensional hollow ceramic nanostructures. Nat Mater. 2013;12:893–8.
125. Bas O, De-Juan-Pardo EM, Chhaya MP, Wunner FM, Jeon JE, Klein TJ, et al. Enhancing structural integrity of hydrogels by using highly organized melt electrospun fiber constructs. Eur Polym J. 2015;72:451–63.
126. Martinez-Marquez D, Delmar Y, Sun S, Stewart RA. Exploring macroporosity of additively manufactured titanium metamaterials for bone regeneration with quality by design: a systematic literature review. Materials. 2020;13:4794.
127. Zhang H, Guo X, Wu J, Fang D, Zhang Y. Soft mechanical metamaterials with unusual swelling behavior and tunable stress-strain curves. Sci Adv. 2018;4:eaar8535.
128. Sydney Gladman A, Matsumoto EA, Nuzzo RG, Mahadevan L, Lewis JA. Biomimetic 4D printing. Nat Mater. 2016;15:413–8.

Correction to: Commonly Used 3D Printing Technologies in Oral Health Science

Ramandeep Kaur Sokhi, Nitika Monga,
Kelly Kathering Achachao Almerco, Dipti Shastri,
Manish Bajracharya, Ujjwal Pyakurel,
Mohammad Khursheed Alam,
and Prabhat Kumar Chaudhari

Correction to:
Chapter 3 in: Prabhat Kumar Chaudhari, Dinesh Bhatia, Jitendra Sharan (Eds.) ***3D Printing in Oral Health Science*****,**
https://doi.org/10.1007/978-3-031-07369-4

This book was inadvertently published with an incorrect author name. This has now been amended throughout the book from Mohammad Mohammed Alam to Mohammad Khursheed Alam.

The updated original version of the book can be found at
https://doi.org/10.1007/978-3-031-07369-4_3

© The Author(s), under exclusive license to Springer Nature Switzerland AG 2022
P. K. Chaudhari et al. (eds.), *3D Printing in Oral Health Science*,
https://doi.org/10.1007/978-3-031-07369-4_15

MIX
Papier aus verantwortungsvollen Quellen
Paper from responsible sources
FSC® C105338

If you have any concerns about our products,
you can contact us on
ProductSafety@springernature.com

In case Publisher is established outside the EU,
the EU authorized representative is:
Springer Nature Customer Service Center GmbH
Europaplatz 3, 69115 Heidelberg, Germany

Printed by Libri Plureos GmbH
in Hamburg, Germany